Strigolactones, Alkamides and Karrikins in Plants

Strigolactones, Alkamides and Karrikins in Plants: Recent Updates and Future Prospects provides comprehensive knowledge on the various aspects of plant growth, physiology, and communication associated with the three potential biomolecules. Strigolactones have gained much importance in the last decade as potential plant growth regulators. Likewise, alkamides are also known for their plant growth regulatory and pharmacological properties. The evolutionary significance of karrikins as a potential signaling molecule in different plant groups has been fascinating to plant physiologists and ecologists. This book enables the reader to gain insights into the myriad role of these biomolecules in plant physiology in normal and challenging environments.

The book offers comprehensive coverage of the most essential topics, including:

- Regulation of strigolactone biosynthesis.
- Strigolactones and plant stress tolerance.
- Strigolactones and parasitic plants.
- Alkamides and plant-microbe interaction in rhizosphere.
- Pharmacological potential of alkamides.
- Molecular associations of strigolactones and karrikins.
- Karrikins in plant biotechnology.
- Commercial realities of karrikins for plant abiotic stress tolerance and biodiversity restoration.

In reference to the recent findings in the field, the authors have provided insights into the role of each biomolecule and analyzed the future prospects of each cover area.

The present book is the first of its kind, in which the three molecules have been integrated as potential regulators of plant signaling, communication, and physiology in adverse conditions. Karrikins in plant biology is an emerging field in which its inter-relationship with strigolactones shall pave the way to future investigations of the molecular mechanisms of plant stress tolerance, regulation of root system architecture, and crop yield.

Strigolactones, Alkamides and Karrikins in Plants

Recent Updates and Future Prospects

Edited by
Soumya Mukherjee, Tariq Aftab

CRC Press
Taylor & Francis Group
Boca Raton London

CRC Press is an imprint of the
Taylor & Francis Group, an **informa** business

First edition published 2023
by CRC Press
6000 Broken Sound Parkway NW, Suite 300, Boca Raton, FL 33487–2742

and by CRC Press
4 Park Square, Milton Park, Abingdon, Oxon, OX14 4RN

CRC Press is an imprint of Taylor & Francis Group, LLC

ISBN: 978-1-032-12433-9 (hbk)
ISBN: 978-1-032-12437-7 (pbk)
ISBN: 978-1-003-22452-5 (ebk)

DOI: 10.1201/9781003224525

Typeset in Times
by Apex CoVantage, LLC

Contents

SECTION II Alkamides: Ecophysiological and Pharmacological Perspectives

SECTION III Karrikins in Plant Biology

Preface

Investigations in the last decade have deciphered the roles of novel plant-growth regulators which are known to exert a plethora of effects on plant growth, morphogenesis, and plant–environment interaction. Apart from the role of conventional phytohormones, strigolactones, karrikins, and alkamides have been reported to function as potential growth regulators in plants. Strigolactones participate in molecular crosstalk pathways with important phytohormones and biomolecules like auxin, ethylene, jasmonic acid, and nitric oxide. Interestingly, strigolactones function as endogenous growth regulators in plants, regulate signaling and communication at the rhizosphere, and act as growth stimulants in parasitic plants. Strigolactones are functional as important signaling molecules in diverse plants groups. Karrikins are a group of organic compounds emitted by wildfire and are known to modulate various processes associated with growth and morphogenesis in plants. Recent investigations have attempted to delineate the mechanism of perception of karrikins by plants. Photomodulation and hormonal crosstalk appear to play pivotal roles in karrikin-mediated regulation of seed dormancy and germination and the modulation of osmotic stress in plants. Alkamides are a class of secondary metabolites present in a wide variety of plants and exhibit precise control over various physiological responses. Alkamides have been known to be present as a chief component of humic acid in rhizosphere, thus regulating root development and architecture. Alkamides exhibit auxin-like activity and are probably involved in crosstalk with jasmonic acid signaling–pathways. Furthermore, alkamides are capable of boosting plant immunity to biotic stress.

In view of the accumulating evidence supporting the signaling role of strigolactones, alkamides and karrikins in plants, it is necessary to collate the recent developments in the field. The new volume shall contain a comprehensive analysis of the recent investigations and future possibilities associated with the physiological roles of strigolactones, alkamides and karrikins in plants. Although various interesting reports have been obtained, it is necessary to summarize the interrelations of these biomolecules with other important plant metabolites. The new volume is comprised of brief evidence supported by analytical schemes and flowcharts depicting the signaling mechanisms of the biomolecules (strigolactones, karrikins, and alkamides) in plant growth, development, and interaction with the environment. Supportive evidence shall unveil the mechanisms of stress tolerance in crops associated with the signaling pathway of the three novel biomolecules. The book shall appear as an informative reference volume to investigators from plant physiology, crop biotechnology, ecology, and plant breeding for crop management. Various investigators associated with the relevant fields are expected to contribute in increasing the value of the volume.

Contributors

Zouhaier Abbes
Field Crop Laboratory (LGC), Carthage University, National Institute for Agricultural Research of Tunisia (INRAT)
Menzah 1, Tunis, Tunisia

Tahseen Afzal
Institute of Soil and Environmental Sciences
University of Agriculture
Faisalabad, Pakistan

Nahid Akhtar
School of Bioengineering and Biosciences
Lovely Professional University
Punjab (144411), India

Muhammad Fraz Ali
Department of Crop Sciences, Faculty of Agricultural Sciences
University of Talca
Talca 3460000, Chile

Moez Amri
AgroBioSciences Program (AgBS)
University Mohammed VI Polytechnic (UM6P)
Ben Guerir, Morocco

Athira KV
Amrita School of Pharmacy
Amrita VishwaVidyapeetham, AIMS Health Sciences Campus
Kochi, 682 041, Kerala, India

Aaliya Batool
Department of Botany
University of Agriculture
Faisalabad 38040, Pakistan

Shabana Bibi
Yunnan Herbal Laboratory, College of Ecology and Environmental Sciences
Yunnan University
Kunming 650091, China

Juan Bueno
Research Center of Bioprospecting and Biotechnology for Biodiversity Foundation (BIOLABB), Colombia

Canellas LP
Núcleo de Desenvolvimento de Insumos Biológicos
Agricultura da Universidade Estadual do Norte
Fluminense Darcy Ribeiro, Campos Dos
Goytacazes-RJ, CEP 28013–602, Brazil

Hitesh Chopra
Chitkara College of Pharmacy
Chitkara University
Punjab 140401, India

Irsa Ejaz
College of Agronomy and Biotechnology,
China Agricultural University,
Beijing 100194, China

Youness En-nahli
AgroBioSciences Program (AgBS), University Mohammed VI Polytechnic (UM6P)
Benguerir, Morocco
and
Materials Science Center, Mohammed V University in Rabat, LPCMIO, Ecole Normale Supérieure, Rabat 10112, Morocco

Javed Iqbal
Department of Agronomy and Horticulture,
University of Nebraska-Lincoln, USA

Ben Guerir, Morocco
Materials Science Center
Mohammed V University in Rabat, LPCMIO, Ecole Normale Supérieure
Rabat 10112, Morocco

Nour Eddine Es-Safi
Materials Science Center
Mohammed V University in Rabat, LPCMIO, Ecole Normale Supérieure
Rabat 10112, Morocco

Gavin R. Flematti
School of Molecular Sciences
The University of Western Australia
Perth, 6009, Australia

Sourbh Suren Garg
School of Bioengineering and Biosciences
Lovely Professional University
Punjab (144411), India

Michel Edmond Ghanem
AGAP Institut, Univ Montpellier
Centre de coopération internationale en recherche agronomique pour le développement (CIRAD)
Montpellier, France
and
AgroBioSciences Program (AgBS)
University Mohammed VI Polytechnic (UM6P)
Ben Guerir, Morocco

Kriti Gupta
Department of Botany
DAV College
Bathinda, Punjab, India 151001

Santosh Kumar Gupta
National Institute of Plant Genome Research
New Delhi, India

Mohammad Mehedi Hasan
Faculty of Life Science, Department of Biochemistry and Molecular Biology
Mawlana Bhashani Science and Technology University
Tangail, Bangladesh

Anju Jose
Amrita School of Pharmacy
Amrita VishwaVidyapeetham, AIMS Health Sciences Campus
Kochi, 682 041, Kerala, India

Chenmala Karthika
Department of Pharmaceutics
JSS College of Pharmacy, J SS Academy of Higher Education and Research
The Nilgiris, Ooty 643001, Tamil Nadu, India

Simranjeet Kaur
Department of Botany
Akal University, Talwandi Sabo
Bathinda, Punjab, India 151302

Ewa Kępczyńska
Institute of Biology
University of Szczecin
Wąska 13, 71–415 Szczecin, Poland

Jan Kępczyński
Institute of Biology
University of Szczecin
Wąska 13, 71–415 Szczecin, Poland

Muhammad Saad Khan
Department of Biosciences
COMSATS University Islamabad, Sahiwal Campus
Sahiwal, Pakistan

Mohamed Kharrat
Field Crop Laboratory (LGC)
Carthage University, National Institute for Agricultural Research of Tunisia (INRAT)
Menzah 1, Tunis, Tunisia

Jitka Kochanek
Australian Institute for Bioengineering and Nanotechnology
The University of Queensland
St Lucia, QLD 4072, Australia

Santi Krisantini
School of Agriculture and Food Sciences
The University of Queensland
St Lucia, QLD 4072, Australia

Avneesh Kumar
Department of Botany, Akal University, Talwandi Sabo,
Bathinda, Punjab, India 151302

William Makaza
AgroBioSciences Program (AgBS) University Mohammed VI Polytechnic (UM6P)
Ben Guerir, Morocco

Shivangi Mathur
Department of Botany
Dayalbagh Educational Institute
Dayalbagh Agra-282005, India

Georgina McGregor
Australian Institute for Bioengineering and Nanotechnology
The University of Queensland
St Lucia, QLD 4072, Australia

Tahir ul Gani Mir
School of Bioengineering and Biosciences
Lovely Professional University
Punjab (144411), India

Sakina Mussarat
Department of Botany
Kohat University
Kohat, Pakistan

Muhammad Asad Naseer
College of Agronomy
Northwest Agriculture and Forestry University
Yangling 712100, Shaanxi, China

Rubina Naseer
Centre of Agricultural Biochemistry and Biotechnology
University of Agriculture
Faisalabad 38040, Pakistan

Ghazala Nawaz
Department of Botany
Kohat University
Kohat, Pakistan

Mereen Nizar
Department of Botany
University of Balochistan
Quetta, Pakistan

Olivares FL
Núcleo de Desenvolvimento de Insumos Biológicos
Agricultura da Universidade Estadual do Norte
Fluminense Darcy Ribeiro, Campos Dos
Goytacazes-RJ, CEP 28013–602, Brazil

Ajit Prakash
Department of Biochemistry and Biophysics
University of North Carolina
Chapel Hill, NC 27599–2760

Farida Rahayu
Indonesian Sweetener and Fiber Crops Research Institute
Jl. Raya Karangploso, KM-4, Malang 65100, Indonesia

Md. Habibur Rahman
Department of Global Medical Science
Wonju College of Medicine, Yonsei University
Wonju 26426, Korea

Mahmood-ur-Rahman
Department of Bioinformatics and Biotechnology
Government College University
Faisalabad-38000, Pakistan

Rajiv Ranjan
Department of Botany
Dayalbagh Educational Institute
Dayalbagh Agra-282005, India

Arshdeep Singh
Department of Agronomy & Horticulture,
University of Nebraska-Lincoln, USA

Abdul Samad
Department of Botany
University of Balochistan
Quetta, Pakistan

Santos MP
Instituto de Biodiversidade e Sustentabilidade
Universidade Federal do Rio de Janeiro
Macaé-RJ, CEP 27965–045, Brazil

Atish Sardar
Jogesh Chandra Chaudhuri College
Kolkata 700033, India

Kanval Shaukat
Department of Botany
University of Balochistan
Quetta, Pakistan

Deeksha Singh
Department of Botany
Dayalbagh Educational Institute
Dayalbagh Agra-282005, India

Rattandeep Singh
School of Bioengineering and Biosciences
Lovely Professional University
Punjab (144411), India

Zhencai Sun
College of Agronomy and Biotechnology
China Agricultural University
Beijing 100194, China

Shrouq H. Sweilam
Faculty of Pharmacy, Department of Pharmacognosy
Prince Sattam Bin Abdulaziz University
Al-Kharj, Saudi Arabia

Tauheed
Department of Botany
Kohat University
Kohat, Pakistan

Imen Trabelsi
Agronomy and Sciences Technology Laboratory, Carthage University
National Institute for Agricultural Research of Tunisia (INRAT)
Menzah 1, Tunis, Tunisia

Kenneth A. Tryggestad
Australian Institute for Bioengineering and Nanotechnology
The University of Queensland
St Lucia, QLD 4072, Australia

Sana Ullah
Institute of Soil and Environmental Sciences,
University of Agriculture Faisalabad
Pakistan

Maryam Valivand
Department of Biology, Falavarjan Branch, Islamic Azad University Isfahan, Iran

Abdul Wahid
Department of Botany
University of Agriculture
Faisalabad 38040, Pakistan

Yunwei Wang
School of Agriculture and Food Sciences; Australian Institute for Bioengineering and Nanotechnology
The University of Queensland
St Lucia, QLD 4072, Australia

Atif Khurshid Wani
School of Bioengineering and Biosciences
Lovely Professional University
Punjab (144411), India

Qudsia Yousafi
Department of Biosciences
COMSATS University Islamabad, Sahiwal Campus
Sahiwal, Pakistan

Noreen Zahra
Department of Botany
University of Agriculture
Faisalabad 38040, Pakistan

Zandonadi DB
Instituto de Biodiversidade e Sustentabilidade
Universidade Federal do Rio de Janeiro
Macaé-RJ, CEP 27965–045, Brazil

Muhammad Aurang Zeb
Key Laboratory of Medicinal Chemistry for Natural Resources
Ministry of Education and Yunnan Province
School of Chemical Science and Technology, Yunnan University
Kunming 650091, People's Republic of China

Editor Biographies

Soumya Mukherjee received his PhD from the Department of Botany, University of Delhi. He is a former faculty of Ramjas College, University of Delhi. Currently, he works as the faculty at the Department of Botany, Jangipur College, University of Kalyani, West Bengal. He has worked as a CSIR research fellow in the area of the abiotic stress physiology of plants. He has published both research and review articles in various peer-reviewed international journals. He is serving as an associate editor on the editorial board of Plant Signaling and Behaviour (Taylor and Francis). He has authored e-learning modules in plant physiology and biochemistry, published by the National Mission on Education through ICT (MHRD Project undertaken by University of Delhi). He has acted as a subject reviewer of some peer-reviewed international journals on plant science. Currently, he is focused on undergraduate teaching to botany and life science students in various fundamental aspects areas of plant physiology. He continues to work with research interests in the area of salt-stress physiology associated with the aspects of hormone signaling and biomolecular crosstalk mechanisms in plants.

Tariq Aftab received his PhD from the Department of Botany at Aligarh Muslim University, India, and is currently an assistant professor there. He is the recipient of a prestigious Leibniz-DAAD fellowship from Germany, the Raman Fellowship from the government of India, and the Young Scientist Award from the state government of Uttar Pradesh (India) and the government of India. Since completing his doctorate, he has worked as research fellow at the National Bureau of Plant Genetic Resources, New Delhi, and as post-doctoral fellow at Jamia Hamdard, New Delhi, India. Dr. Aftab has also worked as visiting scientist at the Leibniz Institute of Plant Genetics and Crop Plant Research (IPK), Gatersleben, Germany, and in the Department of Plant Biology, Michigan State University, USA. He is a member of various scientific associations from India and abroad. He has edited several books with international publishers, including Elsevier Inc., Springer Nature, and CRC Press (Taylor & Francis Group); coauthored several book chapters, and published more than 60 research papers in peer-reviewed international journals. His research interests include physiological, proteomic, and molecular studies on medicinal and aromatic plants.

Introduction

Strigolactones, Alkamides, and Karrikins in Plants: Progress and Evolutionary Significance

Soumya Mukherjee and Tariq Aftab

CONTENTS

Plant metabolites, including volatile organic compounds (VOCs), sesquiterpenoids, and humic acids, are important regulatory molecules in the rhizosphere. Plant-soil interaction is precisely regulated by the release of these growth regulatory biomolecules in the soil. The growth and physiological activity of various plant growth–regulating bacteria (PGRB), mycorrhiza, pathogenic fungi, and parasites are modulated by the activity of these metabolites in the rhizosphere. In the last few decades, we have observed significant progress in the number of investigations of the roles of strigolactones (SLs), karrikins (KKs), and alkamides (AKs) in various aspects of plant growth, development, stress modulation, and soil interactions. An analysis of literature from the PubMed database reveals that a variable number of reports have been observed for the three biomolecules in a time span of last 17 years (i.e., 2005–2022). The total number of reports on strigolactone and plant biology was approximately 730, out of which around 211 studies involved investigations associated with plant stress. Karrikins in plants were represented by around 96 reports while 64 reports depicted the inter-relations of SLs and KKs. The search for alkamides appeared somewhat nonspecific as a significant number of papers discuss the pharmacological effects of plant-derived alkamides in animal systems. However, the total number of reports on alkamides was around 80. Moreover, a surge in the number of publication was observed in the last decade (2012–2022) for investigations pertaining to all the three biomolecules.

Following the discovery of **strigolactones** (SLs) as germination stimulants in the parasitic weed *Striga* (Xie et al., 2010), it has been known to exert multiple physiological and molecular effects on plants. Apart from their role as a phytohormone, SLs are effective ecological communicators among plant and other organisms in the soil (Akiyama et al., 2005). Endogenous SLs exert regulatory effects in shoot and root branching, flowering, nutrient absorption, long-distance signaling, and various other developmental aspects (Yoneyama, 2019). Interestingly, SLs exhibit associative roles in regulating the function of various other phytohormones in normal and challenging environments. In an evolutionary context, the existence of SLs can be traced back in unicellular algae and primitive land plants – namely, mosses and liverworts – in which SLs regulate intercellular communications, protonema differentiation, and competition among the neighbouring organisms (Delaux et al., 2012, Proust et al., 2011). As early in the cretaceous period, during the colonization of the first land plants, SLs were presumed to have facilitated the establishment of plant-fungal symbiosis. Certain plants of the lower group like *Physcomitrella patens* (moss) and *Chara coralline* (alga) are known to employ endogenous SLs in the growth and differentiation of vegetative thallus. Certain breakthrough reports in 2008 led to the characterization of several SL-biosynthetic mutants, and

DOI: 10.1201/9781003224525-1

thus SLs have been known as a new class of plant hormone (Mashiguchi et al., 2021). Investigations have revealed the molecular mechanisms of SL transport, perception, and signaling in plants. Apart from their receptor-mediated activity in various regulatory processes, SLs have been employed as potential agrochemicals in various crop plants for stress mitigation. Moreover, SLs have also shown promising effects on human health as immunomodulatory, antimicrobial, anticancerous and anti-inflammatory molecules (Yoneyama, 2020; Dell'Oste et al., 2021). However, there are challenges associated with the synthesis of various structural derivatives of SLs, and there is also a need to establish the stability and efficacy of exogenous SLs in various formulations required for crop management in the agricultural fields.

Karrikins (KKs) belong to a group of butanolide-related smoke-derived compounds (plant material burns), which were discovered to stimulate seed germination ability in certain plants during forest fires (Nelson et al., 2012). The word *karrikin* was derived from the aboriginal term for smoke – *karrik* – used by the Noongar people of west Australia (Flematti et al., 2009, Nelson et al., 2009). Certain plants called "fire followers" or "fire-ephemerals" are known to exhibit a reduction in seed dormancy immediately after a forest fire, when KKs are likely to be washed by raindrops into the soil of burnt forest beds (Flematti et al., 2015). Thus, KKs possess immense ecological relevance in regulating seed germination, dormancy breakage, and vegetation growth in forest sites. Interestingly, earth fires have been associated with the advent of volcanoes, land plants, and forests where plant materials served as fuels for burning in the presence of oxygen. According to scientists, the existence of burnt wood has been dated back 400 million years ago (Ma) while the fire response of seed plants is assumed to date back to the cretaceous period (65–145 Ma) (He et al., 2012). The presence of the KAI2 gene has been traced back to the early land plants, where its function has been associated with plant growth and development. The function of the KAI2 gene has been determined to operate through the signaling pathways of KK and SL – related compounds. Interestingly, the KAI2 protein from *Selaginella* has been known to complement seedling and leaf development in Arabidopsis (Waters et al., 2015). Interestingly, certain plants have evolved with the sensitivity of the KAI2 protein to forest fire–derived smoke. Subsequent gene duplication of KAI2 has resulted in the existence of DWARF14, where sensitivity to KK is known to be associated with growth, development, and exogenous smoke-induced responses. Inter-relations of KK and SLs have been deciphered in relation to the parasitic plants of Orobanchaceae in which KAI2 exhibits sensitivity to root-exudated SLs (Conn et al., 2015). According to Sun et al. (2020), diversification and duplication of KAI2 receptor proteins has resulted in the ability of the plants to receive diverse chemical signals associated with KKs and SLs. According to Flematti et al. (2015), KKs are more ancient in origin than SLs. Although much remains to be investigated, it is likely that KKs and SLs have co-evolved throughout the ancient lower plant groups – namely, unicellular alga -- and their existence has continued in land plants, including conifers and wide families of angiosperms (Arellano-Saab et al., 2021; Wang et al., 2022).

Alkamides are secondary metabolites comprising of a group of 200 related compounds that are formed by the condensation of unsaturated fatty acid and an amine (Zandonadi et al., 2019). They have evolved as a wide range of compounds with diverse pharmacological effects. Alkamides exert regulatory roles in root hair formation and lateral root elongation. Although soil-based humic substances (HS) have been known to exert auxin-like actions, the specific role of alkamides was deciphered following the discovery of affinins in the early 2000s (Ramirez-Chavez et al., 2004). NMR-spectroscopic analysis has revealed that the primary nitrogen structure associated with HS belongs to the amide functional group. Zandonadi et al. (2019) have reported that alkamide (N-isopropyldecanamide) is associated with elevation in the activity of PM H^+-ATPase activity accompanied by the promotion of root growth. Alkamide-mediated root growth modulation has been determined to be operating through an NO-mediated process (Méndez-Bravo et al., 2010, Zandonadi et al., 2019). Méndez-Bravo et al. (2010) have provided evidence of the involvement of JA and NO in mediating auxin-independent downstream signaling to alkamide action. Furthermore, López-Bucio et al. (2007) have substantiated the role of alkamide to be associated with the cytokinin-signaling pathway accompanying meristem growth and cell division. Further investigations are required to determine the role of alkamides in regulating the signaling response of various phytohormones and other biomolecules associated with plant growth and development. Moreover, it is imperative to perform humeome analysis of various soils to trace the association of alkamide and other plant growth regulators in the rhizosphere, where the molecule might exhibit interaction with other phytohormones or secondary metabolites.

REFERENCES

Akiyama K, Matsuzaki K, Hayashi H. Plant sesquiterpenes induce hyphal branching in arbuscular mycorrhizal fungi. *Nature*. 2005, 435: 824–827. doi: 10.1038/nature03608.

Arellano-Saab A, Bunsick M, Al Galib H, Zhao W, Schuetz S, Bradley JM, Xu Z, Adityani C, Subha A, McKay H, de Saint Germain A, Boyer FD, McErlean CSP, Toh S, McCourt P, Stogios PJ, Lumba S. Three mutations repurpose a plant karrikin receptor to a strigolactone receptor. *Proc Natl Acad Sci U S A*. 2021, 118(30): e2103175118. doi: 10.1073/pnas.2103175118.

Conn CE, Bythell-Douglas R, Neumann D, Yoshida S, Whittington B, Westwood JH, et al. Convergent evolution of strigolactone perception enabled host detection in parasitic plants. *Science*. 2015, 349: 540–543.

Delaux PM, Xie X, Timme RE, Puech-Pages V, Dunand C, Lecompte E, Delwiche CF, Yoneyama K, Bécard G, Séjalon-Delmas N. Origin of strigolactones in the green lineage. *New Phytol*. 2012, 195: 857–871. doi: 10.1111/j.1469–8137.2012.04209.x.

Dell'Oste V, Spyrakis, F, Prandi C. Strigolactones, from plants to human health: Achievements and challenges. *Molecules*. 2021, 26: 4579. doi: 10.3390/molecules26154579.

Flematti GR, Dixon KW, Smith SM. What are karrikins and how were they 'discovered' by plants? *BMC Biol*. 2015, 13(108). doi: 10.1186/s12915-015-0219-0.

Flematti GR, Ghisalberti EL, Dixon KW, Trengove RD. Identification of alkyl substituted 2H-furo[2,3-c] pyran-2-ones as germination stimulants present in smoke. *J Agric Food Chem*. 2009, 57: 9475–9480.

He T, Pausas JG, Belcher CM, Schwilk DW, Lamont BB. Fire-adapted traits of Pinus arose in the fiery Cretaceous. *New Phytol*. 2012, 194: 751–759.

López-Bucio J, Millán-Godínez M, Méndez-Bravo A, Morquecho-Contreras A, Ramírez-Chávez E, Molina-Torres J, Pérez-Torres A, Higuchi M, Kakimoto T, Herrera-Estrella L. Cytokinin receptors are involved in alkamide regulation of root and shoot development in Arabidopsis. *Plant Physiol*. 2007, 145: 1703–1713.

Mashiguchi K, Seto Y, Yamaguchi S. Strigolactone biosynthesis, transport and perception. *Plant J*. 2021, 105(2): 335–350. doi: 10.1111/tpj.15059.

Méndez-Bravo A, Raya-González J, Herrera-Estrella L, López-Bucio J. Nitric oxide is involved in alkamide-induced lateral root development in Arabidopsis. *Plant Cell Physiol*. 2010, 51(10): 1612–1626.

Nelson DC, Flematti GR, Ghisalberti EL, Dixon KW, Smith SM. Regulation of seed germination and seedling growth by chemical signals from burning vegetation. *Annu Rev Plant Biol*. 2012, 63: 107–130.

Nelson DC, Riseborough JA, Flematti GR, Stevens J, Ghisalberti EL, Dixon KW, et al. Karrikins discovered in smoke trigger arabidopsis seed germination by a mechanism requiring gibberellic acid synthesis and light. *Plant Physiol*. 2009, 149: 863–873.

Proust H, Hoffmann B, Xie X, Yoneyama K, Schaefer DG, Yoneyama K, Nogué F, Rameau C. Strigolactones regulate protonema branching and act as a quorum sensing-like signal in the moss Physcomitrella patens. *Development*. 2011, 138: 1531–1539. doi: 10.1242/dev.058495.

Ramirez-Chavez E, Lopez-Bucio J, Herrera-Estrella L, Molina-Torres J. Alkamides isolated from plants promote growth and alter root development in Arabidopsis. *Plant Physiol*. 2004, 134: 1058–1068.

Sun YK, Yao J, Scaffidi, A, et al. Divergent receptor proteins confer responses to different karrikins in two ephemeral weeds. *Nat Commun*. 2020, 11: 1264. doi: 10.1038/s41467-020-14991-w.

Wang Q, Smith SM, Huang J. Origins of strigolactone and karrikin signaling in plants. *Trends Plant Sci*. 2022, 27(5): 450–459. doi: 10.1016/j.tplants.2021.11.009.

Waters MT, Scaffidi A, Moulin ALY, Sun YK, Flematti GR, Smith SM. A Selaginella moellendorfii ortholog of karrikin insensitive2 functions in Arabidopsis development but cannot mediate responses to karrikins or strigolactones. *Plant Cell*. 2015, 27: 1925–1944.

Xie X, Yoneyama K, Yoneyama K. The strigolactone story. *Annu Rev Phytopathol.* 2010, 48: 93–117. doi: 10.1146/annurev-phyto-073009–114453.

Yoneyama K. How do strigolactones ameliorate nutrient deficiencies in plants? *Cold Spring Harb Perspect Biol.* 2019, 11(8): a034686. doi: 10.1101/cshperspect.a034686.

Yoneyama K. Recent progress in the chemistry and biochemistry of strigolactones. *J Pestic Sci.* 2020, 45(2): 45–53. doi: 10.1584/jpestics.D19-084.

Zandonadi DB, Matos CRR, Castro RN, et al. Alkamides: A new class of plant growth regulators linked to humic acid bioactivity. *Chem. Biol. Technol. Agric.* 2019, 6(23). doi: 10.1186/s40538-019-0161-4.

Section I

Strigolactones in Plant Biology

1 Biosynthetic Regulation of Strigolactones in Plants

Maryam Valivand

CONTENTS

1.0 INTRODUCTION

Strigolactones (SLs) are a new group of plant-signaling phytohormones implicated in a range of physiological processes of plants and in agriculture. SLs play important roles in the early plant development, germination of dormant seeds in the root of parasitic plants, control of parasitic weeds, regulation of photomorphogenesis, shoot and root architecture via phytochrome, cryptochrome, and hormonal interactions (Omoarelojie et al., 2019). SLs mediate root architecture by restoration of primary root and root hair elongation and suppression of lateral and new adventitious root formation (Marzec and Melzer, 2018). Moreover, SLs are involved in regulating internode length, stem morphology and thickness, cambial activity, and leaf senescence and structure (Al-Babili and Bouwmeester, 2015). In addition, SLs enhance chlorophyll content and delay abscisic acid (ABA)–dependent stomatal closure (Liu et al., 2015). In the interface of plants with arbuscular mycorrhizal fungi (AMFs), SLs, as a hyphal branching stimulator and an allelochemical agent, are essential for successful root colonization (Lanfranco et al., 2018) and improvement of water uptake and mineral nutrition in plants (Wang et al., 2018). SLs also induce changes in fungal gene expression implicated in transcriptional reprograming, construction of chitin oligosaccharides, respiration (Lanfranco et al., 2018), and production of reactive oxygen species in the mitochondria of the fungus (Belmondo et al., 2016). SLs also play important roles in eliminating nutrient deprivation by increasing the number of nodules and enhancing the progress of infection during the symbiosis nitrogen-fixing bacteria with plant roots (McAdam et al., 2017). The resilience and response of plants to abiotic stresses such as salinity, chilling, and drought (Ma et al., 2017; Cooper et al., 2018; Min et al., 2019) and defense against biotic stresses such as fungal and parasite pathogens are regulated by SLs (Cheng et al., 2017). Moreover, SLs are known to increase nitric oxide and hydrogen peroxide contents and can also influence the activity of other phytohormones such as auxin (IAA), cytokinin (CK), gibberellin (GA), ABA, ethylene (ET), jasmonates, salicylic acid, and brassinosteroids (Lv et al., 2018; Omoarelojie et al., 2019).

Therefore, understanding the regulatory mechanisms of the SLs biosynthesis, which is associated with endogenous SLs levels in plants, is a key factor for determining the functions of these signaling molecules in physiological responses and plant growth. The latest excess of genetic, molecular, and biochemical studies on SLs biosynthesis have contributed to improving our understanding of the regulatory mechanisms related to each gene or enzyme.

DOI: 10.1201/9781003224525-3

1.1 BIOSYNTHETIC PATHWAYS AND RELATED ENZYMES

The SLs biosynthesis pathways have been identified in both eudicotyledons and monocotyledons. Recent studies reported that plant roots produce SLs at high levels, whereas hypocotyls, stems, and leaves create them at low levels. Various types of active SLs are synthesized in different quantities (pico- to nanomolar) by a single plant species and diverse varieties of the same plant species (Mostofa et al., 2018; Faizan et al., 2020). As shown in Figure 1.1, in the carotenoid biosynthesis pathway, β caroten is synthesized from all-trans-lycopene by the function of lycopene β cyclase enzyme in the all-trans configuration. Thus, the synthesis of the strigolactone precursor is in the plastids. The first step of the SLs synthesis pathway is the conversion of β caroten from a trans to a cis configuration by the function of DWARF 27 (D27) enzyme and a β caroten isomerase, and this conversion is reversible (Alder et al., 2012). Subsequently, CAROTENOID CLEAVAGE DIOXYGENASE 7 (CCD7) or (DECREASED APICAL DOMINANCE (DAD3), RAMOSUS (RMS5), MORE AXILLARY GROWTH (MAX3), and DWARF (D17) (studied in *Petunia hybrid*, *Pisum sativum*, *Arabidopsis thaliana*, and *Oryza sativa* respectively) enzyme as a stereospecific for cis configuration, cleaves the C9C10 double bond in the trans moiety β carotene and leads to the production of β ionone and 9-cis-β-apo-10-carotenal compounds. The pathway has been proposed for the synthesis of carlactone (CL), an intermediate product and the last common precursor in SLs biosynthesis, in which CL is derived from the 9-cis-β-apo-10-carotenal (C27 compound) by the function of CAROTENOID CLEAVAGE DIOXYGENASE 8 (CCD8) or (DAD1, RMS1, MAX4, D10) enzyme (Alder et al., 2012). In the cytosolic synthetic pathway, CL is converted to carlactonoic acid (CLA) by CYP 450 enzyme (MAX1) studied in Arabidopsis or by an unidentified CYP 450 enzyme to 5-deoxy-strigol, as a parent molecule of canonical SLs (Zhang et al., 2014). An unknown enzyme methylates CLA to produce methyl carlactonoate (MeCLA). In the next step, MeCLA is renovated to an unknown compound that may be a potent SL by the function of LATERAL BRANCHING OXIDOREDUCTASE (LBO) enzyme (Brewer et al., 2016). MAX1-homologue, Os01g0700900 as a CARLACTONE OXIDASE (CO) enzyme is devoted to converting CL to 4-deoxy-orobanchol (ent-2′-epi-5-deoxystrigol). 4-deoxy-orobanchol by Os01g0701400 as an OROBANCHOL SYNTHASE (OS) enzyme is hydroxylated to form orobanchol in rice (Zhang et al., 2014). Moreover, recent studies have reported direct renovation of CLA to orobanchol by catalyzing the cytochrome P450 CYP722C enzyme in tomato and cowpea (Wakabayashi et al., 2019; Aliche et al., 2020).

Among the factors affecting the regulation of the SLs' biosynthesis pathway, two proteins in the chloroplasts, ORT1 in tomato (Koltai et al., 2011) and PPD5 (a Photosystem II subunit) in Arabidopsis (Roose et al., 2011), are assumed. Both *Slort1* and *Atppd5* mutants have lower SL levels, pushy phenotype, and reduce transcripts of chlorophyll and light-associated genes. SL-deficient phenotype is restored by SL treatment. Liu et al. (2011) also declared two GRAS-type transcription factors important for rhizobium root nodule construction in legumes, NODULATION SIGNALING PATHWAY1 (NSP1/2), have regulatory effects on *D27* gene expression in rice and *Medicago truncatula*.

Two inhibitors have been known in SLs biosynthesis: Abaminc (an ABA biosynthesis inhibitor), which inhibited CCDs enzymes and decreased SL production, and TIS108 (an Azole-type compound), which prevented of oxidation of CL and acts downstream of CL in SL biosynthetic steps (Yoneyama et al., 2019).

The structure of SLs contains an A ring (containing one or two methyl groups and one or more acetylonyl or hydroxyl groups), a B ring (containing one or more acetylonyl or hydroxyl groups), and C and D rings with high conservation. The A, B, and C rings together form the fundamental tricycliclactone that they are connected through an enol ether bridge to the D ring, which shows a butenolide group. Based on the stereochemistry of the BC junction and its orientation, natural SLs can be divided into orobanchol-type (α-oriented) and strigol-type (β-oriented) groups. SLs are also classified into two categories – canonical and non-canonical – by their structures. Canonical SLs

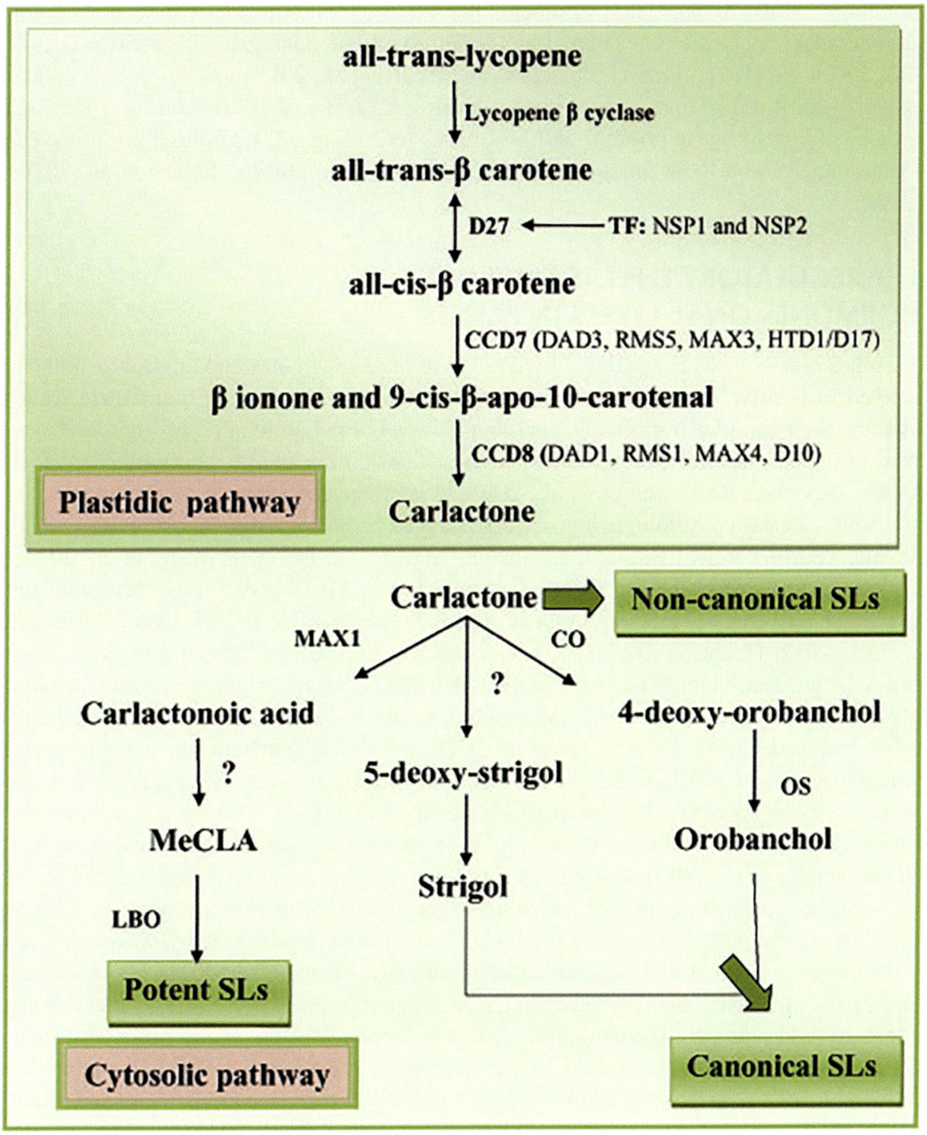

FIGURE 1.1 The model of SLs biosynthesis pathways. In the plastids: β caroten is synthesized from all trans lycopene by lycopene β cyclase enzyme and subsequently during several sequential steps carlactone is synthesized from all-trans-β-carotene by D27 (DWARF 27), CCD7 (carotenoid cleavage dioxygenase 7 encoded by DAD3 in petunia, RMS5 in pea, MAX3 in Arabidopsis, and HTD1/D17 in rice), and CCD8 (carotenoid cleavage dioxygenase8 encoded by DAD1 in petunia, RMS1 in pea, MAX4 in Arabidopsis, and D10 in rice), respectively. Transcription factors NSP1/2 (nodulation signaling pathway) have regulatory effects on D27 enzyme activity in rice and Medicago truncatula. In the cytosol: the functions of MAX1 (more axillary growth 1), an unknown enzyme and LBO (lateral branching oxidoreductase) respectively, transform carlactone into carlactonic acid, MeCLA (methyl carlactonoate) and finally, a potent SLs or by an unidentified CYP 450 enzyme to 5-deoxy-strigol in the case of Arabidopsis whereas in rice the consecutive actions of CO (carlactone oxidase) and OS (orobanchol synthase) convert carlactone into orobanchol.

contain ABC and D rings and non-canonical SLs, which lack some A, B, and C rings but have a methyl furanone (as a D ring) linked to the core structure via an enol ether bond (Banerjee and Roychoudhury, 2018). At present, 18 non-canonical and 23 canonical SLs have been recognized from various plant species (Yoneyama et al., 2019). Synthetic analogs of SL include GR24, GR7, and GR5, which are also commercially accessible (Bharti et al., 2015).

It is worth mentioning that, at the genetic level, *RMS4/MAX2/D3* (encoding a Skp1–Cullin–F-box (SCF) ubiquitin ligase protein) and *D14* genes (encoding α/ß-hydrolase) are involved in SL signal transduction, and their function is downstream of SL synthesis (Faizan et al., 2020; Wani et al., 2020).

1.2 THE REGULATORY EFFECTS OF PLANT HORMONES ON SL LEVELS IN PLANTS

Strigolactones biosynthesis is regulated by a range of plant hormones and environmental cues that are involved as positive and negative regulatory messengers. It is known that SLs-phytohormone crosstalk and their complex hormonal signaling pathway is a dynamic interplay that facilitates plant responses to stress conditions via regulatory nutrient distribution and by modulating developmental and defense processes (Omoarelojie et al., 2019). The prospective role of ABA and SLs has been demonstrated in certain physiological processes, such as plant adaptation to drought and salt stress, by delaying ABA-dependent stomatal closure in plants (Lv et al., 2018; Wang et al., 2018). Some data suggest that SLs synthesis is potentially connected to ABA biosynthesis because of the common precursor (All-trans-β-carotene) that is shared between ABA and SL biosynthesis pathways (Alder et al., 2012; Haider et al., 2018). Lopez-Raez et al. (2010) reported a positive connection between ABA production and SLs levels in the ABA-deficient tomato (*flacca*, *notabilis*, and *sitiens*) mutants and tomato plants treated with AbaminSG, an inhibitor of *NCED* (ABA biosynthetic gene). They also observed lower transcript levels of *CCD7* and *CCD8* genes in the *notabilis* mutant; SL biosynthesis seems to be influenced by ABA. In the work by Ha et al. (2014), ABA was shown to induce transcript levels of *MAX3* and *MAX4* genes in Arabidopsis. This proposes that early ABA synthesis steps might contribute, at least in part, toward the regulation of SL biosynthesis. On the other hand, Haider et al. (2018) declared *D27* transcript levels as a linker and director for regulating ABA and SL homeostasis in rice under stress conditions. Moreover, endogenous SL levels are influenced by the isomerase activity of the HvD27 enzyme in *Hordeum vulgare*, which has a basic role in the directive of ABA and SLs biosynthesis pathways (Wang et al., 2018). However, new studies propose an opposite connection between the SLs contents and ABA synthesis that are organism dependent. In wild type (WT) tomato plants, SLs production was blocked in roots while oversensitivity of SLs-exhausted plants and sensitivity of stomata to ABA led to drought adaptation in shoots (Visentin et al., 2016). As pointed out by Wang et al. (2018), there is an ABA-induced reduction in SL content and SL biosynthesis genes in ABA-deficient barley plants. They also reported a negative association between SL levels and ABA biosynthesis, which organizes tiller formation in barley. In the LOHi lines by silencing of the *HvABA8'*-hydroxylase genes, the ABA level is elevated, which subsequently downregulates the transcript levels of *HvD27*, *HvCCD7/8*, and *HvMAX1* and leads to low SL production. Therefore, at developmental stages, the high-tillering phenotype is observed in this line (Wang et al., 2018). These consequences cumulatively point out that site-specific lowering of SL levels induces local accumulation of ABA, which is essentially useful to engage in abiotic stress. Under osmotic stress, SL levels are low in *Lotus japonicus* roots while expression levels of ABA increased, suggesting the possible role of ABA in the regulation of SL biosynthesis (Liu et al., 2015). However, Ren et al. (2018) observed that ABA upregulated the expression of SL biosynthesis genes in *Sesbania cannabina* seedlings under salt stress.

Lechat et al. (2015) observed that treatment of dormant *Phelipanche ramose* seeds with GR24 and hydroxyurea (methylating agent) hypermethylated PrCYP707A1 (the promoter DNA methylations) of the ABA gene and led to seed dormancy breaking. Therefore, SLs might increase GAs contents, which, in turn, trigger the high accumulation of CK that sets up germination in the later stages. The role of SL as specific regulation of germination has been also shown in SL-defective *Arabidopsis* mutants by preventing the over-expression of the ABA biosynthetic gene *NCED9* during high-temperature stress (Tsuchiya et al., 2010).

The new insight into the crosstalk between GA and SL as the physiological regulatory pathway for SL biosynthesis showed that bioactive GAs in a dose-dependent manner strongly reduced the endogenous levels of 29-epi-5-deoxystriol, (an endogenous SL) in rice roots and root exudates. Furthermore, crosstalk between GA and SLs has been suggested as a promising executive strategy for increasing seed dormancy of the parasitic weed *Striga* (Ito et al., 2017). Gibberellin signaling regulates endogenous SL levels by a GA receptor GIBBERELLIN INSENSITIVE DWARF1 (GID1-DELLA) signaling pathway. The GID and *D14* gene that encode the SL receptor competitively bind to the transcriptional repressor DELLA protein, SLENDER RICE1 (SLR1). The D14 enzyme, by the loss of function of GID1, can dominantly bind to SLR1 and repress SLR1 function. The mechanism is dependent on the negative feedback regulation of the SL biosynthesis genes *D17* and *D10* (Umehara et al., 2010) and independent of SL signaling and *D3* and *D14* genes expression (Ito et al., 2017). Gibberellin treatment downregulated SL biosynthesis genes and maintained their expression levels at the basal level via de novo protein synthesis in roots of rice plants (Tsuchiya et al., 2010). Zou et al. (2019) showed that SL biosynthesis rice mutants exhibited reduced shoot length compared to WT plants. Gibberellin treatment–induced genes expression involved in cell division and elongation ultimately leads to increased GA high bioactive and sensitivity. Moreover, a transcriptional change in an overlapping set of genes was observed in response to GA and GR24 (Lantzouni et al., 2017).

In SL-IAA crosstalk, IAA regulation in an AUXIN RESISTANT1–dependent manner and IAA12 signaling on the SL biosynthesis genes expression under low SL conditions and high IAA content has been shown as a feedback mechanism and a downstream signal. The polar IAA transport stream from apical meristem toward the stems and roots of Arabidopsis promotes *MAX3/4* (and its homolog in pea, *RMS1/5*) transcript levels and leads to SL production. High SL levels reduced IAA export from buds to IAA polar transport, which, in turn, reduced bud outgrowth. On the other hand, low SL levels led to increased IAA content, which, in turn, would feedback upregulate *MAX3/4* genes' expression to increase SL levels. The *MAX3* transcript levels' decrease in Arabidopsis was observed when the apical meristem and young leaves were removed (Hayward et al., 2009). A similar response was reported for the gene expression of *RMS1/5* in the stem of pea plants that were repressed by decapitation but restored by exogenous IAA (Brewer et al., 2009). However, SL signaling has been suggested to be involved in feedback downregulate *MAX3/4* genes expression independent of the IAA pathway (Hayward et al., 2009). Together, these data mean that significant IAA reduction results in increased branching, at least in part by lowering SL synthesis (Brewer et al., 2009). Therefore, IAA and SL signaling could interact in interlocking feedback loops, with each hormone having the possibility to adjust the levels of the other. Moreover, Dun et al. (2012) also depicted the regulatory role of the IAA-dependent AXR1 (AUXIN RESPONSE INCREASED BRANCHING1) pathway on secondary growth in the aerial parts by SL biosynthesis with suppression of the CK biosynthesis genes. On the other hand, Hayward et al. (2009) detected IAA-responsive elements in the promoter sequences of *CCD7, CCD8* genes, and Brewer et al. (2009) showed IAA up-regulation on transcription levels of *CCD7* and *CCD8* genes in Arabidopsis. Rasmussen et al. (2012) also declared IAA induced positive feedback regulation on *MAX3* and *MAX4* genes' expression in the *Arabidopsis d27* mutants and increased transcript levels of *CCD8* gene after exogenous applying by 1-naphthalene acetic acid in Arabidopsis roots. Using CRISPR/Cas9–targeted mutation of the *CCD8SL*-biosynthesis gene, Gao et al. (2018) demonstrated a possible allelic control of SL biosynthesis in *Nicotiana tabacum*. The two mutants' (*Ntccd8a* and *Ntccd8b*) alleles exhibited differential gene expression levels and reaction to IAA treatment, respectively, in root tissues, indicating that SL biosynthesis may be altered at the allelic level. Under heat stress, Hu et al. (2019) showed that the foliar application of GR24 in *Festuca arundinacea* induced leaf elongation that was related to up-regulation of cell cycle-related and down-regulation of IAA transport-related genes expression in elongating leaves. Guillotin et al. (2017) investigated the positive effect of *Sl-IAA27* (an IAA-related gene) on mycorrhization control by regulating SL biosynthesis via *NSP1* in tomatoes. Moreover IAA-SL crosstalk in balancing shoot branching in chrysanthemums under P-deficient conditions has been reported (Xi et al., 2015).

Concerning the role of CK in SL biosynthesis regulation, a negative regulation between CK and SLs has been reported because SLs act as a positive regulator of abiotic stress tolerance, and CK does the opposite (Brewer et al., 2009; Omoarelojie et al., 2019). Using D10-RNAi transgenic lines (in the background of the WT rice plant), Zhang et al. (2010) reported that the application of 10 μM 6-benzylaminopurine in the shoot nodes decreased *D10* gene expression while *HTD1* gene expression was amplified. Furthermore, they found applying 10 μM naphthylacetic acid after decapitation induced *D10* transcription levels to elevate the SL production in rice shoot nodes. They also found that IAA inhibited the genes expression of CK biosynthesis while D10 enzyme promoted CK biosynthesis, revealing that SL regulates CK levels, possibly by reducing IAA levels through a feedback signal. However, Bainbridge et al. (2005) pointed out that CK might affect SL biosynthesis MAX4 promoter::GUS and suggested that CK can inhibit IAA-induced up-regulation of MAX4 SL biosynthesis in transgenic Arabidopsis root tissue.

Both ET and SL signaling participate in definite physiological processes in plants, such as seed germination, root hair elongation, hypocotyl growth, and leaf senescence (Omoarelojie et al., 2019). The relationship between SLs and ET is evident from a study on the Arabidopsis SL-deficient (*MAX1*) mutant that showed ET-induced leaf senescence (Ueda et al., 2015). The authors declared that application of both exogenous GR24 and ET prompted leaf senescence in *MAX1* plants in comparison with ET treatment alone. These results suggest that crosstalk between SLs and ET is involved in delaying leaf senescence in Arabidopsis via suppression of SL-induced ET function.

It is worth noting that recent data pointed to evidence that phytohormones such as jasmonates, salicylic acid, and brassinosteroids are involved in SL signal transduction, and their function is downstream of SL synthesis (Lv et al., 2018; Omoarelojie et al., 2019). Therefore, the exact contribution of other plant hormones in regulation of the SL biosynthesis pathway is to be studied in further detail to understand the significance of hormonal modulation with SL levels.

1.3 THE REGULATORY EFFECTS OF NUTRIENT STARVATION ON SL LEVELS IN PLANTS

The nutrient acquisition approach as limiting factors to plant growth also plays a regulatory role in the SL biosynthesis. Yoneyama et al. (2008) reported that the effects of nutrient deficiency on SL production were extensively dependent on the type of plant, type and degree of nutrient starvation, and uptake strategies. As reported by Umehara et al. (2010), phosphate (Pi) deficiency can negatively regulate the expression of genes in rice roots of WT seedlings that are responsible for changes in SL levels such as *D10*, *D17*, *D27*, and some MAX1-related CYP711A (CYP711A1/MAX1 act downstream of CCDs in Arabidopsis) and leads to a negative correlation between SL concentrations and the number of outgrowing tillers, although SL levels in WT shoot tissues have not been as great as that in the root because of its scarcity. Umehara et al. (2015) also found that P starvation/supplementation differentially regulated *D10*, *D17*, and *D27* genes involved in SL biosynthesis in rice plants. On the other hand, Sun et al. (2014) reported up-regulation of *D3*, *D10*, and *D27* genes expression in both nitrate (N) and Pi starvation in WT rice plants. Several studies also demonstrated that Pi starvation in *Lotus japonicus* (Liu et al., 2015) and in legumes (Pelaez-Vico et al., 2016) or N starvation in sorghum (Yoneyama et al., 2007) elevated SL levels in roots or root exudates. However, Yoneyama et al. (2008) depicted that neither N nor Pi deficiency promoted SL production in white lupin. It seems that plant species, by applying dissimilar strategies, regulate SLs levels in the roots under N/P deficiency. Bonneau et al. (2013) showed in alfalfa up-regulation of 189 genes connected with AM fungal colonization and those integrated SL-biosynthesis genes under Pi/N deficiency. Under similar conditions, they also revealed that in *Medicago truncatula* root, the inspiration of NADPH oxidases led to the regulation of the SLs biosynthesis pathway. Recently, zaxinone (a carotenoid-derived molecule) has been shown to adversely regulate 4-deoxyorobanchol biosynthesis, and a high level of SL has been also confirmed in zaxinone synthase (*zas*) mutant rice seedlings under Pi limiting conditions (Wang et al., 2019). Overall, Pi/N deficiency changed the expression of SL biosynthesis genes, leading to changes in the root architecture, improvement of AM symbioses and colonization, activation and regulation of phosphate starvation marker and Pi transporter genes and reduction of fresh weight in plants (Sun et al., 2014; Ito et al., 2017).

1.4 THE PHOTOMODULATORY REGULATION ON SL LEVELS IN PLANTS

In previous studies, the crosstalk between photoperiod, light intensity and quality, and SL biosynthesis pathways has been mentioned as a required feedback loop for coordinating plant growth and development (Koltai et al., 2011; Bharti et al., 2015). Finlayson et al. (2010) reported in the Arabidopsis phytochrome B (*phyB*) mutant a connection between tall, slender plants with low branch numbers and poor reaction to high red light/far red light conditions. They also observed in Arabidopsis that *MAX4* and *phyB* double mutants repressed the phyB phenotype under earlier conditions and exhibited high branch numbers like the *MAX4* single mutants. These results suggest that SL biosynthesis is controlled by PHYB, and SLs act downstream of PHY B signaling pathways during normal lighting that has an abundance of red wavelength. Similarly, Takagi et al. (2009) observed *Orobanche minor* parasitic plants lacking PHY B function exhibited any more hypocotyls elongating in response to red light and hypothesized that PHY B signaling reduces SL production in these parasitic weeds. A similar regulating role of the light-related gene in the SL synthesis pathway has been also been proven for Arabidopsis plants lacking PPD5 functional (a subunit of Photosystem II) that exhibited increased axillary rosette and root branching. The application of exogenous GR24 treatment restored the SL-defective phenotype (Roose et al., 2011). In the affirmative relationship between SLs biosynthesis and light, Koltai et al. (2011) depicted high light intensity in a time-exposure manner and by changes in IAA efflux carriers, PIN2 distribution and IAA transport can positively enhance the expression of *CCD7* gene in roots of tomatoes. Moreover, they pointed to evidence that light stimulated transcription of the SL biosynthetic genes in the shoots. The positive link between photomodulation, SL accretion, and CCD enzyme activity is demonstrated in the roots and leaves of sunflower seedlings (Bharti et al., 2015) and adventitious root production in the pea mutants (Tsuchiya et al., 2010). This evidence points to the positive regulatory roles of light in SLs production. Rasmussen et al. (2012) showed that Arabidopsis (all *max*) mutants exhibited adventitious root (AR) formation in hypocotyls at etiolation conditions compared with WT plants. Light exposure induced *MAX3/4* genes expression in WT hypocotyls, suggesting that during the process of de-etiolation, local SL biosynthesis is concerned with the regulation of AR formation.

1.5 THE REGULATORY EFFECTS OF ABIOTIC STRESSES ON SL LEVELS IN PLANTS

Recent studies stated the molecular mechanisms involved in SLs responses to abiotic stresses, such as drought (Zhuang et al., 2017; Haider et al., 2018; Min et al., 2019; Visentin et al., 2020), salinity (Ma et al., 2017; Ren et al., 2018), heat (Marzec and Muszynska, 2015), dark chilling (Cooper et al., 2018), freezing (Ramirez et al., 2018), and heavy metals stress (Lovejoy et al., 2021; Mostofa et al., 2021; Qiu et al., 2021). These new findings have contributed to improving our understanding of the regulatory roles of SLs in regulating fine-tuned plant responses and adaptation to abiotic stresses.

A detailed study of *MAX3* and *MAX4* genes expression during drought and salt stress using *Arabidopsis thaliana* showed SLs positively regulate drought response, indicating a clear correlation between MAX mRNA expressions and increased drought tolerance in dehydrated leaf tissues (Ha et al., 2014). The levels of SLs-biosynthetic *LjD27, LjCCD7/CCD8* genes decreased in the roots or root exudates of *Lotus japonicus* under polyethylene glycol–mediated drought stress (Liu et al., 2015). These data were in agreement with the reduction of transcript levels of SLs-biosynthesis *SlCCD7/CCD8* genes (Ruiz-Lozano et al., 2016; Visentin et al., 2016) in the roots of tomatoes under drought that act as a part of the systemic signal of drought stress (Visentin et al., 2016) and a linker molecule between miR156 actions on stomata in tomato leaves (Visentin et al., 2020). Zhuang et al. (2017) observed an association between the inhibition of tiller growth and development with *D17* transcript levels in *Festuca arundinacea* under drought stress, suggesting that drought stress negatively regulates SL production in plants. On the other hand, in rice plants, water-deficit stress upregulated the SLs synthesis *D27* gene expression, designated to play a key role in the plant's adaptation against drought stress (Haider et al., 2018). A similar pattern has been reported for *D10*, *D14*, and *D17* genes expression in the stem base of rice plants (Du et al., 2018). The affirmative effect of SLs in defeating drought stress was also reported in lettuce and tomato root AMF symbiosis (Ruiz-Lozano et al., 2016) and in leaf grapevines (Min et al., 2019). Under salt stress, Ha et al. (2014) showed that Arabidopsis (all *max*) mutants exhibited salt-sensitive phenotypes at vegetative and germination stages compared with WT plants, designating SLs production and signaling their key role in the plant's adaptation against salt stress. Consistently, the increase in SLs contents for induction of salt tolerance helps the symbiosis formation demonstrated in *Sesbania cannabina* seedlings (Ren et al., 2018) and the shoot and root growth of *Brassica napus* plants (Ma et al., 2017).

Under dark-chilling stress, Cooper et al. (2018) showed that Arabidopsis (*MAX4*) and pea (*rms3* and *rms5*) mutants exhibited reduced biomass growth and photosynthetic carbon acclimatization in comparison to WT plants. The existence of GR24 saved the low-temperature sensitivity of the SL-deficient mutants and amplified the dark-chilling tolerance of WT plants. As reported by Ramirez et al. (2018), in Arabidopsis, the recognized gene tbl29/esk1 suppressor mutation decreases the *MAX4* gene expression under freezing stress, which in turn affects the biosynthesis of methyl carlactonoate.

In heat stress conditions, low transcript levels of *D27* and high transcript levels of *MAX3* involved in SLs synthetic pathway were observed in Arabidopsis root tissues, suggesting that SLs act as a plant's survival phytohormones against high-temperature stress (Marzec and Muszynska, 2015).

Recent studies reported that heavy metal–related stresses regulate SLs production in the roots and shoots of dicot and monocot plants. Mostofa et al. (2021) investigated that *d10 and d17* rice mutants exhibited phenotypic abnormalities and reduced chlorosis and biomass compared with WT plants under arsenate stress. Metal stress–induced SL-biosynthetic *D10* and *D17* genes expression in WT shoots ultimately lead to increased metal tolerance through the stimulation of vacuolar sequestration of arsenic and cellular antioxidant defense systems. The favourable effect of SLs on cadmium detoxification and balancing nutrient levels in barley (Qiu et al., 2021) and lead uptake in the roots of sunflowers (Lovejoy et al., 2021) have been reported.

Although, depending on the stress imposed on the plant, the SLs levels in the plant in some cases have been raised. In sum, these interpretations designate the negative regulatory roles of abiotic stress and limiting conditions on gene expression or enzyme activities involved in SL biosynthesis pathways.

1.6 CONCLUSION

As shown in Figure 1.2, it seems that the mechanism of SLs biosynthesis is regulated by complex networks of phytohormones, environmental factors, and abiotic stresses and leads to the physiological role of SLs in changing the architecture (reduced shoot/root ratio and inhibition of shoot branching) and balancing plant growth in response to stress (Omoarelojie et al., 2019; Mostofa et al., 2018). As a result, the expression of SL biosynthetic genes is positively regulated in the roots under N and Pi deficiency. High levels of SLs either negatively affect IAA transport by reduction of PIN2

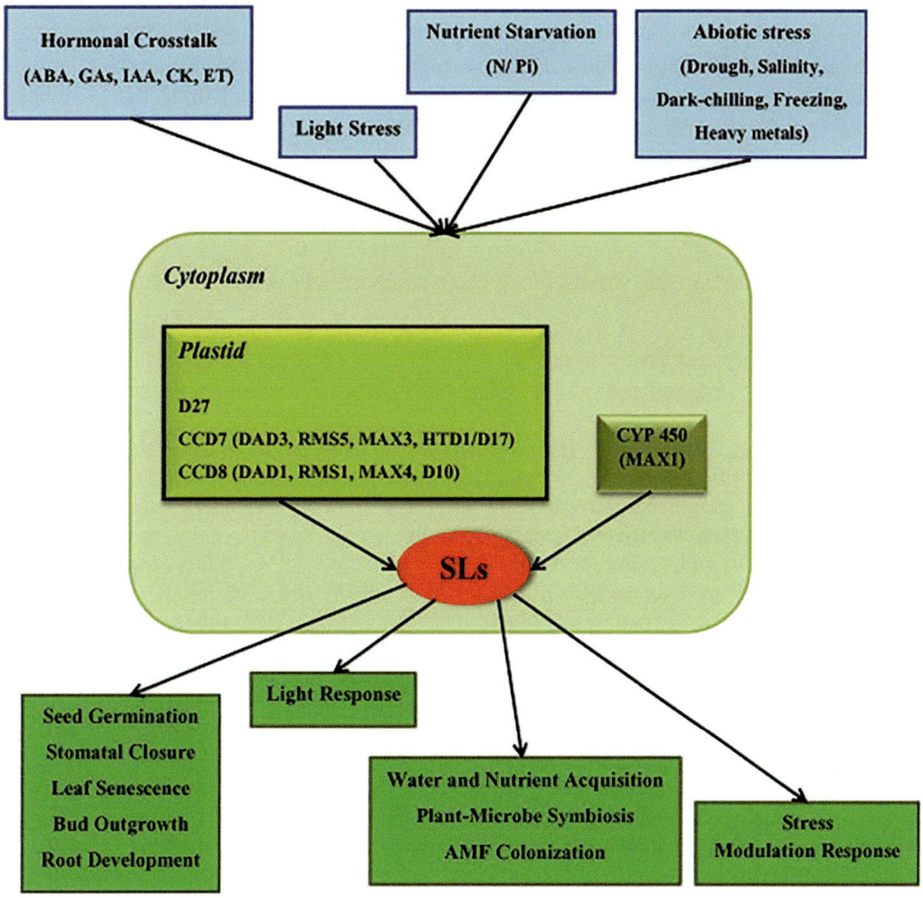

FIGURE 1.2 Schematic diagram of the role of various factors in regulating strigolactones (SLs) biosynthesis. D27 (DWARF 27), CCD7 (carotenoid cleavage dioxygenase7 encoded by DAD3 in petunia, RMS5 in pea, MAX3 in Arabidopsis, and HTD1/D17 in rice), CCD8 (carotenoid cleavage dioxygenase8 encoded by DAD1 in petunia, RMS1 in pea, MAX4 in Arabidopsis, and D10 in rice), and MAX1 (more axillary growth 1) in Arabidopsis.

proteins or regulate the transcriptional level of IAA receptors, namely TIR1 in roots, which, in turn, stimulates the growth of primary and lateral roots and prevents roots elongating (Peret et al., 2014). SLs and ET crosstalk promote the growth of root hair length and density by a recruit of IAA signaling (Omoarelojie et al., 2019). On the other hand, previous studies pointed to evidence that increased SLs levels in the roots not only improve AMFs colonization but also are necessary for complete nodule formation by inducing surface motility in N-fixing bacteria (Ruiz-Lozano et al., 2016; Mostofa et al., 2018). These changes in roots help plants cover larger areas for more efficient macro- and micro-nutrients and water uptake from the soil. Moreover, under environmental stresses such as osmotic, drought, and salinity, SLs production in roots not only decreased to permit an increase in ABA contents in the roots but also may send a systemic signal to aboveground parts to promote SLs synthesis in the local site (Visentin et al., 2016). More ABA flow from root to stem leads to stomatal closure and prevents high transpiration. Less SLs flow from root to stem systemically removes PIN1 from plasma membranes and reduces the IAA transport capacity (Saeed et al., 2017). Therefore, SL functions are downstream of IAA to prevent bud outgrowth as a survival and adaptation strategy to environmental stresses by maintaining local low CK and high SL levels (Brewer et al., 2009).

In sum, the crosstalk between phytohormones and abiotic factors controls key mechanisms at the level of SLs gene expression, synthesis, sensitivity, and transport and regulates plants' accurate responses to environmental conditions. Advance research is still needed to elucidate the particular manners and diverse regulatory mechanisms on genes expression involved in the SL biosynthesis pathways and to detect the effects of probable further factors over physiological and molecular processes involved in the regulation of SLs synthesis.

REFERENCES

Al-Babili S, Bouwmeester HJ. Strigolactones, a novel carotenoid-derived plant hormone. *Annu Rev Plant Biol.* 2015, 66: 161–186.

Alder A, Jamil M, Marzorati M, et al. The path from β-carotene to carlactone, a strigolactone-like plant hormone. *Science.* 2012, 335(6074): 1348–1351.

Aliche EB, Screpanti C, Mesmaeker AD, et al. Science and application of strigolactones. *New Phytol.* 2020, 227: 1001–1011.

Bainbridge K, Sorefan K, Ward S, et al. Hormonally controlled expression of the Arabidopsis *MAX4* shoot branching regulatory gene. *Plant J.* 2005, 44: 569–580.

Banerjee A, Roychoudhury A. Strigolactones: Multi-level regulation of biosynthesis and diverse responses in plant abiotic stresses. *Acta Physiol Plant.* 2018, 40: 86.

Belmondo S, Marschall R, Tudzynski P, et al. Identification of genes involved in fungal responses to strigolactones using mutants from fungal pathogens. *Curr Genet.* 2016, 63(2): 201–213.

Bharti N, Tripathi S, Bhatla SC. Photomodulation of strigolactone biosynthesis and accumulation during sunflower seedling growth. *Plant Signal Behav.* 2015, 10(8): e1049792.

Bonneau L, Huguet S, Wipf D, et al. Combined phosphate and nitrogen limitation generates a nutrient stress transcriptome favorable for arbuscular mycorrhizal symbiosis in Medicago truncatula. *New Phytol.* 2013, 199: 188–202.

Brewer PB, Dun EA, Ferguson BJ, et al. Strigolactone acts downstream of auxin to regulate bud outgrowth in pea and Arabidopsis. *Plant Physiol.* 2009, 150(1): 482–493.

Brewer PB, Yoneyama K, Filardo F, et al. Lateral branching oxidoreductase acts in the final stages of strigolactone biosynthesis in Arabidopsis. *Proc Natl Acad Sci USA.* 2016, 113: 6301–6306.

Cheng X, Floková K, Bouwmeester H, et al. The role of endogenous strigolactones and their interaction with ABA during the infection process of the parasitic weed Phelipanche ramosa in tomato plants. *Front Plant Sci.* 2017, 8: 392.

Cooper JW, Hu Y, Beyyoudh L. Strigolactones positively regulate chilling tolerance in pea and in Arabidopsis. *Plant Cell Environ.* 2018, 41(6): 1298–1310.

Du H, Huang F, Wu N, et al. Integrative regulation of drought escape through ABA-dependent and independent pathways in rice. *Mol Plant.* 2018, 11(4): 584–597.

Dun EA, de Saint Germain A, Rameau C, et al. Antagonistic action of strigolactone and cytokinin in bud outgrowth control. *Plant Physiol.* 2012, 158: 487–449.

Faizan M, Faraz A, Sami F, et al. Role of strigolactones: Signalling and crosstalk with other phytohormones. *Open Life Sci.* 2020, 15: 217–228.

Finlayson SA, Krishnareddy SR, Kebrom TH, et al. Phytochrome regulation of branching in Arabidopsis. *Plant Physiol.* 2010, 152: 1914–1927.

Gao J, Zhang T, Xu B, et al. CRISPR/Cas9-mediated mutagenesis of carotenoid cleavage dioxygenase 8 (CCD8) in tobacco affects shoot and root architecture. *Int J Mol Sci.* 2018, 19: e1062.

Guillotin B, Etemadi M, Audran C, et al. Sl-IAA27 regulates strigolactone biosynthesis and mycorrhization in tomato (var. MicroTom). *New Phytol.* 2017, 213: 1124–1132.

Ha CV, Leyva-Gonzalez MA, Osakabes Y, et al. Positive regulatory role of strigolactone in plant responses to drought and salt stress. *Proc Natl Acad Sci USA.* 2014, 111: 851–856.

Haider I, Andreo-Jimenez B, Bruno M, et al. The interaction of strigolactones with abscisic acid during the drought response in rice. *J Exp Bot.* 2018, 69: 2403–2414.

Hayward A, Stirnberg P, Beveridge C, et al. Interactions between auxin and strigolactone in shoot branching control. *Plant Physiol.* 2009, 151(1): 400–412.

Hu Q, Zhang S, Huang B. Strigolactones promote leaf elongation in tall fescue through upregulation of cell cycle genes and down regulation of auxin transport genes in tall fescue under different temperature regimes. *Int J Mol Sci.* 2019, 20: 1836.

Ito S, Yamagami D, Umehara M, et al. 2017. Regulation of strigolactone biosynthesis by gibberellin signaling. *Plant Physiol.* 2017, 174(2): 1250–1259.

Koltai H, Cohen M, Chesin O, et al. Light is a positive regulator of strigolactone levels in tomato roots. *J Plant Physiol.* 2011, 168: 1993–96.

Lanfranco L, Fiorilli V, Venice F, et al. Strigolactones cross the kingdoms: Plants, fungi, and bacteria in the arbuscular mycorrhizal symbiosis. *J Exp Bot.* 2018, 69: 2175–2188.

Lantzouni O, Klermund C, Schwechheimer C. Largely additive effects of gibberellin and strigolactone on gene expression in Arabidopsis thaliana seedlings. *The Plant Journal.* 2017, 92: 924–938.

Lechat MM, Brun G, Montiel G, et al. Seed response to strigolactone is controlled by abscisic acid-independent DNA methylation in the obligate root parasitic plant, (Phelipanche ramosa L) Pomel. *J Exp Bot.* 2015, 66: 3129–314.

Liu J, He H, Vitali M, et al. Osmotic stress represses strigolactone biosynthesis in Lotus japonicas roots: Exploring the interaction between strigolactones and ABA under abiotic stress. *Planta.* 2015, 241(6): 1435–1451.

Liu W, Kohlen W, Lillo A, et al. Strigolactone biosynthesis in medicago truncatula and rice requires the symbiotic GRAS-type transcription factors NSP1 and NSP2. *Plant Cell.* 2011, 23(12): 3853–3865.

Lopez-Raez JA, Kohlen W, Charnikhova T, et al. Does abscisic acid affect strigolactone biosynthesis? *New Phytol.* 2010, 187: 343–354.

Lovejoy C, Smemo KA. Strigolactone significantly increases lead uptake by dwarf sunflower (Helianthus annuus). *Bioremediat J.* 2021, 25(3): 191–196.

Lv S, Zhang Y, Li C, et al. Strigolactone-triggered stomatal closure requires hydrogen peroxide synthesis and nitric oxide production in an abscisic acid-independent manner. *New Phytol.* 2018, 217(1): 290–304.

Ma N, Hu C, Wan L, et al. Strigolactones improve plant growth, photosynthesis, and alleviate oxidative stress under salinity in rapeseed (Brassica napus L) by regulating gene expression. *Front Plant Sci.* 2017, 27(8): 1671.

Marzec M, Melzer M. Regulation of root development and architecture by strigolactones under optimal and nutrient deficiency conditions. *Int J Mol Sci.* 2018, 19: 1887.

Marzec M, Muszynska A. In silico analysis of the genes encoding proteins that are involved in the biosynthesis of the RMS/MAX/D pathway revealed new roles of strigolactones in plants. *Inte J Mol Sci.* 2015, 16(4): 6757–6782.

McAdam EL, Hugill C, Fort S, et al. Determining the site of action of strigolactones during nodulation. *Plant Physiol.* 2017, 175: 529–542.

Min Z, Li R, Chen L, et al. Alleviation of drought stress in grapevine by foliar-applied strigolactones. *Plant Physiol Biochem.* 2019, 135: 99–110.

Mostofa MG, Ha CV, Rahman Md. M, et al. Strigolactones modulate cellular antioxidant defense mechanisms to mitigate arsenate toxicity in rice shoots. *Antioxidants.* 2021, 10: 1815.

Mostofa MG, Li W, Nguyen KH, et al. Strigolactones in plant adaptation to abiotic stresses: An emerging avenue of plant research. *Plant Cell Environ.* 2018, 41: 2227–2243.

Omoarelojie LO, Kulkarni, MG, Finnie JF, et al. Strigolactones and their crosstalk with other phytohormones. *Annal Bot.* 2019, 124: 749–767.

Pelaez-Vico MA, Bernabéu-Roda L, Kohlen W, et al. Strigolactones in the rhizobium-legume symbiosis: Stimulatory effect on bacterial surface motility and down-regulation of their levels in nodulated plants. *Plant Sci.* 2016, 245: 119–127.

Peret B, Desnos T, Jost R, et al. Root architecture responses: In search of phosphate. *Plant Physiol.* 2014, 166: 1713–1723.

Qiu CW, Zhang C, Wang NH, et al. Strigolactone GR24 improves cadmium tolerance by regulating cadmium uptake, nitric oxide signaling and antioxidant metabolism in barley (Hordeum vulgare L). *Environ Pollut.* 2021, 273: 116486.

Ramirez V, Xiong G, Mashiguchi K, et al. Growth – and stress-related defects associated with wall hypoacetylation are strigolactone dependent. *Plant Direct.* 2018, 2: 1–11.

Rasmussen A, Mason MG, De Cuyper C, et al. Strigolactones suppress adventitious rooting in Arabidopsis and pea. *Plant Physiol.* 2012, 158: 1976–1987.

Ren CG, Kong CC, Xie ZH. Role of abscisic acid in strigolactone induced salt stress tolerance in arbuscular mycorrhizal *Sesbania cannabina* seedlings. *BMC Plant Biol.* 2018, 18: 74.

Roose JL, Frankel LK, Bricker TM. Developmental defects in mutants of the PsbP domain protein 5 in Arabidopsis thaliana. *PLoS One.* 2011, 6(12): e28624.

Ruiz-Lozano JM, Aroca R, Zamarreno AM, et al. Arbuscular mycorrhizal symbiosis induces strigolactone biosynthesis under drought and improves drought tolerance in lettuce and tomato. *Plant Cell Environ.* 2016, 39(2): 441–52.

Saeed W, Nasseem S, Ali Z. Strigolactones biosynthesis and their role in abiotic stress resilience in plants: A critical review. *Front Plant Sci.* 2017, 8: 1487.

Sun H, Tao J, Liu S, et al. Strigolactones are involved in phosphate-and nitrate-deficiency-induced root development and auxin transport in rice. *J Exp Bot.* 2014, 65(22): 6735–6746.

Takagi K, Okazawa A, Wada Y, et al. Unique phytochrome responses of the holoparasitic plant Orobanche minor. *New Phytol.* 2009, 182(4): 965–974.

Tsuchiya Y, Vidaurre D, Toh S, et al. A small-molecule screen identifies new functions for the plant hormone strigolactone. *Nat Chem Biol.* 2010, 6(10): 741.

Ueda H, Kusaba M. Strigolactone regulates leaf senescence in concert with ethylene in Arabidopsis. *Plant Physiol.* 2015, 169: 138–147.

Umehara M, Cao M, Akiyama K, et al. Structural requirements of strigolactones for shoot branching inhibition in rice and Arabidopsis. *Plant Cell Physiol.* 2015, 56(6): 1059–1072.

Umehara M, Hanada A, Magome H, et al. Contribution of strigolactones to the inhibition of tiller bud outgrowth under phosphate deficiency in rice. *Plant Cell Physiol.* 2010, 51: 1118–1126.

Visentin I, Pagliarani C, Deva E, et al. A novel strigolactone-miR156 module controls stomatal behaviour during drought recovery. *Plant Cell Environ.* 2020, 43(7); 1613–1624.

Visentin I, Vitali M, Ferrero M, et al. Low levels of strigolactones in roots as a component of the systemic signal of drought stress in tomato. *New Phytol.* 2016, 212: 954–963.

Wakabayashi T, Hamana M, Mori A, et al. Direct conversion of carlactonoic acid to orobanchol by cytochrome P450 CYP722C in strigolactone biosynthesis. *Sci Adv.* 2019, 5: eaax9067.

Wang H, Chen W, Eggert K, et al. Abscisic acid influences tillering by modulation of strigolactones in barley. *J Exp Bot.* 2018, 69(16): 3883–3898.

Wang JY, Haider I, Jamil M, et al. The apocarotenoid metabolite zaxinone regulates growth and strigolactone biosynthesis in rice. *Nat Commun.* 2019, 10: 810.

Wani KI, Zehra A, Choudhary S, et al. Mechanistic insights into strigolactone biosynthesis, signaling, and regulation during plant growth and development. *J Plant Growth Regul.* 2020, 40: 1836–1852.

Xi L, Wen C, Fang S, et al. Impacts of strigolactone on shoot branching under phosphate starvation in chrysanthemum (Dendranthema grandiflorum cv. Jinba). *Front Plant Sci.* 2015; 6: 694.

Yoneyama K, Xie X, Kusumoto D, et al. Nitrogen deficiency as well as phosphorus deficiency in sorghum promotes the production and exudation of 5-deoxystrigol, the host recognition signal for arbuscular mycorrhizal fungi and root parasites. *Planta.* 2007, 227: 125–132.

Yoneyama K, Xie X, Sekimoto H, et al. Strigolactones, host recognition signals for root parasitic plants and arbuscular mycorrhizal fungi, from Fabaceae plants. *New Phytol.* 2008, 179: 484–494.

Yoneyama K, Xie X, Yoneyama K, et al. Regulation of biosynthesis, perception, and functions of strigolactones for promoting arbuscular mycorrhizal symbiosis and managing root parasitic weeds. *Pest Manage Sci.* 2019, 75(9): 2353–2359.

Zhang S, Li G, Fang J, et al. The Interactions among DWARF10, auxin and cytokinin underlie lateral bud outgrowth in rice. *J Integrat Plant Biol.* 2010, 52(7): 626–638.

Zhang Y, van Dijk AD, Scaffidi A, et al. Rice cytochrome P450 MAX1 homologs catalyze distinct steps in strigolactone biosynthesis. *Nat Chem Biol.* 2014, 10: 1028–103.

Zhuang L, Wang J, Huang B. Drought inhibition of tillering in *Festuca arundinacea* associated with axillary bud development and strigolactone signaling. *Environ Exp Bot.* 2017, 142: 15–23.

Zou X, Wang Q, Chen P, et al. Strigolactones regulate shoot elongation by mediating gibberellin metabolism and signaling in rice (Oryza sativa L). *J Plant Physiol.* 2019, 237: 72–79.

2 Roles of Strigolactone in Plant Growth and Development

Deeksha Singh, Shivangi Mathur, and Rajiv Ranjan

CONTENTS

ABBREVIATIONS:

SLs Strigolactones
GSA Gravitropic set point angle
MAX More axillary growth
CCD Carotenoid cleavage dioxygenase
D14 Dwarf14
LR Lateral root
AR Adventitious root
AMF Arbuscular mycorrhizal fungi

2.0 INTRODUCTION

Life on earth is constantly being challenged by the changing environment. Plants, being sessile, are more vulnerable to environmental fluctuations as they have a lesser option to escape. Usually, they can sustain the mild environmental variations by changing nearly every aspect of their biology in order to thrive, including their morphology, anatomy, physiology, and molecular biology. To maintain uniform growth and maintenance, plants possess finely tuned homeostasis. Despite these

DOI: 10.1201/9781003224525-4

restrictions of limited mobility, they have developed enormous plasticity through pluripotent meristem. All plant meristems are closely coordinated with one another to optimize growth in response of changing environmental conditions. The biochemical processes controlling hormone production and response are thought to provide a genetically determined framework for structural and developmental strategy to optimize the present growth condition while possessing the ability to respond rapidly to the changing environment. Phytohormones are critical to many fundamental aspects of plant biology, including seed germination, flowering time, leaf senescence, fruit ripening, and various other processes of plant development. Plant hormones can typically be classified into five distinct classes: Abscisic acid (ABA), auxins, cytokinins (CKs), gibberellins (GAs), and ethylene. Besides brassinolides and salicylic acid, jasmonates are also widely accepted as plant hormones. Moreover, compounds such as nitric oxide, glucose, systemin, and polyamines are known to have chemical properties that are like plant hormones (McSteen and Zhao, 2008). Many of these are pleiotropic *in vivo* and often work in tandem with other hormones varying chemically.

Strigolactones (SLs) are a new class of endogenous plant hormones that have emerged over the last decade. SLs or their precursors have been shown to behave in a way that controls architectural and developmental processes in plants, especially in inhibiting shoot branching (Umehara et al., 2008; Gomez-Roldan et al., 2008). They are derived from carotenoid compounds, and 80% of terrestrial plants exude them. SLs act as a signaling molecule during symbiotic interaction between plants and arbuscular mycorrhizal fungi and are thought to be a stimulant for parasitic weed seeds to germinate (Akiyama et al., 2005; Besserer et al., 2006). SLs are potential agricultural targets possessing multiple functions, and as an agrochemical, they have the potential to contribute to the maintenance of sustainable agriculture by enhancing the architecture of roots and branching, promoting nutrient uptake, regulating parasitic weeds, enhancing mycorrhization, and reducing droughts. Furthermore, several studies focused on the possible applications of SLs in medicine and the effects they exert on human cells over the last few years. In addition to their anti-apoptotic, anti-inflammatory, and anti-proliferative properties, SLs are found to regulate pathways involved in inflammation and apoptosis (Pollock et al., 2012).

SLs influence factors such as the availability of phosphorus (P) and nitrogen (N) and the intensity of light, droughts, and salinity (Waldie et al., 2014; Sun et al., 2016). Apart from that, these hormones also promote germination of seeds in root parasitic plant species such as *Striga*, *Alectra*, *Phelipanche*, and *Orobanche* (Yoneyama et al., 2013).

2.1 SHORT HISTORY OF SLS RESEARCH

The name *Strigolactone* was given to this hormone group after the discovery of the first candidate in Striga (Cook et al., 1966). Strigol was the first SL elucidated from cotton roots exudates. It was discovered to be a $C_{19}H_{22}O_6$ compound that acts as a germination stimulant for the root parasitic weed, *Striga lutea*; however, it took almost 20 years to fully understand its structure (Coggon et al., 1973; Brooks et al., 1985). Strigol was the only known naturally occurring SL for a long time.

SLs are present in minute amounts in root exudates, providing difficulties in determining their structure (Zwanenburg and Pospíšil, 2013). Based on the potential to stimulate seed germination in root parasitic plants including *Striga*, *Alectra*, *Orobanche*, and *Phelipanche* spp., several structurally similar chemicals have also been identified from root exudates of these plant species (Xie et al., 2010; Yoneyama et al., 2013). After strigol was discovered over 30 years ago, *strigolactone* was coined as a collective term to designate this family of compounds (Butler, 1995).

2.2 CHEMICAL STRUCTURE

There are over 25 SLs identified across the plant kingdom, and each plant species produces different blends of several SLs (Bürger and Chory, 2020). Naturally, there are two major structural classes of SLs: Canonical SLs and non-canonical SLs (Figure 2.1), which differ in whether or not they possess the complete ABC-ring system (Wang and Bouwmeester, 2018; Yoneyama et al., 2018). A conserved

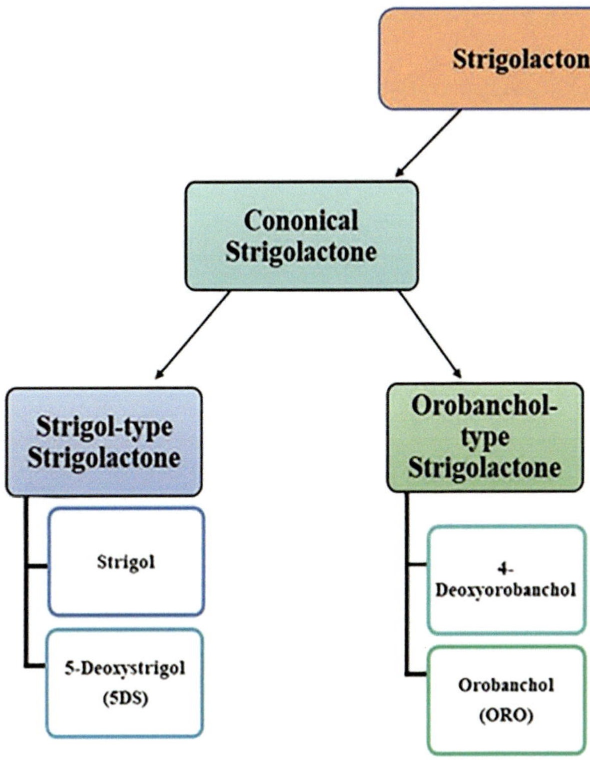

FIGURE 2.1 List of naturally occurring strigolactone.

feature of both canonical and non-canonical structure is the enol-ether bridge and D-ring, which link the ABC core of the molecule (Zwanenburg and Blanco-Ania, 2018). Based on their differing structural characteristics, SLs are characterized by a wide range of functional qualities (Scaffidi et al., 2013). Looking at the structures, it should be noted that the A-ring may contain various oxygen functions at different positions, and the stereochemistry at the B/C junction along with C-2′ is a variable and must be taken into account. Originally, it was thought that the absolute configuration of (+)-strigol (Figure 2.2) would be the only natural configuration (Zwanenburg and Pospíšil, 2013). Flematti et al. (2016) have emphasized the importance of stereochemistry in fine-tuning the biological properties of SLs. These are from the known SLs families: namely, strigol and orobanchol. Based on the stereochemically arranged C-ring, SL can be divided into (-)-orobanchol and (+)-strigol types. In each naturally occurring SL, the butenolide D-ring always occupies the same spatial position at C-2′ (Table 2.1).

Analogues of SLs are essential, having a potential application in the agriculture and biomedicine fields. The replacement of the A-ring by an aromatic ring gives a highly active analogue of strigol known as GR24 (Figure 2.2). Removal of the A-ring leads to analogue GR7, while omitting the A- and B-rings gives compound GR5 (Zwanenburg et al., 2009). GR24 is popular in SL research as it is more stable than natural SLs (Nakamura and Asami, 2014).

2.3 PHYSIOLOGICAL FUNCTIONS

In parasitic plants, strigolactones may have the potential to affect shoot bifurcation, fungal hyphal branching, and seed maturation, which might offer a way to synthesize similar compounds to control these processes. Once parasitic weed seeds germinate, they will die if they cannot reach a host root within a week. A cost-effective chemical analogue of strigolactone that stimulates the germination of these seeds could be synthesized and widely used to trap parasitic weeds in the soil prior to sowing or to stimulate

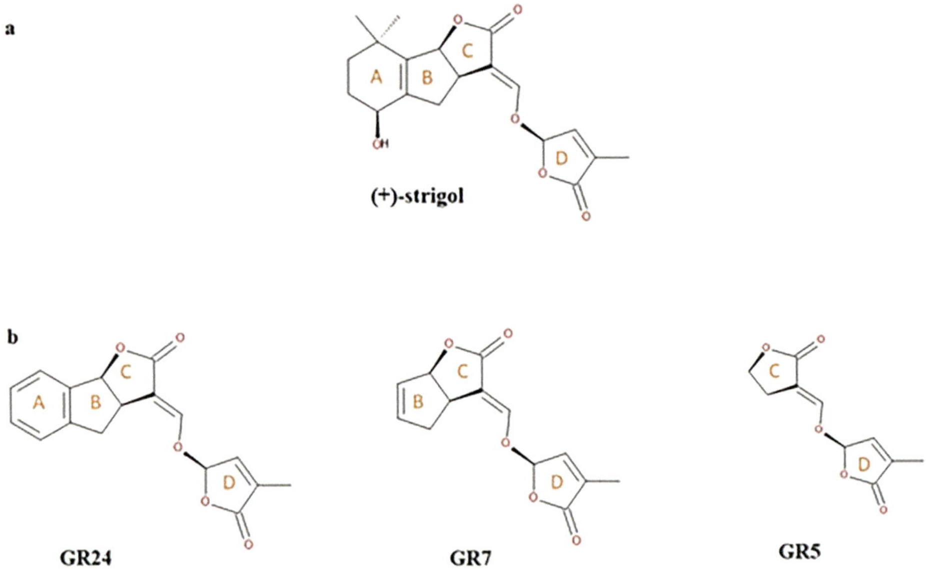

FIGURE 2.2 Structure of strigolactone and their analogues: a. naturally occurring SL: strigol; b. analogs of SL: GR24, GR7, and GR5.

TABLE 2.1
Distribution of SLs in Plant Kingdom

Plant Species	Strigolactone Found	Strigolactone-Responsive Genes	Reference
*Gossypium hirsutum*L. (Cotton)	Strigol and strigyl acetate	D14 (DWARF14)	Wang et al., 2019
Zea mays (Maize)	5-deoxystrigol, Zeapyranolactone	*CCD8 gene* (CAROTENOID CLEAVAGE DIOXYGENASE 8), TEOSINTE BRANCHED1 (TB1)	Guan et al., 2012; Makhzoum et al., 2017; Charnikhova et al., 2018
Oryza sativa (Rice)	4-deoxyorobanchol, orobanchyl acetate, orobanchol, 7-oxoorobanchyl acetate	OsCKX 9 (CYTOKININ OXIDASE/ DEHYDROGENASE9) And D53	Jiang et al., 2013; Zhou et al., 2013; Duan et al., 2019
Lycopersicon esculentum (Tomato)	Orobanchol, 4-deoxyorobanchol, 7-hydroxyorobanchol solanacol	HSP70 (HEAT SHOCK PROTEIN 70), CBF1(CENTROMERE-BINDING PROTEIN 1), and antioxidant-related genes	Chi et al., 2021
Helianthus annuus (Sunflower)	CLA, heliolactone	Salinity tolerance gene	Zulfiqar et al., 2021
Solanum tuberosum (Potato)	Orobancol and orobancyl acetate	CC8 (CAROTENOID CLEAVAGE DIOXYGENASE8)	Roumeliotis et al., 2012; Pasare et al., 2013

(Continued)

TABLE 2.1
(Continued)

Plant Species	Strigolactone Found	Strigolactone-Responsive Genes	Reference
Nicotiana tabacum (Tobacco)	4-deoxyorobanchol, 4-acetoxy-5-deoxystrigol, orobanchol, solanacyl acetate5-deoxystrigol orobanchyl acetate, solanacol, 4-hydroxy-5-deoxystrigol	PDR1 and 6	Xie et al., 2021
Pisum sativum (Pea)	orobanchol, 4-deoxyorobanchol, fabacol, fabacyl acetate, orobanchyl acetate	BRC1 (BRANCHED 1)	Braun et al., 2012
Arabidopsis thaliana (Arabidopsis)	Orobanchol, Strigol, Orobanchyl acetate	MORE AXILLARY GROWTH (MAX3, MAX4), STH7 (SALT TOLERANCE HOMOLOG 7), BRC1(BRANCHED 1), SMXL2, 6, 7, 8, DWARF14-LIKE2 (DLK2), KUF1(KAR-UP F-BOX1), BZS1	Wang et al., 2015; Mashiguchi et al., 2009; Scaffidi et al., 2013; Waters et al., 2015; Lantzouni et al., 2017; Booker et al., 2005
Cucumis sativus (Cucumbers)	7-oxoorobanchyl acetate, 7-hydroxyorobanchyl Acetate, 4-deoxyorobanchol, orobanchol	SPS (SUCROSE PHOSPHATE SYNTHASE) and SS (SUCROSE SYNTHASE)	Xie, 2016; Zhou et al., 2022
Panicum miliaceum (Proso millet)	Strigol	-	Makhzoum et al., 2017
Sorghum bicolor (Sorghum)	Sorgolactone, Strigol, 5-deoxystrigol	D27, CCD7 and CCD8	Abdelhalim et al., 2019; Bhoi et al., 2021

premature germination so that the parasitic weed seedlings would perish before the crop plants could take root. Additionally, strigolactones may also be of great benefit to agriculture due to their ability to inhibit shoot branching. Hence, non-bearing tillers form, which capture the limited nutrients from their seed-bearing counterparts. By applying analogues, it would be possible to reduce non-bearing tillers, keeping nutrients and photosynthesizing directed toward seed production. The development of chemicals that predictably alter plant architecture would also prove beneficial to the horticultural industry.

2.4 STRIGOLACTONE AND SHOOT STRUCTURE REGULATION

2.4.1 SLs Suppress Shoot Branching/Bifurcation

The number of branches in plants must be regulated to maintain plant architecture and produce biomass efficiently. Axillary buds emerge in the axils of numerous leaves, and their expansion causes shoot branching. The results of genetic analysis showed that carotenoid signaling molecules, SLs, were found to be one of the main components repressing the axillary bud process (Figure 2.3). Throughout the past decade, several plant species have developed mutants that display increased shoot branching and in which known hormones, including cytokinin, are not responsible for bud outgrowth. Decreased Apical Dominance1 (DAD1) in *Petunia hybrida*, *Pisum sativum* showed Ramosus1 (rms1)

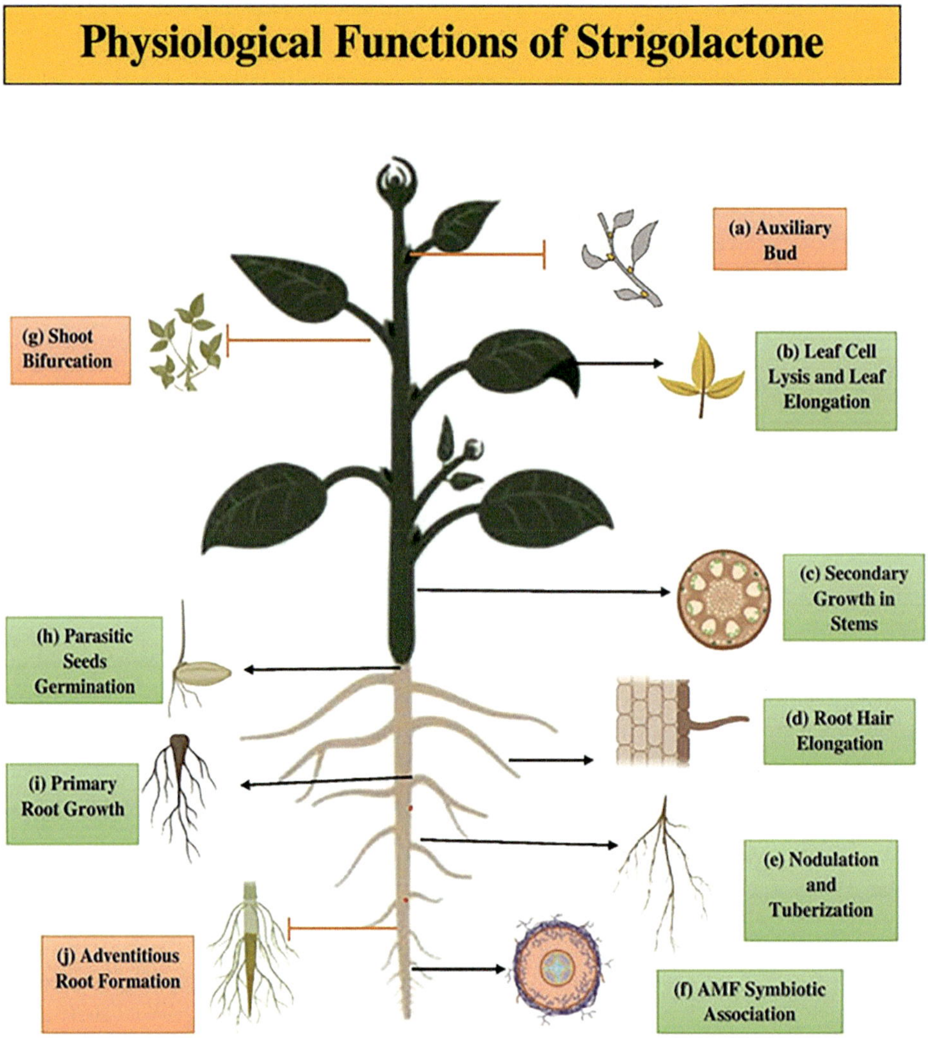

FIGURE 2.3 SLs in plant development: their roles and effects. The SLs controls a variety of plant growth and development processes. In photosynthetic plants, SLs upregulate the leaf elongation and leaf cell lysis (b), secondary growth in stems (c), root hair elongation (d), nodulation and tuberization (e), and primary root growth (i). SLs positively regulate the AMF symbiotic association (f) and germination of the parasitic seeds (h). However, they downregulate axillary bud growth (a), shoot branching/bifurcation (g), and adventitious root formation (j). Black arrows along with green boxes depict the upregulating effects while the T-bars in orange along with red boxes depict downregulating effects of SLs.

to Ramosus5, More Axillary Growth1–4 (MAX1–4) as well as DWARF14 (AtD14) in *Arabidopsis thaliana* and *Oryza sativa* showed dwarf and high-tillering dwarf (d/htd) are among those plant species (Simons et al., 2007; Beveridge et al., 2009; Waldie et al., 2014; Yoshimura et al., 2018). The cloning of the RMS1 or RMS5 and MAX3 or MAX4 genes from various plants, which encode the CCD7 and CCD8, provided a significant clue to the discovery of this unknown branching inhibitor (Booker et al., 2004; Sorefan et al., 2003). Yoshimulactone green (YLG), a fluorescence turn-on probe emits a high fluorescence when hydrolyzed by D14-type strigolactone receptors. By applying YLG *in vitro*, a novel molecule, DL1, was identified as a potential inhibitor of D14. This DL1 competes with endogenous SLs in *Arabidopsis* and rice to increase the number of shoot branches.

Matusova et al. (2005) predicted that SLs would be generated from carotenoids by employing carotenoid-deficient maize mutants and carotenoid biosynthesis inhibitors in cowpea, maize, and sorghum. Exogenous applications of racemic-GR24 (rac-GR24) were used to investigate if it reduced shoot branching, demonstrating that SLs were also the elusive branching inhibitor. In fact, rac-GR24 retarded the development of axillary buds in SL biosynthesis mutants but not in responsive variants, indicating an enigmatic branching regulator, the SL (Umehara et al., 2008; Gomez-Roldan et al., 2008).

Waters et al., 2017 proposed that SLs regulate auxin transport. BRANCHED1 (BRC1), a TCP transcription factor closely related to TEOSINTE BRANCHED1 (TB1) from maize, is considered to work downstream from auxin and SL in *Arabidopsis* (Aguilar-Martínez et al., 2007; Martín-Trillo et al., 2011). The SL signaling pathway is partially mediated by BRC1 and TB1-related genes in the control of axillary bud development (Minakuchi et al., 2010). A genetic study in *Arabidopsis* indicated that the effects of BRC1 and SLs also appear to be partially independent (Seale et al., 2017). In pea, the expression of BRC1 has been found to be suppressed by cytokinin and sucrose, both of which are antagonistic to SLs in the control of axillary bud outgrowth (Mason et al., 2014).

2.4.2 SLs Elevate Internode Elongation and Height

The most conspicuous characteristic of SL-deficient and signaling variants is their strong shoot branching while another notable trait of these genotypes is dwarfism and short internodes. In SL-deficient mutants, axillary buds or branches of pea plants were manually removed to investigate whether this phenotype resulted from energy being diverted to lateral shoots. A cytological examination of internode epidermal cells implied that SLs may alter stem elongation via encouraging cell division. Even when the plants only had one main stem, SL mutants remained stunted, despite the treatment internode length not being altered. This suggests that their short stature was not solely due to the higher branching of SL deficient mutants. Moreover, no further increase in shoot branching in *Medicago truncatula* (barrel medic) was observed, and SL-deficient mutants still had shorter internodes than wild-type plants (Lauressergues et al., 2015). SLs were found to stimulate internode elongation in pea by altering cell division, functioning independent of GA. A combination of mutants with altered SL and GA levels resulted in additive effects on internode elongation.

2.4.3 SLs Regulates Secondary Growth in Stems

Mutants that are related to SL have shorter stem diameters than wild types. Secondary growth is regulated by the cylindrical meristems and vascular cambium, which divide radially, providing stems their thickness. Their cells proliferated by a long-distance signal from the plant hormone auxin. The vascular cambium is responsible for forming the vascular bundles of xylem capillaries and phloem capillaries, and the interfascicular cambium occurs between these two types of cambiums. Cambium activity is reduced in *Arabidopsis* SL mutants compared to wild types: cambium-specific genes are expressed at lower levels, and tissue generated from the interfascicular cambium has less lateral extension in the stems. Additionally, cambium activity is reduced in the pea mutant rmsl–1 (impaired in SL biosynthesis), whereas GR24 treatments positively affect secondary vascular tissue production in *Eucalyptus*. It demonstrated that SL acts as a generally conserved signaling mechanism coordinating the growth of different plants and organs. According to the study of Agusti et al., 2011, the regional rac-GR24 treatment of stems increases cell division in interfascicular areas, indicating that SLs have a significant impact on cambium functioning (Figure 2.3). Decapitation tests, however, not only induce axillary bud development but also inhibit stem secondary growth, indicating that the regulation of branching and secondary growth are interdependent.

2.4.4 SLs Regulates the Leaf Elongation and Its Shape

In leafy crops, leaf elongation is a critical element in shoot biomass production. Cell elongation/expansion and cell division frequency are two developmental factors that affect leaf elongation (Skinner and Nelson, 1994). SL-deficient and response variants in *Arabidopsis* have rounder rosette leaves with smaller petioles (Stirnberg et al., 2002; Soundappan et al., 2015). Moreover, SL-deficient mutants maintained in hydroponic culture treated with rac-GR24 recover leaf curvature (Scaffidi et al., 2013). The degree of serration is reduced in SL mutants. To determine whether the serration phenotype was caused by SLs, the main shoot apex of the SL-deficient mutant was treated with rac-GR24, resulting in sharper serrations than the mock-treated leaflets (Lauressergues et al., 2015). According to Hu et al., 2019, SLs could prevent the retardation of leaf elongation caused by high temperature in *Festuca arundinacea* (tall fescue). In tall fescue, exogenous application of GR24 induced elongation of the leaves and mitigated heat inhibition, accompanied by an increase in cell numbers, upregulation of cell-cycle genes, and reduced apoptosis. The SL-deficient and SL-insensitive *Medicago truncatula* mutants have a different leaf morphology, with fewer serrations near the leaf margin. Exogenous application of SL could cure the serration phenotype of SL-deficient mutants and had no impact on SL-insensitive mutants.

2.4.5 Effect of SLs on Gravitropic Set Point Angle (GSA)

Gravity influences the orientation of plant growth. It is believed that plants can sense gravity and alter their growth orientation accordingly. Shoot architecture and yield are influenced by the angle between the branches and the main stem. In pea and *Arabidopsis*, SL mutants exhibit significantly reduced branch angles (Liang et al., 2016). Negative gravitropism is a fundamental characteristic of shoot organs for keeping branch orientations at non-vertical angles. GSA is controlled by several genes, including members of the LAZY1 gene family. In rice and *Arabidopsis*, LAZY1 mutants have defects in shoot gravitropism and a wide branch angle phenotype. Three LAZY genes are mainly expressed in the endodermal cells of *Arabidopsis* shoots that regulate asymmetric auxin distribution between the upper and lower sides of the stem, which is essential for the gravitropic response (Taniguchi et al., 2017). This auxin biosynthesis repression of SLs was demonstrated in peas (Ligerot et al., 2017). Genetic screening for LAZY1 suppressors in rice plants resulted in the discovery of mutants with SL signaling or biosynthesis defects. However, it has been hypothesized that SL modulation of shoot gravitropism is independent of LAZY1. This asymmetric auxin concentration is compensated by SLs, which suppress auxin levels on the lower side of the shoot (Sang et al., 2014).

2.4.6 SLs Regulate Flowering Time Control and Tuberization

Mutants and transgenic lines of SL-biosynthesis genes (typically CCD7 or CCD8 orthologs) with lower mRNA levels (RNA interference lines, RNAi) have been explored in a variety of species. Reduced SL synthesis causes late-flowering phenotypes in *Petunia*. The DAD1 mutant possessing transposonal insertion in PhCCD8, as well as RNAi lines of DAD1, flowers later than wild types (Snowden et al., 2005). The flower and fruit setting of SL-defective *Lycopersicum esculantum* and *Lotus japonicus* (bird's-foot trefoil) plants are affected greatly (Kohlen et al., 2012; Vogel et al., 2010; Liu et al., 2013). RNAi lines generated from potato StCCD8 genes have provided new insight into SL functions. In addition, these RNAi lines exhibited abnormal tuber formation and development. SLs, along with auxin and GA, regulate stolon architecture by suppressing axillary bud growth (Roumeliotis et al., 2012). The CCD8 gene, which is important in the SL biosynthesis pathway, was silenced in transgenic potato plants through RNAi. As a result, RNAi potato plants had an increased number of lateral branches and reduced stolon development, along with a lack of flowering. Tubers were more numerous and smaller in size, resulting in a total tuber yield up to three times lower in StCCD8 RNAi plants (Pasare et al., 2013).

2.5 SLS AND ROOT STRUCTURE REGULATION

2.5.1 SLs Regulate Root Length

The growth of primary roots could be controlled by SLs (Ruyter-Spira et al., 2011). The main root is slightly shortened in different SL mutants, including *Arabidopsis*, *Oryza sativa*, and *Hordeum vulgare*. According to studies in *Arabidopsis*, the decreased length may be because of a smaller root meristem than in wild types. It is evident from these results that the main root length increases in several plant species when treated with rac-GR24. In contrast, treatment with relatively higher doses of GR24 resulted in an inhibition of primary root (PR) length in a MAX2-independent manner. There is reduction in PR during carbohydrate restriction. Nevertheless, it is important to note that this effect is influenced by phosphate or nitrate concentration in the medium in rice (Sun et al., 2014). However, transgenic *Lotus japonicus* plants containing the silenced SL biosynthesis gene CCD7 produced longer roots than wild-type plants (Liu et al., 2013; Koltai et al., 2010; De Cuyper et al., 2015). Compared to other plant hormones, SLs have a very subtle effect on the main root length, while auxins have a much stronger effect. This is related to their probable action on plant growth (Matthys et al., 2016). The effects of SLs may therefore be masked by other hormones that control the length of the root due to differences in the general hormonal landscape between different species of plants or based on the nutrient conditions under which the plant is grown.

2.5.2 SLs Regulate Development of Lateral Roots

In several studies, SLs have been shown to influence the growth of lateral roots (Ruyter-Spira et al., 2011; De Cuyper et al., 2015; Sun et al., 2014). When SL (and therefore, KAI2-related) signaling is disrupted in *Arabidopsis* MAX2 mutants, the number of lateral roots per length increases. Soundappan et al. (2015) showed that smxl 6, 7, and 8 triple mutants lacking the SL signaling repressors reduce this effect, suggesting that SL may play an important role in this process. Although rice SL biosynthesis mutants show a constant increase in lateral root density, *Arabidopsis* MAX3 and MAX4 biosynthesis mutants show no obvious abnormalities in lateral growth (Kapulnik et al., 2011; Sun et al., 2014; Ruyter-Spira et al., 2011). SL-deficient or insensitive mutants of tomato generated significantly higher LR densities than SL-sensitive mutants (Koltai et al., 2010) while GR24 inhibited LR growth by affecting the MAX2 gene, which is responsible for the LR signaling pathway in *Arabidopsis*. Treatments with rac-GR24 also produce a reduction in lateral root density in wild-type rice plants. The suppression of lateral root development in the upper portion of the root in *Arabidopsis* has been proven to be the main cause of this reaction (Jiang et al., 2016). This phenotypic characteristic could be driven by signaling via the SL or KL pathways. Similarly, both the GR245-DS and GR24ent-DS enantiomers mimic indigenous SL and karrikins, respectively, causing a reduction in lateral root density, implying that both SL and KL play a significant role in lateral root development control (Li et al., 2016).

2.5.3 SLs in Adventitious Root Development

Many plant species have evidence that SLs play a role in adventitious root development. The SLs promote the growth of seminal roots and adventitious roots (ARs), and they repress the lateral roots (LRs) growth (Kapulnik et al., 2011; Kumar et al., 2015). SLs were also linked to the production of adventitious roots in pea and *Arabidopsis*, as well as in rice crown roots, which are developmentally equivalent (Rasmussen et al., 2012). In *Arabidopsis*, SL mutants have shortened primary roots relative to wild-type plants because they have fewer meristem cells than wild-type plants (Ruyter-Spira et al., 2011). GR24 could nullify this effect in mutants but not in SL-insensitive plants. In rice, SL-biosynthesis variants possess shorter and fewer crown roots than the wild type (Sun et al., 2015). It has been noted that rice SL mutants produce fewer ARs at the seedling stage and have a lower number of ARs per tiller during maturity. When the synthetic SL mutant d10 was treated with GR24, the AR number was complemented, but it was not complemented in SL signaling mutants.

2.5.4 Strigolactone Role in Root Hair Development

Root hair density and prolongation are other root developmental processes in which SLs play a crucial role. The root hair density in SL signaling and biosynthesis mutants MAX2 and MAX4 is significantly reduced despite nutrient limitation (Mayzlish-Gati et al., 2012). Additionally, both the SL signaling and biosynthesis mutants MAX2 and MAX1 showed shorter roots under phosphate-deficient conditions (Ito et al., 2015). The roots of wild-type tomato and *Arabidopsis* exhibited increased elongation when exposed to rac-GR24 but showed no effect in *Medicago* (Koltai et al., 2010; Kapulnik et al., 2011; De Cuyper et al., 2015). Studying different enantiomers of synthetic SLs, researchers determined that compounds that act on root hair growth are more potent than for KAI2. However, KAI2 was still involved in root hair development (Li et al., 2016; Artuso et al., 2015). Considering all the evidences, it seems like root hair development may not just be governed by SLs but also by KAI2-dependent signaling.

2.6 STRIGOLACTONES AS GERMINATION STIMULANTS FOR PARASITIC PLANT SEEDS

A strigolactone is a terpenoid-based plant hormone that acts as communication signal within the plant and between plants and fungus, as well as in parasitic plants and their hosts. Plant roots are colonized by parasitic weeds like *Striga* spp. (witchweed) and *Orobanche* spp. (broomrape). The parasites deplete the water and nutrients of the host plant, making agricultural cultivation a severe concern in many regions of the world. Most parasitic weed species are obligatory parasites, so they cannot complete their life cycle without a host. The initial step in their life span is seed germination. Parasitic weed seeds are quite minute (0.2–0.3 mm) and have very limited nutritional reserves. They are fully reliant on the existence of a suitable host after they have germinated. The seedling will die if the developing parasite radicle does not form contact with a host root. Parasitic plant seeds remain latent in the soil until they have been exposed to stimuli produced and secreted by the host roots. Strigolactones are generated and secreted by the roots of almost all higher plants; parasitic weed seedlings receive these molecules as a chemotactic signal. Strigolactones are the only chemical stimulants identified from plant roots (Bouwmeester et al., 2003). The germination-stimulating ability of natural SLs was examined in seeds of *Orobanche minor*, which is an obligate root parasite. The most effective germination stimulant consists of three monohydroxy-SLs: orobanchol, 2-epiorobanchol, and sorgomol, which were able to induce 80% germination of the seeds of *O. minor* (Kim et al., 2010). Strigolactones are highly unstable in soil and serve as a short-range signal molecule; therefore, their existence shows the presence of a host root. Additionally, they cause hyphal branching at low concentrations. Strigolactones are also very efficient against parasitic weed seeds; in the presence of strigolactones at the picogram level, 50% of seeds germinate.

2.7 THE ROLE OF STRIGOLACTONES IN PLANT-ARBUSCULAR MYCORRHIZA INTERACTION

Arbuscular mycorrhizal fungi and plant roots form one of the oldest and most ecologically significant symbiotic relationships between higher plants and microbes. Yet the process by which these fungi detect the presence of plants is poorly understood. During symbiotic interactions between plants and AM fungi, signal molecules are exchanged from both sides. Parniske, 2008 reported SLs to be the branching factors for AM fungi. *Orobanchol* has the most action, followed by 5-deoxystrigol. GR24, a strigolactone analogue, is highly active like strigol, unlike its enantiomer. The strigolactone analogue GR7 was discovered to be a thousand times less active than GR24. Various studies have suggested that SLs need to have all ABC rings to function as active branching factors for fungi. SLs should also possess the same stereochemistry as strigols. In contrast, reports indicated that branching factors can be produced without the B-ring. Plant parasites and AM fungi are synergistic (Bonfante and Requena, 2011). By increasing soil mineral (phosphate, nitrate) uptake, the fungus acts as a biological fertilizer. The growth of arbuscular mycorrhizal (AM) symbiosis is regulated by plant hormones, microRNAs, and secreted peptides in relation to the plant's phosphorus status (Müller and Harrison, 2019). *Rhizophagus* activates the strigolactone biosynthesis genes CCD7 and CCD8 in *Arabidopsis* roots and induces strigolactone biosynthesis gene CCD7 and CCD8 expression in the early stages of association (Fernandez et al., 2019). Through a better understanding of the symbiotic interaction between beneficial fungi and parasitic weeds in agriculture, new experimental methods could be developed for controlling and managing these relationships.

2.8 CONCLUSION

From the discussion in this chapter, it may be concluded that SLs are carotenoid-derived plant hormones and rhizosphere signaling molecules. They facilitate communication between plants and symbiotic mycorrhizal fungi, and they have been deliberately manipulated by root parasitic weeds to trigger seed germination. Several bioactivities of SLs and their analogues have been observed, including the germination of seeds of parasitic weeds, hyphal branching of AM fungi, and inhibition of shoot branching of grasses. These bioactivities have distinct biophysical characteristics, suggesting that each of these functions involves a distinct perception system. Nevertheless, the discovery of strigolactones already opens a variety of novel agricultural and horticultural research opportunities, such as suppressing unwanted vegetative growth after pruning. By learning their transport, regulation, and defense mechanisms against hostile environments, SLs can be made useful for improving crop production under climate change, which can lead to substantial food and nutritional security for the growing population. Given the mobility of SLs both inside and outside the plant, as well as their significance in plant development, a better knowledge of their impact on plant-organism interactions is important for sustainable research. This understanding will open new avenues for biotechnological applications to manipulate plant interactions to promote favourable connections by exploiting them as biofertilizers and bioprotectants while avoiding their negative consequences. Although much has been learned about the behaviour of SLs throughout the years, many more features and practical applications of these fascinating new plant hormones will be discovered in the upcoming years.

REFERENCES

Abdelhalim T, Jannoura R, Joergensen RG. Mycorrhiza response and phosphorus acquisition efficiency of sorghum cultivars differing in strigolactone composition. *Plant and Soil*. 2019; 437(1): 55–63.

Aguilar-Martínez JA, Poza-Carrion C, Cubas P. Arabidopsis branched1 acts as an integrator of branching signals within axillary buds. *The Plant Cell*. 2007; 19(2): 458–472.

Agusti J, Herold S, Schwarz M, et al. Strigolactone signaling is required for auxin-dependent stimulation of secondary growth in plants. *Proceedings of the National Academy of Sciences*. 2011; 108(50): 20242–20247.

Akiyama K, Matsuzaki KI, Hayashi H. Plant sesquiterpenes induce hyphal branching in arbuscular mycorrhizal fungi. *Nature*. 2005; 435(7043): 824–827.

Artuso E, Ghibaudi E, Lace B, et al. Stereochemical assignment of strigolactone analogues confirms their selective biological activity. *Journal of Natural Products*. 2015; 78(11): 2624–2633.

Besserer A, Puech-Pages V, Kiefer P, et al. Strigolactones stimulate arbuscular mycorrhizal fungi by activating mitochondria. *PLoS biology*. 2006; 4(7): e226.

Beveridge CA, Dun EA, Rameau C. Pea has its tendrils in branching discoveries spanning a century from auxin to strigolactones. *Plant Physiology*. 2009; 151(3): 985–990.

Bhoi A, Yadu B, Chandra J, et al. Contribution of strigolactone in plant physiology, hormonal interaction and abiotic stresses. *Planta*. 2021; 254(2): 1–21.

Bonfante P, Requena N. Dating in the dark: how roots respond to fungal signals to establish arbuscular mycorrhizal symbiosis. *Current Opinion in Plant Biology*. 2011; 14(4): 451–457.

Booker J, Auldridge M, Wills S, et al. MAX3/CCD7 is a carotenoid cleavage dioxygenase required for the synthesis of a novel plant signaling molecule. *Current Biology*. 2004; 14(14): 1232–1238.

Booker J, Sieberer T, Wright W, et al. MAX1 encodes a cytochrome P450 family member that acts downstream of MAX3/4 to produce a carotenoid-derived branch-inhibiting hormone. *Developmental Cell*. 2005; 8(3): 443–449.

Bouwmeester HJ, Matusova R, Zhongkui S, et al. Secondary metabolite signalling in host–parasitic plant interactions. Current opinion in plant biology. 2003; 6(4): 358–364.

Braun N, de Saint Germain A, Pillot JP, et al. The pea TCP transcription factor PsBRC1 acts downstream of strigolactones to control shoot branching. *Plant Physiology*. 2012; 158(1): 225–238.

Brooks DW, Bevinakatti HS, Powell DR. The absolute structure of (+)-strigol. *The Journal of Organic Chemistry*. 1985; 50(20): 3779–3781.

Bürger M, Chory J. The many models of strigolactone signaling. *Trends in Plant Science*. 2020; 25(4): 395–405.

Butler LG. Chemical communication between the parasitic weed *Striga* and its crop host: A new dimension in allelochemistry. *American Chemical Society*. 1995; 158–168.

Charnikhova TV, Gaus K, Lumbroso A, et al. Zeapyranolactone – A novel strigolactone from maize. *Phytochemistry Letters*. 2018; 24: 172–178.

Chi C, Xu X, Wang M, et al. Strigolactones positively regulate abscisic acid-dependent heat and cold tolerance in tomato. *Horticulture Research*. 2021; 1: 8.

Coggon P, Luhan PA, McPhail AT. Crystal and molecular structure of the germination stimulant strigol by X-ray analysis. *Journal of the Chemical Society, Perkin Transactions 2*. 1973; (4): 465–469.

Cook CE, Whichard LP, Turner B, et al. Germination of witchweed (*Striga lutea* Lour.): isolation and properties of a potent stimulant. *Science*. 1966; 154(3753): 1189–1190.

De Cuyper C, Fromentin J, Yocgo RE, et al. From lateral root density to nodule number, the strigolactone analogue GR24 shapes the root architecture of *Medicago truncatula*. *Journal of Experimental Botany*. 2015; 66(1): 137–146.

Duan J, Yu H, Yuan K, et al. Strigolactone promotes cytokinin degradation through transcriptional activation of cytokinin oxidase/dehydrogenase 9 in rice. *Proceedings of the National Academy of Sciences*. 2019; 116(28): 14319–14324.

Fernandez I, Cosme M, Stringlis IA, et al. Molecular dialogue between arbuscular mycorrhizal fungi and the nonhost plant Arabidopsis thaliana switches from initial detection to antagonism. *New Phytologist*. 2019; 223(2): 867–881.

Flematti GR, Scaffidi A, Waters MT, et al. Stereospecificity in strigolactone biosynthesis and perception. *Planta*. 2016; 243(6): 1361–1373.

Gomez-Roldan V, Fermas S, Brewer PB, et al. Strigolactone inhibition of shoot branching. *Nature*. 2008; 455(7210): 189–194.

Guan JC, Koch KE, Suzuki M, et al. Diverse roles of strigolactone signaling in maize architecture and the uncoupling of a branching-specific subnetwork. *Plant Physiology*. 2012; 160(3): 1303–1317.

Hu Q, Zhang S, Huang B. Strigolactones promote leaf elongation in tall fescue through upregulation of cell cycle genes and downregulation of auxin transport genes in tall fescue under different temperature regimes. *International Journal of Molecular Sciences*. 2019; 20(8): 1836.

Ito S, Nozoye T, Sasaki E, et al. Strigolactone regulates anthocyanin accumulation, acid phosphatases production and plant growth under low phosphate condition in *Arabidopsis*. *PLoS One*. 2015; 10(3): e0119724.

Jiang L, Liu X, Xiong G, et al. DWARF 53 acts as a repressor of strigolactone signalling in rice. *Nature*. 2013; 504(7480): 401–405.

Jiang L, Matthys C, Marquez-Garcia B, et al. Strigolactones spatially influence lateral root development through the cytokinin signaling network. *Journal of Experimental Botany*. 2016; 67(1): 379–389.

Kapulnik Y, Resnick N, Mayzlish-Gati E, et al. Strigolactones interact with ethylene and auxin in regulating root-hair elongation in *Arabidopsis*. *Journal of Experimental Botany*. 2011; 62(8): 2915–2924.

Kim HI, Xie X, Kim HS, et al. Structure–activity relationship of naturally occurring strigolactones in *Orobanche* minor seed germination stimulation. *Journal of Pesticide Science*. 2010; 35(3): 344–347.

Kohlen W, Charnikhova T, Lammers M, et al. The tomato carotenoid cleavage dioxygenase 8 (S l CCD 8) regulates rhizosphere signaling, plant architecture and affects reproductive development through strigolactone biosynthesis. *New Phytologist*. 2012; 196(2): 535–547.

Koltai H, Dor E, Hershenhorn J, et al. Strigolactones' effect on root growth and root-hair elongation may be mediated by auxin-efflux carriers. *Journal of Plant Growth Regulation*. 2010; 29(2): 129–136.
Kumar M, Pandya-Kumar N, Kapulnik Y, et al. Strigolactone signaling in root development and phosphate starvation. *Plant Signaling & Behavior*. 2015; 10(7): e1045174.
Lantzouni O, Klermund C, Schwechheimer C. Largely additive effects of gibberellin and strigolactone on gene expression in *Arabidopsis thaliana* seedlings. *The Plant Journal*. 2017; 92(5): 924–938.
Lauressergues D, Andre O, Peng J, et al. Strigolactones contribute to shoot elongation and to the formation of leaf margin serrations in *Medicago truncatula* R108. *Journal of Experimental Botany*. 2015; 66(5): 1237–44.
Li S, Chen L, Li Y, et al. Effect of GR24 stereoisomers on plant development in *Arabidopsis*. *Molecular plant*. 2016; 9(10): 1432–1435.
Liang Y, Ward S, Li P, et al. Smax1-Like7 signals from the nucleus to regulate shoot development in *Arabidopsis* via partially EAR motif-independent mechanisms. *The Plant Cell*. 2016; 28(7): 1581–601.
Ligerot Y, de Saint Germain A, Waldie T, et al. The pea branching RMS2 gene encodes the PsAFB4/5 auxin receptor and is involved in an auxin-strigolactone regulation loop. *PLoS Genetics*. 2017; 13(12): e1007089.
Liu J, Novero M, Charnikhova T, et al. Carotenoid cleavage dioxygenase 7 modulates plant growth, reproduction, senescence, and determinate nodulation in the model legume Lotus japonicus. *Journal of Experimental Botany*. 2013; 64(7): 1967–1981.
Makhzoum A, Yousefzadi M, Malik S, et al. Strigolactone biology: genes, functional genomics, epigenetics and applications. *Critical Reviews in Biotechnology*. 2017; 37(2): 151–62.
Martín-Trillo M, Grandío EG, Serra F, et al. Role of tomato BRANCHED1-like genes in the control of shoot branching. *The Plant Journal*. 2011; 67(4): 701–714.
Mashiguchi K, Sasaki E, Shimada Y, et al. Feedback-regulation of strigolactone biosynthetic genes and strigolactone-regulated genes in *Arabidopsis*. *Bioscience, Biotechnology, and Biochemistry*. 2009; 73(11): 2460–2465.
Mason MG, Ross JJ, Babst BA, et al. Sugar demand, not auxin, is the initial regulator of apical dominance. *Proceedings of the National Academy of Sciences*. 2014; 111(16): 6092–6097.
Matthys C, Walton A, Struk S, et al. The whats, the wheres and the hows of strigolactone action in the roots. *Planta*. 2016; 243(6): 1327–1337.
Matusova R, Rani K, Verstappen FW, et al. The strigolactone germination stimulants of the plant-parasitic *Striga* and *Orobanche* spp. are derived from the carotenoid pathway. *Plant Physiology*. 2005; 139(2): 920–934.
Mayzlish-Gati E, De-Cuyper C, Goormachtig S, et al. Strigolactones are involved in root response to low phosphate conditions in *Arabidopsis*. *Plant Physiology*. 2012; 160(3): 1329–1341.
McSteen P, Zhao Y. Plant hormones and signaling: common themes and new developments. *Developmental Cell*. 2008; 14(4): 467–73.
Minakuchi K, Kameoka H, Yasuno N, et al. Fine Culm1 (FC1) works downstream of strigolactones to inhibit the outgrowth of axillary buds in rice. *Plant and Cell Physiology*. 2010; 51(7): 1127–1135.
Müller LM, Harrison MJ. Phytohormones, miRNAs, and peptide signals integrate plant phosphorus status with arbuscular mycorrhizal symbiosis. *Current Opinion in Plant Biology*. 2019; 50: 132–139.
Nakamura H, Asami T. Target sites for chemical regulation of strigolactone signaling. *Frontiers in Plant Science*. 2014; 5: 623.
Parniske M. Arbuscular mycorrhiza: the mother of plant root endosymbioses. *Nature Reviews Microbiology*. 2008; 6(10): 763–775.
Pasare SA, Ducreux LJ, Morris WL, et al. The role of the potato (*Solanum tuberosum*) CCD8 gene in stolon and tuber development. *New Phytologist*. 2013; 198(4): 1108–1120.
Pollock CB, Koltai H, Kapulnik Y, et al. Strigolactones: a novel class of phytohormones that inhibit the growth and survival of breast cancer cells and breast cancer stem-like enriched mammosphere cells. *Breast Cancer Research and Treatment*. 2012; 134(3): 1041–1055.
Rasmussen A, Mason MG, De Cuyper C, et al. Strigolactones suppress adventitious rooting in *Arabidopsis* and pea. *Plant Physiology*. 2012; 158(4): 1976–1987.
Roumeliotis E, Kloosterman B, Oortwijn M, et al. The effects of auxin and strigolactones on tuber initiation and stolon architecture in potato. *Journal of Experimental Botany*. 2012; 63(12): 4539–4547.
Ruyter-Spira C, Kohlen W, Charnikhova T, et al. Physiological effects of the synthetic strigolactone analog GR24 on root system architecture in *Arabidopsis*: another below-ground role for strigolactones? *Plant Physiology*. 2011; 155(2): 721–734.

Sang D, Chen D, Liu G, et al. Strigolactones regulate rice tiller angle by attenuating shoot gravitropism through inhibiting auxin biosynthesis. *Proceedings of the National Academy of Sciences*. 2014; 111(30): 11199–11204.

Scaffidi A, Waters MT, Ghisalberti EL, et al. Carlactone-independent seedling morphogenesis in *Arabidopsis*. *The Plant Journal*. 2013; 76(1): 1–9.

Seale M, Bennett T, Leyser O. BRC1 expression regulates bud activation potential but is not necessary or sufficient for bud growth inhibition in *Arabidopsis*. *Development*. 2017; 144(9): 1661–1673.

Simons JL, Napoli CA, Janssen BJ, et al. Analysis of the decreased apical dominance genes of *Petunia* in the control of axillary branching. *Plant Physiology*. 2007; 143(2): 697–706.

Skinner RH, Nelson CJ. Epidermal cell division and the coordination of leaf and tiller development. *Annals of Botany*. 1994; 74(1): 9–15.

Snowden KC, Simkin AJ, Janssen BJ, et al. The decreased apical dominance1/petunia hybrida carotenoid cleavage dioxygenase8 gene affects branch production and plays a role in leaf senescence, root growth, and flower development. *The Plant Cell*. 2005; 17(3): 746–759.

Sorefan K, Booker J, Haurogné K, et al. MAX4 and RMS1 are orthologous dioxygenase-like genes that regulate shoot branching in *Arabidopsis* and pea. *Genes & Development*. 2003; 17(12): 1469–1474.

Soundappan I, Bennett T, Morffy N, et al. Smax1-Like/D53 family members enable distinct MAX2-dependent responses to strigolactones and karrikins in *Arabidopsis*. *The Plant Cell*. 2015; 27(11): 3143–3159.

Stirnberg P, van De Sande K, Leyser HO. MAX1 and MAX2 control shoot lateral branching in *Arabidopsis*. *Development*. 2002; 1131–1141.

Sun H, Tao J, Gu P, et al. The role of strigolactones in root development. *Plant Signaling & Behavior*. 2016; 11(1): e1110662.

Sun H, Tao J, Hou M, et al. A strigolactone signal is required for adventitious root formation in rice. *Annals of Botany*. 2015; 115(7): 1155–1162.

Sun H, Tao J, Liu S, et al. Strigolactones are involved in phosphate- and nitrate-deficiency-induced root development and auxin transport in rice. *Journal of Experimental Botany*. 2014; 65(22): 6735–6746.

Taniguchi M, Furutani M, Nishimura T, et al. The Arabidopsis LAZY1 family plays a key role in gravity signaling within statocytes and in branch angle control of roots and shoots. *The Plant Cell*. 2017; 29(8): 1984–1999.

Umehara M, Hanada A, Yoshida S, et al. Inhibition of shoot branching by new terpenoid plant hormones. *Nature*. 2008; 455(7210): 195–200.

Vogel JT, Walter MH, Giavalisco P, et al. SlCCD7 controls strigolactone biosynthesis, shoot branching and mycorrhiza-induced apocarotenoid formation in tomato. *The Plant Journal*. 2010; 61(2): 300–311.

Waldie T, McCulloch H, Leyser O. Strigolactones and the control of plant development: lessons from shoot branching. *The Plant Journal*. 2014; 79(4): 607–622.

Wang L, Wang B, Jiang L, et al. Strigolactone signaling in Arabidopsis regulates shoot development by targeting D53-like SMXL repressor proteins for ubiquitination and degradation. *The Plant Cell*. 2015; 27(11): 3128–3142.

Wang P, Zhang S, Qiao J, et al. Functional analysis of the GbDWARF14 gene associated with branching development in cotton. *PeerJ*. 2019; 7: e6901.

Wang Y, Bouwmeester HJ. Structural diversity in the strigolactones. *Journal of Experimental Botany*. 2018; 69(9): 2219–2230.

Waters MT, Gutjahr C, Bennett T, et al. Strigolactone signaling and evolution. *Annual Review of Plant Biology*. 2017; 68: 291–322.

Waters MT, Scaffidi A, Moulin SL, et al. A Selaginella moellendorffii ortholog of Karrikin Insensitive2 functions in *Arabidopsis* development but cannot mediate responses to karrikins or strigolactones. *The Plant Cell*. 2015; 27(7): 1925–1944.

Xie X. Structural diversity of strigolactones and their distribution in the plant kingdom. *Journal of Pesticide Science*. 2016; J16–02.

Xie X, Cao P, Wang Z, et al. Genome-wide characterization and expression profiling of the PDR gene family in tobacco (*Nicotiana tabacum*). *Gene*. 2021; 788: 145637.

Xie X, Yoneyama K, Yoneyama K. The strigolactone story. *Annual Review of Phytopathology*. 2010; 48: 93–117.

Yoneyama K, Kisugi T, Xie X, et al. Chemistry of strigolactones: why and how do plants produce so many strigolactones. *Molecular Microbial Ecology of the Rhizosphere*. 2013; 1: 373–379.

Yoneyama K, Xie X, Yoneyama K, et al. Which are the major players, canonical or non-canonical strigolactones? *Journal of Experimental Botany*. 2018; 69(9): 2231–2239.

Yoshimura M, Sato A, Kuwata K, et al. Discovery of shoot branching regulator targeting strigolactone receptor DWARF14. *ACS Central Science*. 2018; 4(2): 230–234.

Zhou F, Lin Q, Zhu L, et al. D14–SCFD3-dependent degradation of D53 regulates strigolactone signalling. *Nature*. 2013; 504(7480): 406–410.

Zhou X, Tan Z, Zhou Y, et al. Physiological mechanism of strigolactone enhancing tolerance to low light stress in cucumber seedlings. *BMC Plant Biology*. 2022; 22(1): 1–17.

Zulfiqar H, Shahbaz M, Ahsan M, et al. Strigolactone (GR24) induced salinity tolerance in sunflower (*Helianthus annuus* L.) by ameliorating morpho-physiological and biochemical attributes under in vitro conditions. *Journal of Plant Growth Regulation*. 2021; 40(5): 2079–2091.

Zwanenburg B, Blanco-Ania D. Strigolactones: new plant hormones in the spotlight. *Journal of Experimental Botany*. 2018; 69(9): 2205–2218.

Zwanenburg B, Mwakaboko AS, Reizelman A, et al. Structure and function of natural and synthetic signalling molecules in parasitic weed germination. *Pest Management Science: Formerly Pesticide Science*. 2009; 65(5): 478–491.

Zwanenburg B, Pospíšil T. Structure and activity of strigolactones: new plant hormones with a rich future. *Molecular Plant*. 2013; 6(1): 38–62.

3 Strigolactones: Regulators of Root Development

Kanval Shaukat, Mahmood-ur-Rahman, Abdul Wahid, Noreen Zahra, Abdul Samad, Mereen Nizar, Rubina Naseer, and Aaliya Batool

CONTENTS

LIST OF ABBREVIATIONS

SLs	Strigolactones
RSM	Rhizosphere signaling molecules
PR	Primary root
PRM	Primary root meristem
WT	Wild type
AM	Arbuscular mycorrhizal
LR	Lateral root
LB	Lateral branches
AMF	Arbuscular mycorrhizal fungi
HB	Hyphal branches
RH	Root hairs
LRD	Lateral root development
RHE	Root hair elongation
tir1	Transport inhibitor response 1
EIN2	Ethylene insensitive 2
ETR1	Ethylene responsive 1
SCR	SCARECROW
AR	Adventitious roots
CR	Crown root

DOI: 10.1201/9781003224525-5

CjCCD7	Carotenoid cleavage dioxygenase 7 modulate
ABA	Abscisic acid
CYK	Cytokinin
AUX	Auxins
GA	Gibberellic acid
ZN	Zaxinone
JA	Jasmonic acid
SA	Salicylic acid
P	Phosphate
N	Nitrogen

3.0 INTRODUCTION

Root is the major organ that ensures plant survival by providing strong anchorage in the soil, ensuring uptake of water and nutrients, and being the mediator of beneficial interactions with soil microflora. Root development and growth is highly regulated processes under various abiotic and biotic stresses, mainly by hormonal pathways. Apart from auxin (AUX) and cytokinin (CYK), which are well-known regulators of root development, a novel hormone, Strigolactone (SL), identified in 2008 (Gomez-Roldan et al., 2008), is known to play a key role in maintaining and regulating the architecture of the root (Kapulnik et al., 2011a; Ruyter-Spria et al., 2011; Rasmussen et al., 2012; Sun et al., 2014). SLs are known as a new class of phytohormones, carotenoid-derived, rhizosphere-signaling molecules (RSM) that help to regulate the growth and development of plants. SLs in apocarotenoid form are produced by several enzymes in plastid and cytosol (Matusova et al., 2005, Hossain et al., 2021).

The SLs are well known to initiate seed germination of various parasitic plant roots (Cook et al., 1966). The SLs are synthesized by many plant species. Several kinds of SLs are synthesized by particular species; however, distinct varieties of a single species may synthesize a mixture of SLs (Xie et al., 2010; Yoneyama et al., 2013). SLs are mainly involved in various developmental processes by influencing the morpho-physiological attributes of plants. Mainly, they are synthesized in the roots; however, their synthesis is not just restricted to roots, and other plant parts like epicotyl and internodes also synthesize SLs (Koltai and Beveridge, 2013; Yoneyama et al., 2013). They control the development of shoot and root, not only in vascular plants but also in moss Caulonema in response to various environmental stresses (Foo et al., 2013; Umehara et al., 2008; Hoffmann et al., 2014). They are also known to be synthesized in primitive plants such as Charales and Embryophyta (Kyozuka et al., 2022). Kohlen et al. (2011) showed that SLs are transported through xylem from root to shoot in *Arabidopsis*. Moreover, they also control the root and shoot architecture of plants (Jiu et al., 2022). They also are responsible for the development of root hairs (Koltai et al., 2010; Arite et al., 2012; Kapulnik et al., 2011; Koltai and Kapulnik, 2011), the induction of nodulation (Foo and Davies, 2011), and mesocotyl elongation in plants (Hu et al., 2010). SLs are suggested to suppress the growth of axillary buds, thus producing dwarf or stunted plants (Gomez-Roldan et al., 2008; Umehara et al., 2008).

Several studies documented that exogenous application of synthetic SL promotes shoot and root development (Umehara et al., 2008; Kapulnik et al., 2011) under various stresses like nutrient availability (Kohlen et al., 2011; Czarnecki et al., 2013), drought (Ruiz-Lozano et al., 2016), salinity (Ma et al., 2017), and heat (Omoarelojie et al., 2020). Root hairs of *Arabidopsis* were positively influenced by the application of SLs (Kapulnik et al., 2011). In rice, SLs are responsible for crown root stimulation (Arite et al., 2012). However, it has been well documented that they inhibit the adventitious root development in various plants: e.g., *Pisum sativum* (Urquhart et al., 2015) and *Solanum lycopersicum* (Rasmussen et al., 2013). Root genes and their transcriptional activities get enhanced by the application of GR24, a synthetic analogue of Strigolactone (Ma et al., 2017). GR24 application alleviated salinity stress by regulating the enzymatic antioxidants system (Ma et al., 2017). It nullified and reduced ion toxicity under salt stress by improving seedling and shoot growth with overexpression of particular genes (Ma et al., 2017). Likewise, *Oryza sativa* (rice) seedling growth was improved by

GR24 application, which enhanced the photosynthetic attributes several folds (Ling et al., 2020). This chapter's emphasis on how SLs impact root development and its regulation.

3.1 STRIGOLACTONES REGULATES ROOT DEVELOPMENT

Strigolactones' involvement in modulating root architecture is well documented in several plants like *Arabidopsis*, tomato, pea, rice, and *Medicago truncatula*. However, prime evidence that SLs play significant role in the development of the root in *Arabidopsis* mutants as compared to the wild type (WT) (Kapulnik et al., 2011a; Ruyter-Spira et al., 2011). SLs impact root architecture – i.e., the development of adventitious roots (AR), primary root (PR), and lateral root (LR) – as well as the elongation of root hairs (RH), which are discussed next. (See Table 3.1.)

TABLE 3.1
Impact of Synthetic SL on Roots of Various Plant Species

Plant Species	Mutant	Impact of SLs	Reference
	Primary root (PR) seminal root (SR) length		
Arabidopsis thaliana L	Mutants of SL signaling (*MAX2–1*) and SL-biosynthesis (*MAX3-11*, *MAX4-1*)	Under optimum conditions they have no impact and under carbohydrates starvation they promotes the length of PR	Ruyter-Spira et al. (2011)
Oryza sativa L.	SL-biosynthesis (*d10*, *d17*, *d27*) and SL-signaling (*d3*, *d14*) mutants	Lack of effect on SR length	Arite et al. (2012)
Hordeum vulgare L.	SL-insensitive mutants (*hvd14*)	Promotion of SR growth	Marzec et al. (2016)
Lotus japonicas L.	RNAi transgenic line for *LjCCD7*	Inhibition of PR	Liu et al. (2013)
	Lateral root Density (LRD)		
A. thaliana	Mutants of SL signaling (*MAX2-1*) and SL-biosynthesis (*MAX3-11*, *MAX4-1*)	Inhibition of LRD	Kapulnik et al. (2011)
A. thaliana	Mutants of SL signaling (*MAX2-1*) and SL-biosynthesis (*MAX3-11*, *MAX4-1*)	Inhibition of LRD	Ruyter-Spira et al. (2011)
A. thaliana	Mutants of SL signaling (*MAX2*)	Inhibition of primary outgrowth of LR	Jiang et al. (2016)
O. sativa	SLs synthesis (*d10*) and signaling (*d14*) mutants	No impact on density of LR	Arite et al. (2012)
H. vulgare	SL-insensitive mutants (*hvd14.d*)	Inhibition of LRD	Marzec et al. (2016)
	Crown root (CR) length		
O. sativa	SLs-biosynthesis (*d10*) and Signaling (*d14*) mutants	Promotes the growth of CR	Arite et al. (2012)

(Continued)

TABLE 3.1
(Continued)

		Adventitious root (AR) development	
A. thaliana	SL-biosynthesis (*MAX1*, *MAX3*, *MAX4*) and signaling mutant (*MAX2*)	Negative effect of AR formation	Rasmussen et al. (2012)
Pisum sativum L.	SL-biosynthesis (*rms1*, *rms5*) and signaling mutant (*rms4*)	Negative effect of AR formation	Rasmussen et al. (2012)
O. sativa	Mutants of SL-signaling (*d3*, *d14*) and SL-biosynthesis (*d10*, *d17*, *d27*)	Promotes the development of AR	Sun et al. (2015)

3.1.1 Development of Primary Root

Strigolactones' effects on primary root (PR) development depend upon plant species and growth conditions. Treating seedlings of *Arabidopsis* with rac-GR24 (a synthetic and bioactive form of SLs) increased PR length by increasing the cortex cells of the primary root meristem (PRM). Signaling mutant MAX2 and biosynthetic mutants MAX4 and MAX1 had shorter PR than WT, with limited/fewer cortex cells in PRM, validating that endogenous SL controls root development and growth (Ruyter-Spira et al., 2011). Moreover, a concentration greater than 2.5μM of rac-GR24 inhibits PR. Sun et al. (2014) documented related outcomes in rice, in which the seminal roots of SL mutant was shorter than WT under low nitrate and phosphate conditions. However, rac-GR24 application resulted in PR elongation in WT and SL mutants except the *d3* mutant (signaling mutant). On the other hand, GR24 does not have any impact on seminal roots (Arite et al., 2012). For tomato, combinational treatment with AUX was shown to have an effect on PR length; the inhibitory impact of AUX on the root is reduced by adding rac-GR24– amplified concentrations (Koltai et al., 2010). Contrarily, in *Lotus japonicus*, on the silencing of the MAX3 ortholog, shorter PR has been documented (Liu et al., 2013). Thus, it can be suggested that SLs are indeed the positive regulators of PR, depending on the plant species, growth conditions, and varying endogenous SL hormonal concentrations.

3.1.2 Development of Lateral Roots

The impact of SLs on LR is more pronounced than that of PR. Arabidopsis mutants have more LR than WT. Treating *Arabidopsis* seedlings with rac-GR24 suppresses lateral root formation in the wild type and SL-synthetic mutant (MAX3 and MAX4) but not in the SL-response mutant (MAX2), suggesting its regulatory role (Kapulnik et al., 2011; Ruyter-Spira et al., 2011). Spatio-temporal analysis through LR marker *GATA23* exposed that rac-GR24's effect on LR is complex and dual as GR24 application results in less effective priming, the early step of LRD (lateral root development) and the main impact on the development later in the root region, close to the junction of shoot and root (Jiang et al., 2016). In *Arabidopsis*, formation of the lateral root was revealed to be adversely regulated by SLs under sufficient phosphate (Kapulnik et al., 2011a).

The impact of rac-GR24 on the development of the lateral root of *Arabidopsis* is stimulated by AUX (Keret et al., 2022). At lower supplementation of AUX, the addition of rac-GR24 resulted in reduced development of LR while under elevated AUX concentrations, an increase in LRD occurred

(Ruyter-Spira et al., 2011). To explain this phenomenon, rac-GR24 may cause AUX to sink below optimum for LRD by decreasing AUX flow-adapted PIN recycling on the xylem cells membrane (xylem parenchyma cells), causing reduced LRD. Nonetheless, at greater AUX concentrations, the rac-GR24 prompted a decrease in AUX flow that caused the total AUX content material to move toward the given optimum, thereby increasing LRD (Shinohara et al., 2013). However, further studies are required to unveil the exact mechanism.

3.1.3 Root Hair Elongation

Root hairs (RH) play an important role in nutrient uptake from soil (Lopez-Bucio et al., 2003) while the SLs also affect the development of RH. Application of rac-GR24 to *Arabidopsis* improved length of RH in both SL-deficient mutants (MAX3 and MAX4) and wild types but not in the SL-insensitive mutant (MAX2), even if MAX2 is treated with greater concentrations of GR24 (Cohen et al., 2013). Moreover, AUX and ethylene (ET) signaling is essential for the positive impact of SLs on root hair elongation (RHE). Thus, SLs have a positive impact, which is mediated by MAX2. The crosstalk among AUX and SL signaling in RHE is much less clear, despite the fact that each hormone enhances the elongation of root hairs. AUX treatment enhances the response of RH to rac-GR24, and the MAX2 mutant stays receptive to AUX in the phenotype of root hair (Kapulnik et al., 2011b), though some crosstalk might follow. Certainly, the AUX receptor mutant *tir1* is much less responsive to rac-GR24 (Kapulnik et al., 2011b). SLs no longer seem to be vital for RH formation to ET; however, each ethylene-signaling mutant ET response1 mutant (*etr1*) and ET insensitive2 mutant (*ein2*) displayed much less response to the rac-GR24 phenotype of RH, signifying that ET is epistatic to SLs. The impact of exogenous supplementation of SLs becomes evident within 24 hours of application. However, the impact gets reduced under high doses of GR24 (Kapulnik et al., 2011b). Moreover, as RHE is regulated in the epidermis or, specifically, the epiblema of roots, the efficiency of MAX2 expression under the control of the SCARECROW (SCR) promotor, which is primarily in quiescent centre and endodermis of roots, for sensitivity to GR24 documents that SLs act on non-cells autonomously at a smaller range (Kapulnik et al., 2011b).

3.1.4 Development of Adventitious Roots

Strigolactones (SLs) regulate adventitious root (AR) formation both positively or negatively, depending up plant species and experimental and growth conditions. In pea and *Arabidopsis*, both mutants SL-biosynthesis and SL-signaling showed greater AR than WT, validating that endogenous SL suppresses AR formation (Rasmussen et al., 2012). Pea SL-biosynthesis mutants grown in dark displayed reduced a number of AR, showing that SLs play the autonomous role of MAX2 in AR synthesis in dark conditions (Urquhart et al., 2015). Sun et al. (2015) elucidated that SL mutants of rice had minimum AR at the seedling stage and a lesser number of AR tiller^{-1} at the mature stage.

Furthermore, crown roots (CR) in rice of all SL mutants displayed smaller and fewer AR roots than WT (Arite et al., 2012; Sun et al., 2014). Hence, it can be suggested that in rice, SLs regulate CR development positively.

3.2 STRIGOLACTONES REGULATE ROOT DEVELOPMENT UNDER ENVIRONMENTAL STRESSES

The SLs control the development of root architecture under environmental stress too. Several abiotic and biotic stresses along with nutrient deficiency influence SLs concentration and initiate/stimulate its signaling, thereby bringing about various morpho-physiological alteration/modifications not only in the root structure but also in other plant organs such as shoot, buds, etc.

3.2.1 Abiotic Stress

Voluminous evidences validates the role of SLs in root modulations in several abiotic stress conditions (Van Ha et al., 2014; Marzec et al., 2016; Ruiz-Lozano et al., 2016; Ling et al., 2020, Raza et al., 2021), although the effects may differ from species to species and depending on growth circumstances. For instance, LRD may be subdued under optimal growth conditions by SLs; however, it is enhanced under nutrient stress to facilitate nutrient uptake (Ruyter-Spira et al., 2011, Marzec and Melzer, 2018).

SLs are well documented for their effectiveness in the promotion of seed germination (Lechat et al., 2015). SLs facilitate seed germination by maintaining the balance of various phytohormones (i.e., lowering ABA biosynthesis and increasing cytokinin levels. GR24 application to *Arabidopsis* mutants stimulated germination at high temperatures whilst reducing ABA to GA levels and increasing CK levels slightly (Tsuchiya et al., 2010). However, it can further be determined that the downregulation of biosynthetic genes by exposure to SLs reduced the SLs production in the roots of plants, which moderately affected the architecture of the roots (Liu et al., 2015).

The variations in SLs biosynthesis were also reported in root development among various species under different stress conditions. In tomato, parallel consequences were reported under water deficit conditions that had negative affect on the SL expression in biosynthesis of SlCCD7 gene in root tissues (Ruiz-Lozano et al., 2016). In rice, contradictory effects were also reported in roots under drought conditions (Haider et al., 2018), thus showing that in monocot and dicot roots, the SLs responded differently to stress environment. Genes for abscisic acid (ABA) biosynthesis are identified for their regulation process in many plants' initial and development stages by providing timely response to abiotic stress (Sah et al., 2016; Visentin et al., 2020. Further, the exogenously applied zaxinone (ZN) greatly elevated the ABA transcription level and SLs biosynthetic genes in the root tissues of *Arabidopsis* seedlings. ZN acts as a crucial SL repressor by improving the formation of roots, leading to the assumption that ZN and SL work oppositely in various plants under environmental cues. As a result of its capacity to upregulate SL and ABA synthetic gene expression under various abiotic stress conditions, the role of ZN is still debatable (Ablazov et al., 2020).

Furthermore, SLs have been recognized to regulate the tolerance for chilling stress in *Arabidopsis* and *Pisum sativum* plants (Cooper et al., 2018). SL-signaling pea mutants (*rms3, rms4* and *rms5*) showed a greater number of branches with increased chlorophyll and carotenoid contents than WT. However, *Arabidopsis* MAX mutants had less biomass than that of WT. Interestingly, GR24 supplementation decreases the leaf area of WT and mutants MAX3-9 and MAX4-1, but no reduction was detected in mutant MAX2-1 (Cooper et al., 2018). Hu et al. (2018) highlighted SLs' positive influence on crown root development and its elongation under different temperatures, which implicates the regulation of cell division and intervention in the transport gene expression of AUX. Osmotic stress decreases SLs concentration in *Lotus japonicas* roots. The reduction is due to the downregulation of SL-biosynthetic genes such as *LjCCD7, LjD27*, and *LjCCD8*. ABA production is necessary for better root adaptation in *L. japonicus* (Liu et al., 2015). Furthermore, SL-responsive mutants are more susceptible to drought stress than WT (Zhang et al., 2018b). Visentin et al. (2016) documented that reduction in SL production in tomato roots under water stress increases the hypersensitivity of tomatoes to water stress. Moreover, *Arabidopsis* seedlings of mutant MAX2 showed hypersensitivity to salinity and drought stress. Thus, SLs' interaction with various phytohormones contributes to eventual root architectural modification under environmental stresses.

3.2.2 Nutrient Deficiency

Phosphate (P) and nitrogen (N) are the most significant nutrients essential for the development and growth of plants (Forde and Lorenzo, 2001; Lopez-Bucio et al., 2003). SLs possess a very significant role in adaptive responses under N and P scarcities (Yoneyama et al., 2007; Xu et al., 2015).

For instance, SLs promote symbiotic relations with *Arbuscular mycorrhizal* (AM) fungi via hyphae ramification and also regulate shoot architecture that inhibits the outgrowth of tiller bud in order to better adapt during the deficiencies imposed by P and N (Umehara et al., 2008). Mayzlish-Gati et al. (2012) observed that SLs regulate responses to P-limited conditions. Mutants of SL-signaling in *Arabidopsis* along with SL-biosynthesis showed a reduction in response to P-deficient conditions as compared to WT plants. However, in MAX4-1 plants, the reduction in response to P was compensated for by GR24 application. Interestingly, evidence from recent studies suggested that SLs were responsible for root growth under P and N deficits (Arite et al., 2012; Sun et al., 2014). The GR24 application to WT plants, along with SL mutants under nutrient deficient conditions, leads to the overall recovery of seminal root length and LRD as compared to WT plants under P- and N-deficient conditions. This confirmed that the level of SL was elevated under conditions of nutrient limiting, leading to the D3-dependent induction and elongation of seminal roots and reduction in LRD (Sun et al., 2014).

3.2.3 Biotic Stress

Under biotic stress, jasmonic acid (JA) and salicylic acid (SA), along with ABA, play key roles in SLs-mediated regulations. The communication between parasites and plants depends on SL molecules that are exuded from the roots into the rhizosphere (Joel and Bar, 2013). Synthetic GR24 is 100 times less effective and active than natural SLs (Kim et al., 2010). In tomato, SLs' role has been well documented as SL-biosynthetic mutants, which displayed resistance against root-knot nematode and mycological pathogens, e.g., *Botrytis cinerea* and *Alternaria alternate* (Xu et al., 2019). However, no involvement of SLs in pea plants against fungal species (e.g., *Fusarium oxysporum* and *Pythium irregulare*) has been observed (Foo et al., 2016). By using a reverse-genetic approach, the main component identified in response to plant defense was MAX2. Thus, the mutants has increased stomatal conductance, which promoted pathogen entry in the apoplast, and susceptibility to various pathogens increased (Piisila et al., 2015).

3.3 STRIGOLACTONES SIGNALING FOR PLANT INTERACTIONS

3.3.1 Mycorrhizal Symbiosis

The first role of SLs discovered was that they induced seed germination in parasitic plants (Cook et al., 1966). Later, it was documented that these molecules play a role in triggering the symbiotic relationship between fungi (*Arbuscular mycorrhizal* fungi [AMF]) and plants (Akiyama et al., 2005; Parniske, 2008). The mutualism between AMF and plants has been shown in 80% of plants and almost 450 million years ago (Parniske, 2008). In legume plants, the SLs concern the initiation of nodule and root bacteria swarming with the interaction of fungal, bacterial, and oomycetes pathogens (Akiyama and Hayashi, 2006; Mcadam et al., 2017). The function of SLs is mostly reported with respect to the symbiotic interaction with the AMF. SLs released by the root trigger the pre-symbiotic branches of AMF and also initiate metabolism and enhances spore germination (Besserer et al., 2008; Mori et al., 2016). The AMF hyphal branches reach the rhizosphere, absorb water and nutrients from soil, and transfer them to the roots in exchange of photosynthesis (Smith and Read, 2010; Yoneyama et al., 2007a). This provides extra water and minerals to the plants (Yoneyama et al., 2007a). Under P deficiency, the roots release more SLs to increase the AMF symbiosis (Yoneyama et al., 2012). SLs signaling components are required by AMF for root penetration of pea and rice while, in rice, the karrikin receptor complex is required for AMF perception (Yoshida et al., 2012; Foo et al., 2013). AMF symbiotic characters or gene expression and colony is suppressed under high doses of exogenous P (Breuillin et al., 2010). It may limit AMF symbiosis unless high levels of nutrients are required by plants. It is observed that SLs did not respond to phosphate regulation in plants. Moreover, the pea mutants were impaired in SL sensing and colonized poorly in AM compared to the wild type, which showed that mycorrhizal symbiosis was controlled by the phosphate supply (Foo et al., 2013). Non-symbiotic plants such as *Arabidopsis* (Goldwasser et al., 2008) and *Lupinus* spp (Yoneyama et al., 2008) roots release SL, which indicates processes other than the mycorrhizal symbiotic association.

3.3.2 Symbiotic Interaction for the Formation of Root Nodules

Nitrogen-fixing bacteria play a key role in root nodule development. Thus, a symbiotic interaction on the roots of legumes, clover, and alfalfa can be observed. This symbiotic association involves signal exchange between the plant root and the bacteria to form the nodule (Markmann and Parniske, 2009). SLs' signal initiates symbiotic association with the bacteria, thereby increasing the motility of the bacteria to the root surface to build an interaction (Pelaez-Vico et al., 2016). The SLs' capability to adjust the development of meristem unveiled SL participation in the symbiotic relationship.

The ability of SLs to govern nodulation was substantiated by the use of the RMS1 mutant in pea (Foo and Davies, 2011). The study confirmed that endogenously synthesized SLs are controllers of nodule formation. By using the root tissue and root exudates of the RMS1 mutants that are deficient in SLs, Foo and Davies. (2011) confirmed a 40% decrease in nodules relative to WT that contained SL. GR24 application to RMS1 plants caused an increased number of nodules in mutants compared to WT (devoid of exogenous supplementation of SLs). GR24 is shown to increase nodule formation in the WT of alfalfa, pea, and *L. japonicus* (Soto et al., 2010; Liu et al., 2013; Rehman et al., 2022). Additionally, it was proved that SLs are derived from roots rather than shoots, which helps to modulate the nodule formation (Foo et al., 2014). SLs do not directly interact with the rhizobium (Soto et al., 2010). The mechanism governing SLs enhancement under nodule formation is still a paradox.

3.4 CONCLUSION

Roots are the strong foundations of plants. Development of roots and their ability to withstand various environmental cues decides the fate of plant survival. Literature reveals that strigolactones are mainly recognized for their crucial role in plant architecture, especially shoot and root. SLs regulate primary, adventitious, and lateral root development along with various other phytohormones. Exogenous supplementation of synthetic SL improves root development in plants. Moreover, SLs help plants combat the negative impact of various environmental stresses and biotic constraints. They promote symbiotic association, specifically arbuscular mycorrhizal association, and nitrogen-fixing bacteria to form nodules in leguminous plants. Further studies are required that may unveil mechanisms and SLs-associated metabolic pathways that influence root development under biotic and environmental cues.

REFERENCES

Ablazov A, Mi J, Jamil M, Jia K-P, Wang JY, Feng Q, Al-Babili S. (2020). The apocarotenoid zaxinone is a positive regulator of strigolactone and abscisic acid biosynthesis in Arabidopsis roots. *Front Plant Sci* 11: 578.

Akiyama K, Hayashi H. (2006). Strigolactones: Chemical signals for fungal symbionts and parasitic weeds in plant roots. *Ann. Bot-London* 97: 925–931.

Akiyama K, Matsuzaki K, Hayashi H. (2005). Plant sesquiterpenes induce hyphal branching in arbuscular mycorrhizal fungi. *Nature* 435, 824–827.

Arite T, Kameoka H, Kyozuka J. (2012). Strigolactone positively controls crown root elongation in rice. *J Plant Growth Regul* 31: 165–172.

Besserer A, Becard G, Jauneau A, Roux C, Sejalon-Delmas N. (2008). GR24, a synthetic analog of strigolactones, stimulates the mitosis and growth of the arbuscular mycorrhizal fungus Gigaspora rosea by boosting its energy metabolism. *Plant Physiology* 148: 402–413.

Breuillin F, Schramm J, Hajirezaei M, Ahkami A, Favre P, Druege U, Hause B, Bucher M, Kretzschmar T, Bossolini E, Kuhlemeier C, Martinoia E, Franken P, Scholz U, Reinhardt D. (2010). Phosphate systemically inhibits development of arbuscular mycorrhiza in Petunia hybrida and represses genes involved in mycorrhizal functioning. *Plant J* 64: 1002–1017.

Cohen M, Prandi C, Occhiato EG, Tabasso S, Wininger S, Resnick N, Steinberger Y, Koltai H, Kapulnik Y. (2013). Structure-function relations of strigolactone analogs: Activity as plant hormones and plant interactions. *Mol Plant* 6: 141–152.

Cook CE, Whichard LP, Turner B, Wall ME. (1966). Germination of witchweed (Striga lutea Lour) – Isolation and properties of a potent stimulant. *Science* 154: 1189–1190.

Cooper JW, Hu Y, Beyyoudh L, Yildiz Dasgan H, Kunert K, Beveridge CA, Foyer CH. (2018). Strigolactones positively regulate chilling tolerance in pea and in *Arabidopsis*. *Plant Cell Environ* 41: 1298–1310.

Czarnecki O, Yang J, Weston DJ, Tuskan GA, Chen JG. (2013). A dual role of strigolactones in phosphate acquisition and utilization in plants. *Int J Mol Sci* 14: 7681–701, PMID: 23612324. http://dx.doi.org/10.3390/ijms14047681.

Foo E, Blake SN, Fisher BJ, Smith JA, Reid JB. (2016). The role of Strigolactones during plant interactions with the pathogenic fungus Fusarium oxysporum. *Planta* 243: 1387–1396.

Foo E, Davies NW. (2011). Strigolactones promote nodulation in pea. *Planta* 234: 1073–1081.

Foo E, Ferguson BJ, Reid JB. (2014). The potential roles of strigolactones and brassinosteroids in the autoregulation of nodulation pathway. *Annals of Botany* 113(6): 1037–1045.

Foo E, Yoneyama K, Hugill CJ, Quittenden LJ., Reid JB. (2013). Strigolactones and the regulation of pea symbioses in response to nitrate and phosphate deficiency. *Mol. Plant* 6: 76–87.

Forde B, Lorenzo H. (2001). The nutritional control of root development. *Plant Soil* 232: 51–68.

Goldwasser Y, Yoneyama K, Xie XA, Yoneyama K. (2008). Production of strigolactones by Arabidopsis thaliana responsible for Orobanche aegyptiaca seed germination. *Plant Growth Regul* 55: 21–28.

Gomez-Roldan V, Fermas S, Brewer PB, Puech-Pagès V, Dun EA, Pillot J-P, Letisse F, Matusova R, Danoun S, Portais JC. (2008). Strigolactone inhibition of shoot branching. *Nature* 455: 189–194.

Haider I, Andreo-Jimenez B, Bruno M, Bimbo A, Floková K, Abuauf H, Ntui VO, Guo X, Charnikhova T, Al-Babili S. (2018). The interaction of strigolactones with abscisic acid during the drought response in rice. *J Exp Bot* 69: 2403–2414.

Hoffmann B, Proust H, Belcram K, Labrune C, Boyer F-D, et al. (2014). Strigolactones inhibit caulonema elongation and cell division in the Moss *Physcomitrella patens*. *PLoS ONE* 9(6): e99206.

Hossain A, Raza A, Maitra S, Asaduzzaman M, Islam MR, Hossain MJ, . . . Aftab T. (2021). Strigolactones: A novel carotenoid-derived phytohormone–Biosynthesis, transporters, signalling, and mechanisms in abiotic stress. In *Plant Growth Regulators* (pp. 275–303). Springer, Cham.

Hu L, Xie Y, Fan S, Wang Z, Wang F, Zhang B, . . . Kong L. (2018). Comparative analysis of root transcriptome profiles between drought-tolerant and susceptible wheat genotypes in response to water stress. *Plant Science* 272: 276–293.

Hu Z, Haifang Yan, Jinghua Yang, Shinjiro Yamaguchi, Masahiko Maekawa, Itsuro Takamure, Nobuhiro Tsutsumi, Junko Kyozuka, Mikio Nakazono. (2010). Strigolactones negatively regulate mesocotyl elongation in rice during germination and growth in darkness. *Plant and Cell Physiology* 51(7): 1136–1142.

Jiang L, Matthys C, Marquez-Garcia B, De Cuyper C, Smet L, De Keyser A, Boyer FD, Beeckman T, Depuydt S, Goormachtig S. (2016). Strigolactones spatially influence lateral root development through the cytokinin signaling network. *J Exp Bot* 67: 379–389.

Jiu S, Xu Y, Xie X, Wang J, Xu J, Liu X, Zhang C. (2022). Strigolactones affect the root system architecture of cherry rootstock by mediating hormone signaling pathways. *Environ. Exp. Bot* 193.

Joel DM, Bar H. (2013). The seed and the seedling. In Joel DM, Gressel J, Musselman LJ, eds, *Parasitic Orobanchaceae* (pp. 147–165). Springer, Berlin.

Kapulnik Y, Delaux PM, Resnick N, Mayzlish-Gati E, Wininger S, Bhattacharya C, Séjalon-Delmas N, Combier JP, Bécard G, Belausov E, Beeckman T, Dor E, Hershenhorn J, Koltai H. (2011a). Strigolactones affect lateral root formation and root-hair elongation in *Arabidopsis*. *Planta* 233: 209–216.

Kapulnik Y, Resnick N, Mayzlish-Gati E, Kaplan Y, Wininger S, Hershenhorn J, Koltai H. (2011b). Strigolactones interact with ethylene and auxin in regulating root-hair elongation in Arabidopsis. *J Exp Bot* 62: 2915–2924.

Keret R, Nakhooda M, Jones NB, Hills PN. (2022). Optimisation of micropropagation protocols for temperate eucalypt hybrids in South Africa, with a focus on auxin transport proteins. *Southern Forests: A Journal of Forest Science*: 1–10.

Kim HI, Xie X, Kim HS, Chun JC, Yoneyama K, Nomura T, Takeuchi Y, Yoneyama K. (2010). Structure-activity relationship of naturally occurring strigolactones in Orobanche minor seed germination stimulation. *J. Pestic Sci* 35: 344–347.

Kohlen W, Charnikhova T, Liu Q, Bours R, Domagalska MA, Beguerie S, Verstappen F, Leyser O, Bouwmeester H, Ruyter-Spira C. (2011). Strigolactones are transported through the xylem and play a key role in shoot architectural response to phosphate deficiency in nonarbuscular mycorrhizal host Arabidopsis. *Plant Physiol* 155: 974–987.

Koltai H, Beveridge CA. (2013). Strigolactones and the coordinated development of shoot and root. In *Long-Distance Systemic Signaling and Communication in Plants* (pp. 189–204). Springer, Berlin.

Koltai H, Dor E, Hershenhorn J, Joel DM, Weininger S, Lekalla S, Kapulnik Y. (2010). Strigolactones' effect on root growth and root-hair elongation may be mediated by auxin-efflux carriers. *Journal of Plant Growth Regulation* 29(2), 129–136.

Koltai H, Kapulnik Y. (2011). Strigolactones as mediators of plant growth responses to environmental conditions. *Plant Signaling & Behavior* 6: 37–41.

Kyozuka J, Nomura T, Shimamura M. (2022). Origins and evolution of the dual functions of strigolactones as rhizosphere signaling molecules and plant hormones. *Curr. Opin. Plant Biol* 65.

Lechat MM, Brun G, Montiel G, Véronési C, Simier P, Thoiron S, Pouvreau JB, Delavault P. (2015). Seed response to strigolactone is controlled by abscisic acid-independent DNA methylation in the obligate root parasitic plant, Phelipanche Ramosa L. *Pomel. J Exp Bot* 66: 3129–3140.

Ling F, Su Q, Jiang H, Cui J, He X, Wu Z, Zhang Z, Liu J, Zhao Y. (2020). Effects of strigolactone on photosynthetic and physiological characteristics in salt-stressed rice seedlings. *Sci Rep* 10: 1–8.

Liu JW, He H, Vitali M, Visentin I, Charnikhova T, Haider I, Schubert A, Ruyter-Spira C, Bouwmeester HJ, Lovisolo C. (2015). Osmotic stress represses strigolactone biosynthesis in Lotus japonicus roots: Exploring the interaction between strigolactones and ABA under abiotic stress. *Planta* 241: 1435–1451.

Liu JW, Novero M, Charnikhova T, Ferrandino A, Schubert A, Ruyter-Spira C, Bonfante P, Lovisolo C, Bouwmeester HJ, Cardinale F. (2013). Carotenoid cleavage dioxygenase 7 modulates plant growth, reproduction, senescence, and determinate nodulation in the model legume Lotus japonicus. *J. Exp. Bot* 64: 1967–1981.

Lopez-Bucio J, Cruz-Ramırez A, Herrera-Estrella L. (2003). The role of nutrient availability in regulating root architecture. *Curr Opin Plant Biol* 6: 280–287.

Ma N, Hu C, Wan L, Hu Q, Xiong J, Zhang C. (2017). Strigolactones improve plant growth, photosynthesis, and alleviate oxidative stress under salinity in rapeseed (Brassica napus L) by regulating gene expression. *Front Plant Sci* 8: 1671.

Markmann K, Parniske M. (2009). Evolution of root endosymbiosis with bacteria: How novel are nodules? *Trends Plant Sci* 14: 77–86.

Marzec M, Gruszka D, Tylec P, Szarejko I. (2016). Identification and functional analysis of the HvD14 gene involved in strigolactone signaling in Hordeum vulgare. *Physiologia Plantarum* 158(3): 341–355.

Marzec M, Melzer M. (2018). Regulation of root development and architecture by strigolactones under optimal and nutrient deficiency conditions. *International Journal of Molecular Sciences* 19(7): 1887.

Matusova R, Rani K, Verstappen FWA, Franssen MCR, Beale MH, Bouwmeester HJ. (2005). The strigolactone germination stimulants of the plant parasitic striga and orobanche spp. are derived from the carotenoid pathway. *Plant Physiol* 139: 920–934.

Mayzlish-Gati E, De-Cuyper C, Goormachtig S, Beeckman T, Vuylsteke M, Brewer PB, Beveridge CA, Yermiyahu U, Kaplan Y, Enzer Y. (2012). Strigolactones are involved in root response to low phosphate conditions in Arabidopsis. *Plant Physiol* 160: 1329–1341.

McAdam EL, Hugill C, Fort S, Samain E, Cottaz S, Davies NW, Reid JB, Foo E. (2017). Determining the site of action of strigolactones during nodulation. *Plant Physiol* 175: 529–542.

Mori N, Nishiuma K, Sugiyama T, Hayashi H, Akiyama K. (2016). Carlactone-type strigolactones and their synthetic analogues as inducers of hyphal branching in arbuscular mycorrhizal fungi. *Phytochemistry* 130: 90–98.

Omoarelojie LO, Kulkarni MG, Finnie JF, Pospíšil T, Strnad M, Van Staden J. (2020). Synthetic strigolactone (rac-GR24) alleviates the adverse effects of heat stress on seed germination and photosystem II function in lupine seedlings. *Plant Physiology and Biochemistry* 155: 965–979.

Parniske M. (2008). Arbuscular mycorrhiza: the mother of plant root endosymbioses. *Nat. Rev. Microbiol* 6: 763–775.

Pelaez-Vico MA, Bernabeu-Roda L, Kohlen W, Soto MJ, Lopez-Raez JA. (2016). Strigolactones in the Rhizobium-legume symbiosis: Stimulatory effect on bacterial surface motility and down-regulation of their levels in nodulated plants. *Plant Sci* 245: 119–127.

Piisila M, Keceli MA, Brader G, Jakobson L, Joesaar I, Sipari N, Kollist H, Palva ET, Kariola T. (2015). The F-box protein MAX2 contributes to resistance to bacterial phytopathogens in Arabidopsis thaliana. *BMC Plant Biol* 15: 53.

Rasmussen A, Depuydt S, Goormachtig S, Geelen D. (2013). Strigolactones fine-tune the root system. *Planta* 238: 615–626.

Rasmussen A, Mason MG, De Cuyper C, Brewer PB, Herold S, Agusti J, Geelen D, Greb T, Goormachtig S, Beeckman T. (2012). Strigolactones suppress adventitious rooting in Arabidopsis and pea. *Plant Physiol* 158: 1976–1987.

Raza A, Javed R, Zahid Z, Sharif R, Hafeez MB, Ghouri MZ, . . . Siddiqui MH. (2021). Strigolactones for sustainable plant growth and production under adverse environmental conditions. In *Plant Performance Under Environmental Stress* (pp. 129–166). Springer, Cham.

Rehman N, Khan FU, Imran M, Rajput SA, Li Y, Ullah I, Azhar MT. (2022). Knockdown of GmD53a confers strigolactones mediated rhizobia interaction and promotes nodulation in soybean. *Peer J* 10: e12815.

Ruiz-Lozano JM, Aroca R, Zamarreño ÁM, Molina S, Andreo-Jiménez B, Porcel R, García – Mina JM, Ruyter-Spira C, López-Ráez JA. (2016). Arbuscular mycorrhizal symbiosis induces strigolactone biosynthesis under drought and improves drought tolerance in lettuce and tomato. *Plant Cell Environ* 39: 441–452.

Ruyter-Spira C, Kohlen W, Charnikhova T, van Zeijl A, van Bezouwen L, de Ruijter N, Cardoso C, Lopez-Raez JA, Matusova R, Bours R. (2011). Physiological effects of the synthetic strigolactone analog GR24 on root system architecture in *Arabidopsis*: another below ground role for strigolactones? *Plant Physiol* 155: 721–734.

Sah SK, Reddy KR, Li J. (2016). Abscisic acid and abiotic stress tolerance in crop plants. *Front Plant Sci* 7: 571.

Shinohara N, Taylor C, Leyser O. (2013). Strigolactone can promote or inhibit shoot branching by triggering rapid depletion of the auxin efflux protein PIN1 from the plasma membrane. *PLoS Biol* 11.

Smith SE, Read DJ. (2010). *Mycorrhizal Symbiosis*. Cambridge: Academic Press.

Soto MJ, Fernandez-Aparicio M, Castellanos-Morales V, Garcia-Garrido JM, Ocampo JA, Delgado MJ, Vierheilig H. (2010). First indications for the involvement of strigolactones on nodule formation in alfalfa (Medicago sativa). *Soil Biol. Biochem* 42: 383–385.

Sun H, Tao J, Hou M, Huang S, Chen S, Liang Z, . . . Zhang Y. (2015). A strigolactone signal is required for adventitious root formation in rice. *Annals of Botany* 115(7): 1155–1162.

Sun H, Tao J, Liu S, Huang S, Chen S, Xie X, Yoneyama K, Zhang Y, Xu G. (2014). Strigolactones are involved in phosphate-and nitrate-deficiency-induced root development and auxin transport in rice. *J. Exp. Bot* 65: 6735–6746.

Tsuchiya Y, Vidaurre D, Toh S, Hanada A, Nambara E, Kamiya Y, Yamaguchi S, McCourt P. (2010). A small-molecule screen identifies new functions for the plant hormone strigolactone. *NatChem Biol* 6: 741–749.

Umehara M, Hanada A, Yoshida S, Akiyama K, Arite T, Takeda-Kamiya N, Magome H, Kamiya Y, Shirasu K, Yoneyama K. (2008). Inhibition of shoot branching by new terpenoid plant hormones. *Nature* 455: 195–200.

Urquhart S, Foo E, Reid JB. (2015). The role of strigolactones in photomorphogenesis of pea is limited to adventitious rooting. *Physiol Plant* 153: 392–402.

Van Ha C, Leyva-González MA, Osakabe Y, Tran UT, Nishiyama R, Watanabe Y, Tanaka M, Seki M, Yamaguchi S, Van Dong N. (2014). Positive regulatory role of strigolactone in plant responses to drought and salt stress. *Proc. Natl. Acad. Sci* 111(2): 851–856.

Visentin I, Pagliarani C, Deva E, Caracci A, Turečková V, Novák O, Lovisolo C, Schubert A, Cardinale F. (2020). A novel strigolactone-miR156 module controls stomatal behaviour during drought recovery. *Plant Cell Environ* 43(7): 1613–1624.

Visentin I, Vitali M, Ferrero M, Zhang Y, Ruyter-Spira C, Novák O, Strnad M, Lovisolo C, Schubert A, Cardinale F. (2016). Low levels of strigolactones in roots as a component of the systemic signal of drought stress in tomato. *New Phytol* 212(4): 954–963.

Xie X, Yoneyama K, Yoneyama K. (2010). The strigolactone story. *Annu Rev Phytopathol* 48: 93–117.

Xu J, Zha M, Li Y, Ding Y, Chen L, Ding C, Wang S. (2015). The interaction between nitrogen availability and auxin, cytokinin, and strigolactone in the control of shoot branching in rice (Oryza sativa L). *Plant Cell Rep* 34: 1647–1662.

Xu X, Fang P, Zhang H, Chi C, Song L, Xia X, Shi K, Zhou Y, Zhou J, Yu J. (2019). Strigolactones positively regulate defense against root-knot nematodes in tomato. *J. Exp. Bot* 70: 1325–1337.

Yoneyama K, Kisugi T, Xie X, Yoneyama K. (2013). Chemistry of strigolactones: why and how do plants produce so many strigolactones. *J Mol Microbial Ecol Rhizosphere* 1: 373–379.

Yoneyama K, Xie XN, Kim HI, Kisugi T, Nomura T, Sekimoto H, Yokota T, Yoneyama K. (2012). How do nitrogen and phosphorus deficiencies affect strigolactone production and exudation? *Planta* 235: 1197–1207.

Yoneyama K, Xie XN, Kusumoto D, Sekimoto H, Sugimoto Y, Takeuchi Y, Yoneyama K. (2007). Nitrogen deficiency as well as phosphorus deficiency in sorghum promotes the production and exudation of 5-deoxystrigol, the host recognition signal for arbuscular mycorrhizal fungi and root parasites. *Planta* 227(1): 125–132.

Yoneyama K, Xie X, Sekimoto H, Takeuchi Y, Ogasawara S, Akiyama K, . . . Yoneyama K. (2008). Strigolactones, host recognition signals for root parasitic plants and arbuscular mycorrhizal fungi, from Fabaceae plants. *New Phytologist* 179(2): 484–494.

Yoshida S, Kameoka H, Tempo M, Akiyama K, Umehara M, Yamaguchi S, Hayashi H, Kyozuka J, Shirasu K. (2012). The D3 F-box protein is a key component in host strigolactone responses essential for arbuscular mycorrhizal symbiosis. *New Phytol* 196: 1208–1216.

Zhang Y, Lv S, Wang G. (2018b). Strigolactones are common regulators in induction of stomatal closure in planta. *Plant Signal Behav* 13: e1444322.

4 Regulatory Roles of Strigolactones in Drought Tolerance in Plants

Kriti Gupta, Simranjeet Kaur, Muhammad Asad Naseer, and Avneesh Kumar

CONTENTS

4.0 INTRODUCTION

During development and growth, plants undergo various environmental stresses under natural and agricultural conditions. Drought, commonly called water stress, is one of the most detrimental, severe environmental stresses affecting plants' growth and productivity. The plant body comprises 80–95% water, which plays a crucial role in various aspects of plant growth, development, and metabolism (Brodersen et al., 2019; Abbasi and Abbasi, 2010). Due to this, few scientists regarded desiccation as the primary environmental stress globally, one of the hypercritical threats to world food security and the catalyst of famines in the past, especially in areas vulnerable to desiccation (Anjum et al., 2011; Diatta et al., 2020). Crops can experience water limitations during the growing season at all developmental stages. Drought at the vegetative stage is termed intermittent drought and in the reproductive phase is known as terminal drought. Intermittent drought can occur in the growing season while terminal drought usually affects crops during the reproductive phase and continues till harvest. It is known as pseudo-drought or physiological drought when the plant cannot uptake water from the soil, even with an adequate water supply (Salehi-Lisar and Bakhshayeshan-Agdam, 2020). Several factors, such as unequal and unreliable rainfall patterns, evapotranspiration, and the water holding capacity of the roots, are responsible for the uncertain drought (Passioura and Angus, 2010; Devincentis, 2020).

Drought stress impairs the various molecular, biochemical, physiological, morphological, and ecological functions and processes of the plants. Plants, therefore, can lower resource use, improve their growth, and adopt different mechanisms such as drought tolerance, avoidance, and escape to survive under abiotic stresses such as drought (Osakabe et al., 2014).

Strigolactones (SLs) are a class of plant hormones that protect the plants from adverse environmental conditions. They were first discovered from the exudates of the roots of cotton plants around 60 years ago (Cook et al., 1966). SLs are well-founded host and root parasitic transmitters in

DOI: 10.1201/9781003224525-6

weeds of the genera *Striga*, *Orobanche*, *Phelipanche*, and *Alectra* (Yoneyama et al., 2010). SLs act as chemical signals upon excretion from the roots of host plants. They make non-dormant seeds of parasitic plants germinate, thereby allowing them to infect hosts to take their nutrients, water, and other assimilates. SLs confer hyphal-branching signals with respect to arbuscular mycorrhizal (AM) fungi (Besserer et al., 2006; Akiyama et al., 2005). SLs were further designated as a novel class of branch-inhibiting phytohormones, which modulate the general structure of plants (Umehara et al., 2008; Gomez-Roldan et al., 2008) by playing a dual role as signaling molecules in the rhizosphere.

In the past few years, several scholarly investigations have especially explored the functions of SLs in plants and their interplay with root symbionts and parasitic weeds. Here we wish to report the recent breakthroughs in the area of SL functions in plant resilience under drought stress.

4.1 BIOSYNTHESIS OF STRIGOLACTONES

Strigolactones have been grouped into two categories on the basis of the C-ring orientation and ABC-ring system. Category 1 C-ring orientation includes β-oriented strigol-type SL and α-oriented orobranchol-type SL. Generally, the majority of plants produce only one type of SL, except *Tobacco*, in which both α and β are produced. Category 2 ABC-ring systems include canonical SL; another one lacking canonical A, B, and C rings is the non-canonical SL. Avenaol, heliolactone, zealactone, and lotuslactone are some of the non-canonical SLs isolated from root exudates of wild oat, sunflower, maize, and *Lotus japonicum*, respectively, as seen in Figure 4.1 (Yoneyama, 2020).

SLs belong to apocarotenoid-class compounds derived from carotenoids by the action of certain enzymes such as carotenoid isomerase D27, carotenoid cleavage dioxygenases CCD7 and CCD8, and cytochrome P450 monooxygenases localized in cytosol and plastids. The biosynthesis of SLs initiates from all-trans-β-carotene, which is converted to 9-cis-β-carotene by D27 isomerase, followed by chronological reactions catalyzed by CCD7 and CCD8 convert 9-cis-β-carotene to carlactone (CL) exhibiting A- and D-ring structure (Figure 4.2). Carlactone is an endogenous precursor of both canonical and non-canonical SLs in *Arabidopsis* and rice (Seto et al., 2014). Carotenoid isomerase, CCD7, and CCD8 convert all-trans-β-carotene into CL and 3-hydroxy-carlactone (3-OH-CL) in maize (Baz et al., 2018). In terms of stability, these non-canonical SLs are less stable than canonical SLs (Yoneyama, 2020), which thwart further investigations into the biological functions of non-canonical SLs.

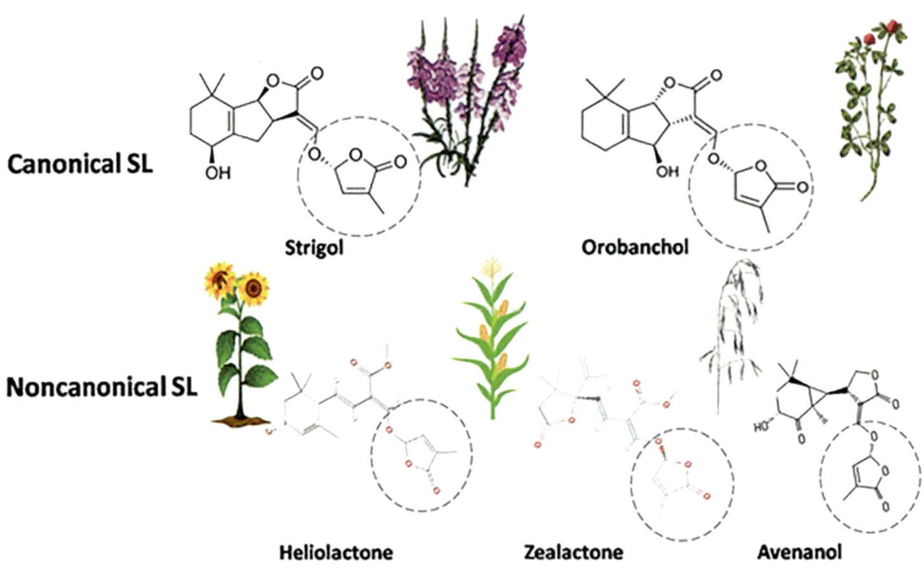

FIGURE 4.1 Structures of canonical and non-canonical strigolactones.

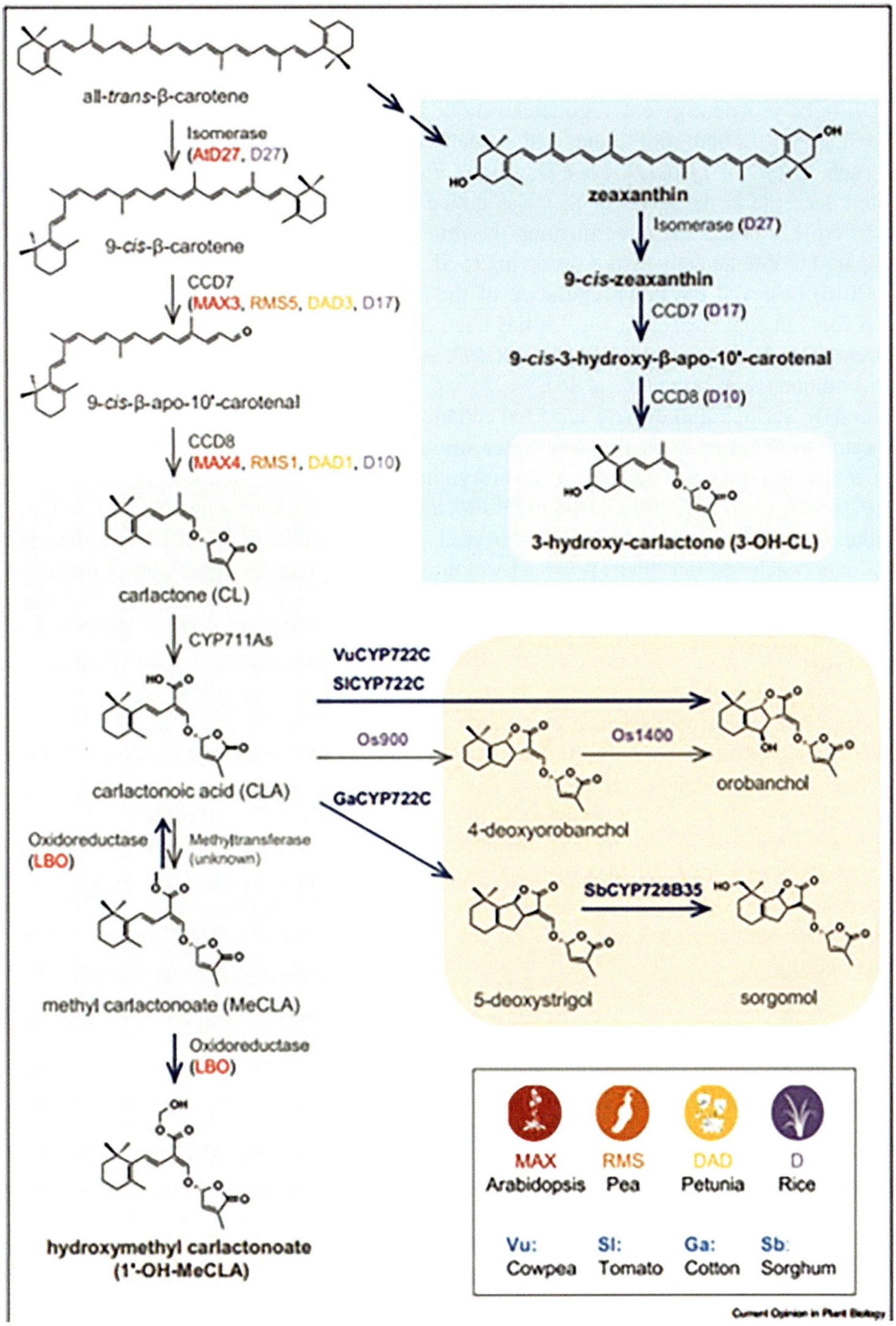

FIGURE 4.2 Schematic representation of biosynthesis of strigolactones.

Source: Yoneyama et al., 2021

4.1.1 SL Production under Drought Stress

There is a lot of discrepancy in SL production in terms of both upregulation and downregulation in roots subjected to diverse types of drought treatments at various stages.

Under osmotic stress, the production of SLs has been negatively regulated in the roots of dicot plants such as tomato, *L. japonicus*, and lettuce (Ruiz-Lozano et al., 2016; Liu et al., 2015; Aroca et al., 2013). Polyethylene glycol–regulated osmotic stress in *L. japonicus* was responsible for reducing the SLs levels in both root tissues and exudates of the plant. The expression of SL-biosynthetic genes such as *LjD27, LjCCD7, LjCCD8*, and *LjPDR1* was reduced further. Further, this PEG-mediated decrease in the levels of SLs was noticed in both phosphorus (P) supplementation and P-starvation (Liu et al., 2015), confirming the prominent PEG-mediated osmotic stress responses as compared to P-deficiency responses owing to SL production in *L. japonicus* roots. Ruiz-Lozano et al. (2016) observed the downregulation of the SL-biosynthetic gene *SlCCD7* in the drought-stressed roots of tomatoes. Likewise, it has been demonstrated that there is a downregulation of genes related to SL-biosynthesis such as *SlCCD7* and *8* and SL contents in tomato roots in water deficit conditions (Visentin et al., 2016).

Contrarily, various studies have reported enhanced SL levels endogenously in rice roots following a water withholding–mediated low water stress (Haider et al., 2018). There was an upregulation of transcription factors related to SL-biosynthetic genes MAX1-homologues *Os01g0701400, Os01g0700900, Os02g0221900, Os06g0565100*, and *D27* in rice shoots and *SlCCD7* and *SlCCD8* upregulation in the roots of tomato (Haider et al., 2018; Visentin et al., 2016). Further, some of the findings concluded that there is an activation of SL biosynthesis in the aboveground parts of plants with regard to osmotic stress. In *Arabidopsis* leaves, there was an enhanced transcription of SL-biosynthetic genes such as MAX3 and MAX4 subjected to desiccation (Ha et al., 2014). The SL biosynthetic *FaD17* gene and *D10, D17*, and *D27* genes have been upregulated in *Festuca arundinacea* crowns and the base of rice stem, respectively, under low water stress (Zhuang et al., 2017; Du et al., 2018). Unfortunately, in present times, SLs cannot be investigated much in the shoots because of their significantly reduced content endogenously in shoots (Liu et al., 2015; Xie et al., 2015; Umehara et al., 2010).

It can be concluded that the endogenous level of SL is often low in dicot roots, which may transmit a systemic signal to the aboveground parts in order to activate the localized synthesis of SLs (Visentin et al., 2016; Cardinale et al., 2018). This stimulation of genes involved in SL-biosynthesis in aboveground parts of diverse plant species suggested playing their an active role in the osmotic-stress tolerance, which can enhance plant survivability in adverse environmental conditions.

4.2 PLANTS RESPONSES AND ADAPTATIONS TO DROUGHT STRESS UNDER APPLICATION OF SL

There is increasing evidence of ameliorating drought stress with the supplementation of SLs in plants. SLs play a significant role in the closing of stomata in plants. SLs further crosstalk with several phytohormones, especially ABA (Ha et al., 2014). Several studies have been reported in reference to this. In *Lotus japonicus*, the ABA-SL crosstalk shows the ABA insensitivity of the SL-biosynthesis mutant *Ljccd7* (Liu et al., 2015). In rice, there were enhanced endogenous levels of ABA in both SL-deficient (*d10* and *d17*) and SL-insensitive (*d3*) mutants in the shoot, resulting in drought tolerance (Haider et al., 2018). The overexpression of *OsD27* is highly expressed in SL-deficient insensitive mutants and involved in SL-ABA crosstalk. Further investigations reported that in *Arabidopsis* signaling mutants, single and double mutants (MAX2), biosynthesis of ABA (*aba2*), and signaling in guard cells by ABA (*ost1*) showed greater stomatal conductance in the double mutants than in the single mutants, pointing to a possible ABA-independent/MAX2-mediated stress response. This ABA-independent (MAX2-mediated) stress response shows the concerted role of the strigolactones-karrikins paralogous pathways since both molecules signal through MAX2 (Kalliola et al., 2019).

In agricultural fields, the foliar spray of SLs played many physiological roles in stomatal regulation and antioxidant response in mitigating drought stress. In support of this, Davidson (2015) noticed that the application of SLs in maize fields led to more effective drought tolerance than non-treated fields. During foliar application of synthetic SLs (e.g., rac-GR 24) on two-year-old grape seedlings (*Vitis vinifera* L.), SLs were in crosstalk with ABA, which lessened the effects of drought stress by stomatal closing, modulated the chlorophyll components and photosynthesis, and activated the antioxidant defense system in shoots and roots of grapes under drought stress (Min et al., 2019). In winter wheat, GR24 application showed increased yield and photosynthesis under drought stress by providing tolerance to osmotic stress (Sedaghat et al., 2021). In barley, it was concluded that the drought response by a specific gene (the SL-specific receptor HvD14) is linked to SL signaling and ABA-dependent osmotic response (Marzec et al., 2020). There is a report showing that exogenous application of SLs improved the water relations, increased photosynthetic pigments and gas exchange parameters, and enhanced antioxidant enzymatic activities to impart drought stress tolerance in maize seedlings (Sattar et al., 2022). SLs in tomato act as a molecular link between drought and miR156, and miR156 acts as a mediator of the ABA-dependent effect of SLs on the aftereffects of drought on stomata (Visentin et al., 2020).

4.2.1 Effect of SLs on the Root Architecture and Its Adaptations

An additional role of SLs is to make complexes with other hormones such as auxin (IAA), cytokinin, ethylene, and karrikin pathways, enhancing nutrient uptake in nutrient-poor soils with root initiation and nutrient uptake. SLs inhibit the formation of lateral roots under optimal conditions, but during nutrient stress, they facilitate nutrient uptake in *Arabidopsis* (Ruyter-Spira et al., 2011; Marzec and Melzer, 2018). In *Arabidopsis* roots, SLs interact with auxins and positively regulate primary root biomass and root hair length and density (Ruyter-Spira et al., 2011; Kapulnik and Koltai, 2014). Further studies concluded that optimized SL transportation could extricate feedback inhibition of SL biosynthetic pathway, enhance root growth, and induce root hair elongation by overexpressing the PLEIOTROPIC DRUG RESISTANCE1 (*PDR1*) genes in *Petunia hybrida* (Liu et al., 2018).

4.2.2 Effect of SLs on the Shoot Architecture

SLs play a significant role in shoot architectural modifications with the help of D-14, a signaling receptor, and *SMXLs*. *SMXL6,7,* and *8* and MAX2 quadruple mutants in *Arabidopsis* accelerate the mobilization of auxin and accretion of PIN1 in stems, recommending the roles of SLs in modulating shoot branching via interacting with auxins (Soundappan et al., 2015). They combine with auxins to control branching patterns in shoots (Hayward et al., 2009; Bennett et al., 2016; Ligerot et al., 2017). The expression of BRC1 in axillary buds is repressed by SMXLs (Soundappan et al., 2015). During SL signaling pathways in pea and *Arabidopsis*, BRC1 proceeds downstream, thus suppressing the axillary growth of buds (Braun et al., 2012; Aguilar-Martinez et al., 2007). Under field conditions, the exogenous SL supplementation to seedlings influenced shoot architecture and metabolite portioning to the yield organ. In relation to this, SL foliar application increased by 20% the yield in maize in various field studies (Davidson, 2015). In rapeseed (*Brassica napus*), GR 24 application significantly increased plant biomass in seven days (Ma et al., 2017). Further, in agricultural fields, synthetic SL and ABA in combination hamper the ripening of farm products, thus enhancing the shelf life of grapevine (Ferrero et al., 2018). All the responses of various crops are listed in Table 4.1.

TABLE 4.1
Response of Various Crops with an Application of SLs

Plant System	Response	Reference
Lotus japonicus	ABA-SL crosstalk helped in synthesizing SL-producing genes	Liu et al., 2015
Rice	Enhanced endogenous levels of ABA in both SL-deficit and SL-insensitive mutants	Haider et al., 2018
Arabidopsis	Increased stomatal conductance and ABA-SL mediated response	Kalliola et al., 2019
Maize	Foliar application increased yield by 20%	Davidson, 2015
Vitis vinifera	SL-ABA crosstalk led to the closing of stomata and activated antioxidant defense system under water stress	Min et al., 2019
Winter wheat	Increased yield and photosynthesis	Sedaghat et al., 2021
Barley	SL-ABA crosstalk activated the SL-specific receptor HvD14	Marzec et al., 2020
Maize	SLs improved water relations and increased photosynthetic pigments and antioxidant enzyme activity	Sattar et al., 2022
Tomato	miR-156 acted as a molecular link between ABA, and SL provided tolerance against osmotic stress	Visentin et al., 2020
Petunia hybrida	Overexpression of *PLEIOTROPIC DRUG RESISTANCE1* (*PDR1*) genes enhanced root growth with SL application	Liu et al., 2018
Brassica napus	GR-24 application increased plant biomass	Ma et al., 2017
Grapes	SL and ABA crosstalk enhanced the shelf life of grapes	Ferrero et al., 2018
Arabidopsis	SLs facilitated nutrient uptake, interacted with auxins and promoted root biomass, root hair length, and density. SLs with auxins modulated shoot branching	Ruyter-Spira et al., 2011; Kapulnik and Koltai, 2014; Soundappan et al., 2015; Hayward et al., 2009; Bennett et al., 2016; Ligerot et al., 2017

4.3 SIGNALING UNDER DROUGHT STRESS

Various studies reported that strigolactones showed growth improvement and development under various abiotic stresses, particularly drought stress. In recent times, functional analysis of genes associated with the biosynthetic pathway and signaling of SL revealed that SLs impart protection against water deficit conditions. Progressive mutant screening and physiological and biochemical experiments in petunia (*Petunia hybrida*), pea (*Pisum sativum*), rice (*Oryza sativa*), and *Arabidopsis* established the involvement of strigolactones in the activation of stress-responsive proteins (Mostofa et al., 2018). Genetic study of the two crucial carotenoid cleavage dioxygenases in SL biosynthesis MAX3 and MAX4 (*MORE AXILLARY GROWTH 3, 4*) mutants in *Arabidopsis* provided the primary evidence of SLs as a positive regulator of drought stress (Ha et al., 2014). This discovery is corroborated by more discrete studies, revealing water-stress susceptibility in SL-depleted MAX1 *Arabidopsis* lines (Zhang et al., 2018) and SL-deficient *Lotus japonicus LjCCD7*-silenced and tomato *SlCCD7*-silenced plants (Visentin et al., 2016; Liu et al., 2015).

Li et al. (2016), established that SL acts as a negative regulator in drought resistance by analyzing the drought-tolerant phenotype of the triple *smxl6,7,8* mutant plants. These mutated *Arabidopsis* plants showed greater leaf surface temperature, lowered cuticle permeability, decreased drought-induced water deficit, and increased electrolyte leakage due to cell membrane damage than wild-type plants. It was found that SLs reduce the stomatal conductance and stomatal density (Ha et al., 2014) under water deficit conditions (Zhang et al., 2018; Visentin et al., 2016; Liu et al., 2015), which is associated with SL-mediated overexpression of genes associated with drought response in the shoots (Visentin et al., 2016; Ha et al., 2014).

Bu and his coworkers in 2014 suggested the role of MAX2 (*MORE AXILLARY GROWTH 2*) in salinity and mannitol and desiccation resistance during seed germination and evidenced thinner cuticle and larger stomata aperture in MAX2 mutant *Arabidopsis* plants under water stress conditions. Microarray analysis in MAX2 and wild-type plants discovered that the SL network plays a pivotal role in abiotic stress resistance, comprising formerly characterized genes responsible for abiotic stress tolerance and phytohormone signaling, particularly with ABA and cytokinins.

SL signaling has also been associated with ABA production; for instance, SL-deficient mutants have shown diminished ABA responsiveness and less synthesis of anthocyanin but improved shoot branching and a reduced root-to-shoot ratio (Mostofa et al., 2018; Ha et al., 2014; Visentin et al., 2016). MAX2 mutant plants show impaired stomatal function by downregulating the ABA import genes *ABCG22* and *ABCG40* in well-watered and water-stressed MAX2 leaves, respectively, compared to the wild type, representing that *ABCG22* and *ABCG40* mutants are susceptible to drought (Kuromori et al., 2011; Kang et al., 2010). Additionally, in tomato via LC-MS/MS, enzyme-specific inhibitors, and ABA-deficient mutants (for instance *notabilis, sitiens*, and *flacca*) showed the relationship between ABA and SL signaling under drought conditions. Plants treated with the NCED-inhibitor abamine SG and the non-treated ABA-deficient mutants display lower ABA and SL contents. Furthermore, downregulation of *LeCCD7* and *LeCCD8* transcripts was demonstrated via performing the expression analysis in ABA deficient mutants (López-Ráez et al., 2010).

Numerous SL biosynthesis genes (e.g., MAX3 and MAX4) have been downregulated by drought in *ahp2,3,5* mutants as compared to the wild type (Nishiyama et al., 2013), signifying that SL biosynthesis is dependent on CK levels. Besides repression of four of seven cytokinin catabolite genes, such as *CKX1, CKX2, CKX3*, and *CKX5*, subsequent water deficit is noticed in MAX2 mutant (SL biosynthesis gene) relative to wild-type plants (Ha et al., 2014). These results emphasize the multifaceted homeostatic adjustment of one hormone by another in plants subjected to water deficit. This study proposes an adverse consequence of SL signaling on the CK level while an exact interpretation could be drawn by discovering SL-specific signaling mutants systematically since MAX2 is found to participate in karrikin signaling (Mostofa et al., 2018).

Quantitative real-time reverse transcription PCR (qRT-PCR) assays illustrated that there is downregulation of ABA-induced marker genes, comprising *KIN1 (COLD INDUCIBLE), RD29A (RESPONSIVE TO DEHYDRATION 29A), RD29B (RESPONSIVE TO DEHYDRATION 29B*), responsive to ABA and *COR47 (COLD-REGULATED PROTEIN47*) genes, which are found to be participating in the ABA biosynthetic transportation signaling pathways and their catabolism with *ABA Insensitive1 (ABI1), NCED3, Hypersensitive to ABA1, 707A3, CytochromeP450*, and *ABCG22*.

4.3.1 Transgenics

In *Arabidopsis* cis-acting motifs, *ACGTATERD1* and *MYBIAT* mainly bind to drought-responsive TFs were identified by silico analysis of the promoter sequences of SL biosynthetic genes (Marzec and Muszynska, 2015). Subsequently, a complementary approach confirms the evolutionary conservation of MAX2 activity between parasitic and non-parasitic plants (Li et al., 2016). It has been discovered that there is a recovery of loss-of-function mutants of *AtMAX2* in *Arabidopsis* transgenics by transferring the *OaMAX2* gene from the parasitic plant, *Orobanche aegyptiaca*, under water-deficit conditions (Li et al., 2016). Additionally, group 1 glycosyltransferase, *SDG8i* from *Sporobolus stapfanus* (resurrection grass), when overexpressed in *Arabidopsis*, showed high auxin content, which proceeded downstream of ABA to curtail low water-induced senescence. This illustrated

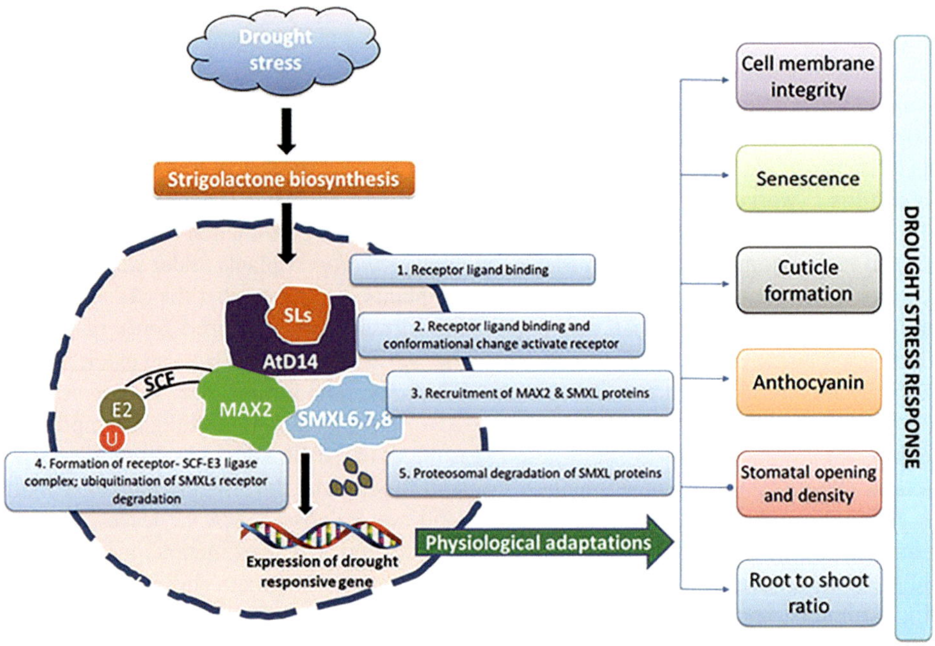

FIGURE 4.3 In *Arabidopsis*, five enzymes enciphered by *AtD27*, MAX3, MAX4, MAX1, and *LBO* have shown involvement in the biosynthesis of SL. Genes responsible for the catabolism of SL are yet to be known. However, D14 may catabolize by performing both hydrolase and receptor functions. Following SLs synthesis, it binds to the receptor D14 (for instance, AtD14 in *Arabidopsis*) to commence SL signaling supported by the F-box protein MAX2. This MAX2-dependent SCF (i.e., Skp1-Cullin-F-box) E2 ubiquitin ligase complex disrupts the transcription of the repressor proteins such as *SMXL6*, *SMXL7*, and *SMXL8* in *Arabidopsis*, thereby upregulating the expression of SL-mediated drought-responsive genes. SL-mediated drought adaptations improve SL signaling, resulting in enhanced physiological adaptations under water deficit conditions.

Note: Solid arrows indicate positive regulation, and solid blunt bars signify negative regulation. Abbreviations: AtD14, 27 (Arabidopsis thaliana DWARF14, 27); E2 (ubiquitin-conjugating enzyme); LBO (LATERAL BRANCHING OXIDOREDUCTASE); MAX2, 3, 4 (MORE AXILLARY GROWTH2, 3, 4); SMXL6 (SUPPRESSOR OF MAX21-LIKE6); (U, ubiquitin).

an immediate connection between the *SDG8i* and SL-signaling pathways and showed improved growth and yield under water deficit compared to control conditions (Islam et al., 2013). The expression of the rice *cystatin*, *Oryza cystatin-I* (*OCI*) demonstrated improved drought tolerance traits in *Glycine max* and *Arabidopsis thaliana* plants by regulating the strigolactone biosynthetic pathway. It has been observed that inhibition of OCI-dependent cysteine proteases results in better desiccation tolerance in soybean and *A. thaliana* photosynthetic assimilation of CO_2 than OCI-expressing soybean lines. It was evident from enhanced densities of lateral roots in MAX2-1 mutants (defective in strigolactone signaling), aside from MAX3-9 mutants (defective in strigolactone synthesis) in *E64*- (cysteine protease inhibitor) treated *A. thaliana* plants (Quain et al., 2014). Huang et al. (2021) reported the improved drought tolerance in apples by upregulating the strigolactone (SL) biosynthesis genes such as *MdD27*, *MdCCD7*, *MdCCD8a*, *MdCCD8b*, and *MdMAXa* genes from apples on AM colonization. Furthermore, it has been illustrated by elevated relative water content and osmolyte levels, better gas exchange capacity, increased antioxidative ability, enhanced chlorophyll fluorescence rate, and reduced relative electrolytic leakage in overexpression (OE) lines *MdIAA24* under drought stress conditions. Recently, drought stress acclimatization by SL-mediated early maize silk emergence to improve yield stability has been evaluated by overexpressing the carboxylesterase 20 (*AtCXE20)* in *Arabidopsis* and maize.

4.4 CONCLUSION AND FUTURE PROSPECTS

In agriculture, phytohormone-based technologies are well adapted to mitigate stress and the development of crops. SL emerged as an exciting target in adapting plants to abiotic stress with several applications in sustainable agriculture. Most importantly, SLs alleviate drought stress, enhance nutrient assimilation, and promote the development of crops in early stages.

Though several studies have involved SLs in drought mitigation, more efforts are needed to explore the underlying mechanisms behind this stress adaptation. Foliar application of synthetic analogues of SL (e.g., GR-24) provides evidence that SLs mitigate the adverse effects of drought. In terms of signaling, SL, in coordination with ABA, leads to stomatal closure in many plants, thus ameliorating the severe effects of drought. Several studies proved that natural or synthetic SLs influence the architecture of roots and promote plant growth in several plant species. They also played a vital role in the absorption of P in nutrientless soils.

The positive effects of SLs make them an interesting target for developing crop-enhancement products. New opportunities in sequencing technologies and analysis of root microbiome, rhizobiology, will lead to the development of new transgenics and improved crop varieties. Future and continued R&D investments will employ more in the successful development and use of phytohormone technology in agriculture. These new ventures, collaborations, bring many SL-based technological applications to the market and, hence, to the farmers.

REFERENCES

Abbasi T, Abbasi SA. (2010). Biomass energy and the environmental impacts associated with its production and utilization. *Renewable and Sustainable Energy Reviews*. 14: 919–937.

Aguilar-Martinez JA, Poza-Carrion C, Cubas P. (2007). *Arabidopsis* BRANCHED1 acts as an integrator of branching signals within axillary buds. *The Plant Cell*. 19: 458–472.

Akiyama K, Matsuzaki KI, Hayashi H. (2005). Plant sesquiterpenes induce hyphal branching in arbuscular mycorrhizal fungi. *Nature*. 435: 824–827.

Anjum SA, Xie XY, Wang LC, et al. (2011). Morphological, physiological, and biochemical responses of plants to drought stress. *African Journal of Agricultural Research*. 6: 2026–2032.

Aroca R, Ruiz-Lozano JM, Zamarreño ÁM, et al. (2013). Arbuscular mycorrhizal symbiosis influences strigolactone production under salinity and alleviates salt stress in lettuce plants. *Journal of Plant Physiology*. 170: 47–55.

Baz L, Mori N, Mi J, et al. (2018). 3-Hydroxycarlactone, a novel product of the strigolactone biosynthesis core pathway. *Molecular Plant*. 11: 1312–1314.

Bennett T, Liang Y, Seale M, et al. (2016). Strigolactone regulates shoot development through a core signaling pathway. *Biology Open*. 5: 1806–1820.

Besserer A, Puech-Pagès V, Kiefer P, et al. (2006). Strigolactones stimulate arbuscular mycorrhizal fungi by activating mitochondria. *PLoS Biology*. 4: e226.

Braun N, de Saint Germain A, Pillot JP, et al. (2012). The pea TCP transcription factor PsBRC1 acts downstream of strigolactones to control shoot branching. *Plant Physiology*. 158: 225–238.

Brodersen CR, Roddy AB, Wason JW, et al. (2019). Functional status of xylem through time. *Annual Review of Plant Biology*. 70: 407–433.

Cardinale F, Korwin-Krukowski P, Schubert A, et al. (2018). Strigolactones: Mediators of osmotic stress responses with a potential for agrochemical manipulation of crop resilience. *Journal of Experimental Botany*. 23: 2291–2303.

Cook CE, Whichard LP, Turner B, et al. (1966). Germination of witchweed (Striga lutea Lour.): Isolation and properties of a potent stimulant. *Science*. 154: 1189–1190.

Davidson E. (2015). Enhancing drought tolerance in maize using strigolactones. *USDA*. [WWW document] https://portal.nifa.usda.gov/web/crisprojectpages/1002887-enhancing-drought-tolerance-in-maize-using-strigolactones.html.

Diatta AA, Fike JH, Battaglia ML, et al. (2020). Effects of biochar on soil fertility and crop productivity in arid regions: A review. *Arabian Journal of. Geosciences*. 13: 595.

Du H, Huang F, Wu N, et al. (2018). Integrative regulation of drought escape through ABA-dependent and independent pathways in rice. *Molecular Plant*. 11: 584–597.

Ferrero M, Pagliarani C, Novak O, et al. (2018). Exogenous strigolactone interacts with an abscisic acid-mediated accumulation of anthocyanins in grapevine berries. *Journal of Experimental Botany*. 69: 2391–2401.

Gomez-Roldan V, Fermas S, Brewer PB, et al. (2008). Strigolactone inhibition of shoot branching. *Nature*. 455: 189–194. doi: 10.1038/nature07271.

Haider I, Andreo-Jimenez B, Bruno M, et al. (2018). The interaction of strigolactones with abscisic acid during the drought response in rice. *Journal of Experimental Botany*. 69: 2403–2414.

Hayward A, Stirnberg P, Beveridge C, et al. (2009). Interactions between auxin and strigolactone in shoot branching control. *Plant Physiology*. 151: 400–412.

Huang D, Wang Q, Jing G, et al. (2021). Overexpression of MdIAA24 improves apple drought resistance by positively regulating strigolactone biosynthesis and mycorrhization. *Tree Physiology*. 41(1): 134–146.

Islam S, Grifths CA, Blomstedt CK, et al. (2013). Increased biomass, seed yield, and stress tolerance is conferred in Arabidopsis by a novel enzyme from the resurrection grass *Sporobolus stapfanus* that glycosylates the strigolactone analogue GR24. *PLoS One*. 8: e80035.

Kalliola M, Jakobson L, Davidsson P, et al. (2019). The role of strigolactones in the regulation of stomatal conductance and plant-pathogen interactions in Arabidopsis thaliana. *Plant Stomata Encyclopedia*.

Kang J, Hwang J-U, Lee M, Kim Y-Y, Assmann SM, Martinoia E, Lee Y. (2010). PDR-type ABC transporter mediates cellular uptake of the phytohormone abscisic acid. *Proceedings of the National Academy of Sciences*. 107(5): 2355–2360.

Kapulnik Y, Koltai H. (2014). Strigolactone involvement in root development, response to abiotic stress, and interactions with the biotic soil environment. *Plant Physiology*. 166: 560–569.

Kuromori T, Sugimoto E, Shinozaki K. (2011). Arabidopsis mutants of AtABCG22, an ABC transporter gene, increase water transpiration and drought susceptibility. *Plant Journal*. 67(5): 885–894.

Li W, Nguyen KH, Watanabe Y, et al. (2016). OaMAX2 of *Orobanche aegyptiaca* and Arabidopsis AtMAX2 share conserved functions in both development and drought responses. *Biochemistry and Biophysics Research Communication*. 478: 521–526.

Ligerot Y, de Saint GA, Waldie T, et al. (2017). The pea branching RMS2 gene encodes the PsAFB4/5 auxin receptor and is involved in an auxin-strigolactone regulation loop. *PLoS Genetics*. 13: e1007089.

Liu G, Pfeifer J, de Brito Francisco R, Emonet A, Stirnemann M, Gübeli C, Hutter O, Sasse J, Mattheyer C, Stelzer E, Walter A, Martinoia E, Borghi L. (2018, Jan). Changes in the allocation of endogenous strigolactone improve plant biomass production on phosphate-poor soils. *New Phytologist*. 217(2): 784–798. doi: 10.1111/nph.14847.

Liu G, Stirnemann M, Gübeli C, Egloff S, Courty PE, Aubry S, Borghi L. (2019). Strigolactones play an important role in shaping exodermal morphology via a KAI2-dependent pathway. *iScience*. 17: 144–154.

Liu J, He H, Vitali M, Visentin I, Charnikhova T, Haider I, Cardinale F. (2015). Osmotic stress represses strigolactone biosynthesis in *Lotus japonicus* roots: Exploring the interaction between strigolactones and ABA under abiotic stress. *Planta*. 241: 1435–1451.

López-Ráez JA, Kohlen W, Charnikhova T, Mulder P et al. (2010). Does abscisic acid affect strigolactone biosynthesis? *New Phytologist*. 187(2): 343–354.

Ma N, Hu C, Wan L, et al. (2017). Strigolactones improve plant growth, photosynthesis, and alleviate oxidative stress under salinity in rapeseed (*Brassica napus* L.) by regulating gene expression. *Frontiers in Plant Science*. 8: e1671.

Marzec M, Daszkowska-Golec A, Collin A, et al. (2020). Barley strigolactone signaling mutant hvd14.d reveals the role of strigolactones in ABA-dependent response to drought. *Plant, Cell & Environment*. 43: 2239–2253.

Marzec M, Melzer M. (2018). Regulation of root development and architecture by Strigolactones under Optimal and Nutrient Deficiency Conditions. *International Journal of Molecular Sciences*. 19(7): 1887

Marzec M, Muszynska A. (2015). In silico analysis of the genes encoding proteins that are involved in the biosynthesis of the RMS/MAX/D pathway revealed new roles of strigolactones in plants. *International Journal of Molecular Sciences*. 16: 6757–6782.

Min Z, Li R, Chen L, et al. (2019). Alleviation of drought stress in grapevine by foliar-applied strigolactones. *Plant Physiology and Biochemistry*. 135: 99–110.

Mostofa MG, Li W, Nguyen KH, Fujita M, Tran LP. (2018, Oct.). Strigolactones in plant adaptation to abiotic stresses: An emerging avenue of plant research. *Plant, Cell & Environment*. 41(10): 2227–2243. doi: 10.1111/pce.13364.

Nishiyama R, Watanabe Y, Leyva-Gonzalez MA, Ha CV, Fujita Y, Tanaka M, Seki M, Yamaguchi-Shinozaki K, Shinozaki K, Herrera-Estrella L, Tran LS (2013, Mar 19). *Arabidopsis* AHP2, AHP3, and AHP5 histidine phosphotransfer proteins function as redundant negative regulators of drought stress response. *Proceedings of the National Academy of Sciences USA*. 110(12): 4840–4845. doi: 10.1073/pnas.1302265110.

Osakabe Y, Osakabe K, Shinozaki K, et al. (2014). Response of plants to water stress. *Frontiers in Plant Sciences*. 5: 86.

Passioura JB, Angus JF. (2010). Improving productivity of crops in water-limited environments. In *Advances in Agronomy* (106), Academic Press: Cambridge, pp. 37–75.

Quain MD, Makgopa ME, Márquez-García B, et al. (2014). Ectopic phytocystatin expression leads to enhanced drought stress tolerance in soybean (Glycine max) and *Arabidopsis thaliana* through effects on strigolactone pathways and can also result in improved seed traits. *Plant Biotechnology Journal*. 12(7): 903–913.

Ruiz-Lozano JM, Aroca R, Zamarreño ÁM, et al. (2016). Arbuscular mycorrhizal symbiosis induces strigolactone biosynthesis under drought and improves drought tolerance in lettuce and tomato. *Plant, Cell & Environment*. 39: 441–452.

Ruyter-Spira C, Kohlen W, Charnikhova T, et al. (2011). Physiological effects of the synthetic strigolactone analog GR24 on the root system architecture in Arabidopsis: another belowground role for strigolactones? *Plant Physiology*. 155: 721–734.

Salehi-Lisar SY, Bakhshayeshan-Agdam H. (2020). Agronomic crop responses, and tolerance to drought stress. In *Agronomic Crops*, Springer: Berlin/Heidelberg, pp. 63–91.

Sattar A, Ul-Allah S, Ijaz M, et al. (2022). Exogenous application of strigolactone alleviates drought stress in maize seedlings by regulating the physiological and antioxidants defense mechanisms. *Cereal Research Communications*. 50: 263–272.

Sedaghat M, Emam Y, Mokhtassi-Bidgoli A, et al. (2021). The potential of the synthetic strigolactone analogue gr24 for the maintenance of photosynthesis and yield in winter wheat under drought: Investigations on the mechanisms of action and delivery modes. *Plants*. 10(6): 1223. doi: 10.3390/plants10061223.

Seto Y, Sado A, Asami K, et al. (2014). Carlactone is an endogenous biosynthetic precursor for strigolactones. *Proceedings of the National Academy of Sciences*. 111: 1640–1645.

Soundappan I, Bennett T, Morffy N, et al. (2015). SMAX1-LIKE/D53 family members enable distinct MAX2-dependent responses to strigolactones and karrikins in Arabidopsis. *Plant Cell*. 27: 3143–3159.

Umehara M, Hanada A, Magome H, et al. (2010). Contribution of strigolactones to the inhibition of tiller bud outgrowth under phosphate deficiency in rice. *Plant & Cell Physiology*. 51: 1118–1126.

Umehara M, Hanada A, Yoshida S, et al. (2008). Inhibition of shoot branching by new terpenoid plant hormones. *Nature*. 455: 195–200. doi: 10.1038/nature07272.

Visentin I, Pagliarani C, Deva E, et al. (2020). A novel strigolactone-miR156 module controls stomatal behavior during drought recovery. *Plant, Cell and Environment*. 43: 1613–1624.

Visentin I, Vitali M, Ferrero M, et al. (2016). Low levels of strigolactones in roots as a component of the systemic signal of drought stress in tomato. *New Phytologist*. 212: 954–963.

Xie X, Yoneyama K, Kisugi T, et al. (2015). Strigolactones are transported from roots to shoots, although not through the xylem. *Journal of Pesticide Science*. 40: 214–216.

Yoneyama K. (2020). Recent progress in the chemistry and biochemistry of strigolactones. *Journal of Pesticide Science*. 45: 45–53.

Yoneyama K, Awad AA, Xie X, et al. (2010). Strigolactones as germination stimulants for root parasitic plants. *Plant & Cell Physiology*. 51: 1095–110.

Yoneyama K, Brewer PB. (2021). Strigolactones, how are they synthesized to regulate plant growth and development? *Current Opinion in Plant Biology*. 63: 102072. doi: 10.1016/j.pbi.2021.102072.

Zhang Y, Lv S, Wang G. (2018). Strigolactones are common regulators in induction of stomatal closure in planta. *Plant Signalling and Behaviour*. 13(3): e1444322. doi: 10.1080/15592324.2018.1444322.

Zhuang L, Wang J, Huang B. (2017). Drought inhibition of tillering in *Festuca arundinacea* associated with axillary bud development and strigolactone signaling. *Environmental and Experimental Botany*. 142: 15–23.

5 Strigolactones in Plants
Web Networking with Phytohormonal Crosstalk

Atif Khurshid Wani, Nahid Akhtar, Sourbh Suren Garg, Farida Rahayu, Tahir ul Gani Mir, Ajit Prakash, and Rattandeep Singh

CONTENTS

5.0 INTRODUCTION

SL plant hormones are an essential class of phytohormones that manages many dimensions of root and shoot growth (Figure 5.1). They were isolated first from cotton plants in 1966 (Cook et al., 1966). However, their significance regarding germination was determined later (Bunsick et al., 2020). Similar compounds like alectrol and sorgolactone, both having lactone groups, were proven to play the same role and were thereby classified as SLs (Sato et al., 2003). They are structurally diverse chemical compounds synthesized by the rhizosphere that play an essential role in various plant physiological processes. SLs are vital for establishing a mutualistic union between the symbiotic fungi and the rhizosphere that helps in nutrient accumulation (Gamir et al., 2020). SLs are a vast group of plant hormones, but chemically, they have a common core in which hydroxymethyl butanolide is linked to a tricyclic lactone (Yoneyama, 2020). SLs bolster parasitic organisms by promoting their germination in the roots of the plant host like *Striga lutea* (Cook et al., 1966). As SLs are concerned with germination, signaling pathways of parasitic species like *Striga sp.*, they are supposed to control pests (Miyakawa, 2020). Another process in which SLs play their role is the prevention of excess bud development in stems, thus acting as inhibitory hormones (Dun et al., 2012). SLs differ from each other because of the functional group variation, but their melting points range from 200–202°C, with high solubility in polar solvents like benzene, and acetone (Halouzka et al.,

DOI: 10.1201/9781003224525-7

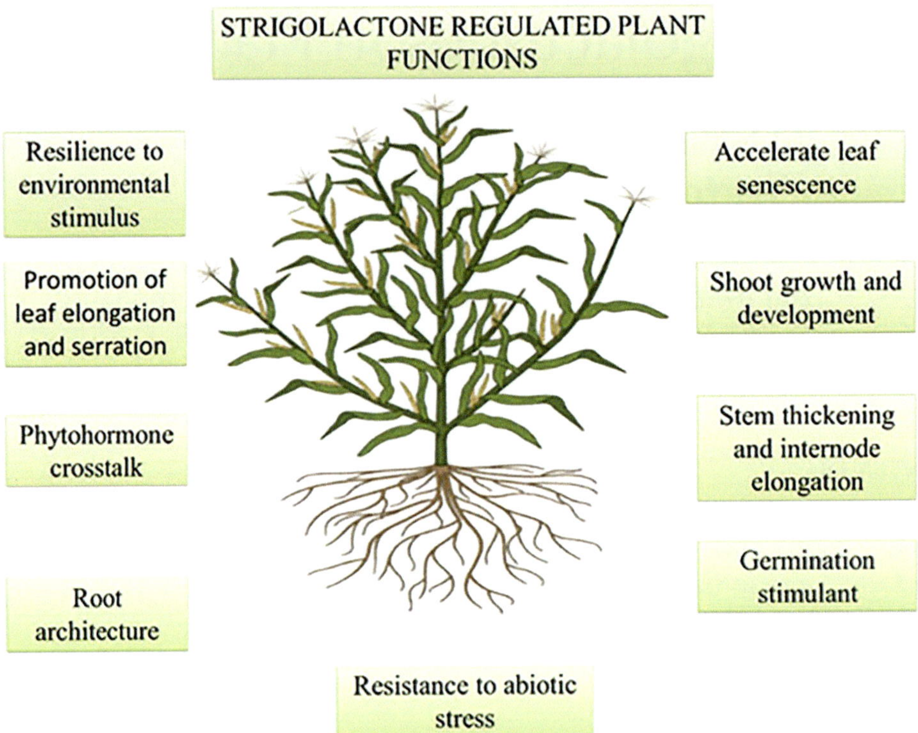

FIGURE 5.1 Summary Diagram

2020). Besides being pivotal to shoot branching, SLs are imperative for anthocyanin accumulation (Ferrero et al., 2018), plant height (Brewer et al., 2013), adaptation to phosphate starvation and drought (Ling et al., 2020), root architecture, and shoot branching (Xie et al., 2020). SLs in plants are perceived with the help of DWARF14 (D14) hydrolase. D14 has a low substrate turnover; thus, it requires a catalytic triad for biological functioning (Zhou et al., 2013). Molecular studies have reported the flexibility of the ligand binding pocket; thus, the catalytic triad is essential for ligand positioning and binding (Bürger and Chory, 2020; Carbonnel et al., 2020; Zhao et al., 2015). D14 hydrolase has been characterized as an SL receptor in *Arabidopsis, Oryza sativa, and Pisum sativum* (Yao et al., 2016). SLs are also well known for their usefulness in the plant-fungi association. A study on *Lotus japonicus* reported that the root-extracted compounds are vital for arbuscular mycorrhizal fungi (Xie et al., 2019). This was found true for many other plants, such as sorghum and maize (Roesler et al., 2021; Kaori Yoneyama et al., 2013). The isolated compounds of the SL family include strigol, 5-deoxystrigol, and sorgolactone. Since there exists a symbiotic relationship between arbuscular mycorrhiza and most gymnosperms and angiosperms, the existence of SLs in plants is always expected. Despite the fact that SLs are abundantly present in a wide variety of plants, a significantly smaller number of plants have been explored regarding SLs signaling because of the difficulties in the extraction process and early disintegration of SLs. Since the hormonal function discovery of SLs, there have been several research breakthroughs about the molecular perspectives of SL biosynthesis, perception, and transport (Brewer et al., 2013; Koltai, 2015). The SL-associated hormonal function discovery helped in understanding the long-sought SL biosynthesis through the involved genes. In this chapter, we summarize the molecular perspectives of SL biosynthesis with a focus on its role in plant development and stress response via hormonal crosstalk.

5.1 STRIGOLACTONE BIOSYNTHESIS: A MOLECULAR PERSPECTIVE

There have been 20 SL and SL-like molecules that have been characterized from the root exudates of different plants (Dell'Oste et al., 2021). They share a characteristic tricyclic lactone structure (ABC rings) connected through an enol-ether bridge to the D ring (unsaturated furanone moiety) (Cook et al., 1966).

5.1.1 Synthesis via Carlactone: The Carotenoid Pathway

The exact molecular details of SL biosynthesis through the carotenoid pathway are not clear yet, but several studies of the sequence of events in SL synthesis are somewhat interesting. These are supported by a study in which two rice mutants, *d17* and *d10*, were unable to cause the cleavage of carotenoid dioxygenase 7 (CCD7) and CCD8, respectively (Wang et al., 2019). Besides, cytochrome P450 and more axillary growth 1 (MAX1) have also been reported to be helpful in SL biosynthesis (Challis et al., 2013; Zhang et al., 2014). D27, a Fe-carrying protein, is also regarded as a biosynthetic component in SL synthesis (Lin et al., 2009). Carlactone, carrying a carbon skeleton similar to SL, has been identified as a resultant product of a series of reactions by CCD7, CCD8, and D27 utilizing trans-β-carotene as an *in vitro* substrate (Seto and Yamaguchi, 2014). During these sequence-off reactions, D27 was reported to mediate the isomerization of trans-β-carotene, which is a reversible reaction. This occurs at the C-9 position, and the resultant product is 9-cis-β-carotene. CCD7 causes preferential cleavage of 9-cis-β-carotene to form 9-cis-β-apo-10′-carotenal, which undergoes cleavage mediated by CCD8 giving rise to carolactone (Alder et al., 2012). Finally, intramolecular rearrangement occurs to form the characteristic D-ring of SLs. Figure 5.2 illustrates the carotenoid pathway of SL-biosynthesis through carlactone. Interestingly, carlactone causes the stimulation of *Striga hermonthica* germination as well as the suppression of shoot branching in biosynthetic mutants of rice like *d27*, therefore suggesting CL as the main precursor molecular for SLs (Seto and Yamaguchi, 2014; Alder et al., 2012).

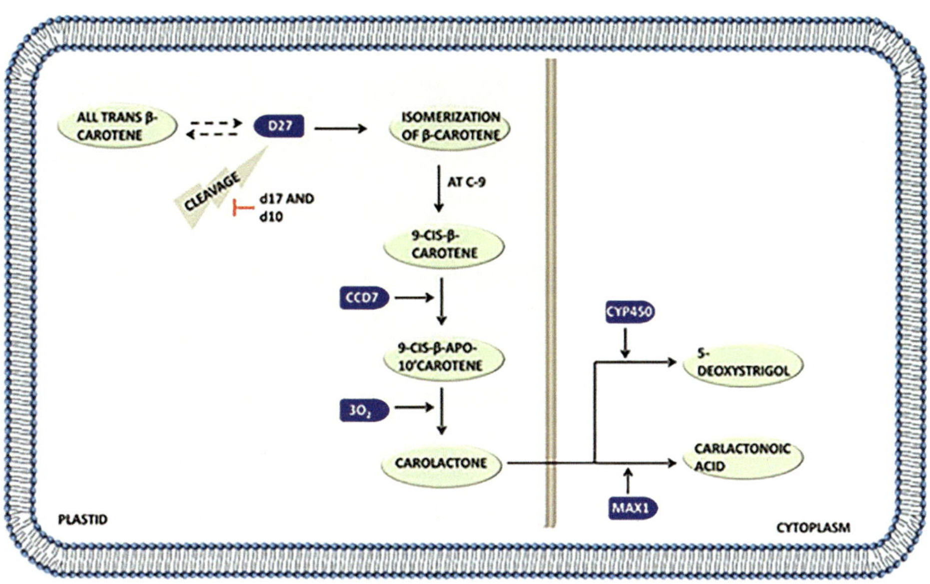

FIGURE 5.2 SL-biosynthesis through the carotenoid pathway.

5.1.2 SL and Abscisic Acid (ABA): A Biosynthetic Relation

Before the discovery of carlactone, SL biosynthesis was believed to occur by the coupling of the D-ring and other ring parts. Both portions were predicted to be synthesized separately, although it has been suggested that ABA helps in the regulation of SL (Koltai et al., 2010). β-carotene hydroxylase (HYD) catalyzes the conversion of all-trans-β-carotene to zeaxanthin. The formed product catalyzes the formation of violaxanthin when the reaction is catalyzed with epoxidase. Then, violaxanthin is converted into 9-cis-violaxanthin; which enzyme mediates this step is still unknown. In addition, NCED (or 9-cis-epoxycarotenoid dioxygenases) converts the 9-cis-violaxanthin to xanthoxin. This reaction proceeds, and xanthoxin forms the ABA-aldehyde in the presence of SDR (or short-chain-dehydrogenase/reductase). Finally, oxidation of ABA-aldehyde takes place with the help of an enzyme called AAO (or abscisic aldehyde oxidase), which gives the abscisic acid as an end product 9 (Figure 5.3). The mutant plants lacking or carrying defective NCED genes were found to have either extremely low or negligible levels of ABA besides having low SL levels in roots. Several correlations have been established between ABA and SL in rhizosphere extracts, but the exact mechanism remains obscure (J. Liu et al., 2015; López-Ráez et al., 2010).

5.2 SLS AND SEQUENTIAL SIGNALING FOR PLANT DEVELOPMENT

SLs were initially identified for symbiotic and parasitic relationships in the roots. Recent research has established their significance in shoot branching, arbuscular mycorrhiza germination, and other plant developments through a cascade of biochemical reactions.

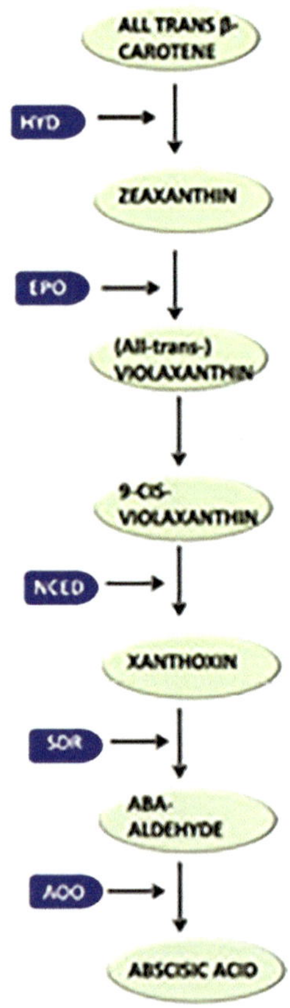

FIGURE 5.3 ABA biosynthesis and relation with SLs.

5.2.1 Shoot Branching

SLs are known for their repressive action on shoot branching. In pea and rice, SLs act on transcription factor (TF) (OsTB1/PsBRC1) to suppress bud outgrowth (Braun et al., 2012; Yamada et al., 2014). It is important to note that this TF acts as a connecting link between various pathways like the sucrose signaling pathway and cytokinin pathway (Wang et al., 2019). TB1, an ortholog of this gene in maize, encodes this TF act in an SL-independent manner to quell the shoot branching (Guan et al., 2012). IPA1/OsSPL14 and MADS57 TFs involved in rice shoot branching are also linked with SL signaling, but their downstream existence is still unknown (Song et al., 2017). It has been reported that SL mutants show additional branching other than *brc1* mutants (Chevalier et al., 2014). In *Arabidopsis*, auxin efflux removal from stem xylem depends on SLs (Shinohara et al., 2013). This increases the competition among buds in order to export transfer auxin into the principal auxin stream. It is therefore a proven fact that secondary plant growth is controlled by auxin levels (Milhinhos et al., 2013). However, auxin secretion is concurrently regulated by SLs, thus indirectly controlling plant development (Koltai, 2015). With regard to stem buds, SLs can cause a transport protein blockade, which is important for the auxin movement within the buds (Zhang et al., 2020). Therefore, the presence of higher levels of PIN1 protein in SL mutants is no exception.

5.2.2 Root Development

The role of SLs in root development is illustrated by the fact that *Arabidopsis* SL mutants exhibit more lateral roots than the wild type (Kumar et al., 2015). However, GR24 (biologically active synthetic-SL) supplementation causes the suppression of lateral root development in both SL mutants and the wild type, suggesting negative regulation of root formation by SLs (Sun et al., 2019). SLs have also been studied for their role in the hair length of roots, root length, and meristem cell numbers (Santoro et al., 2020). The supplementation of synthetic SLs induces *Arabidopsis* root elongation in SL mutants as well as the wild type (Mayzlish-Gati et al., 2012). D14 plays an important role in SL perception, mediating the conversion of SLs active forms. When D14 interacts with SL, it increases the chances of D14 interaction with D53, forming D14-SL-D53 complex, which, in turn, increases the chances of D53 interaction with the F-box subset (SCF D3/MAX2). This eventually causes D53 protein's polyubiquitination, thereby degrading through the proteosome pathway (Seto and Yamaguchi, 2014; Nakamura et al., 2013). Thus, it is proposed that the SL induces D14 degradation through proteosomes. Recent studies have reported that the epidermis is an essential mediator in SL-regulated root architecture (Koltai, 2013).

5.2.3 Fruit Development and Ripening

The majority of the phytohormones – including auxins, gibberellins, jasmonic acid, abscisic acid, brassinolide, ethylene, and cytokinins – participate in fruit development and ripening (Dubois et al., 2018; Huang et al., 2017; Verma et al., 2016; Wani et al., 2022). However, the role of SLs in fruit development has largely remained unclear. Wu et al. carried out an expression analysis of SL-signaling genes and revealed the involvement of SL in *Fragaria vesca* (woodland strawberry) development. The study showed higher expression levels in developing anther, style, and carpel (Wu et al., 2019). Notably, there was a significant decrease in receptacles during receptacle development, but they increased after pollination. Lee and Yoon reported elevation in ethylene biosynthesis in *Arabidopsis* seedlings, thereby influencing fruit ripening, senescence, and germination. SL does this by increasing the acetyl-CoA carboxylase transcript levels, thereby amplifying ethylene biosynthesis (Lee and Yoon, 2020). However, it is pertinent to note that SLs don't influence the expression and stability of acetyl-CoA synthases (Shi and Tu, 2015).

5.3 SLS-KEY PHYTOHORMONES IN ABIOTIC STRESS RESPONSE

The SL discovery has opened a new gateway for the exploration of phytohormone regulation during plant development and adaptation during abiotic stresses. The hormonal crosstalk largely assists in developing plant response under strenuous conditions like cold, heat, salinity, and nutrient starvation by regulating nutrient supply and other growth parameters (Table 5.1).

TABLE 5.1
Importance of SLs in Plant Development by Improving Tolerance Levels When Subjected to Abiotic Stresses

Plant	Stress	Effect/importance of SL	References
Lolium perenne (perennial ryegrass)	Cadmium	Controls bud outgrowth by transcriptional regulation	(Niu et al., 2021)
Solanum lycopersicum (tomato)	Drought	Enhances stomata sensitivity via ABA pathway	(Visentin et al., 2016)
Arabidopsis thaliana (Thale cress)	Salinity	Increases photosynthesis and stomatal conductance	(Ha et al., 2014)
Festuca arundinacea	Heat	Accelerates *D3* and *D14* expression	(Hu et al., 2018)
Malus hupehensis	KCl	Decreases chlorosis and increases transpiration rate	(Zheng et al., 2021)
Oryza sativa	Nutrient (P)	Restores leaf senescence	(Yamada et al., 2014)
Helianthus annus	Salinity	Increases superoxide dismutases and callus biomass	(Zulfiqar et al., 2020)
Sporobolus stapfianus	Cold	Overexpression of SDG8i gene encoding glycosyltransferase	(Islam et al., 2013)
Panicum virgatum	Cadmium	Decreases electrolyte leakage thus increasing shoot biomass and water content	(Tai et al., 2017)
Vitis vinifera	Drought	Lessens stomata opening and electrolyte leakage	(Min et al., 2019)
Lotus japonicus	Osmotic	Decreases ABA level	(Liu et al., 2011)
Arabidopsis thaliana	Salinity	Increases leaf stomatal density	(Ha et al., 2014)
Brassica napus	Salinity	Increases stomatal conductance	(N. Ma et al., 2017)
Lupinus	Heat	Increases proline content and reduces in peroxidation of lipids	(Omoarelojie et al., 2020)
Potentilla saundersiana	Low oxygen, UV exposure, and temperature	Increases primary metabolites and antioxidant proteins	(Ma ct al., 2015)
Arabidopsis thaliana	Low light or darkness	Accelerates leaf senescence when applied with ethylene	(Ueda and Kusaba, 2015)
Salvia nemorosa	Salinity	Decreases electrolyte leakage and increases photosynthetic rate	(Sharifi and Bidabadi, 2020)
Hordeum vulgare	Drought	Accelerates stomatal conductance	(Marzec et al., 2020)

5.3.1 Nutrient Starvation

Nitrogen (N) and phosphate (P) scarcity in the soil is a common but serious abiotic stress that plants encounter (Mu and Chen, 2021). Since the rhizosphere is the main P acquisition site, several changes occur in roots during its deficiency, which include an increase in root growth, shoot branching inhibition, reduced root elongation, and large-scale growth of root hairs and lateral roots (Rana et al., 2020; Xomphoutheb et al., 2020). It has been reported that SL production increases during P starvation (de Souza Campos et al., 2019). Yoneyama *et al.* studied the exudation of orobanchol from *Trifolium pratense* L. (red clover) due to P and K deficiency. The quantification of red clover–derived SL showed a 20-fold increase in orobanchol when subjected to P starvation. In sorghum, a 30-fold increase of SL 5-deoxystrigol was quantified under low N and P (Kaori Yoneyama, Xie et al., 2007; Kaori Yoneyama, Yoneyama et al., 2007). The majority of plants showed increased SL exudation as a result of N and P scarcity, except *Medicago sativa* (alfalfa) and *Solanum lycopersicum* L (tomato) (Saeed et al., 2017). While the surge in SLs has been reported in many plants, the stress signaling mechanism leading to overexpression of SL-associated proteins is still debatable. Umehara *et al.* reported that the genes in rice (D10, D17, D27) involved in SL biosynthesis were regulated by P supplementation and starvation (Umehara et al., 2015). Additionally, the gene responsible for SL exudation and translocation in soil from roots up to shoots in *Petunia hybrida* is accumulated during P scarcity (Kretzschmar et al., 2012).

5.3.2 Drought and Salinity Stress

The role of SLs in plants encountering water stress has been demonstrated by SL-ABA interplay. It has been reported that the NCED inhibitor–treated ABA mutants show reduced SL and ABA levels (Cheng et al., 2017). The SL-deficient mutants show hypersensitivity to drought, and their physiological effect is well observed in shoots (Li et al., 2019). Other studies on SL response mutants (MAX2) and SL synthesis (*MAX4* and *MAX3*) show that SL treatment can rescue germination under water-scarce conditions in synthesis mutants (Conn and Nelson, 2015; Toh et al., 2012). The comparative study on SL-depleted and wild-type *Lotus japonicus* subjected them to osmotic stress and P starvation. The SL-depleted exhibited higher stomatal conductance and showed lesser resistance to delayed stomatal closure in response to ABA (J. Liu et al., 2015). It has been established that arbuscular mycorrhiza symbiosis increases drought stress resistance in several plants (Begum et al., 2019). It has also been observed that drought influences the hormonal balance of lettuce and tomato plants (Ruiz-Lozano et al., 2016). This favours strigolactone accumulation, which thereby influences symbiosis, which is pivotal in coping with drought stress (Saeed et al., 2017).

The arbuscular mycorrhiza symbiosis is known to reduce the negative implications of salt stress on several plants, such as lettuce (Evelin et al., 2019). One of the studies conducted by Aroca and colleagues suggested that salinity stress instigated SL production in lettuce plants and further induces fungal (arbuscular mycorrhiza) growth, thereby helping in resisting salt stress (Aroca et al., 2013). Since Arabidopsis mutants (*MAX3* and *MAX4)* are sensitive toward salinity and drought stress, they show slow stomatal closure induced by ABA in comparison to the wild type (Ha et al., 2014). Ma et al. reported improvement in photosynthesis and reduction in oxidative stress under salt stress conditions in *Brassica napus* L. using synthetic GR24 SL (Ma et al., 2017). Moreover, the application of GR24 also causes a decrease in non-photochemical quenching and an increase in the quantum yield of photosystem II (Lu et al., 2019, 24).

5.3.3 Temperature Stress

Physiological and biochemical plant processes require optimal temperature, and any temperature fluctuation may subject them to stress conditions (Hasanuzzaman et al., 2013). For example, seed germination is highly dependent on temperature, and higher temperature usually inhibits the seed germination process (Roberts, 1988, 19). It is well known that GA- and CK-like phytohormones help in germination while ABA harms it (Miransari and Smith, 2014). SL causes the lowering of ABA and the increase of CK under temperature stress conditions; therefore, SL has a positive effect on seed germination (Mishra et al., 2016; Hu et al., 2018). Tsuchiya *et al.* reported that, on applying GR24 to SL-defective *Arabidopsis* under temperature stress conditions, the seed germination is stimulated (Tsuchiya et al., 2010). SL also causes seed rescue in dormant *Philipanche ramose* through ABA reduction (Lechat et al., 2015). Zhang *et al.* studied the effect of GR24 on the development of *Brassica rapa* L. (Rapa) at 4°C. They reported the inhibition of seedlings at 4°C, and the application of GR24 reduced the impact of the low temperature by increasing soluble protein content, improving photosynthesis, and increasing viability of the cells (Zhang et al., 2020).

5.3.4 Light Stress

Several studies have revealed the relationship between light and SLs. One of the studies on *Arabidopsis* revealed genes related to light signaling are instigated after exposing them to GR24, whereas, in SL mutants, the same genes underwent downregulation**.** The use of GR24 has been reported to increase the levels of chlorophyll and genes involved in the light-harvesting complex (Mayzlish-Gati et al., 2012). Sternberg *et al.* correlated SLs with light by carrying out the phenotypic evaluation of *Arabidopsis* growing under crowded conditions and low light intensity and found the plants with slender stems and elongated leaves. On the other hand, SL mutants such as *MAX1* and *MAX2* were found to have short and round leaves (Stirnberg et al., 2002). Low light is among the principal parameters that cause a decrease in plant yield, such as tomato, making stress resistance imperative. Lu et al. were able to improve plant growth during low light conditions in tomatoes using GR24. They reported improved seedling development via the reduction of oxidative stress and photosynthetic inhibition (Lu et al., 2019). Cooper et al. explored the significance of SL in *Arabidopsis* and pea defective in SL biosynthesis or signaling during dark chilling and reported that SL positively regulates chilling tolerance in both the plants (Cooper et al., 2018).

5.4 SLS AND OTHER PHYTOHORMONES: A PARADIGM OF HORMONAL CROSSTALK

Plant hormones are signal molecules synthesized by plants in very low concentrations. They control important characteristics of the plant, including embryogenesis (Jenik and Barton, 2005), organ size (Hepworth and Lenhard, 2014), stress tolerance (Xia et al., 2015), reproductive development (Terceros et al., 2020), and pathogen defense (Pieterse et al., 2012). ABA, a universal plant stress hormone, mediates stomatal closure and leads to signal activation during abiotic stresses like drought, salinity, heat, and nutrient starvation (Danquah et al., 2014). Since both SL and ABA are carotenoid derivatives, they are equally vital in developing stress tolerance in plants. The importance of ABA in SL biosynthesis has been already elucidated, and we explained the role of the NCED enzyme (Seto and Yamaguchi, 2014). In *Solanum lycopersicum*, alleviated SL concentration was reported when subjected to NCED treatment, resulting in the blockage of ABA biosynthesis (Finkelstein, 2013). Auxin, a growth-promoting phytohormone, induces the expression of genes associated with SL. These genes include *CCD7* and *CCD8 in Pisum sativa, Arabidopsis thaliana, Chrysanthemum morifolium*, and *Oryza sativa* (Lee and Yoon, 2020; Liang et al., 2010; W. Liu et al., 2011). The extended branching *of Petunia hybrida* (*dad* mutants), *Arabidopsis thaliana* (*MAX* mutants), and *Pisum sativum* (*rms* mutants) and the use of exogenous SLs like GR24 inhibit branching through auxin transport regulation (J. Zhang et al., 2020; Brewer et al., 2015). PIN protein (auxin transporter) has been reported to regulate the route and amount of auxin transport. The increased auxin content in MAX mutants ensures positive feedback control (*MAX* pathway) and increases the expressions of *MAX3* and *MAX4*. Thus, low SL increases auxin levels, which further helps in SL biosynthesis (Zhang et al., 2020a). Zhang et al. studied the effect of endogenous and exogenous SLs in *Arabidopsis thaliana* and *Pisum sativum* and reported that SLs interact through auxin feedback (Zhang et al., 2020). Three plant hormones – auxin, cytokinin, and SLs – control the bud outgrowth. Auxin, alongside SLs, inhibits the outgrowth of bud while cytokinins promote it. Braun *et al.* reported the positive regulation of axillary bud initiation and outgrowth by cytokinins (Braun et al., 2012). Sarina *et al.* observed that elevation of SL concentration in *Zantedeschia aethiopica* may result in axillary bud inhibition (Sarina et al., 2017). The report was supported by similar results obtained in *Oryza sativa*, wherein cytokinin triggered the bud growth, whereas SL (GR24) inhibited it (Xu et al., 2015). Several studies have been conducted to establish a relationship between SL and gibberellic acid (GA) in the control of shoot branching. It has been observed that SL biosynthesis is negatively affected by GA-bioactive forms like GA4, GA3, and GA1 in *Lotus japonicus* and *Oryza sativa* (Marzec et al., 2020; Ito et al., 2017).

5.5 CONCLUSION

Over the years, SLs have drawn significant attention because of their role in plant growth mechanisms like root development, shoot branching, seed germination, leaf senescence, etc. Since abiotic stresses such as extreme temperatures, mineral toxicity, drought, nutrient starvation, etc. negatively impact plant development and seed quality, thereby decreasing the yield, SLs play a remarkable role in plant adaptation during abiotic stresses by controlling a number of the physiological and molecular processes through hormonal crosstalk. This illustrates the influence of novel endogenous and exogenous phytohormones on agriculture. This opens a new gateway for the exploration of new phytohormones, technologies, and approaches for sustainable agriculture.

REFERENCES

Alder, Adrian, Muhammad Jamil, Mattia Marzorati, Mark Bruno, Martina Vermathen, Peter Bigler, Sandro Ghisla, Harro Bouwmeester, Peter Beyer, and Salim Al-Babili. 2012. "The Path from β-Carotene to Carlactone, a Strigolactone-like Plant Hormone." *Science (New York, N.Y.)* 335 (6074): 1348–51. https://doi.org/10.1126/science.1218094.

Aroca, Ricardo, Juan Manuel Ruiz-Lozano, Angel María Zamarreño, José Antonio Paz, José María García-Mina, María José Pozo, and Juan Antonio López-Ráez. 2013. "Arbuscular Mycorrhizal Symbiosis Influences Strigolactone Production under Salinity and Alleviates Salt Stress in Lettuce Plants." *Journal of Plant Physiology* 170 (1): 47–55. https://doi.org/10.1016/j.jplph.2012.08.020.

Begum, Naheeda, Cheng Qin, Muhammad Abass Ahanger, Sajjad Raza, Muhammad Ishfaq Khan, Muhammad Ashraf, Nadeem Ahmed, and Lixin Zhang. 2019. "Role of Arbuscular Mycorrhizal Fungi in Plant Growth Regulation: Implications in Abiotic Stress Tolerance." *Frontiers in Plant Science* 10: 1068. https://doi.org/10.3389/fpls.2019.01068.

Braun, Nils, Alexandre de Saint Germain, Jean-Paul Pillot, Stéphanie Boutet-Mercey, Marion Dalmais, Ioanna Antoniadi, Xin Li, et al. 2012. "The Pea TCP Transcription Factor PsBRC1 Acts Downstream of Strigolactones to Control Shoot Branching." *Plant Physiology* 158 (1): 225–38. https://doi.org/10.1104/pp.111.182725.

Brewer, Philip B., Elizabeth A. Dun, Renyi Gui, Michael G. Mason, and Christine A. Beveridge. 2015. "Strigolactone Inhibition of Branching Independent of Polar Auxin Transport." *Plant Physiology* 168 (4): 1820–29. https://doi.org/10.1104/pp.15.00014.

Brewer, Philip B., Hinanit Koltai, and Christine A. Beveridge. 2013. "Diverse Roles of Strigolactones in Plant Development." *Molecular Plant* 6 (1): 18–28. https://doi.org/10.1093/mp/sss130.

Bunsick, Michael, Shigeo Toh, Cynthia Wong, Zhenhua Xu, George Ly, Christopher S.P. McErlean, Gianni Pescetto, et al. 2020. "SMAX1-Dependent Seed Germination Bypasses GA Signalling in Arabidopsis and Striga." *Nature Plants* 6 (6): 646–52. https://doi.org/10.1038/s41477-020-0653-z.

Bürger, Marco, and Joanne Chory. 2020. "In-Silico Analysis of the Strigolactone Ligand-Receptor System." *Plant Direct* 4 (9): e00263. https://doi.org/10.1002/pld3.263.

Carbonnel, Samy, Debatosh Das, Kartikye Varshney, Markus C. Kolodziej, José A. Villaécija-Aguilar, and Caroline Gutjahr. 2020. "The Karrikin Signaling Regulator SMAX1 Controls Lotus Japonicus Root and Root Hair Development by Suppressing Ethylene Biosynthesis." *Proceedings of the National Academy of Sciences of the United States of America* 117 (35): 21757–65. https://doi.org/10.1073/pnas.2006111117.

Challis, Richard J., Jo Hepworth, Céline Mouchel, Richard Waites, and Ottoline Leyser. 2013. "A Role for MORE AXILLARY GROWTH1 (MAX1) in Evolutionary Diversity in Strigolactone Signaling Upstream of MAX2[C][W][OA]." *Plant Physiology* 161 (4): 1885–902. https://doi.org/10.1104/pp.112.211383.

Cheng, Xi, Kristýna Floková, Harro Bouwmeester, and Carolien Ruyter-Spira. 2017. "The Role of Endogenous Strigolactones and Their Interaction with ABA during the Infection Process of the Parasitic Weed Phelipanche Ramosa in Tomato Plants." *Frontiers in Plant Science* 8: 392. https://doi.org/10.3389/fpls.2017.00392.

Chevalier, Florian, Kaisa Nieminen, Juan Carlos Sánchez-Ferrero, María Luisa Rodríguez, Mónica Chagoyen, Christian S. Hardtke, and Pilar Cubas. 2014. "Strigolactone Promotes Degradation of DWARF14, an α/β Hydrolase Essential for Strigolactone Signaling in Arabidopsis[W]." *The Plant Cell* 26 (3): 1134–50. https://doi.org/10.1105/tpc.114.122903.

Conn, Caitlin E., and David C. Nelson. 2015. "Evidence That KARRIKIN-INSENSITIVE2 (KAI2) Receptors May Perceive an Unknown Signal That Is Not Karrikin or Strigolactone." *Frontiers in Plant Science* 6: 1219. https://doi.org/10.3389/fpls.2015.01219.

Cook, C.E., L.P. Whichard, B. Turner, M.E. Wall, and G.H. Egley. 1966. "Germination of Witchweed (Striga Lutea Lour.): Isolation and Properties of a Potent Stimulant." *Science (New York, N.Y.)* 154 (3753): 1189–90. https://doi.org/10.1126/science.154.3753.1189.

Cooper, James W., Yan Hu, Leila Beyyoudh, H. Yildiz Dasgan, Karl Kunert, Christine A. Beveridge, and Christine H. Foyer. 2018. "Strigolactones Positively Regulate Chilling Tolerance in Pea and in Arabidopsis." *Plant, Cell & Environment* 41 (6): 1298–310. https://doi.org/10.1111/pce.13147.

Danquah, Agyemang, Axel de Zelicourt, Jean Colcombet, and Heribert Hirt. 2014. "The Role of ABA and MAPK Signaling Pathways in Plant Abiotic Stress Responses." *Biotechnology Advances* 32 (1): 40–52. https://doi.org/10.1016/j.biotechadv.2013.09.006.

Dell'Oste, Valentina, Francesca Spyrakis, and Cristina Prandi. 2021. "Strigolactones, from Plants to Human Health: Achievements and Challenges." *Molecules (Basel, Switzerland)* 26 (15): 4579. https://doi.org/10.3390/molecules26154579.

Dubois, Marieke, Lisa Van den Broeck, and Dirk Inzé. 2018. "The Pivotal Role of Ethylene in Plant Growth." *Trends in Plant Science* 23 (4): 311–23. https://doi.org/10.1016/j.tplants.2018.01.003.

Dun, Elizabeth A., Alexandre de Saint Germain, Catherine Rameau, and Christine A. Beveridge. 2012. "Antagonistic Action of Strigolactone and Cytokinin in Bud Outgrowth Control." *Plant Physiology* 158 (1): 487–98. https://doi.org/10.1104/pp.111.186783.

Evelin, Heikham, Thokchom Sarda Devi, Samta Gupta, and Rupam Kapoor. 2019. "Mitigation of Salinity Stress in Plants by Arbuscular Mycorrhizal Symbiosis: Current Understanding and New Challenges." *Frontiers in Plant Science* 10: 470. https://doi.org/10.3389/fpls.2019.00470.

Ferrero, Manuela, Chiara Pagliarani, Ondrej Novák, Alessandra Ferrandino, Francesca Cardinale, Ivan Visentin, and Andrea Schubert. 2018. "Exogenous Strigolactone Interacts with Abscisic Acid-Mediated Accumulation of Anthocyanins in Grapevine Berries." *Journal of Experimental Botany* 69 (9): 2391–401. https://doi.org/10.1093/jxb/ery033.

Finkelstein, Ruth. 2013. "Abscisic Acid Synthesis and Response." *The Arabidopsis Book/American Society of Plant Biologists* 11 (November): e0166. https://doi.org/10.1199/tab.0166.

Gamir, Jordi, Rocío Torres-Vera, Carlos Rial, Estefanía Berrio, Pedro M. de Souza Campos, Rosa M. Varela, Francisco A. Macías, María J. Pozo, Victor Flors, and Juan A. López-Ráez. 2020. "Exogenous Strigolactones Impact Metabolic Profiles and Phosphate Starvation Signalling in Roots." *Plant, Cell & Environment* 43 (7): 1655–68. https://doi.org/10.1111/pce.13760.

Guan, Jiahn Chou, Karen E. Koch, Masaharu Suzuki, Shan Wu, Susan Latshaw, Tanya Petruff, Charles Goulet, Harry J. Klee, and Donald R. McCarty. 2012. "Diverse Roles of Strigolactone Signaling in Maize Architecture and the Uncoupling of a Branching Specific Subnetwork1[C][W][OA]." *Plant Physiology* 160 (3): 1303–17. https://doi.org/10.1104/pp.112.204503.

Ha, Chien Van, Marco Antonio Leyva-González, Yuriko Osakabe, Uyen Thi Tran, Rie Nishiyama, Yasuko Watanabe, Maho Tanaka, et al. 2014. "Positive Regulatory Role of Strigolactone in Plant Responses to Drought and Salt Stress." *Proceedings of the National Academy of Sciences of the United States of America* 111 (2): 851–56. https://doi.org/10.1073/pnas.1322135111.

Halouzka, Rostislav, Sanja Ćavar Zeljković, Bořivoj Klejdus, and Petr Tarkowski. 2020. "Analytical Methods in Strigolactone Research." *Plant Methods* 16: 76. https://doi.org/10.1186/s13007-020-00616-2.

Hasanuzzaman, Mirza, Kamrun Nahar, Md. Mahabub Alam, Rajib Roychowdhury, and Masayuki Fujita. 2013. "Physiological, Biochemical, and Molecular Mechanisms of Heat Stress Tolerance in Plants." *International Journal of Molecular Sciences* 14 (5): 9643–84. https://doi.org/10.3390/ijms14059643.

Hepworth, Jo, and Michael Lenhard. 2014. "Regulation of Plant Lateral-Organ Growth by Modulating Cell Number and Size." *Current Opinion in Plant Biology* 17 (February): 36–42. https://doi.org/10.1016/j.pbi.2013.11.005.

Hu, Qiannan, Shuoxin Zhang, and Bingru Huang. 2018. "Strigolactones and Interaction with Auxin Regulating Root Elongation in Tall Fescue under Different Temperature Regimes." *Plant Science: An International Journal of Experimental Plant Biology* 271 (June): 34–39. https://doi.org/10.1016/j.plantsci.2018.03.008.

Huang, Huang, Bei Liu, Liangyu Liu, and Susheng Song. 2017. "Jasmonate Action in Plant Growth and Development." *Journal of Experimental Botany* 68 (6): 1349–59. https://doi.org/10.1093/jxb/erw495.

Islam, Sharmin, Cara A. Griffiths, Cecilia K. Blomstedt, Tuan-Ngoc Le, Donald F. Gaff, John D. Hamill, and Alan D. Neale. 2013. "Increased Biomass, Seed Yield and Stress Tolerance Is Conferred in Arabidopsis by a Novel Enzyme from the Resurrection Grass Sporobolus Stapfianus That Glycosylates the Strigolactone Analogue GR24." *PloS One* 8 (11): e80035. https://doi.org/10.1371/journal.pone.0080035.

Ito, Shinsaku, Daichi Yamagami, Mikihisa Umehara, Atsushi Hanada, Satoko Yoshida, Yasuyuki Sasaki, Shunsuke Yajima, et al. 2017. "Regulation of Strigolactone Biosynthesis by Gibberellin Signaling." *Plant Physiology* 174 (2): 1250–59. https://doi.org/10.1104/pp.17.00301.

Jenik, Pablo D., and M. Kathryn Barton. 2005. "Surge and Destroy: The Role of Auxin in Plant Embryogenesis." *Development (Cambridge, England)* 132 (16): 3577–85. https://doi.org/10.1242/dev.01952.

Koltai, Hinanit. 2013. "Strigolactones Activate Different Hormonal Pathways for Regulation of Root Development in Response to Phosphate Growth Conditions." *Annals of Botany* 112 (2): 409–15. https://doi.org/10.1093/aob/mcs216.

———. 2015. "Cellular Events of Strigolactone Signalling and Their Crosstalk with Auxin in Roots." *Journal of Experimental Botany* 66 (16): 4855–61. https://doi.org/10.1093/jxb/erv178.

Koltai, Hinanit, Evgenia Dor, Joseph Hershenhorn, Daniel M. Joel, Smadar Weininger, Sivarama Lekalla, Hagit Shealtiel, et al. 2010. "Strigolactones' Effect on Root Growth and Root-Hair Elongation May Be Mediated by Auxin-Efflux Carriers." *Journal of Plant Growth Regulation* 29 (2): 129–36. https://doi.org/10.1007/s00344-009-9122-7.

Kretzschmar, Tobias, Wouter Kohlen, Joelle Sasse, Lorenzo Borghi, Markus Schlegel, Julien B. Bachelier, Didier Reinhardt, Ralph Bours, Harro J. Bouwmeester, and Enrico Martinoia. 2012. "A Petunia ABC Protein Controls Strigolactone-Dependent Symbiotic Signalling and Branching." *Nature* 483 (7389): 341–44. https://doi.org/10.1038/nature10873.

Kumar, Manoj, Nirali Pandya-Kumar, Yoram Kapulnik, and Hinanit Koltai. 2015. "Strigolactone Signaling in Root Development and Phosphate Starvation." *Plant Signaling & Behavior* 10 (7): e1045174. https://doi.org/10.1080/15592324.2015.1045174.

Lechat, Marc-Marie, Guillaume Brun, Grégory Montiel, Christophe Véronési, Philippe Simier, Séverine Thoiron, Jean-Bernard Pouvreau, and Philippe Delavault. 2015. "Seed Response to Strigolactone Is Controlled by Abscisic Acid-Independent DNA Methylation in the Obligate Root Parasitic Plant, Phelipanche Ramosa L. Pomel." *Journal of Experimental Botany* 66 (11): 3129–40. https://doi.org/10.1093/jxb/erv119.

Lee, Han Yong, and Gyeong Mee Yoon. 2020. "Strigolactone Elevates Ethylene Biosynthesis in Etiolated Arabidopsis Seedlings." *Plant Signaling & Behavior* 15 (11): 1805232. https://doi.org/10.1080/15592324.2020.1805232.

Li, Weiqiang, Luis Herrera-Estrella, and Lam-Son Phan Tran. 2019. "Do Cytokinins and Strigolactones Crosstalk during Drought Adaptation?" *Trends in Plant Science* 24 (8): 669–72. https://doi.org/10.1016/j.tplants.2019.06.007.

Liang, Jianli, Liangjun Zhao, Richard Challis, and Ottoline Leyser. 2010. "Strigolactone Regulation of Shoot Branching in Chrysanthemum (Dendranthema Grandiflorum)." *Journal of Experimental Botany* 61 (11): 3069–78. https://doi.org/10.1093/jxb/erq133.

Lin, Hao, Renxiao Wang, Qian Qian, Meixian Yan, Xiangbing Meng, Zhiming Fu, Cunyu Yan, et al. 2009. "DWARF27, an Iron-Containing Protein Required for the Biosynthesis of Strigolactones, Regulates Rice Tiller Bud Outgrowth." *The Plant Cell* 21 (5): 1512–25. https://doi.org/10.1105/tpc.109.065987.

Ling, Fenglou, Qingwang Su, Hao Jiang, Jingjing Cui, Xiaoliang He, Zhihai Wu, Zhian Zhang, Juan Liu, and Yongjun Zhao. 2020. "Effects of Strigolactone on Photosynthetic and Physiological Characteristics in Salt-Stressed Rice Seedlings." *Scientific Reports* 10 (1): 6183. https://doi.org/10.1038/s41598-020-63352-6.

Liu, Junwei, Hanzi He, Marco Vitali, Ivan Visentin, Tatsiana Charnikhova, Imran Haider, Andrea Schubert, et al. 2015. "Osmotic Stress Represses Strigolactone Biosynthesis in Lotus Japonicus Roots: Exploring the Interaction between Strigolactones and ABA under Abiotic Stress." *Planta* 241 (6): 1435–51. https://doi.org/10.1007/s00425-015-2266-8.

Liu, Wei, Wouter Kohlen, Alessandra Lillo, Rik Op den Camp, Sergey Ivanov, Marijke Hartog, Erik Limpens, et al. 2011. "Strigolactone Biosynthesis in Medicago Truncatula and Rice Requires the Symbiotic GRAS-Type Transcription Factors NSP1 and NSP2." *The Plant Cell* 23 (10): 3853–65. https://doi.org/10.1105/tpc.111.089771.

López-Ráez, Juan A., Wouter Kohlen, Tatsiana Charnikhova, Patrick Mulder, Anna K. Undas, Martin J. Sergeant, Francel Verstappen, et al. 2010. "Does Abscisic Acid Affect Strigolactone Biosynthesis?" *The New Phytologist* 187 (2): 343–54. https://doi.org/10.1111/j.1469-8137.2010.03291.x.

Lu, Tao, Hongjun Yu, Qiang Li, Lin Chai, and Weijie Jiang. 2019. "Improving Plant Growth and Alleviating Photosynthetic Inhibition and Oxidative Stress From Low-Light Stress With Exogenous GR24 in Tomato (Solanum Lycopersicum L.) Seedlings." *Frontiers in Plant Science* 10 (April): 490. https://doi.org/10.3389/fpls.2019.00490.

Ma, Lan, Xudong Sun, Xiangxiang Kong, Jose Valero Galvan, Xiong Li, Shihai Yang, Yunqiang Yang, Yongping Yang, and Xiangyang Hu. 2015. "Physiological, Biochemical and Proteomics Analysis Reveals the Adaptation Strategies of the Alpine Plant Potentilla Saundersiana at Altitude Gradient of the Northwestern Tibetan Plateau." *Journal of Proteomics* 112 (January): 63–82. https://doi.org/10.1016/j.jprot.2014.08.009.

Ma, Ni, Chao Hu, Lin Wan, Qiong Hu, Junlan Xiong, and Chunlei Zhang. 2017. "Strigolactones Improve Plant Growth, Photosynthesis, and Alleviate Oxidative Stress under Salinity in Rapeseed (Brassica Napus L.) by Regulating Gene Expression." *Frontiers in Plant Science* 8: 1671. https://doi.org/10.3389/fpls.2017.01671.

Marzec, Marek, Agata Daszkowska-Golec, Anna Collin, Michael Melzer, Kai Eggert, and Iwona Szarejko. 2020. "Barley Strigolactone Signalling Mutant Hvd14.d Reveals the Role of Strigolactones in Abscisic Acid-Dependent Response to Drought." *Plant, Cell & Environment* 43 (9): 2239–53. https://doi.org/10.1111/pce.13815.

Mayzlish-Gati, Einav, Carolien De-Cuyper, Sofie Goormachtig, Tom Beeckman, Marnik Vuylsteke, Philip B. Brewer, Christine A. Beveridge, et al. 2012. "Strigolactones Are Involved in Root Response to Low Phosphate Conditions in Arabidopsis[W][OA]." *Plant Physiology* 160 (3): 1329–41. https://doi.org/10.1104/pp.112.202358.

Milhinhos, Ana, Jakob Prestele, Benjamin Bollhöner, Andreia Matos, Francisco Vera-Sirera, José L. Rambla, Karin Ljung, et al. 2013. "Thermospermine Levels Are Controlled by an Auxin-Dependent Feedback Loop Mechanism in Populus Xylem." *The Plant Journal: For Cell and Molecular Biology* 75 (4): 685–98. https://doi.org/10.1111/tpj.12231.

Min, Zhuo, Runyu Li, Li Chen, Yang Zhang, Ziyu Li, Min Liu, Yanlun Ju, and Yulin Fang. 2019. "Alleviation of Drought Stress in Grapevine by Foliar-Applied Strigolactones." *Plant Physiology and Biochemistry: PPB* 135 (February): 99–110. https://doi.org/10.1016/j.plaphy.2018.11.037.

Miransari, Mohammad, and D.L. Smith. 2014. "Plant Hormones and Seed Germination." *Environmental and Experimental Botany* 99 (March): 110–21. https://doi.org/10.1016/j.envexpbot.2013.11.005.

Mishra, Sonal, Swati Upadhyay, and Rakesh K. Shukla. 2016. "The Role of Strigolactones and Their Potential Cross-Talk under Hostile Ecological Conditions in Plants." *Frontiers in Physiology* 7: 691. https://doi.org/10.3389/fphys.2016.00691.

Miyakawa, Takuya, Yuqun Xu, and Masaru Tanokura. 2020. "Molecular Basis of Strigolactone Perception in Root-Parasitic Plants: Aiming to Control Its Germination with Strigolactone Agonists/Antagonists." *Cellular and Molecular Life Sciences: CMLS* 77 (6): 1103–13. https://doi.org/10.1007/s00018-019-03318-8.

Mu, Xiaohuan, and Yanling Chen. 2021. "The Physiological Response of Photosynthesis to Nitrogen Deficiency." *Plant Physiology and Biochemistry: PPB* 158 (January): 76–82. https://doi.org/10.1016/j.plaphy.2020.11.019.

Nakamura, Hidemitsu, You-Lin Xue, Takuya Miyakawa, Feng Hou, Hui-Min Qin, Kosuke Fukui, Xuan Shi, et al. 2013. "Molecular Mechanism of Strigolactone Perception by DWARF14." *Nature Communications* 4 (1): 2613. https://doi.org/10.1038/ncomms3613.

Niu, Kuiju, Ran Zhang, Ruiting Zhu, Yong Wang, Dan Zhang, and Huiling Ma. 2021. "Cadmium Stress Suppresses the Tillering of Perennial Ryegrass and Is Associated with the Transcriptional Regulation of Genes Controlling Axillary Bud Outgrowth." *Ecotoxicology and Environmental Safety* 212 (April): 112002. https://doi.org/10.1016/j.ecoenv.2021.112002.

Omoarelojie, Luke O., Manoj G. Kulkarni, Jeffrey F. Finnie, Tomáš Pospíšil, Miroslav Strnad, and Johannes Van Staden. 2020. "Synthetic Strigolactone (Rac-GR24) Alleviates the Adverse Effects of Heat Stress on Seed Germination and Photosystem II Function in Lupine Seedlings." *Plant Physiology and Biochemistry: PPB* 155 (October): 965–79. https://doi.org/10.1016/j.plaphy.2020.07.043.

Pieterse, Corné M.J., Dieuwertje Van der Does, Christos Zamioudis, Antonio Leon-Reyes, and Saskia C.M. Van Wees. 2012. "Hormonal Modulation of Plant Immunity." *Annual Review of Cell and Developmental Biology* 28: 489–521. https://doi.org/10.1146/annurev-cellbio-092910-154055.

Rana, Muhammad Shoaib, Xuecheng Sun, Muhammad Imran, Shafaqat Ali, Muhammad Shaaban, Mohamed G. Moussa, Zaid Khan, et al. 2020. "Molybdenum-Induced Effects on Leaf Ultra-Structure and Rhizosphere Phosphorus Transformation in Triticum Aestivum L." *Plant Physiology and Biochemistry: PPB* 153 (August): 20–9. https://doi.org/10.1016/j.plaphy.2020.05.010.

Roberts, E.H. 1988. "Temperature and Seed Germination." *Symposia of the Society for Experimental Biology* 42: 109–32.

Roesler, Keith, Cheng Lu, Jill Thomas, Qingzhang Xu, Peter Vance, Zhenglin Hou, Robert W. Williams, Lu Liu, Michaela A. Owens, and Jeffrey E. Habben. 2021. "Arabidopsis Carboxylesterase 20 Binds Strigolactone and Increases Branches and Tillers When Ectopically Expressed in Arabidopsis and Maize." *Frontiers in Plant Science* 12: 639401. https://doi.org/10.3389/fpls.2021.639401.

Ruiz-Lozano, Juan Manuel, Ricardo Aroca, Ángel María Zamarreño, Sonia Molina, Beatriz Andreo-Jiménez, Rosa Porcel, José María García-Mina, Carolien Ruyter-Spira, and Juan Antonio López-Ráez. 2016. "Arbuscular Mycorrhizal Symbiosis Induces Strigolactone Biosynthesis under Drought and Improves Drought Tolerance in Lettuce and Tomato." *Plant, Cell & Environment* 39 (2): 441–52. https://doi.org/10.1111/pce.12631.

S, Manandhar, Funnell Ka, Woolley Dj, and Cooney Jm. 2017. "Interaction Between Strigolactone and Cytokinin on Axillary and Adventitious Bud Development in Zantedeschia." *Journal of Plant Physiology & Pathology* 6 (1). https://doi.org/10.4172/2329-955X.1000172.

Saeed, Wajeeha, Saadia Naseem, and Zahid Ali. 2017. "Strigolactones Biosynthesis and Their Role in Abiotic Stress Resilience in Plants: A Critical Review." *Frontiers in Plant Science* 8: 1487. https://doi.org/10.3389/fpls.2017.01487.

Santoro, Veronica, Michela Schiavon, Francesco Gresta, Andrea Ertani, Francesca Cardinale, Craig J. Sturrock, Luisella Celi, and Andrea Schubert. 2020. "Strigolactones Control Root System Architecture and Tip Anatomy in Solanum Lycopersicum L. Plants under P Starvation." *Plants (Basel, Switzerland)* 9 (5): E612. https://doi.org/10.3390/plants9050612.

Sato, Daisuke, Ayman A. Awad, Sang Heon Chae, Takao Yokota, Yukihiro Sugimoto, Yasutomo Takeuchi, and Koichi Yoneyama. 2003. "Analysis of Strigolactones, Germination Stimulants for Striga and Orobanche, by High-Performance Liquid Chromatography/Tandem Mass Spectrometry." *Journal of Agricultural and Food Chemistry* 51 (5): 1162–68. https://doi.org/10.1021/jf025997z.

Seto, Yoshiya, and Shinjiro Yamaguchi. 2014. "Strigolactone Biosynthesis and Perception." *Current Opinion in Plant Biology* 21 (October): 1–6. https://doi.org/10.1016/j.pbi.2014.06.001.

Sharifi, Parisa, and Siamak Shirani Bidabadi. 2020. "Strigolactone Could Enhances Gas-Exchange through Augmented Antioxidant Defense System in Salvia Nemorosa L. Plants Subjected to Saline Conditions Stress." *Industrial Crops and Products* 151 (September): 112460. https://doi.org/10.1016/j.indcrop.2020.112460.

Shi, Lei, and Benjamin P. Tu. 2015. "Acetyl-CoA and the Regulation of Metabolism: Mechanisms and Consequences." *Current Opinion in Cell Biology* 33 (April): 125–31. https://doi.org/10.1016/j.ceb.2015.02.003.

Shinohara, Naoki, Catherine Taylor, and Ottoline Leyser. 2013. "Strigolactone Can Promote or Inhibit Shoot Branching by Triggering Rapid Depletion of the Auxin Efflux Protein PIN1 from the Plasma Membrane." *PLoS Biology* 11 (1): e1001474. https://doi.org/10.1371/journal.pbio.1001474.

Song, Xiaoguang, Zefu Lu, Hong Yu, Gaoneng Shao, Jinsong Xiong, Xiangbing Meng, Yanhui Jing, et al. 2017. "IPA1 Functions as a Downstream Transcription Factor Repressed by D53 in Strigolactone Signaling in Rice." *Cell Research* 27 (9): 1128–41. https://doi.org/10.1038/cr.2017.102.

Souza Campos, Pedro M. de, Pablo Cornejo, Carlos Rial, Fernando Borie, Rosa M. Varela, Alex Seguel, and Juan Antonio López-Ráez. 2019. "Phosphate Acquisition Efficiency in Wheat Is Related to Root:Shoot Ratio, Strigolactone Levels, and PHO2 Regulation." *Journal of Experimental Botany* 70 (20): 5631–42. https://doi.org/10.1093/jxb/erz349.

Stirnberg, Petra, Karin van De Sande, and H.M. Ottoline Leyser. 2002. "MAX1 and MAX2 Control Shoot Lateral Branching in Arabidopsis." *Development (Cambridge, England)* 129 (5): 1131–41.

Sun, Huwei, Fugui Xu, Xiaoli Guo, Daxia Wu, Xuhong Zhang, Manman Lou, Feifei Luo, Quanzhi Zhao, Guohua Xu, and Yali Zhang. 2019. "A Strigolactone Signal Inhibits Secondary Lateral Root Development in Rice." *Frontiers in Plant Science* 10: 1527. https://doi.org/10.3389/fpls.2019.01527.

Tai, Zhenglan, Xinqiang Yin, Zhigang Fang, Gaoling Shi, Laiqing Lou, and Qingsheng Cai. 2017. "Exogenous GR24 Alleviates Cadmium Toxicity by Reducing Cadmium Uptake in Switchgrass (Panicum Virgatum) Seedlings." *International Journal of Environmental Research and Public Health* 14 (8): 852. https://doi.org/10.3390/ijerph14080852.

Terceros, Giada Callizaya, Francesca Resentini, Mara Cucinotta, Silvia Manrique, Lucia Colombo, and Marta A. Mendes. 2020. "The Importance of Cytokinins during Reproductive Development in Arabidopsis and Beyond." *International Journal of Molecular Sciences* 21 (21): E8161. https://doi.org/10.3390/ijms21218161.

Toh, Shigeo, Yuji Kamiya, Naoto Kawakami, Eiji Nambara, Peter McCourt, and Yuichiro Tsuchiya. 2012. "Thermoinhibition Uncovers a Role for Strigolactones in Arabidopsis Seed Germination." *Plant & Cell Physiology* 53 (1): 107–17. https://doi.org/10.1093/pcp/pcr176.

Tsuchiya, Yuichiro, Danielle Vidaurre, Shigeo Toh, Atsushi Hanada, Eiji Nambara, Yuji Kamiya, Shinjiro Yamaguchi, and Peter McCourt. 2010. "A Small-Molecule Screen Identifies New Functions for the Plant Hormone Strigolactone." *Nature Chemical Biology* 6 (10): 741–49. https://doi.org/10.1038/nchembio.435.

Ueda, Hiroaki, and Makoto Kusaba. 2015. "Strigolactone Regulates Leaf Senescence in Concert with Ethylene in Arabidopsis." *Plant Physiology* 169 (1): 138–47. https://doi.org/10.1104/pp.15.00325.

Umehara, Mikihisa, Mengmeng Cao, Kohki Akiyama, Tomoki Akatsu, Yoshiya Seto, Atsushi Hanada, Weiqiang Li, Noriko Takeda-Kamiya, Yu Morimoto, and Shinjiro Yamaguchi. 2015. "Structural Requirements of Strigolactones for Shoot Branching Inhibition in Rice and Arabidopsis." *Plant & Cell Physiology* 56 (6): 1059–72. https://doi.org/10.1093/pcp/pcv028.

Verma, Vivek, Pratibha Ravindran, and Prakash P. Kumar. 2016. "Plant Hormone-Mediated Regulation of Stress Responses." *BMC Plant Biology* 16 (April): 86. https://doi.org/10.1186/s12870-016-0771-y.

Visentin, Ivan, Marco Vitali, Manuela Ferrero, Yanxia Zhang, Carolien Ruyter-Spira, Ondřej Novák, Miroslav Strnad, Claudio Lovisolo, Andrea Schubert, and Francesca Cardinale. 2016. "Low Levels of Strigolactones in Roots as a Component of the Systemic Signal of Drought Stress in Tomato." *The New Phytologist* 212 (4): 954–63. https://doi.org/10.1111/nph.14190.

Wang, Jian You, Imran Haider, Muhammad Jamil, Valentina Fiorilli, Yoshimoto Saito, Jianing Mi, Lina Baz, et al. 2019. "The Apocarotenoid Metabolite Zaxinone Regulates Growth and Strigolactone Biosynthesis in Rice." *Nature Communications* 10 (1): 810. https://doi.org/10.1038/s41467-019-08461-1.

Wani, Atif Khurshid, Nahid Akhtar, Reena Singh, Chirag Chopra, Prachi Kakade, Mahesh Borde, Jameel M. Al-Khayri, Penna Suprasanna, and Saurabha B. Zimare. 2022. "Prospects of Advanced Metagenomics and Meta-Omics in the Investigation of Phytomicrobiome to Forecast Beneficial and Pathogenic Response." *Molecular Biology Reports* (September). https://doi.org/10.1007/s11033-022-07936-7.

Wu, Han, Huihui Li, Hong Chen, Qi Qi, Qiangqiang Ding, Juan Xue, Jing Ding, Xiangning Jiang, Xilin Hou, and Yi Li. 2019. "Identification and Expression Analysis of Strigolactone Biosynthetic and Signaling Genes Reveal Strigolactones Are Involved in Fruit Development of the Woodland Strawberry (Fragaria Vesca)." *BMC Plant Biology* 19 (1): 73. https://doi.org/10.1186/s12870-019-1673-6.

Xia, Xiao-Jian, Yan-Hong Zhou, Kai Shi, Jie Zhou, Christine H. Foyer, and Jing-Quan Yu. 2015. "Interplay between Reactive Oxygen Species and Hormones in the Control of Plant Development and Stress Tolerance." *Journal of Experimental Botany* 66 (10): 2839–56. https://doi.org/10.1093/jxb/erv089.

Xie, Xiaonan, Narumi Mori, Kaori Yoneyama, Takahito Nomura, Kenichi Uchida, Koichi Yoneyama, and Kohki Akiyama. 2019. "Lotuslactone, a Non-Canonical Strigolactone from Lotus Japonicus." *Phytochemistry* 157 (January): 200–5. https://doi.org/10.1016/j.phytochem.2018.10.034.

Xie, Yurong, Yang Liu, Mengdi Ma, Qin Zhou, Yongping Zhao, Binbin Zhao, Baobao Wang, Hongbin Wei, and Haiyang Wang. 2020. "Arabidopsis FHY3 and FAR1 Integrate Light and Strigolactone Signaling to Regulate Branching." *Nature Communications* 11 (1): 1955. https://doi.org/10.1038/s41467-020-15893-7.

Xomphoutheb, Thidaphone, Shuai Jiao, Xinxin Guo, Frank Stephano Mabagala, Biao Sui, Hongbin Wang, Lanpo Zhao, and Xingmin Zhao. 2020. "The Effect of Tillage Systems on Phosphorus Distribution and Forms in Rhizosphere and Non-Rhizosphere Soil under Maize (Zea Mays L.) in Northeast China." *Scientific Reports* 10 (1): 6574. https://doi.org/10.1038/s41598-020-63567-7.

Xu, Junxu, Manrong Zha, Ye Li, Yanfeng Ding, Lin Chen, Chengqiang Ding, and Shaohua Wang. 2015. "The Interaction between Nitrogen Availability and Auxin, Cytokinin, and Strigolactone in the Control of Shoot Branching in Rice (Oryza Sativa L.)." *Plant Cell Reports* 34 (9): 1647–62. https://doi.org/10.1007/s00299-015-1815-8.

Yamada, Yusuke, Soya Furusawa, Seiji Nagasaka, Koichiro Shimomura, Shinjiro Yamaguchi, and Mikihisa Umehara. 2014. "Strigolactone Signaling Regulates Rice Leaf Senescence in Response to a Phosphate Deficiency." *Planta* 240 (2): 399–408. https://doi.org/10.1007/s00425-014-2096-0.

Yao, Ruifeng, Zhenhua Ming, Liming Yan, Suhua Li, Fei Wang, Sui Ma, Caiting Yu, et al. 2016. "DWARF14 Is a Non-Canonical Hormone Receptor for Strigolactone." *Nature* 536 (7617): 469–73. https://doi.org/10.1038/nature19073.

Yoneyama, Kaori, Xiaonan Xie, Takaya Kisugi, Takahito Nomura, and Koichi Yoneyama. 2013. "Nitrogen and Phosphorus Fertilization Negatively Affects Strigolactone Production and Exudation in Sorghum." *Planta* 238 (5): 885–94. https://doi.org/10.1007/s00425-013-1943-8.

Yoneyama, Kaori, Xiaonan Xie, Dai Kusumoto, Hitoshi Sekimoto, Yukihiro Sugimoto, Yasutomo Takeuchi, and Koichi Yoneyama. 2007. "Nitrogen Deficiency as Well as Phosphorus Deficiency in Sorghum Promotes the Production and Exudation of 5-Deoxystrigol, the Host Recognition Signal for Arbuscular Mycorrhizal Fungi and Root Parasites." *Planta* 227 (1): 125–32. https://doi.org/10.1007/s00425-007-0600-5.

Yoneyama, Kaori, Koichi Yoneyama, Yasutomo Takeuchi, and Hitoshi Sekimoto. 2007. "Phosphorus Deficiency in Red Clover Promotes Exudation of Orobanchol, the Signal for Mycorrhizal Symbionts and Germination Stimulant for Root Parasites." *Planta* 225 (4): 1031–38. https://doi.org/10.1007/s00425-006-0410-1.

Yoneyama, Koichi. 2020. "Recent Progress in the Chemistry and Biochemistry of Strigolactones." *Journal of Pesticide Science* 45 (2): 45–53. https://doi.org/10.1584/jpestics.D19-084.

Zhang, Jing, Ewa Mazur, Jozef Balla, Michelle Gallei, Petr Kalousek, Zuzana Medveďová, Yang Li, et al. 2020. "Strigolactones Inhibit Auxin Feedback on PIN-Dependent Auxin Transport Canalization." *Nature Communications* 11 (1): 3508. https://doi.org/10.1038/s41467-020-17252-y.

Zhang, Yanxia, Aalt D.J. van Dijk, Adrian Scaffidi, Gavin R. Flematti, Manuel Hofmann, Tatsiana Charnikhova, Francel Verstappen, et al. 2014. "Rice Cytochrome P450 MAX1 Homologs Catalyze Distinct Steps in Strigolactone Biosynthesis." *Nature Chemical Biology* 10 (12): 1028–33. https://doi.org/10.1038/nchembio.1660.

Zhao, Li-Hua, X. Edward Zhou, Wei Yi, Zhongshan Wu, Yue Liu, Yanyong Kang, Li Hou, et al. 2015. "Destabilization of Strigolactone Receptor DWARF14 by Binding of Ligand and E3-Ligase Signaling Effector DWARF3." *Cell Research* 25 (11): 1219–36. https://doi.org/10.1038/cr.2015.122.

Zheng, Xiaodong, Yuqi Li, Xiangli Xi, Changqing Ma, Zhijuan Sun, Xueqing Yang, Xiangyang Li, Yike Tian, and Caihong Wang. 2021. "Exogenous Strigolactones Alleviate KCl Stress by Regulating Photosynthesis, ROS Migration and Ion Transport in Malus Hupehensis Rehd." *Plant Physiology and Biochemistry: PPB* 159 (February): 113–22. https://doi.org/10.1016/j.plaphy.2020.12.015.

Zhou, Feng, Qibing Lin, Lihong Zhu, Yulong Ren, Kunneng Zhou, Nitzan Shabek, Fuqing Wu, et al. 2013. "D14-SCF(D3)-Dependent Degradation of D53 Regulates Strigolactone Signalling." *Nature* 504 (7480): 406–10. https://doi.org/10.1038/nature12878.

Zulfiqar, Hira, Muhammad Shahbaz, Muhammad Ahsan, Muhammad Nafees, Hammad Nadeem, Muhammad Akram, Ambreen Maqsood, et al. 2020. "Strigolactone (GR24) Induced Salinity Tolerance in Sunflower (Helianthus Annuus L.) by Ameliorating Morpho-Physiological and Biochemical Attributes Under In Vitro Conditions." *Journal of Plant Growth Regulation* (October). https://doi.org/10.1007/s00344-020-10256-4.

6 Strigolactone in Abiotic Stress Tolerance of Crop Plants

Ghazala Nawaz, Sakina Mussarat, and Tauheed

CONTENTS

6.0 INTRODUCTION

By 2050, the world's population is expected to rise from 7.3 billion to 9.7 billion people (Ackland et al., 2017). To keep up with population growth, food production will need to expand by 50% by 2030 and by 70–100% by 2050 for a well-fed world population (Ackland et al., 2017). Crop production is affected by global climate change not only because of altered weather patterns but also because of increased environmental stresses, both abiotic and biotic (Prandecki, 2016). It is anticipated that these pressures will diminish staple food crop yields by up to 70% (Prosekov and Ivanova, 2018). Plants are sessile and cannot avoid abiotic pressure, but they can evolve sophisticated mechanisms to adapt to the changing environment. As a result, gaining a better understanding of plant responses to abiotic stresses is critical for developing biorational strategies to improve crop plants. Various omics techniques are being used to shed light on the role of phytohormones in a variety of ways of improving plant tolerance against abiotic stress. Various plant hormones have been shown to cooperatively regulate adaptive responses to abiotic stressors. This chapter highlights the recent updates to our understanding of the strigolactones against various abiotic stresses, inducing tolerance in crop plants.

Strigolactones (SLs) are phytohormones, a growth stimulant in the parasitic weed of family Orobanchaceae, discovered approximately a half century ago in root exudates of cotton. SLs help in regulating developmental processes and have an important role in combating abiotic stresses in crop plants. Strigol, the first strigolactone, was isolated from cotton root exudates in 1966 (Cook et al., 1966). Another function of SLs is to stimulate the symbiotic kinship between plants and arbuscular mycorrhizal (AM) (Fiorilli et al., 2019). Around 80% of terrestrial plants get benefit from this symbiotic interaction, especially in nutritional stress areas (Smith et al., 2011). SL also plays an important role in biotic stress mitigation in plants against pests and insects. One of such kind of role is studied in *A. thaliana*, where it produces resistance against bacterial infection like *Rhodococcus fascians*, *Pectobacterium carotovorum*, and *Pseudomonas syringae* (Stes et al., 2015, Piisilä et al., 2015).

Recently, more than 25 SLs have been discovered across the plant kingdom naturally, classified as canonical or non-canonical based on the presence or absence of the full ABC-ring structure,

DOI: 10.1201/9781003224525-8

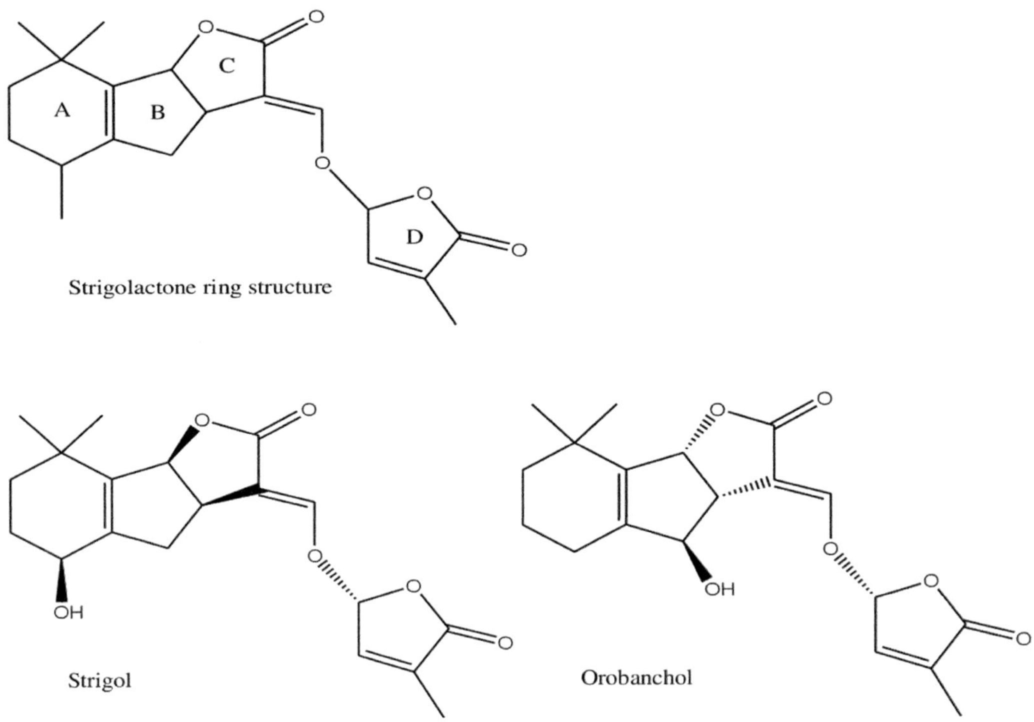

FIGURE 6.1 Structure of strigolactones.
Source: Cardinale et al., 2018

respectively (Wang and Bouwmeester, 2018, Bürger and Chory, 2020). The D ring is a distinguishing characteristic of both canonical and non-canonical SLs, but other structural changes have been found, including differences in stereochemistry and the absence of the standard ABC ring in non-canonical SLs (Wang and Bouwmeester, 2018). Most, but not all, of the SLs studied so far have a four-ring structure, with the A, B, and C rings condensed in a tricyclic lactone and the D ring being a butenolide connected to ring C by an enol ether bridge and considered essential for its activity (Lumba et al., 2017, Al-Babili and Bouwmeester, 2015). Substitution on the A ring and isomerism in the B-C adjunction account for the major diversity within the strigolactone family, like strigol and orobanchol compounds with β- and α-oriented C rings, respectively, and share the R orientation of C-2′ (Figure 6.1).

6.1 ABIOTIC STRESS IN CROP PLANTS

In agriculture, stress is a condition that reduces crop productivity or causes biomass to be destroyed (Grime, 1979). Crop plants' yields are mostly limited by abiotic stresses like heat and cold, water, salt, heavy metals, and nutrient deficiency; all have an impact on the growth and productivity of crops (Zafar et al., 2018, Waqas et al., 2019). Previous studies have reported that 90% of cultivable land is susceptible to one or more of these stresses, resulting in 70% yield losses of major food crops, leading to serious consequences in food security (Dos Reis et al., 2012; Mantri et al., 2012). During the last two decades (1990–2013), the level of salt stress has increased up to 37% in irrigated farmland (Ghassemi et al., 1995, Qadir et al., 2014). As precipitation patterns have changed and evapotranspiration has increased as a result of global warming, drought stress has become more common and severe. Desiccation, which is caused by salt, drought, and heat stress, is a crucial element in plant growth, development, and production. Plants activate different enzymes, complicated gene interactions, and crosstalk with molecular pathways to respond to biotic stress.

6.2 STRESS TOLERANCE IN CROP PLANTS THROUGH PHYTOHORMONES

Different factors, such as the photosynthetic process and metabolism involved in the growth and development of crops, can help acclimate them to environmental stress. Plants activate different enzymes and complex gene interactions to cope with abiotic stresses. Plant growth regulators (PGRs) are a class of chemical substances that regulate plant responses to worse conditions. Phytohormones are necessary for acclimation to negative changes and the production of sophisticated responses that enhance plant growth and development under stress (Abdelrahman et al., 2017, Choudhary et al., 2012). Auxin is a growth regulator that helps seedlings, increases dry weight, and balances ionic concentration in plants during salt stress (Iqbal and Ashraf, 2007). Under salt stress, foliar-applied gibberellic acid improves salt tolerance by boosting the growth of the seedling, crop yield, and efficiency of antioxidant enzymes in wheat crops (Tabatabaei, 2013).

Strigolactones have recently gotten a lot of attention because of their critical involvement in plant adaptation by producing changes in physiological and molecular processes during abiotic stress. According to the literature, the synthesis of strigolactones in plants is related to the types of stresses that plants face at different phases of growth. Plant stress tolerance is controlled by transcriptional activation or repression.

6.3 ROLE OF STRIGOLACTONES IN ABIOTIC STRESS

SLs are produced and secreted in very minute amounts, and even their concentration at pico- to nanomolar amounts is active and reported in different plant species (Andreo-Jimenez et al., 2015). A single plant species or its different varieties can synthesize and produce various types of SLs, mainly in roots, in different quantities according the abiotic stress (Ruyter-Spira et al., 2011,

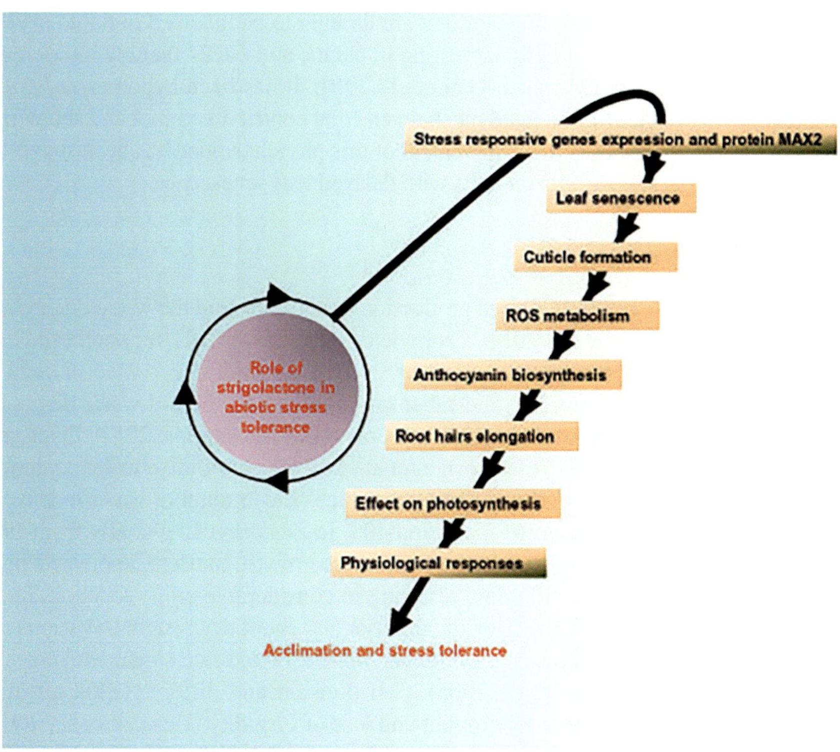

FIGURE 6.2 Mechanisms of abiotic stress tolerance mediated by strigolactone signaling in plants.

Mostofa et al., 2018). According to genetic studies, various abiotic stimuli, such as nutrient restriction, drought, and salt, are also known to affect the production of strigolactones (Du et al., 2018, Zhuang et al., 2017).

6.3.1 Temperature

During important periods of plant development, temperature stress has significant effects on crops (Waqas et al., 2019). By modifying molecular pathways, heat and cold stress decrease the plant's optimal biochemical and physiological role (Djanaguiraman et al., 2018). For example, *Jatropha curcas* is an important bioenergy crop, but its potential to create biofuel is generally limited when it is exposed to freezing temperatures (Wang et al., 2013). Cold stress reduces grain yield but also has an impact on grain quality (Dreccer et al., 2018). Both high and cold stresses limit crop productivity; for example, global wheat productivity declines by 6% for each degree Celsius rise in temperature (Asseng et al., 2015). *Zea mays* L. crops are sensitive to high temperature (>3 °C), and experience a strong reduction in yield if exposed for longer durations. It is estimated that until 2050, 45% of the maize production area is likely to be exposed to average temperatures of>35°C during the reproductive stage annually (Gourdji et al., 2013).

Seed germination was improved in *Philipanche ramose* at high temperature by SL application (Lechat et al., 2015). Seed germination was also improved in SL-deficient *Arabidopsis* mutants when GR-24 was applied under heat stress (Tsuchiya et al., 2010). It assumes that seed with lowering SL levels enhances the ABA/GA ratio, in turn increasing response to thermo inhibition while SL application decreases the ABA/GA ratio, which results in alleviating seed thermo inhibition. In *P. sativum* and *A. thaliana*, SL is involved in dark cold tolerance of photosynthesis. The shoot fresh weights of *P. sativum* and *A. thaliana* lines decreased after dark chilling. In the rms5-3, rms3-1, and MAX4-1 mutants, dry biomass buildup was drastically reduced. The data suggest that SL protects shoots against the deleterious effects of chilling on the buildup of dry biomass (Cooper et al., 2018). The GR-24 application significantly increased the expression of genes that are thought to be involved in light harvesting. SLs are potentially positive regulators of light harvesting in plants, and GR24 therapy interferes with the root's reaction to IAA treatment (Mayzlish-Gati et al., 2010). It has been hypothesized that SLs could be positive regulators of other light-related mechanisms that contribute to root and shoot growth (Koltai and Kapulnik, 2011). In addition, the involvement of this phytohormone has been suggested in leaf senescence, as observed in several SLs mutants with delayed leaf senescence.

6.3.2 Water Stress

Drought is a severe danger to the viability of food production systems around the world due to rapidly changing climate dynamics (Kogan et al., 2019). Crop yields have decreased over southern Europe; for example, the expected rate for spring-sown crops by 2050 is generally decreased in yield (e.g., legumes –30 to +5%, sunflower –12 to +3%, and tuber crops –14 to +7%), and water demand has increased (e.g., maize +2 to +4%, potato +6 to +10%) (Skuras and Psaltopoulos, 2012). Plant growth and output are both affected by drought stress (Bartlett et al., 2019). Leaf relative water content is an index used to measure the water state in tissues. Drought reduces the amount of water in leaves and reduced yields by up to 21% in *Triticum aestivum* L. and 40% in *Zea mays* L. globally from 1980 to 2015 (Dias et al., 2018, Daryanto et al., 2016). Drought disrupts growth, nutrient, and water interactions; photosynthesis; and assimilate partitioning, resulting in considerable crop reduction. Drought stress has a negative impact on germination, growth, and root and shoot dry weight in various crops, such as *Pisum sativum* L., *Medicago sativa* L., *Helianthus annuus* L, and rice *Oryza sativa* L. (Kaya et al., 2006, Farooq et al., 2014, Okçu et al., 2005). Cell division and differentiation are mainly responsible for the growth that is impaired by drought and loss of turgidity (Hussain et al., 2008).

Strigolactones act as a positive regulator in drought response via ABA signaling, as evidenced by all *Arabidopsis max mutants* that showed lower sensitivity to ABA under drought stress. SL and ABA crosstalk has been shown to be critical in integrating stress signals in order to control stomata growth and function. The SLs application improved drought tolerance in wild-type plants and recovered the drought-sensitive phenotype of SL-deficient mutants but not of SL-response mutants, verified the SL role in the stress response (Van Ha et al., 2014). In a study, tolerance has been improved in lettuce and tomato when *arbuscular mycorrhizal* symbiosis stimulates SL biosynthesis under drought conditions (Ruiz-Lozano et al., 2016). Application of GR-24 has regulated the opening and closure of stomata via ABA or ROS, modulation of chlorophyll components and photosynthesis, and activation of the antioxidant defense capability and helped alleviate the negative impacts of water stress in grapevines. Crosstalk with other hormones, particularly ABA, has also been suggested as a key factor in this process (Min et al., 2019). SLs could cause stomatal closure by increasing hydrogen peroxide and nitric oxide generation and activating the SLOW ANION CHANNEL-ASSOCIATED 1 (SLAC1) anion channel, inducing plant resistance to stress (Lv et al., 2018). Due to their common carotenoid origin, strigolactones and abscisic acid (ABA) interact and show their expression and production in *O. sativa* plants during drought stress, assisting in drought tolerance (Haider et al., 2018). Due to a combination of drought resistance and escape characteristics, multiple means of GR-24 application resulted in improved photosynthesis and yield under stress. GR-24 has the potential to improve winter wheat tolerance and drought resistance, demonstrating that the delivery route can impact the severity and efficacy of physiological and biochemical responses to drought stress (Sedaghat et al., 2021). This should be explored in future study of the practical applications of this new phytohormone.

6.3.3 Nutrition and Heavy Metal Stress

Soil solution is the major reservoir for plants to gain nutrient for their optimal growth. Almost all the other environmental stress factors relate to nutritional stress: e.g., heavy metal contamination, water stress, temperature, and salinity. Increased concentration of heavy metals in cultivable land is one of the prominent causes of the loss of productivity of crops (Proshad et al., 2018). Poor irrigation practices, extra use of synthetic nutrients, and chemical fertilizers can add metal toxicity to the crops (Pasala et al., 2016). Heavy metal pollution occurs through a variety of channels, including home garbage disposal in water resources, industrial and sewage waste disposal, urban runoff, and solid and liquid fuel combustion. Heavy metals, such as Ni, Cu, Co, Zn, Se, and Fe, are potentially toxic but also necessary for optimal plant function and cause toxicity when piled up in large amounts in soil solution (Narendrula-Kotha et al., 2019). A very minute quantity of nonessential elements, like arsenic (As), cadmium (Cd), cesium (Cs), and lead (Pb), accumulated in the soil can reduce agricultural yield as well as posing a health risk for both humans and animals (Khalid et al., 2018). Cd stress affects the growth of plants, leaf count, and the dry weight of root and shoot (Shah et al., 2019).

SL stimulates root growth in *Oryza sativa* L. under nutrient deficiency stress (phosphate and nitrate limitation) (Sun et al., 2014). D3 and modulation of auxin transport from shoots to roots are involved in this regulation during stress. Their synthesis and exudation increase in response to low phosphate levels, and they're linked to root responses. Protein interactions and ubiquitin-dependent repressor degradation are part of their signaling cascade in plants (Kapulnik and Koltai, 2014). SLs have a wide range of effects on the development of plants, letting the plants respond to nutrient deprivation (Marzec et al., 2013). In phosphate-deprived conditions, the synthesis and exudation level of SLs increase, causing alternation in shoot and root growth. These changes are likely to be linked to variations in auxin transport and sensitivity. Strigolactones may have a favourable effect on plant-mycorrhiza interactions, promoting phosphate acquisition by the plant. SLs could help plants fine-tune their development patterns in the face of phosphate deficiency (Kapulnik and Koltai, 2016).

Strigolactones play a role in root architecture regulation in response to nutritional circumstances. Despite the presence of adequate Pi, strigolactones inhibit the development of lateral roots (Koltai and Kapulnik, 2011), but when Pi is constrained, they switch from negative to positive regulation, expressing their positive role in nutrient-deficient stress (Ruyter-Spira et al., 2011).

6.3.4 Salinity

Worldwide, about 45 million hectares of farmland is exposed to salt stress, and it is estimated that by the year 2050, more than half of cultivable land might be salt affected, making salinity a key limiting issue for crop plants leading to food security (Munns and Tester, 2008). Salinity affects 20% of the world's arable land and is steadily growing due to climate change and anthropogenic activity, resulting in 50% loss in crop productivity (Arora, 2019), and 45% loss in yield has been reported in salt-stressed wheat (Skalicky et al., 2021). Salinity increases the production of reactive oxygen species (ROS), which spiflicates biomolecules and interrupts redox equilibrium (Kundu et al., 2018). Salinity also affects the ultrastructural components of cells, disrupts the photosynthetic mechanism, amends the cell membrane, increases reactive oxygen species formation, and lowers the activity of enzymes, all of which limit crop development and yield (Hasanuzzaman et al., 2014). Plants use various mechanisms to perceive, respond, and adapt to overcome the salt stress, including changes in morpho-physiological features as well as concentration of ions and molecular metabolisms that could be improved further by application of phytohormones.

Salt tolerance is induced by SL in arbuscular mycorrhizal (AM) *Sesbania cannabina* seedlings and regulated by abscisic acid (ABA) (Ren et al., 2018). GR-24, a synthesized strigolactone, showed improvement in the growth of shoots and roots of *Brassica napus* L., a worldwide edible oil crop, under salt stress. Application of GR24 also reverses the salinity-induced effects on photosynthetic rates, conductivity of stomata, intercellular concentration of CO_2, and rate of transpiration (Ma et al., 2017). Under salt stress conditions, GR-24 application increased and decreased photosystem II quantum productivity and non-photochemical quenching, respectively. Due to the application of GR-24, the activities of peroxidase and superoxide dismutase were enhanced and the lipid peroxidation rate reduced. Transcriptome analysis of shoot and root showed 166 special differentially expressed genes (DEGs) after GR-24 treatment under salinity stress. Gene expression of tryptophan metabolism, signal transduction of phytohormone, and photosynthesis were all linked to stress reduction (Ma et al., 2017). Photosynthetic system dysfunction is frequently associated with the growth retardation found in most plants exposed to NaCl toxicity (Loreto et al., 2003). Salt stress was used to investigate the effects of synthetic strigolactone (GR-24) on rice seedlings. Rice growth was decreased by NaCl stress by about 14% and 40%, respectively, in terms of plant height and root length. The negative effects of salt stress on rice seedling growth, leaf photosynthesis, and physiological/biochemical indices were reduced in the GR24 treatment group compared to the control group. The seedlings' plant height and root length increases as the GR24 concentration increased. Rice seedlings were resistant to the negative effects of high salt stress when the GR24 concentration was 1 M. As a result, adding the right amount of GR24 to could boost rice yields in saline-alkali land and produce tolerance (Ling et al., 2020). In lettuce plants cultivated under high NaCl concentrations, SLs symbioses with AMF may boost crop resilience to salt stress (Aroca et al., 2013). Plants boost strigolactone production in these unfavourable conditions to promote symbiosis formation and deal with salt stress (Table 6.1).

6.4 CONCLUSIONS AND FUTURE PERSPECTIVES

Strigolactones (SLs) are the latest-known phytohormone and derivatives thereof shown to have an important role in plant development. Under normal and stressful situations, SLs can govern plant-arbuscular mycorrhizal fungi symbiotic relationships, impact plant development, and control plant metabolism. They are thought to have evolved in plants as essential regulators in harsh environmental conditions and are helpful in the agricultural sector. The application of this hormone is eco-friendly and prevents the crop loss. More research is needed on the diversified structures of SL and its chemical synthesis for the agriculture sector. Research is still needed to test the effect of SLs on cold and heavy metal stresses in crop varieties. Moreover, research is also needed on the

TABLE 6.1
Production and Role of Strigolactones (SL) in Plants under Abiotic Stress Conditions

Plant	Stress Type	Site of SL Detection	SL Compound	SL Role and Effects Studied	References
Temperature stress					
A. thaliana	Heat, UV	Root tissues	NA	Abiotic stresses responses	(Marzec and Muszynska, 2015)
Solanum lycopersicum	Light	Root tissues	(↑) Solanacol, orobanchol, and didehydro-orobanchol isomer	Tolerance to light	(Koltai et al., 2011)
Glycine max	Dark chilling	Root tissues	NA	Chilling tolerance	(Cooper et al., 2018)
Water stress					
S. lycopersicum	Drought	Root extracts	(↑) Solanacol and didehydro-orobanchol isomers 1 and 2	Low levels of SLs act as components of the systemic signal of drought response	(Ruiz-Lozano et al., 2016, Visentin et al., 2016)
Festuca arundinacea	Drought	Crowns	(↑) Total SLs	SL accumulation and signaling; drought responses and axillary bud development	(Zhuang et al., 2017)
O. sativa	Drought	Stem base	NA	Drought escape and tillering	(Du et al., 2018)
Lotus japonicus	Low P		NA	Drought resistance	(Liu et al., 2015)
O. sativa	Drought		NA	Interferes with ABA formation	(Haider et al., 2018)
Nutrition stress					
Sorghum bicolor	Low P, N	Roots	(↑)5-deoxystrigol	Germination	(Yoneyamaet al., 2007)
Trifolium pretense	Low P	Roots	(↑) Orobanchol	Germination	(Yoneyama et al., 2007)
Solanum lycopersicum	Low P		(↑) Orobanchol, solanacol, and didehydroorobanchol isomers		(López-Ráez, 2016)
Oryza sativa	Low P	Roots	(↑) 2′-epi-5-deoxystrigol	Inhibition of tiller bud outgrowth	(Umehara et al., 2010)

(Continued)

TABLE 6.1 (Continued)

Plant	Stress Type	Site of SL Detection	SL Compound	SL Role and Effects Studied	References
Oryza sativa	Low P and N	Root exudates	(↑) Orobanchol, 2′-epi-5-deoxystrigol	Produces root growth under P and N limitation by decreasing auxin transport from shoots to roots and helps in germination	(Jamil et al., 2011)
Arabidopsis thaliana	Low P	Root and xylem sap	(↑) Orobanchol	Regulates the response of plants to low Pi; axillary shoot branching	(Kohlen et al., 2011)
Pisum sativum	High P	Root exudates	(↓) Fabacyl acetate and orobanchyl acetate	Regulation of AMF symbiosis	(Balzergue et al., 2011)
Medicago sativa	Low P	Root exudates	(↑) Orobanchol and orobanchyl acetate	Plant growth	(Yoneyama et al., 2012)
Astragalus sinicus	Low P and N	Root exudates	(↑) Sorgomol and 5-deoxystrigol	Plant growth	(Yoneyama et al., 2012)
Tagetes patula	Low P and N	Root exudates	(↑) Orobanchol and orobanchyl acetate	Plant growth	(Yoneyama et al., 2012)
Triticum aestivum	Low P and N	Root exudates	(↑) Orobanchol	Plant growth	(Yoneyama et al., 2012)
Solanum lycopersicum	Low P	Root exudates	(↑) Orobanchol	Plant growth	(Yoneyama et al., 2012)
Lactuca sativa	Low P and N	Root exudates	(↑) Orobanchol and orobanchyl acetate	Plant growth	(Yoneyama et al., 2012)
Lotus japonicus	Low P	Root exudates and root tissues	(↑) 5-deoxystrigol	Osmotic stress response	
Dendranthema grandiflorum	Low P	Root exudates	Chrysanthemum SL analogues	Shoot branching	(Xi et al., 2015)
M. sativa	Low P and High N	Root extracts	(↑) Orobanchol and orobanchyl acetate	Rhizobium-legume symbiosis	(Peláez-Vico et al., 2016)
A. thaliana	Low N	Root tissues	NA	Leaf senescence and plant growth	(Ito et al., 2016)

Plant	Stress Type	Site of SL Detection	SL Compound	SL Role and Effects Studied	References
O. sativa	Low N and Low P	Roots	NA	NO-activated elongation of seminal root	(Sun et al., 2014)
Salinity stress					
A. thaliana	Dehydration and salinity	Leaf tissues	NA	SLs positively regulate plant responses to stress; osmotic stress responses	(Van Ha et al., 2014)
A. thaliana	Salinity	Root tissues	NA	Salinity responses	(Marzec and Muszynska, 2015)
A. thaliana	Salinity and mannitol	Leaf tissues	NA	Osmotic stress on stomata closure	(Lv et al., 2018)
Sesbania cannabina	Salinity	Root extracts	AMF (↑) Total SLs (by crosstalk with H_2O_2 and salt tolerance germination activity	Enhances salinity tolerance	(Kong et al., 2017)
Brassica napus	Salinity	Root and leaf tissues	NA	GR24 improves the shoots and roots growth	(Ma et al., 2017)

Key: NA=not available, P=phosphorus, N=nitrogen

concentration, mode of application, and time of application on developmental stage of the crops. Crosstalk of SL with other phytohormones is the lacking area of research and needs to be explored. The mechanism underlying the role of SL in oxidative stress and salinity balances should be investigated. There is still a need to explore how SL interacts with heavy metal stress that is due to soil pollution and plants growing in contaminated soils.

REFERENCES

Abdelrahman M, El-Sayed M, Jogaiah S, et al. The "STAY-GREEN" trait and phytohormone signaling networks in plants under heat stress. Plant Cell Rep. 2017; 36:1009–1025.

Ackland P, Resnikoff S, and Bourne R. World blindness and visual impairment: Despite many successes, the problem is growing. Commu Eye health. 2017; 30:71.

Al-Babili S, and Bouwmeester HJ. Strigolactones, a novel carotenoid-derived plant hormone. Annu Rev Plant Biol. 2015; 66:161–186.

Andreo-Jimenez B, Ruyter-Spira C, Bouwmeester HJ, et al. Ecological relevance of strigolactones in nutrient uptake and other abiotic stresses, and in plant-microbe interactions below-ground. Plant Soil. 2015; 394:1–19.

Aroca R, Ruiz-Lozano JM, Zamarreño ÁM, et al. Arbuscular mycorrhizal symbiosis influences strigolactone production under salinity and alleviates salt stress in lettuce plants. J Plant Physiol. 2013; 170:47–55.

Arora NK. *Impact of climate change on agriculture production and its sustainable solutions*. Springer, 2019.

Asseng S, Ewert F, Martre P, et al. Rising temperatures reduce global wheat production. Nat Climate Change. 2015; 5:143–147.

Balzergue C, Puech-Pagès V, Bécard G, et al. The regulation of arbuscular mycorrhizal symbiosis by phosphate in pea involves early and systemic signalling events. J Exp Bot. 2011; 62:1049–1060.

Bartlett MK, Detto M, and Pacala SW. Predicting shifts in the functional composition of tropical forests under increased drought and CO 2 from trade-offs among plant hydraulic traits. Ecol Lett. 2019; 22:67–77.

Bürger M, and Chory J. The many models of strigolactone signaling. Trends Plant Sci. 2020; 25:395–405.

Cardinale F, Korwin Krukowski P, Schubert A, et al. Strigolactones: mediators of osmotic stress responses with a potential for agrochemical manipulation of crop resilience. J Exp Bot 2018; 69:2291–2303.

Choudhary SP, Yu JQ, Yamaguchi-Shinozaki K, et al. Benefits of brassinosteroid crosstalk. Trends Plant Sci. 2012; 17:594–605.

Cook C, Whichard L P, Turner B, et al. Germination of witchweed (Striga lutea Lour.): Isolation and properties of a potent stimulant. Sci. 1966; 154:1189–1190.

Cooper JW, Hu Y, Beyyoudh L, et al. Strigolactones positively regulate chilling tolerance in pea and in Arabidopsis. Plant Cell Environ. 2018; 41:1298–1310.

Daryanto S, Wang L, and Jacinthe PA. Global synthesis of drought effects on maize and wheat production. PLoS One. 2016; 11:e0156362.

Dias MC, Correia S, Serôdio J, et al. Chlorophyll fluorescence and oxidative stress endpoints to discriminate olive cultivars tolerance to drought and heat episodes. Scient Horticul. 2018; 231:31–35.

Djanaguiraman M, Perumal R, Jagadish S, et al. Sensitivity of sorghum pollen and pistil to high-temperature stress. Plant Cell Environ. 2018; 41:1065–1082.

Dos Reis SP, Lima AM, and De Souza CRB. Recent molecular advances on downstream plant responses to abiotic stress. Inter J Mol Sci. 2012; 13:8628–8647.

Dreccer MF, Fainges J, Whish J, et al. Comparison of sensitive stages of wheat, barley, canola, chickpea and field pea to temperature and water stress across. Aust Agricul Forest Meteorol. 2018; 248:275–294.

Du H, Huang F, Wu N, et al. Integrative regulation of drought escape through ABA-dependent and-independent pathways in rice. Molecul plant. 2018; 11:584–597.

Farooq M, Hussain M, and Siddique KH. Drought stress in wheat during flowering and grain-filling periods. Crit. Rev. Plant Sci. 2014; 33:331–349.

Fiorilli V, Wang JY, Bonfante P, et al. Apocarotenoids: old and new mediators of the arbuscular mycorrhizal symbiosis. Front Plant Sci. 2019; 10:1186.

Ghassemi F, Jakeman AJ, and Nix HA. *Salinisation of land and water resources: Human causes, extent, management and case studies*. CAB Inter, 1995.

Gourdji SM, Sibley AM, and Lobell DB. Global crop exposure to critical high temperatures in the reproductive period: Historical trends and future projections. Environmental Res Lett. 2013; 8:024041.

Grime J. *Plant strategies and vegetation processes*. J. Wiley Publishing, 1979, 222.

Haider I, Andreo-Jimenez B, Bruno M, et al. The interaction of strigolactones with abscisic acid during the drought response in rice. J Exp Bot. 2018; 69:2403–2414.

Hasanuzzaman M, Alam M, Rahman A, et al. Exogenous proline and glycine betaine mediated upregulation of antioxidant defense and glyoxalase systems provides better protection against salt-induced oxidative stress in two rice (Oryza sativa L) varieties. BioMed Res Inter. 2014.

Hussain M, Malik M, Farooq M, et al. Improving drought tolerance by exogenous application of glycinebetaine and salicylic acid in sunflower. J Agro Crop Sci 2008; 194:193–199.

Iqbal M, and Ashraf M. Seed treatment with auxins modulates growth and ion partitioning in salt-stressed wheat plants. J Integr Plant Biol. 2007; 49:1003–1015.

Ito S, Ito K, Abeta N, et al. Effects of strigolactone signaling on Arabidopsis growth under nitrogen deficient stress condition. Plant Signaling & Behav. 2016; 11:e1126031.

Jamil M, Charnikhova T, Cardoso C, et al. Quantification of the relationship between strigolactones and Striga hermonthica infection in rice under varying levels of nitrogen and phosphorus. Weed Res. 2011; 51:373–385.

Kapulnik Y, and Koltai H. Fine-tuning by strigolactones of root response to low phosphate. J. Integr Plant Biol. 2016; 58:203–212.

Kapulnik Y, and Koltai H. Strigolactone involvement in root development, response to abiotic stress, and interactions with the biotic soil environment. Plant Physiol. 2014; 166:560–569.

Kaya MD, Okçu G, Atak M, et al. Seed treatments to overcome salt and drought stress during germination in sunflower (Helianthus annuus L). Eur J Agron. 2006; 24:291–295.

Khalid S, Shahid M, Bibi I, et al. A review of environmental contamination and health risk assessment of wastewater use for crop irrigation with a focus on low and high-income countries. Int. J. Environ. Res. Public Health. 2018; 15:895.

Kogan F, Guo W, and Yang W. Drought and food security prediction from NOAA new generation of operational satellites. Geomatics, Natural Hazards and Risk. 2019; 10:651–666.

Kohlen W, Charnikhova T, Liu Q, et al. Strigolactones are transported through the xylem and play a key role in shoot architectural response to phosphate deficiency in nonarbuscular mycorrhizal host Arabidopsis. Plant Physiol. 2011; 155:974–987.
Koltai H, Cohen M, Chesin O, et al. Light is a positive regulator of strigolactone levels in tomato roots. J. Plant Physiol. 2011; 168:1993–1996.
Koltai H, and Kapulnik Y. Strigolactones as mediators of plant growth responses to environmental conditions. Plant signaling Behavior. 2011; 6:37–41.
Kong CC, Ren CG, Li RZ, et al. Hydrogen peroxide and strigolactones signaling are involved in alleviation of salt stress induced by arbuscular mycorrhizal fungus in Sesbania cannabina seedlings. J. Plant Growth Regul. 2017; 36:734–742.
Kundu P, Gill R, Ahlawat S, et al.. "Targeting the redox regulatory mechanisms for abiotic stress tolerance in crops." In *Biochemical, physiological and molecular avenues for combating abiotic stress tolerance in plants*. Elsevier, 2018, 151–220.
Lechat MM, Brun G, Montiel G, et al. Seed response to strigolactone is controlled by abscisic acid-independent DNA methylation in the obligate root parasitic plant, Phelipanche ramosa L. Pomel. J. Exp. Bot. 2015; 66:3129–3140.
Ling F, Su Q, Jiang H, et al. Effects of strigolactone on photosynthetic and physiological characteristics in salt-stressed rice seedlings. Sci Rep. 2020; 10:1–8.
Liu J, He H, Vitali M, et al. Osmotic stress represses strigolactone biosynthesis in Lotus japonicus roots: exploring the interaction between strigolactones and ABA under abiotic stress. Planta. 2015; 241:1435–1451.
López-Ráez JA. How drought and salinity affect arbuscular mycorrhizal symbiosis and strigolactone biosynthesis? Planta. 2016; 243:1375–1385.
Loreto F, Centritto M, and Chartzoulakis K. Photosynthetic limitations in olive cultivars with different sensitivity to salt stress. Plant Cell Environ. 2003; 26:595–601.
Lumba S, Bunsick M, and McCourt P. Chemical genetics and strigolactone perception. F1000Research. 2017; 6.
Lv S, Zhang Y, Li C, et al. Strigolactone-triggered stomatal closure requires hydrogen peroxide synthesis and nitric oxide production in an abscisic acid-independent manner. New Phytol. 2018; 217:290–304.
Ma N, Hu C, Wan L, et al. Strigolactones improve plant growth, photosynthesis, and alleviate oxidative stress under salinity in rapeseed (Brassica napus L) by regulating gene expression. Front Plant Sci. 2017; 8:1671.
Mantri N, Patade V, Penna S, et al. "Abiotic stress responses in plants: present and future." In *Abiotic stress responses in plants*. Springer, 2012, 1–19.
Marzec M, and Muszynska A. In silico analysis of the genes encoding proteins that are involved in the biosynthesis of the RMS/MAX/D pathway revealed new roles of strigolactones in plants. Inter J Mol Sci. 2015; 16:6757–6782.
Marzec M, Muszynska A, and Gruszka D. The role of strigolactones in nutrient-stress responses in plants. Inter J Mole. Sci. 2013; 14:9286–9304.
Mayzlish-Gati E, LekKala S P, Resnick N, et al. Strigolactones are positive regulators of light-harvesting genes in tomato. J. Exp. Bot. 2010; 61:3129–3136.
Min Z, Li R, Chen L, et al. Alleviation of drought stress in grapevine by foliar-applied strigolactones. Plant Physiol. Biochem. 2019; 135:99–110.
Mostofa M G, Li W, Nguyen K H, et al. Strigolactones in plant adaptation to abiotic stresses: An emerging avenue of plant research. Plant Cell Environ. 2018; 41:2227–2243.
Munns R, and Tester M, Mechanisms of salinity tolerance. Annu. Rev. Plant Biol. 2008; 59:651–681.
Narendrula-Kotha R, Theriault G, Mehes-Smith M, et al. Metal toxicity and resistance in plants and microorganisms in terrestrial ecosystems. Rev Environ Contaminat Toxicol. 2019; 249:1–27.
Okçu G, Kaya MD, and Atak M. Effects of salt and drought stresses on germination and seedling growth of pea (Pisum sativum L). Turkish J Agricul Forest. 2005; 29:237–242.
Pasala RK, Khan MIR, Minhas PS, et al. Can plant bioregulators minimize the crop productivity losses caused by drought, heat stress and salinity? J. Appl. Bot. and Food Qual. 2016; 89:113–125.
Peláez-Vico MA, Bernabéu-Roda L, Kohlen W, et al. Strigolactones in the Rhizobium-legume symbiosis: stimulatory effect on bacterial surface motility and down-regulation of their levels in nodulated plants. Plant Sci. 2016; 245:119–127.
Piisilä M, Keceli MA, Brader G, et al. The F-box protein MAX2 contributes to resistance to bacterial phytopathogens in Arabidopsis thaliana. BMC Plant Biol. 2015; 15:1–17.
Prandecki K. Factors Affecting the Availability of Food in 2050. Przyszłość. Świat-Europa-Polska. 2016:208–233.
Prosekov AY, and Ivanova SA. Food security: The challenge of the present. Geoforum. 2018; 91:73–77.
Proshad R, Kormoker T, Mursheed N, et al. Heavy metal toxicity in agricultural soil due to rapid industrialization in Bangladesh: A review. Inter J Advan Geosci. 2018; 6:83–88.

Qadir M, Quillérou E, Nangia V, et al. Economics of salt-induced land degradation and restoration. Nat Resour. Forum. 2014; 38(4):282–295.
Ren CG, Kong CC, and Xie ZH. Role of abscisic acid in strigolactone-induced salt stress tolerance in arbuscular mycorrhizal Sesbania cannabina seedlings. BMC Plant Biol. 2018; 18:1–10.
Ruiz-Lozano JM, Aroca R, Zamarreño ÁM, et al. Arbuscular mycorrhizal symbiosis induces strigolactone biosynthesis under drought and improves drought tolerance in lettuce and tomato. Plant Cell Environ. 2016; 39:441–452.
Ruyter-Spira C, Kohlen W, Charnikhova T, et al. Physiological effects of the synthetic strigolactone analog GR24 on root system architecture in Arabidopsis: another belowground role for strigolactones? Plant Physiol. 2011; 155:721–734.
Sedaghat M, Emam Y, Mokhtassi-Bidgoli A, et al. The potential of the synthetic strigolactone analogue GR24 for the maintenance of photosynthesis and yield in winter wheat under drought: Investigations on the mechanisms of action and delivery modes. Plants. 2021; 10:1223.
Shah K, Nahakpam S, Chaturvedi V, et al. "Cadmium-induced anatomical abnormalities in plants." In *Cadmium Toxicity and Tolerance in Plants*. Elsevier, 2019, 111–139.
Skalicky M, Raza M, Singh K, et al. Salinity stress in wheat (*Triticum aestivum* L) in the changing climate: Adaptation and management strategies. Front. Agron. 2021; 3:661932.
Skuras D, and Psaltopoulos D. A broad overview of the main problems derived from climate change that will affect agricultural production in the Mediterranean area. Building resilience for adaptation to climate change in the agriculture sector. Environ. Sci. 2012; 23:217.
Smith SE, Jakobsen I, Grønlund M, et al. Roles of arbuscular mycorrhizas in plant phosphorus nutrition: interactions between pathways of phosphorus uptake in arbuscular mycorrhizal roots have important implications for understanding and manipulating plant phosphorus acquisition. Plant Physiol. 2011; 156:1050–1057.
Stes E, Depuydt S, De Keyser A, et al. Strigolactones as an auxiliary hormonal defence mechanism against leafy gall syndrome in Arabidopsis thaliana. J Exp Bot. 2015; 66:5123–5134.
Sun H, Tao J, Liu S, et al. Strigolactones are involved in phosphate-and nitrate-deficiency-induced root development and auxin transport in rice. J. Exp. Bot. 2014; 65:6735–6746.
Tabatabaei S, The effect of salicylic acid and gibberellin on enzyme activity and germination characteristics of wheat seeds under salinity stress conditions. Inter J Agricul Crop Sci. 2013; 6:236–240.
Tsuchiya Y, Vidaurre D, Toh S, et al. A small-molecule screen identifies new functions for the plant hormone strigolactone. Nat Chem Biol. 2010; 6:741–749.
Umehara M, Hanada A, Magome H, et al. Contribution of strigolactones to the inhibition of tiller bud outgrowth under phosphate deficiency in rice. Plant Cell Physiol. 2010; 51:1118–1126.
Van Ha C, Leyva-González MA, Osakabe Y, et al. Positive regulatory role of strigolactone in plant responses to drought and salt stress. Proceed Nat Acad Sci. 2014; 111:851–856.
Visentin I, Vitali M, Ferrero M, et al. Low levels of strigolactones in roots as a component of the systemic signal of drought stress in tomato. New Phytol. 2016; 212:954–963.
Wang H, Zou Z, Wang S, et al. Global analysis of transcriptome responses and gene expression profiles to cold stress of Jatropha curcasL. PLoS One. 2013; 8:e82817.
Wang Y, and Bouwmeester H J, Structural diversity in the strigolactones. J Exp Bot. 2018; 69:2219–2230.
Waqas MA, Kaya C, Riaz A, et al. Potential mechanisms of abiotic stress tolerance in crop plants induced by thiourea. Front Plant Sci. 2019; 10:1336.
Xi L, Wen C, Fang S, et al. Impacts of strigolactone on shoot branching under phosphate starvation in chrysanthemum (Dendranthema grandiflorum cv. Jinba). Front Plant Sci. 2015; 6:694.
Yoneyama K, Xie X, Kim HI, et al. How do nitrogen and phosphorus deficiencies affect strigolactone production and exudation? Planta. 2012; 235:1197–1207.
Yoneyama K, Xie X, Kusumoto D, et al. Nitrogen deficiency as well as phosphorus deficiency in sorghum promotes the production and exudation of 5-deoxystrigol, the host recognition signal for arbuscular mycorrhizal fungi and root parasites. Planta 2007; 227:125–132.
Yoneyama K, Yoneyama K, Takeuchi Y, et al. Phosphorus deficiency in red clover promotes exudation of orobanchol, the signal for mycorrhizal symbionts and germination stimulant for root parasites. Planta. 2007; 225:1031–1038.
Zafar SA, Noor MA, Waqas MA, et al. Temperature extremes in cotton production and mitigation strategies. In Mehboob-Ur-Rahman M, Zafar Y (eds.) *Past, Present and Future Trends in Cotton Breeding [Internet]*. IntechOpen, 2018 [cited 2023 Jan 04]. Available from: https://www.intechopen.com/chapters/59774. doi: 10.5772/intechopen.74648.
Zhuang L, Wang J, and Huang B. Drought inhibition of tillering in Festuca arundinacea associated with axillary bud development and strigolactone signaling. Environ Exp Bot. 2017; 142:15–23.

7 Host–Parasitic Plant Interaction

The Key Role of Strigolactones

Youness En-nahli, Zouhaier Abbes, Imen Trabelsi, William Makaza, Mohamed Kharrat, Michel Edmond Ghanem, Nour Eddine Es-Safi, and Moez Amri

CONTENTS

7.0 INTRODUCTION

Approximately 1% of angiosperms (about 4,000 species belonging to 17 plant families) have evolved the ability to parasitize other plants and sink all or a large part of their requirements of water, assimilates, and nutrients (Abbes et al., 2020). Such relation differs, depending on the type of parasitism (Bouwmeester et al., 2007). All these parasitic plants require the presence of a host plant to survive. Some of them are completely dependent on their host and are called obligate parasites or holoparasites, such as *Alectra*, *Striga* spp. (witch weeds), *Orobanche* spp., and *Phelipanche* spp. (Bouwmeester et al., 2003). Other parasitic plants, called facultative parasites or *hemi parasites*, do not require host-derived germinating stimulus and have retained some of their photosynthetic ability, such as *Triphysaria versicolor* (Fisch. & C.A. Mey) and *Phtheirospermum japonicum* (Thunb.) Kanitz (Saucet and Shirasu, 2016). In the case of obligate parasites, seeds have limited resources. Consequently, it is crucial that the embryo's radicle reaches the host root within a few millimeters of the seed. This makes the perception of host-derived germination factors crucial in the selection of a potential host and prohibits seed germination in the absence of the host (Raupp and Spring, 2013).

The first seed-germination stimulant of *Striga lutea*, known as the strigol, was identified in 1966 (Cook et al., 1966; Zwanenburg et al., 2009; de Saint Germain et al., 2019). Since then, the interest in compounds inducing parasitic weed germination has increased and led to the discovery of a new class of plant hormones known as strigolactones (SLs) (de Saint Germain et al., 2016) (Figure 7.1). To date, more than 30 natural SLs have been identified from different plants as germination stimulants for parasitic weeds. As an example, sorgolactone was derived from the roots of sorghum, which is a true host for *S. hermonthica* and *S. asiatica* (Hauck et al., 1992). Almost at the same time as sorgolactone, alectrol was isolated from cowpea, known as a host for *S. gesnerioides* (Müller et al., 1992). Orobanchol was isolated from red clover (*Trifolium pretense* L.) root, which is a host

DOI: 10.1201/9781003224525-9

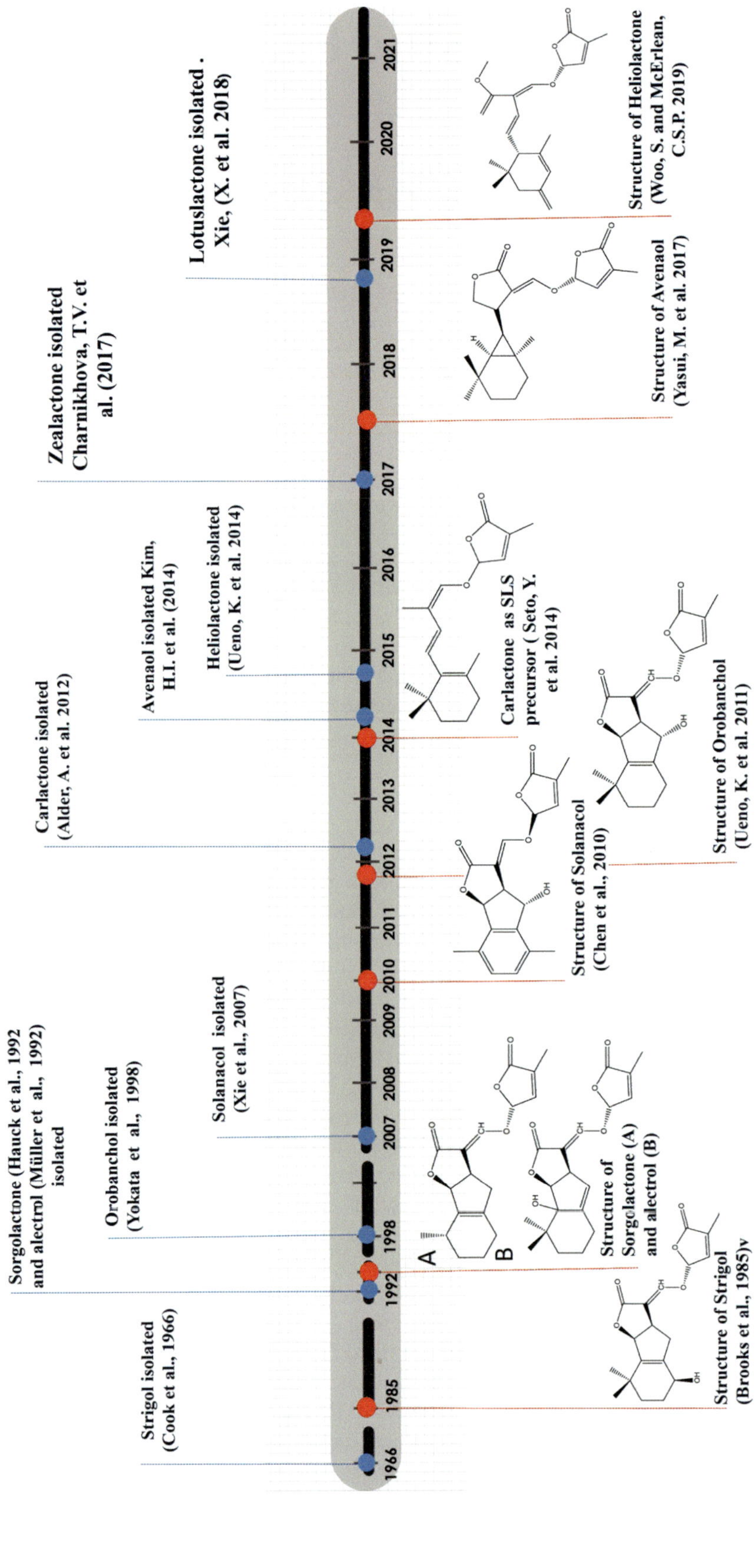

FIGURE 7.1 Timeline of parasitic weed stimulant discovery.

for *Orobanche minor* (Yokota et al., 1998). Alder et al. (2012) isolated carlactone, which was identified as a germination stimulant of *S. hermonthica* and found to be a biosynthetic intermediate that leads to SLs production. Kim et al. (2014) tested the stimulation properties of aveneol, which was extracted from black oat (*Avena strigose* Schreb.) and reported to have an allelopathic effect on parasitic plants. The authors found that aveneol is a potent stimulant to *Phelipanche ramosa* seed germination against a low stimulation effect on *S. hermonthica* and *O. minor*. Heliolactone, which was isolated from sunflower (*Helianthus annuus* L.) root exudates, was also reported to stimulate the seed germination of a wide range of parasitic plant species including *O. crenata*, *O. minor*, *O. cumuna*, *P. aegyptiaca*, and *S. hermonthica* (Brun et al., 2018). In this chapter, we focus on (i) the biosynthesis, structural diversity, and production of the SLs involved in the interaction between parasitic weeds and their host plant and (ii) the use of these SLs in parasitic plant management in many potential food crop species.

7.1 STRUCTURAL DIVERSITY OF STRIGOLACTONES

SLs that are terpenoid lactones play a crucial role in the growth and development of plants as well as the chemical communications with soil microorganisms and root parasitic plants (Yoneyama, 2020; Wang et al., 2020). A β-unsaturated furanone ring called the D ring, which is linked by an enol ether bridge to a tricyclic lactone (ABC rings), characterizes a natural SL. The region connecting the core (ABC) with the D ring connection is conserved in the 2′R configuration, and it is required for biological activities (Wani et al., 2021). The configuration between the Bring and C ring separates SLs into two major classes: strigol types (C-ring in β-orientation/up) produced by rice and tomato and orobanchol (C-ring in α-orientation/down) types produced by tobacco (Bürger and Chory, 2020). Both classes (strigol and orobanchol) have A, B, C, and D rings, and they are called canonical SLs. Other non-canonical SLs lack A, B, and/or C rings but have the D ring bonded to the rest of the molecule (Wang et al., 2021; Wani et al., 2021) (Figure 7.2).

7.2 STRIGOLACTONES BIOSYNTHESIS

SLs biosynthesis takes place in both plant root and shoot and involves many biosynthetic genes: *D27*, *D27-like1 (D27L1)*, and *D27-like2 (D27L2)* (Waters et al., 2017) (Figure 7.3). Putative homologs of the previously mentioned genes are found in some cyanobacteria, chlorophyte, charophyte, and algae, which suggests that the pathway of SLs biosynthesis is conserved throughout the whole plant kingdom (Wang et al., 2021). In plant cells, SLs are synthetized in the plastids and involves several enzymes such as beta-carotene-isomerase (D27), which converts all-transβ-carotene into 9-cis-β-carotene (Li et al., 2020). Sequential actions catalyzed by carotenoid cleavage dioxygenases (CCD7/MAX3 and CCD8/MAX4) convert 9-cis-β-carotene to 9-cis-β-apo-10′-carotenal and then to carlactone having the A and D ring structure, which is the last common precursor of the canonical and non-canonical SLs. In *Arabidopsis*, the plastidal enzymes cytochrome P450 (MAX1) catalyze the oxidation of carlactone to carlactonoic acid methyl, which is methylated after that to carlactonoate (MeCLA) by an unknown methyl transferase (Alder et al., 2012). Then a lateral branching oxidoreductase (LBO) converts MeCLA to hydroxymethyl carlactonoate (1′-OH-MeCLA) (Kolbert, 2019). In rice, five homologs of MAX1 have been found (Brun et al., 2018). One of these homologues (Os01g0700900) catalyzes the conversion of carlactone into 4-deoxyorobanchol, which is considered a precursor to the biosynthesis of orobanchol-type SLs. Then the4-deoxyorobanchol will be converted to orobanchol through the activity of the Os01g0701400 enzyme. However, the other enzymes involved in the diversification of this type of SLs are still unidentified (Waters et al., 2017). Conversely to rice, tomato plants are unable to convert 4-deoxyorobanchol to orobanchol because they have only one MAX1 homolog (CYP711A21). In cowpea, carlactonoic acid (CLA) is directly converted into orobanchol with no 4-deoxyorobanchol intermediate through the activity of CYP722C (Yoneyama and Brewer, 2021). In sorghum, it has been demonstrated that the conversion

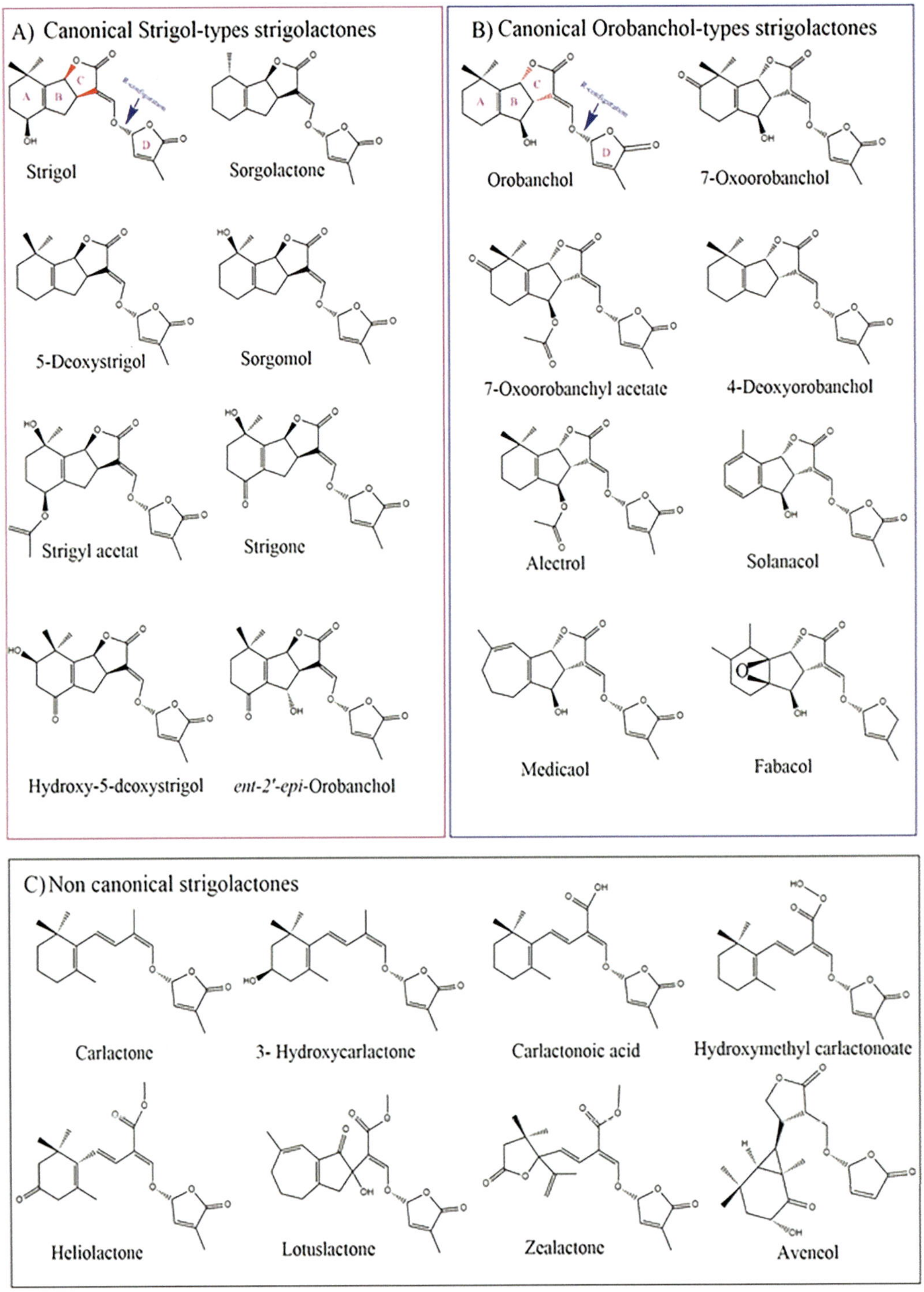

FIGURE 7.2 Chemical structure of strigolactones (SLs).

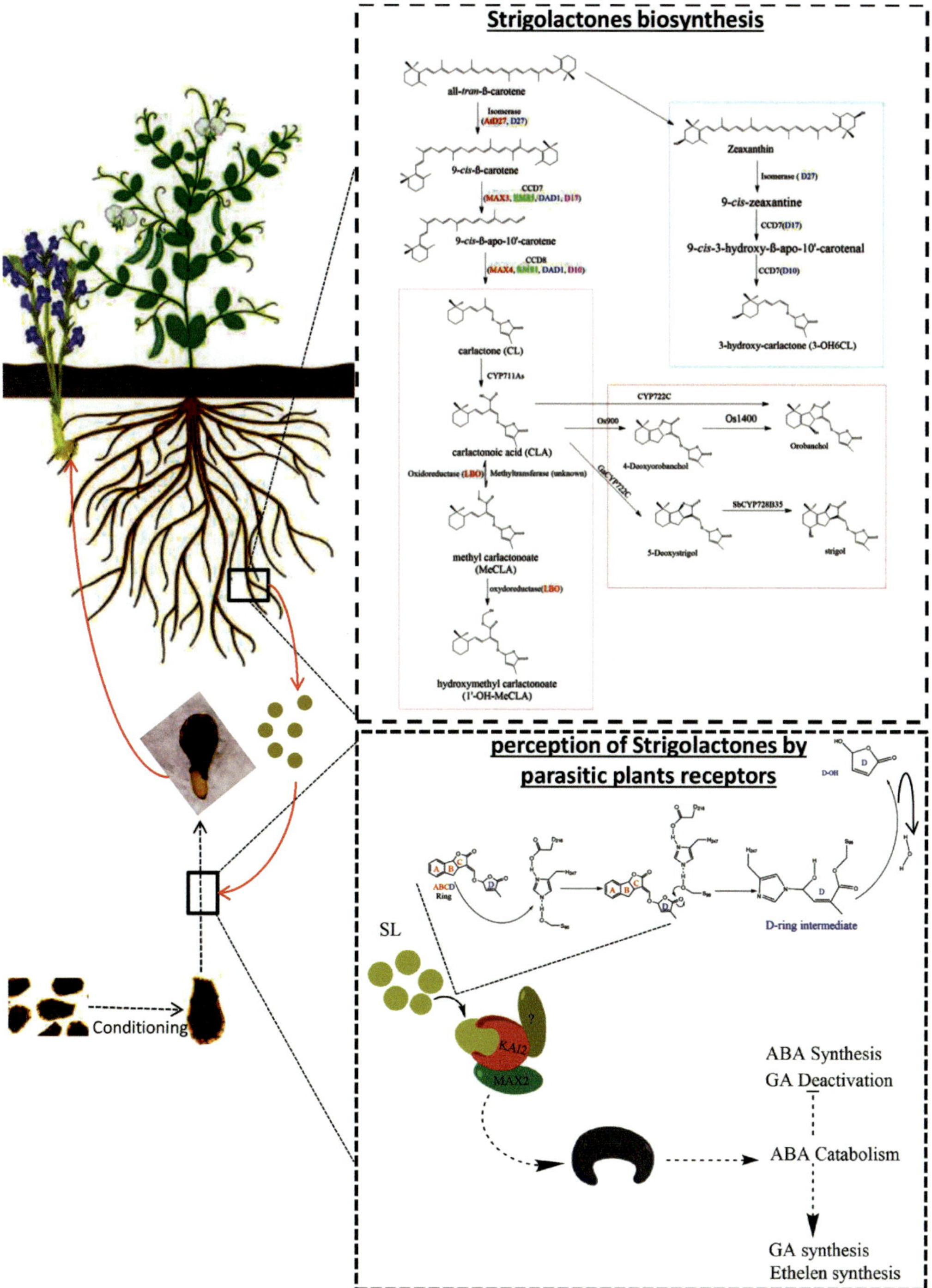

FIGURE 7.3 Host-parasite interaction: biosynthesis, signaling, and perception of strigolactones by parasitic weed seeds.

of canonical SLs' precursor 4-deoxyorobanchol and 4-deoxystrigol into sorgomol and *ent*-2′-epi-sorgomol, as well as the production and the diversification of SLs might be established through the several enzymatic reactions of cytochrome P450 (Brun et al., 2018). In sunflower, methylcarlactonoate acid (MeCLA) has been identified as the precursor to the synthesis of non-canonical SLs such as heliolactone, but not for avenaol and zealactone (Wakabayashi et al., 2020). These non-canonical SLs are synthetized through the conversion of all-trans-b-carotene into carlactone (CL) and 3-hydroxy-carlactone (3-OH-CL) via zeaxanthin and with the involvement of upstream enzymes such as carotenoid, CCD7, and CCD8 (Baz et al., 2018).

7.3 HOST–PARASITE INTERACTION: THE SIGNALING ROLE OF STRIGOLACTONES

Throughout their evolution, parasitic weeds such as *Striga*, *Orobanche*, and *Phelipanche* have developed an adapted perception and signaling mechanisms to use the existing SLs in the rhizosphere of host root system as germination triggers. Understanding parasitic weeds' perception to SLs remains very challenging because of the complexity of the host-parasite interaction pathway and interface (Yaoet al., 2017). However, the identification of the KAI2 gene in parasitic weeds compared to non-parasitic ones constitutes a big step forward in our understanding of SLs perception and signaling in parasitic weeds (Bürger, 2021). The current evolutionary model shows that KAI2 genes are grouped in the lamiales order into three groups: conserved (KAI2c), intermediate (KAI2i), and divergent (KAI2d) (Conn et al., 2015). KAI2 genes were expressed in the KAI2 mutant of *A. thaliana* to elucidate the functions of these parasitic proteins in SLs perception in parasitic weeds. The results indicated that the function of the conserved and intermediary forms of KAI2 are the same in autotrophic and parasitic plants, which suggests that KAI2 proteins might have evolved to become sensitive to natural SLs and to be involved in SLs perception in parasitic weeds (Conn and Nelson, 2016; Xu et al., 2016).

7.4 STRIGOLACTONES AND HOST SPECIFICITY

Different types and quantities of SLs could be produced by a single plant species. However, different genotypes within the same species could produce different SLs combinations (Cardoso et al., 2011). The stereochemistry and structural diversity of SLs influence their biological activities (Kolbert, 2019). Bouwmeester et al. (2007) showed that parasitic weeds have preferences for the host plant's root exudates. In studies conducted on SLs' effect on parasitic weed seeds germination, it has been reported that the natural SLs show different levels of germination activity in parasitic plant species. For example, the 2′epiorobanchol has been found to be more active in eliciting *O. minor* and *P. ramosa* germination than orobanchol, strigol, or solanacol (Xie et al., 2007). Furthermore, orobanchyl acetate (alectrol) was found to be very active in stimulating *S. gesnerioides* germination (Matusova et al., 2005). However, the SLs analogue GR4 showed no effect on seed germination for either *O. foetida* or *S. gesnerioides* against high germination stimulation observed for the other parasitic species (Yoneyama et al., 2009; Trabelsi et al., 2017). The small variation in A, B, and/or C rings of the natural SLs also affect the sensitivity to root exudates within the same parasitic plant species (Yoneyama and Brewer, 2021; Zwanenburg and Tomáš Pospíšil, 2013). Parasitic species such as *O. densiflora*, *O. gracilis*, and *O. hederae* were found to have a strong response only to their host root exudates. However, *P. aegyptiaca*, *P. ramosa*, and *O. minor* were found to be elicited on contact with most plant exudates (Fernández-Aparicio et al., 2009). Brun et al. (2018) suggested that alteration in SLs structures can reduce their stimulation effect on parasite seed germination and could be used to control parasitic weeds and reduce the infestation level. Brun et al. (2021) showed that host specificity is not only determined by the wide structural diversity of SLs but is also defined by other factors such as the haustorium-inducing factors, the incompatibility reactions, and/or the presence of toxic compounds, especially at later development stages.

7.5 STRIGOLACTONES PRODUCTION LEVEL AND RESISTANCE TO PARASITIC WEEDS

Many studies have demonstrated that SLs are likely to be critical for triggering interactions between parasitic and host plants (Brun et al., 2018; Tsuchiya et al., 2018). However, the role of SLs in resistance to parasitic weeds is still far less understood. Qualitative and quantitative differences in SLs production by different host plants and their potential parasitic seeds' germination capacity raise the question whether the resistance or susceptibility in plants correlate with the level of production and/or the structural diversity of SLs.

Sorghum (*Sorghum bicolor* (L.) Moench), maize (*Zea mays* L.), and pearl millet (*Pennisetum typhoideum* Rich.) were found to produce different germination stimulants for *Striga* and *Orobanche* (Awad et al., 2006). It was demonstrated that 5-deoxystrigol was the major SL produced by Striga-susceptible sorghum cultivars, whereas sorgomol was mainly exuded by Striga-resistant cultivars (Yoneyamaet al., 2010). Other studies conducted on *Pisum sativum* carrying *CCD7* or *CCD8* mutation genes showed that the root exudates of mutant plants promoted less germination of parasitic seeds as well as fewer infections by *Orobanche* plants than non-mutant plants (Gomez-Roldan et al., 2008). The same authors showed that these ccd7 and ccd8 mutations were associated with the genes responsible for the cleavage of the apocarotenoid carbon backbone in the strigolactone biosynthetic pathway. In line with these observations, several studies reported that reduced exudation and synthesis of germination stimulants could be involved in plant resistance to parasitic weed attacks (Trabelsi et al., 2017; Mohemed et al., 2018; Badu-Apraku et al., 2020). Trabelsi et al. (2017) showed that susceptible fava bean (*Vicia faba* L.) genotypes produced larger amounts of both orobanchol and orobanchyl acetate than resistant genotypes, resulting in less infestation with both parasitic weeds *O. foetida* and *O. crenata*. In addition, it was reported that root exudates simulating *P. aegyptiaca* germination in resistant tomato species were significantly lower than those of susceptible cultivars (El-Halmouch et al., 2006). Several mutants in tomato and maize were found to be less susceptible to *P. aegytiaca* (López-Ráez et al., 2009). Such resistance was correlated with low production of SLs compared to non-mutant wild types. Furthermore, Dor et al. (2011) reported that the fast-neutron-mutagenized tomato SL-ORT1 showed a high resistance level to *Phelipanche* and *Orobanche* spp. compared to the susceptible tomato cv. M-82. The same authors showed that root exudates analyses of both tomato genotypes revealed the presence of orobanchol, solanacol, and didehydro-orobanchol isomer in the susceptible tomato cv. M-82 and their absence in the mutant tomato SL-ORT1. In chickpea (*Cicer arietinum* L.), the Kabuli mutant line CCD7M14 was reported to have high resistance to both *P. aegyptiaca* and *O. crenata* (Galiliet al., 2021). Root exudates' characterization revealed the presence of orobanchol, orobanchyl acetate, and didehydro-orobancholin non-mutant/wild line, but nothing was detected in the mutant CCD7M14 root exudates, which could be associated with the resistance to broomrape observed for the CCD7M14 mutant line. In addition, Pavan et al. (2016) conducted a large screening for resistance to *O. crenata* using local pea (*P. sativum* L.) genetic resources and were able to identify and select some genotypes with high resistance levels. The root exudates analysis and *in vitro* broomrape germination bioassay of these resistant genotypes showed that the resistance was highly associated and correlated with the low production of SLs in the host plant root rhizosphere (Pavan et al., 2016).

In fact, disabling SLs biosynthesis by mutation is another alternative approach that has been explored to improve crop resistance to parasitic weeds and mitigate the negative effects of the parasite (Fukui et al., 2013). Dor et al. (2011) reported that the ability to induce *P. ramosa* seed germination was reduced by 90% in the mutant tomato SL-ORT1 compared to the non-mutant wild genotype. Similar results were also observed with the rice mutant d10 and *S. hermonthica* (Umehara et al., 2008). In addition, the inhibition and/or manipulation of carotenoid biosynthesis pathway in some host crops such as sorghum and cowpea results in a significant decrease in their ability to stimulate the seed germination of both *S. hermonthica* and *O. crenata* (Matusova et al., 2005). Even though this approach showed promising results in reducing the infestation and the severity of the parasite attack, it could affect the plants' development. This occurs as the SLs play pivotal roles in plant physiology, which includes the regulation of plant shoot and root architecture as well as the modulation of its interaction with the beneficial AM fungi (Cardoso et al., 2011). It has been proven that different SL compounds exuded by rice have different effects on parasitic seed germination and AM fungi (Cardoso et al., 2014). Yoneyama et al. (2015) conducted another study on maize, showing that there were no significant differences in AM symbioses between resistant and susceptible cultivars. Such results indicated that altering SLs exudation and profile could be a better strategy than completely suppressing SLs production. This could guide breeders toward developing parasite-resistant host plants without losing the SLs exudation's beneficial effects on host plant development.

7.6 STRIGOLACTONES AND THEIR POTENTIAL USE FOR PARASITIC WEED CONTROL

During the last decades, research activities on parasitic weed control and management have been intensified, covering all aspects, including mainly those related to breeding for resistance, control methods, and host-parasite interaction. More interest and focus were given to the development of resistant genotypes in major host crops such as tomato, peas, fava bean, chickpea, grass pea, lentils, rice, sorghum, and maize (Matusova et al., 2005; El-Halmouch et al., 2006; Jamil et al., 2011; Nefzi et al., 2016; Amri et al., 2019, 2021; Abdallah et al., 2021; En-nahli et al., 2021). For some of these genotypes, resistance to parasitic weeds was associated with reduced exudation and production of germination stimulants (SLs) (Pavan et al., 2016; Trabelsi et al., 2017; Brun et al., 2018). Such mechanism of resistance with the variation in SLs production and the stimulation of parasitic weed seed germination make breeding for cultivars with low germination stimulants a feasible, attractive, and promising strategy to fight parasitic weed attacks.

Suicidal germination is another strategy for parasitic weed control. Application of a germinating agent in the soil such as SL analogues (Figure 7.4A) stimulates the seed germination of the parasitic weeds and results the death of the germinated seeds in the absence of any suitable host crop (Kgosi et al., 2012). Johnson et al. (1976) reported the first successful use of the suicidal germination approach in controlling *Striga* and *Orobanche* infestation. They found that using SL analogue GR7 in soil at 10 mg L^{-1} led to a considerable reduction in germination of *S. asiatica* and *P. ramosa* seeds. Many other SLs analogues were generated and tested (Figure 7.4B), such as GR24, which was used to reduce up to 50% the *S. asiatica* seed bank in the soil (Musselman, 1987). Also, Fukui et al. (2013) reported that 4-Br debranone (4BD), which is a chemical compound that mimics SL functions, showed less stimulating activity on *S. hermonthica* seeds than GR24 and could be used as an SL mimic for specific target functions in host plants. It is important to mention that successful suicidal methods using SLs analogues depends on some restrictions related to their hydrolyzation in the soil that should happen under the most suitable soil conditions to insure the maximum germination rate of parasitic weed seeds (Figure 7.4C). Several studies demonstrated that the germination potential of SL analogues was strongly influenced by soil moisture, temperature, and pH (Babiker and Hamdoun, 1982; Babiker et al., 1987, 1988). This make the suicidal germination approach to parasitic weed control very difficult and sometimes not a realistic option because of the untimely decomposition, the hydrolysis, and the low stability of the stimulants in soil (Zwanenburg and Tomáš Pospíšil, 2013).

Futhermore, in the last decade, research has implicated SLs in nutrient signaling pathways, and they have been proposed as mediators of plant responses to environmental conditions. SLs production has been demonstrated to be regulated by nutrient availability in plants (Yoneyama et al., 2007; López-Ráez et al., 2008; Umehara et al., 2008; Chiou and Lin, 2011). For example, leguminous plants show an increase in SLs levels in response to both nitrogen- (N) and phosphorus- (P) deficient conditions, and SLs act as a germination stimulant for parasitic weeds in this case (Yoneyama

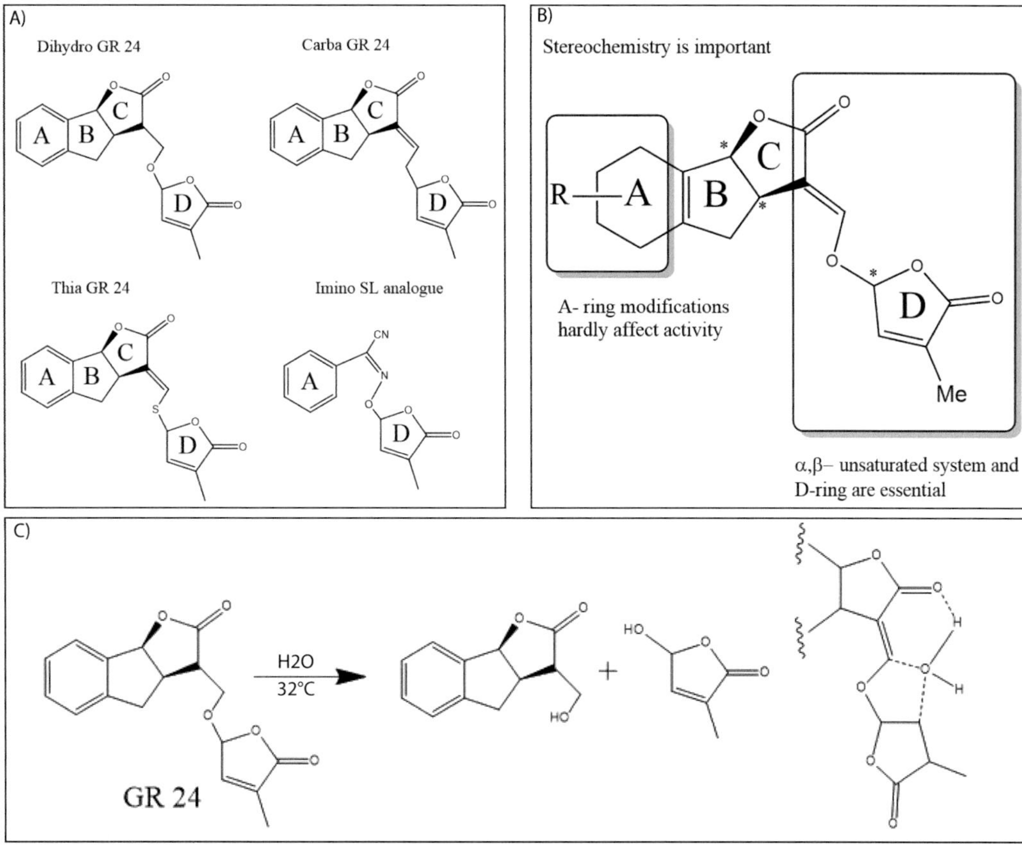

FIGURE 7.4 A) Some modified SL analogues, B) Model for designing novel SL analogues, C) Hydrolysis of GR24 analogue in the soil.

et al., 2011; Foo et al., 2013; Sun et al., 2016). Similar results were reported for other species, such as rice (Umehara et al., 2010; Jamil et al., 2011), sorghum (Yoneyama et al., 2007), Chinese milk vetch, marigold, lettuce, and wheat (Yoneyama et al., 2012). In other plant species, such as red clover (Yoneyama et al., 2007), tomato (López-Ráez et al., 2008; Yoneyama et al., 2012), and alfalfa (Yoneyama et al., 2012), only P deficiency enhanced SL production. Trabelsi et al. (2017) showed that P and N deficiency enhanced germination of *Orobanche*spp. seeds parasitizing fava beans. It has also been demonstrated inStriga that the application of N and P fertilizer in soil deficient in both nutrients limited strigolactone exudation, leading to lower Striga seeds germination (Czarnecki et al., 2013; Yoneyama, 2020; Cechin and Press, 1993; Yoneyama et al., 2007; Jamil et al., 2012, 2014). It was also reported that Striga takes advantages from the production of SLs to attract symbiotic arbuscular mycorrhizal fungi (AMF) in the rhizosphere for phosphorus assimilation in poorly fertile soil to induce seed germination, but SLs exudation still remains a function of the deficiency of macronutrients N and P in the soil (Yoneyama et al., 2010). Other studies also demonstrated that AM fungi colonization induced a clear reduction in strigolactone production in mycorrhizal plants (Lendzemo et al., 2007). It was shown that AM fungi colonization in pea induced less germination in *Orobanche* and *Phelipanche* species than non-colonized plants, suggesting a reduction in strigolactones in those plants (López-Ráez et al., 2011). It was also shown that AM symbiosis leads to a reduction in the germination stimulation of *P. ramosa* seeds by root exudates of tomato; this reduction was caused by a decrease in SLs production. (López-Raez et al., 2011).

7.7 CONCLUSION

Manipulation of SLs signaling in host plants has a strong potential and could significantly contribute to addressing the problem of parasitic weeds in many strategic food and feed crops worldwide. Low SLs production could significantly reduce the severity of the parasitic attack and increase crop production and productivity. Combining interdisciplinary research approaches that could bring together multiple research areas, including breeding advanced sequencing, genetic engineering and gene editing (CRISPR-Cas9 technology), biochemistry, and synthetic biology technologies, is the best way to move forward in parasitic weed management. Such an approach will allow better understanding of the interaction between the host plant and the parasite and decrypt the relationship between SLs structure and their functions. From plant-breeding perspectives, the new emergent CRISPR-Cas9 technology could open new opportunities for further options to fight the spread of the parasitic weeds targeting both the parasite and the host plant.

REFERENCES

Abbes, Zouhaier, Amal Bouallegue, Imen Trabelsi et al. "Investigation of some biochemical mechanisms involved in the resistance of faba bean (*Vicia faba* L.) varieties to *Orobanche* spp." *Plant Protection Science* 56 no. 4 (2020): 317–328.

Abdallah, Fadoua, Shiv Kumar, Ahmed Amri et al. "Wild *Lathyrus* species as a great source of resistance for introgression into cultivated grass pea (Lathyrus sativus L) against broomrape weeds (*Orobanche crenata* Forsk. and *Orobanche foetida* Poir)." *Crop Science* 61 no. 1 (2021): 263–276.

Alder, Adrian, Muhammad Jamil, Mattia Marzorati et al. "The path from β-carotene to carlactone, a strigolactone-like plant hormone." *Science* 335 no. 6074 (2012): 1348–1351.

Amri, Moez, Zouhaier Abbes, Imen Trabelsi, Michel Edmond Ghanem, Rachid Mentag, and Mohamed Kharrat. "Chlorophyll content and fluorescence as physiological parameters for monitoring *Orobanche foetida* Poir. infection in faba bean." *Plos one* 16 no. 5 (2021): e0241527.

Amri, Moez, Imen Trabelsi, Zouhaier Abbes, and Mohamed Kharrat. "Release of a new faba bean variety" Chourouk" resistant to the parasitic plants *Orobanche foetida* and *O. crenata* in Tunisia." *International Journal of Agriculture and Biology* 23 (2019): 499–505.

Awad, Ayman A., Daisuke Sato, Dai Kusumoto, Hiroaki Kamioka, Yasutomo Takeuchi, and Koichi Yoneyama. "Characterization of strigolactones, germination stimulants for the root parasitic plants *Striga* and *Orobanche*, produced by maize, millet and sorghum." *Plant Growth Regulation* 48 no. 3 (2006): 221–227.

Babiker, A. G. T., and A. M. Hamdoun. "Factors affecting the activity of GR7 in stimulating germination of *Striga hermonthica* (Del.) Benth." *Weed Research* 22 no. 2 (1982): 111–115.

Babiker, A. G. T., A. M. Hamdoun, A. Rudwan, N. G. Mansi, and H. H. Faki. "Influence of soil moisture on activity and persistence of the strigol analogue GR 24." *Weed Research* 27 no. 3 (1987): 173–178.

Babiker, A. G. T., Nadia E. Ibrahim, and W. G. Edwards. "Persistence of GR7 and Striga germination stimulant (s) from *Euphorbia aegyptiaca* Boiss. in soils and in solutions." *Weed Research* 28 no. 1 (1988): 1–6.

Badu-Apraku, Baffour, Samuel Adewale, Agre Paterne, Melaku Gedil, and Robert Asiedu. "Identification of QTLs controlling resistance/tolerance to *Striga hermonthica* in an extra-early maturing yellow maize population." *Agronomy* 10 no. 8 (2020): 1168.

Baz, Lina, Narumi Mori, Jianing Mi et al. "3-Hydroxycarlactone, a novel product of the strigolactone biosynthesis core pathway." *Molecular Plant* 11 no. 10 (2018): 1312–1314.

Bouwmeester, Harro J., Matusova Radoslava, Zhongkui Sun, and Michael H. Beale. "Secondary metabolite signalling in host–Parasitic plant interactions." *Current Opinion in Plant Biology* 6 no. 4 (2003): 358–364.

Bouwmeester, Harro J., Christophe Roux, Juan Antonio Lopez-Raez, and Guillaume Becard. "Rhizosphere communication of plants, parasitic plants and AM fungi." *Trends in Plant Science* 12 no. 5 (2007): 224–230.

Brun, Guillaume, Lukas Braem, Séverine Thoiron, Kris Gevaert, Sofie Goormachtig, and Philippe Delavault. "Seed germination in parasitic plants: what insights can we expect from strigolactone research?." *Journal of Experimental Botany* 69 no. 9 (2018): 2265–2280.

Brun, Guillaume, Thomas Spallek, Philippe Simier, and Philippe Delavault. "Molecular actors of seed germination and haustoriogenesis in parasitic weeds." *Plant Physiology* 185 no. 4 (2021): 1270.

Bürger, Marco. "Insights into the evolution of strigolactone signaling." *Plant Cell* 33 no. 11 (2021): 3389–3390.

Bürger, Marco, and Joanne Chory. "The many models of strigolactone signaling." *Trends in Plant Science* 25 no. 4 (2020): 395–405.

Cardoso, Catarina, Carolien Ruyter-Spira, and Harro J. Bouwmeester. "Strigolactones and root infestation by plant-parasitic Striga, *Orobanche* and *Phelipanche* spp." *Plant Science* 180 no. 3 (2011): 414–420.
Cardoso, Catarina, Yanxia Zhang, Muhammad Jamil et al. "Natural variation of rice strigolactone biosynthesis is associated with the deletion of two MAX1 orthologs." *Proceedings of the National Academy of Sciences* 111 no. 6 (2014): 2379–2384.
Cechin, I., and M. C. Press. "Nitrogen relations of the sorghum-*Striga hermonthica* hostparasite association: Growth and photosynthesis." *Plant, Cell & Environment* 16 no. 3 (1993): 237–247.
Chiou, Tzyy-Jen, and Shu-I Lin. "Signaling network in sensing phosphate availability in plants." *Annual Review of Plant Biology* 62 (2011): 185–206.
Conn, Caitlin E., Rohan Bythell-Douglas, Drexel Neumann et al. "Convergent evolution of strigolactone perception enabled host detection in parasitic plants." *Science* 349 no. 6247 (2015): 540–543.
Conn, Caitlin E., and David C. Nelson. "Evidence that Karrikin-Insensitive2 (KAI2) receptors may perceive an unknown signal that is not karrikin or strigolactone." *Frontiers in Plant Science* 6 (2016): 1219.
Cook, C. E., Leona P. Whichard, Beverly Turner, Monroe E. Wall, and Grant H. Egley. "Germination of witchweed (*Striga lutea* Lour.): Isolation and properties of a potent stimulant." *Science* 154 no. 3753 (1966): 1189–1190.
Czarnecki, Olaf, Jun Yang, David J. Weston, Gerald A. Tuskan, and Jin-Gui Chen. "A dual role of strigolactones in phosphate acquisition and utilization in plants." *International Journal of Molecular Sciences* 14 no. 4 (2013): 7681–7701.
de Saint Germain, Alexandre, Guillaume Clavé, Marie-Ange Badet-Denisot et al. "An histidine covalent receptor and butenolide complex mediates strigolactone perception." *Nature Chemical Biology* 12 no. 10 (2016): 787–794.
de Saint Germain, Alexandre, Pascal Retailleau, Stéphanie Norsikian et al. "Contalactone, a contaminant formed during chemical synthesis of the strigolactone reference GR24 is also a strigolactone mimic." *Phytochemistry* 168 (2019): 112112.
Dor, Evgenia, Koichi Yoneyama, Smadar Wininger et al. "Strigolactone deficiency confers resistance in tomato line SL-ORT1 to the parasitic weeds *Phelipanche* and *Orobanchespp*." *Phytopathology* 101 no. 2 (2011): 213–222.
El-Halmouch, Y., H. Benharrat, and P. Thalouarn. "Effect of root exudates from different tomato genotypes on broomrape (*O. aegyptiaca*) seed germination and tubercle development." *Crop Protection* 25 no. 5 (2006): 501–507.
En-nahli, Youness, Hicham El Arroussi, Shiv Kumar et al. "Resistance to *Orobanche crenata* Forsk. in lentil (Lens culinaris Medik.): Exploring some potential altered physiological and biochemical defense mechanisms." *Journal of Plant Interactions* 16 no. 1 (2021): 321–331.
Fernández-Aparicio Mónica, Fernando Flores, and Diego Rubiales. "Recognition of root exudates by seeds of broomrape (*Orobanche* and *Phelipanche*) species." *Annals of Botany* 103 no. 3 (2009): 423–431.
Foo, Eloise, Kaori Yoneyama, Cassandra J. Hugilla, Laura J. Quittenden, and James B. Reid. "Strigolactones and the Regulation of Pea Symbioses in Response to Nitrate and Phosphate Deficiency." *Molecular Plant* 6 no. 1 (2013): 76–87.
Fukui, Kosuke, Shinsaku Ito, and Tadao Asami. "Selective mimics of strigolactone actions and their potential use for controlling damage caused by root parasitic weeds." *Molecular Plant* 6 no. 1 (2013): 88–99.
Galili, Shmuel, Joseph Hershenhorn, Evgeny Smirnov et al. "Characterization of a Chickpea Mutant Resistant to Phelipanche aegyptiaca Pers and Orobanche crenata Forsk." *Plants* 10 no. 12 (2021): 2552.
Gomez-Roldan, Victoria, Soraya Fermas, Philip B. Brewer et al. "Strigolactone inhibition of shoot branching." *Nature* 455 no. 7210 (2008): 189–194.
Hauck, Christian, Müller Sigrid, and Schildknecht Hermann. "A germination stimulant for parasitic flowering plants from *Sorghum bicolor*, a genuine host plant." *Journal of Plant Physiology* 139 no. 4 (1992): 474–478.
Jamil, M., T. Charnikhova, B. Houshyani, A. van Ast, and H. J. Bouwmeester. "Genetic variation in strigolactone production and tillering in rice and its effect on Striga hermonthica infection." *Planta* 235 no. 3 (2012): 473–484.
Jamil, M., T. Charnikhova, C. Cardoso, T. Jamil, K. Ueno, F. Verstappen, T. Asami, and H. J. Bouwmeester. "Quantification of the relationship between strigolactones and Striga hermonthica infection in rice under varying levels of nitrogen and phosphorus." *Weed Research* 51 no. 4 (2011): 373–385.
Jamil, M., T. Charnikhova, F. Verstappen, Z. Ali, H. Wainwright, and H. J. Bouwmeester. "Effect of phosphate-based seed priming on strigolactone production and *Striga hermonthica* infection in cereals." *Weed Research* 54 no. 3 (2014): 307–313.

Johnson, A. W., G. Rosebery, and C. Parker. "A novel approach to *Striga* and *Orobanche* control using synthetic germination stimulants." *Weed Research* 16 no. 4 (1976): 223–227.

Kgosi, R. L., B. Zwanenburg, A. S. Mwakaboko, and A. J. Murdoch. "Strigolactone analogues induce suicidal seed germination of Striga spp. in soil." *Weed Research* 52 no. 3 (2012): 197–203.

Kim, Hyun I., Takaya Kisugi, Pichit Khetkam et al. "Avenaol, a germination stimulant for root parasitic plants from Avena strigosa." *Phytochemistry* 103 (2014): 85–88.

Kolbert, Zsuzsanna. "Strigolactone-nitric oxide interplay in plants: The story has just begun." *Physiologia Plantarum* 165 no. 3 (2019): 487–497.

Lendzemo, Venasius W., Thomas W. Kuyper, Radoslava Matusova, Harro J. Bouwmeester, and A. Van Ast. "Colonization by arbuscular mycorrhizal fungi of sorghum leads to reduced germination and subsequent attachment and emergence of *Striga hermonthica*." *Plant Signaling & Behavior* 2 no. 1 (2007): 58–62.

Li, Weiqiang, Kien Huu Nguyen, Cuong Duy Tran et al. "Negative roles of strigolactone-related SMXL6, 7 and 8 proteins in drought resistance in Arabidopsis." *Biomolecules* 10 no. 4 (2020): 607.

López-Ráez, Juan Antonio, Tatsiana Charnikhova, Ivan Fernández, Harro Bouwmeester, and Maria J. Pozo. "Arbuscular mycorrhizal symbiosis decreases strigolactone production in tomato." *Journal of Plant Physiology* 168 no. 3 (2011): 294–297.

López-Ráez, Juan Antonio, Tatsiana Charnikhova, Victoria Gómez-Roldán et al. "Tomato strigolactones are derived from carotenoids and their biosynthesis is promoted by phosphate starvation." *New Phytologist* 178 no. 4 (2008): 863–874.

López-Ráez, Juan Antonio, Radoslava Matusova, Catarina Cardoso et al. "Strigolactones: ecological significance and use as a target for parasitic plant control." *Pest Management Science: Formerly Pesticide Science* 65 no. 5 (2009): 471–477.

Matusova, Radoslava, Kumkum Rani, Francel WA Verstappen, Maurice CR Franssen, Michael H. Beale, and Harro J. Bouwmeester. "The strigolactone germination stimulants of the plant-parasitic Striga and Orobanche spp. are derived from the carotenoid pathway." *Plant Physiology* 139 no. 2 (2005): 920–934.

Mohemed, Nasreldin, Tatsiana Charnikhova, Emilie F. Fradin, Juriaan Rienstra, Abdelgabar GT Babiker, and Harro J. Bouwmeester. "Genetic variation in Sorghum bicolor strigolactones and their role in resistance against Striga hermonthica." *Journal of Experimental Botany* 69 no. 9 (2018): 2415–2430.

Müller, Sigrid, Christian Hauck, and Hermann Schildknecht. "Germination stimulants produced by Vigna unguiculata Walp cv Saunders Upright." *Journal of Plant Growth Regulation* 11 no. 2 (1992): 77–84.

Musselman, Lytton J. "Taxonomy of witch weeds." *Taxonomy of Witch weeds* (1987): 3–12.

Nefzi, Fatma, Imen Trabelsi, Moez Amri, Emna Triki, Mohamed Kharrat, and Zouhaier Abbes. "Response of some chickpea (*Cicer arietinum* L.) genotypes to *Orobanche foetida* Poir. parasitism." *Chilean Journal of Agricultural Research* 76 no. 2 (2016): 170–178.

Pavan, Stefano, Adalgisa Schiavulli, Angelo Raffaele Marcotrigiano et al. "Characterization of low-strigolactone germplasm in pea (*Pisum sativum* L.) resistant to crenate broomrape (*Orobanche crenata* Forsk.)." *Molecular Plant-Microbe Interactions* 29 no. 10 (2016): 743–749.

Raupp, Frank M., and Otmar Spring. "New sesquiterpene lactones from sunflower root exudate as germination stimulants for *Orobanche cumana*." *Journal of Agricultural and Food Chemistry* 61 no. 44 (2013): 10481–10487.

Saucet, Simon B., and Ken Shirasu. "Molecular parasitic plant–Host interactions." *PLoS Pathogens* 12 no. 12 (2016): e1005978.

Sun, H., J. Tao, M. Hou, S. Huang, S. Chen, Z. Liang, T. Xie, Y. Wei, X. Xie, K. Yoneyama, G. Xu, and Y. Zhang. "A strigolactone signal is required for adventitious root formation in rice." *Annals of Botany* 115 no. 7 (2016): 1155–1162.

Trabelsi, Imen, Kaori Yoneyama, Zouhaier Abbes et al. "Characterization of strigolactones produced by *Orobanche foetida* and *Orobanche crenata* resistant faba bean (Vicia faba L.) genotypes and effects of phosphorous, nitrogen, and potassium deficiencies on strigolactone production." *South African Journal of Botany* 108 (2017): 15–22.

Tsuchiya, Yuichiro, Masahiko Yoshimura, and Shinya Hagihara. "The dynamics of strigolactone perception in Striga hermonthica: A working hypothesis." *Journal of Experimental Botany* 69 no. 9 (2018): 2281–2290.

Umehara, Mikihisa, Atsushi Hanada, Hiroshi Magome, Noriko Takeda-Kamiya, and Shinjiro Yamaguchi. "Contribution of strigolactones to the inhibition of tiller bud outgrowth under phosphate deficiency in rice." *Plant and Cell Physiology* 51 no. 7 (2010): 1118–1126.

Umehara, Mikihisa, Atsushi Hanada, Satoko Yoshida et al. "Inhibition of shoot branching by new terpenoid plant hormones." *Nature* 455 no. 7210 (2008): 195–200.

Wakabayashi, Takatoshi, Hikaru Shinde, Nanami Shiotani et al. "Conversion of methyl carlactonoate to heliolactone in sunflower." *Natural Product Research* (2020): 1–8.

Wang, Jian Y., Pei-Yu Lin, and Salim Al-Babili. "On the biosynthesis and evolution of apocarotenoid plant growth regulators." *Seminars in Cell & Developmental Biology* (Academic Press)109 (2021): 3–11.

Wang, Lei, Qian Xu, Hong Yu et al. "Strigolactone and karrikin signaling pathways elicit ubiquitination and proteolysis of SMXL2 to regulate hypocotyl elongation in Arabidopsis." *The Plant Cell* 32 no. 7 (2020): 2251–2270.

Wani, Kaiser I., Andleeb Zehra, Sadaf Choudhary et al. "Mechanistic insights into strigolactone biosynthesis, signaling, and regulation during plant growth and development." *Journal of Plant Growth Regulation* 40 no. 5 (2021): 1836–1852.

Waters, Mark T., Caroline Gutjahr, Tom Bennett, and David C. Nelson. "Strigolactone Signaling and Evolution."*Annual Review of Plant Biology* 68 no. 1 (2017): 291–322.

Xie, Xiaonan, Dai Kusumoto, Yasutomo Takeuchi, Kaori Yoneyama, Yoichi Yamada, and Koichi Yoneyama. "2′-Epi-orobanchol and solanacol, two unique strigolactones, germination stimulants for root parasitic weeds, produced by tobacco." *Journal of Agricultural and Food Chemistry* 55 no. 20 (2007): 8067–8072.

Xu, Yuqun, Takuya Miyakawa, Hidemitsu Nakamura et al. "Structural basis of unique ligand specificity of KAI2-like protein from parasitic weed *Striga hermonthica*." *Scientific Reports* 6 no. 1 (2016): 1–9.

Yao, Ruifeng, Fei Wang, Zhenhua Ming et al. "ShHTL7 is a non-canonical receptor for strigolactones in root parasitic weeds." *Cell Research* 27 no. 6 (2017): 838–841.

Yokota, Takao, Hidenori Sakai, Keiji Okuno, Koichi Yoneyama, and Yasutomo Takeuchi. "Alectrol and orobanchol, germination stimulants for Orobanche minor, from its host red clover." *Phytochemistry* 49 no. 7 (1998): 1967–1973.

Yoneyama, Kaori, and Philip B. Brewer. "Strigolactones, how are they synthesized to regulate plant growth and development?." *Current Opinion in Plant Biology* 63 (2021): 102072.

Yoneyama, Kaori, Ryota Arakawa, Keiko Ishimoto et al. "Difference in Striga-Susceptibility is reflected in strigolactone secretion profile, but not in compatibility and host preference in arbuscular mycorrhizal symbiosis in two maize cultivars." *New Phytologist* 206 no. 3 (2015): 983–989.

Yoneyama, Kaori, Xiaonan Xie, Dai Kusumoto et al. "Nitrogen deficiency as well as phosphorus deficiency in sorghum promotes the production and exudation of 5-deoxystrigol, the host recognition signal for arbuscular mycorrhizal fungi and root parasites." *Planta* 227 no. 1 (2007): 125–132.

Yoneyama, Kaori, Xiaonan Xie, Hyun Il Kim et al. "How do nitrogen and phosphorus deficiencies affect strigolactone production and exudation?." *Planta* 235, no. 6 (2012): 1197–1207.

Yoneyama, Koichi. "Recent progress in the chemistry and biochemistry of strigolactones." *Journal of Pesticide Science* (2020): D19– D084.

Yoneyama, Koichi, X. Xie, T. Kisugi, T. Nomura, H. Sekimoto, T. Yokota, and K. Yoneyama. "Characterization of strigolactones exuded by Asteraceae plants." *Plant Growth Regulation* 65 (2011): 495–504.

Yoneyama, Koichi, Ayman A. Awad, Xiaonan Xie, Kaori Yoneyama, and Yasutomo Takeuchi. "Strigolactones as germination stimulants for root parasitic plants." *Plant and Cell Physiology* 51, no. 7 (2010): 1095–1103.

Yoneyama, Koichi, Xiaonan Xie, Kaori Yoneyama, and Yasutomo Takeuchi. "Strigolactones: structures and biological activities." *Pest Management Science: Formerly Pesticide Science* 65 no. 5 (2009): 467–470.

Zwanenburg, Binne, Alinanuswe S. Mwakaboko, Anat Reizelman, Gopinathan Anilkumar, and Divakaramenon Sethumadhavan. "Structure and function of natural and synthetic signalling molecules in parasitic weed germination." *Pest Management Science: Formerly Pesticide Science* 65 no. 5 (2009): 478–491.

Zwanenburg, Binne, and Tomáš Pospíšil. "Structure and activity of strigolactones: new plant hormones with a rich future." *Molecular Plant* 6 no. 1 (2013): 38–62.

Section II

Alkamides: Ecophysiological and Pharmacological Perspectives

8 Alkamides and Humic Acids
Insights into Soil Fertility

Tahseen Afzal, Arshdeep Singh, Javed Iqbal, and Muhammad Sanaullah

CONTENTS

8.0 INTRODUCTION

Numerous organic amendments – manures, biofertilizers, composts, biosolids, crop residues, and humic products – are applied to soil with the intent to ameliorate soil fertility and nutrition efficiency. Crop lands with <1% soil organic matter (SOM) are deficient in humus, those having 1–4% SOM are humic, while those containing >8% SOM are abundant in humus (Larcher, 2003). Soil humification is the main pathway through which the transformation of SOM to humic matter takes place in soil as consequence of the consistent activities of soil biota on SOM, resulting in the formation of a macro-molecule (humus) of a very persistent nature as an end result (Guggenberger, 2005; Sollins et al., 2007). Humic substances are likely to attain molecular weights of several hundred Daltons (Da) (Zandonadi et al., 2019). In general, these are categorized into three main classes; humic acids (alkaline soluble), fulvic acids (acid and alkaline soluble), and insoluble humin (DiDonato et al., 2016). When the alkaline-mediated extracts from the humic matter are acidified, it leads to the formation of precipitates (humic acids). These humic materials possess a homogenous mixture of oxidized organic components with various molecular moieties and functional groups (Stevenson, 1994); thereby humic acids usually solubilize in soil at alkaline soil pH, and the fulvic acids are soluble in soil at all pH ranges (MacCarthy, 2001). Humic acids are formed by the unspecified assemblage of micro-molecules of completely degraded organic materials (Piccolo et al., 2002), which is so intricate that no one structure or set of structures can define it (MacCarthy, 2001). However, it was evinced from the ultrahigh resolution analysis of the humic substances that humic acids isolated from various sources showed the similar structural configurations (DiDonato et al., 2016).

The use of humic products is increasing among farmers due to their pragmatic ability to induce favourable changes in soil health and soil physicochemical characteristics (Van Oosten et al., 2017). It is estimated that the international trade dealing with humic products may potentially surpass over

DOI: 10.1201/9781003224525-11

US $1 billion by 2024 (Olk et al., 2018). In addition to being used an soil conditioning agents, humic products can be used as additives in the formulations of different synthetic and organic fertilizers (Omoarelojie et al., 2021). Due to the pragmatic effects on the soil properties, nutrients availability, and root growth (improved root architecture), humic acids may likely minimize the burgeoning demand of different synthetic fertilizers (Khaled and Fawy, 2011), which may be due to increased fertilizer usage efficiencies. Plant scientists regard humic acids as bio-stimulants since they can regulate plant growth processes, confer tolerance against abiotic stresses, and enhance plant nutrient use efficiencies, which may likely improve the quality of produce (Calvo et al., 2014; Canellas et al., 2015; Shah et al., 2018).

The term "Alkamides" in general refers to a group of secondary metabolites (almost 200 compounds), and all of them have a unique structural framework that develops from the condensation of an unsaturated fatty acid and an amine group (Ramírez-Chávez et al., 2004). Specifically, these compounds develop when different amine parts are condensed with unsaturated fatty acids through an amide link (Greger, 1984). The unbound fraction of humic acid isolated from cow-dung vermicompost (N-iso-propyldecanamide) is considered to be related to the bioactive properties of humic matter (Zandonadi et al., 2019). The humic acids can alter the cellular electrical environment via modulation of the plasma membrane proton pump (PM-ATPase) (Ramos et al., 2015). Improved functionality of PM-ATPase enhances the electrochemical gradient across the cellular membrane, which helps in increasing nutrients uptake (Morsomme and Boutry, 2000). Humic acids in this way acidifies the soil environment (enhanced PM-ATPase activity) which likely improves the cell mitosis and root elongation, may likely improving the nutrients uptake from the soil (Zandonadi et al., 2019), especially micro-nutrients and soil P availability. This chapter illustrates the possible effects of humic acids and Alkamides on root development (improved root architecture), nutrients uptake efficiency, improved nutrients cycling in soil, high quality produce, and crop yields (Figure 8.1).

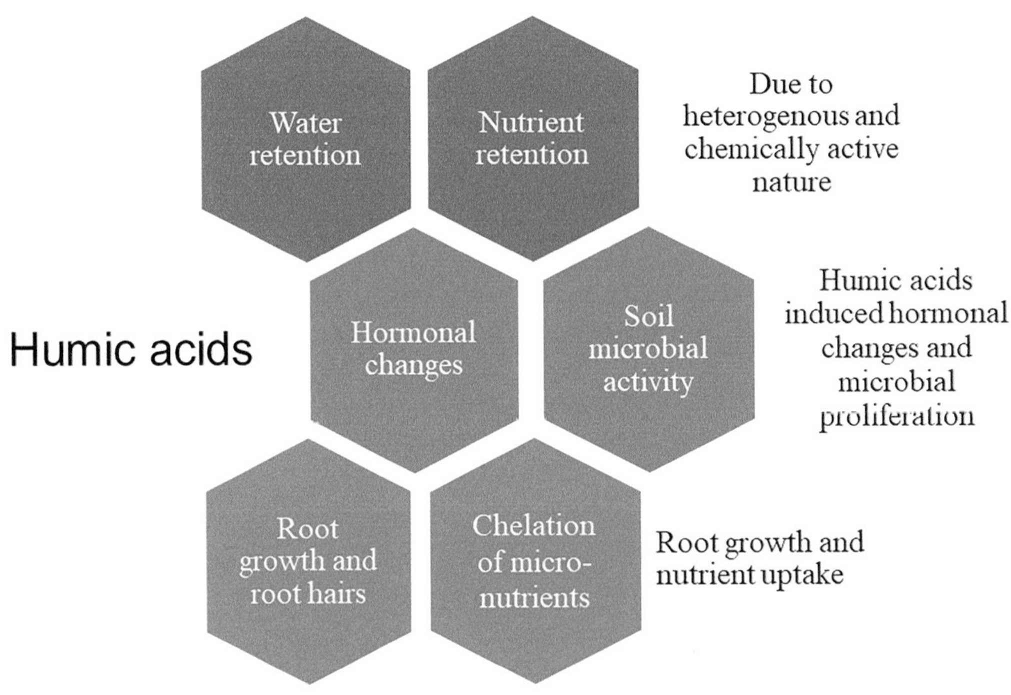

FIGURE 8.1 Humic acids and their effects on soil conditioning and plant nutrients uptake.

8.1 HUMIC ACIDS AND SOIL CHARACTERISTICS

Humic acids are soil conditioners that have significant effects on soil's physical, chemical, and biological properties. This is largely attributed to their unique intrinsic nature and bioactivity and the presence of chemical functional moieties in them. As being soil conditioners, humic acids can potentially improve the cation exchange capacity (CEC) of soil, which may contribute to improving the mineral nutrients retention and efficiencies. The application of K-humate at a high rate caused a significant increase in SOM status, total nitrogen (N), base saturation, soil pH, and CEC (Yilmaz, 2011). The K-humate improved the soil aggregate formation (soil aggregation) due to enhanced flocculation and cementation (organo-mineral complexations) and reduced the soil bulk density (Haynes and Swift, 1990), an indicative of the improved functionality of humic acids (Glaser et al., 2002). Under the presence of 2:1 clay minerals in soil (smectite), humic substances were found to be more efficacious in improving adherence between soil mineral and organic particles, resulting in improved soil structural changes (Yamaguchi et al., 2004). Before adding humic materials to soil, the characterization of soil mineral suite seems important in terms of evaluating the effects of humic acids on improving soil structure. The soil mineral suite shares a strong bonding with positive-charged surfaces, which predominantly influences the soil pH and electrolyte concentrations of soil solution (Spark et al., 1997a). In sum up, the humic materials are generally considered recalcitrant to microbial activities and processes, and act as soil conditioning agents (Albiach et al., 2001; Imbufe et al., 2005). The active humic groups important for soil conditioning and plant growth stimulation constitute the soil humus (Benedetti et al., 1996).

8.2 HUMIC ACIDS AND ROOT DEVELOPMENT

Roots regulate the influx of nutrients and water into and from the plants. The variations in root architecture and morphology lead to the differences in nutrients acquisition patterns from the soil among the plant species (Sattelmacher et al., 1993). Humic substances are considered involved in triggering the changes in root growth and affecting the outgrowth of lateral roots and root hairs (Nardi et al., 2002).

8.2.1 Hormonal Effects of Humic Acids

The effects of humic acids on root nutrients uptake are largely attributed to their ameliorative effects on soil health since the intrinsic nutritional content (mineral composition) of humic acids is negligible. The effective root development caused by a low-dose humic acid application generally supports this notion. As the humic acid application at 50 ppm (mg kg^{-1} soil) promoted the root length and density, however, these were found to be reduced when humic acid was applied at 1,000 ppm (mg kg^{-1} soil) (Mylonas and McCants, 1980), which indicates that the root development is sensitive to the high dose humic acid application (Sharif et al., 2002). Due to the presence of quinone functional groups in humic substances, the intensification of mineral intrusions into root cells under humic acids applications may be related to its involvement in plant metabolic processes and plant growth regulations (Petronio et al., 1982).

The Alkamides (N-iso-propyldecanamide) in the macro-molecular structure of humic acids is the main cause of hormonal changes under humic acid additions, including auxin and nitric oxide, that presumably regulate root growth and architecture (Zandonadi et al., 2019). Under humic acid additions, induction of auxins (Indole-3-acetic acid; IAA) stimulates the lateral roots proliferation and retards the primary root growth, leading to the formation of an extensive and surficial root system (Benková et al., 2003). In addition, humic acids also influence the induction of cytokinin, polyamines, and abscisic acids in plants (Mora et al., 2010). In parallel, higher PM-ATPase activity levels promote the acid-intensification of the cellular environment by high proton (H^+) efflux, the acidified apoplast in turn helps in loosening the cell wall structure. The slackness of the cell wall paves the way for the formation of new mitotic sites, triggering root elongation and root hair formation (Ramos et al., 2015). Intensified H^+ extrusion to apoplast and enhanced root growth both reflect an improvement in root absorption capacity that indicates the improvement in plant nutrition (Morsomme et al., 2000). Zandonadi et al. (2019) reported that N-iso-propyldecanamide in calcium chloride ($CaCl_2$) growth medium triggered the formation of new mitotic sites and high PM-ATPase activity levels, which stimulated an increase in root biomass and lateral root formation. High PM-ATPase activity in root tissues signals its crucial role in nutrient uptake and root growth (Palmgren, 2001). Zandonadi et al. (2010) illustrated the bioactivity of humic acids through IAA signaling, nitric oxide (NO) scavengers, and inductors, which had an influence on lateral root development and increased PM-ATPase activities. Studies focused on mRNA genes encoding PM-ATPase activity also supported the specific modulating influence of humic acids in nutrient acquisition and plant bioenergetics (Mora et al., 2010). The role of humic substances in regulation of plant metabolism is a complicated phenomenon and demands focused scientific investigations (Canellas et al., 2008).

In addition to the hormonal bioactivities, some evidence signals that macro-molecular configuration of humic acids can pose direct physiological influences, being independent of the hormonal molecule contained in its macro-molecular assembly (Mora et al., 2012). Chen et al. (2004) reported that the phytohormones generally are not sourced from the humic substances, because the rigorous isolation and purification processes mediated to isolate the bioactive humate fractions could not extricate phytohormones. In this context, the physiological functions of humic acids might be related to the presence of different functional groups (moieties) in the macro-molecular structure. This discussion underlines the fact that physiological functions of humic acids in soil conditioning and plant growth regulation may likely be associated with various mechanisms; e.g., PM-ATPase activity levels and hormonal influences, however, detailed scientific explorations are needed to pinpoint the origins of humic acid bioactivity.

8.2.2 Improved Plant Nutrition Mediated by Root Growth

The addition of humic acids to soil strongly stimulates root growth (root hair development), which increases the plant nutrients uptake (Atiyeh et al., 2002). The humic acid additions to soil stimulate the root growth more than shoot growth (Nardi et al., 2002). The source (origin), concentrations, and molecular weight of humic matter are the main factors which affects the root response to humic acid applications. The very small-size (<28 nm in general) humic acids can easily penetrate the plasmalemma of plant cells, stimulate nutrient uptake, and modify the release of rhizo-deposits into soil (Fahramand et al., 2014).

8.3 MINERAL AVAILABILITY IN SOIL

Humic substances can induce notable changes in nutrients' availability in the soil (Zandonadi et al., 2013; Canellas and Olivares, 2014; Rose et al., 2014). Soil application of humic acids promoted the N and phosphorus (P) uptake by corn (Khaled and Fawy, 2011). Both humic and fulvic acids tended to impede the phosphate occlusion with sesquioxide minerals (Al and Fe), which enhanced the P availability in soil and mitigated the extent of cations bridging in soil (Hiemstra et al., 2013), by the formation of humate-cation-phosphate complexes. Wang et al. (2015) reported a significant increase in P availability (a decline in mineral-attached P) and crop yield when phosphatic fertilizers were co-applied with humic acids. The functional groups, e.g., carboxyl and hydroxyl, in large contribute to the affinity of humic acids for cations, which may prevent the P precipitation (occlusion) in soil through several physico-chemical and biological ways and promote the transformation of mineral-associated P to the bioavailable P in soil (Urrutia et al., 2014a). Thus, it looks plausible that synthetic phosphatic fertilizers should be co-applied with different humic substances to improve their usage efficiency. Different humic acids include the artificial humates (Yang et al., 2019a), natural humic substances (Quan-Xian et al., 2008), and geo-chemically modified sewage mud (Zhang et al., 2020). Sharif et al. (2002) reported that low dose addition of humic acids ameliorated the soil biochemical conditions, such as microbial activities, CEC, water retention in soil, and increased corn yield. The structural composition of humic acids indicates that these contain organic carbon (OC) (51–57%), N (4–6%), and P (0.2–1%), which may be bioavailable for plant uptake over time following soil additions (Brannon and Sommers, 1985). Very low intrinsic N of humic acids indicates their stable nature, slow decay (Nisar and Mir, 1989), N assimilation (Mora et al., 2010; Zanin et al., 2018), limited microbial C metabolism (glycolysis and Krebs cycle) (Trevisan et al., 2011), and synthesis of secondary metabolites (e.g., phenylpropanoids).

8.3.1 Humic Acids as Chelating Agents

Humate substances tend to form chelates when they come in contact with cations, which may potentially overcome the mineral disturbances in soil by improving the bioavailable portions of different cations (Sharif et al., 2002) and micro-nutrients in soil solution (Tipping, 2002). Amine and carboxyl functional groups are typically involved in the chelation affinity of humic acids for cations. The formation of chelates prolongs the nutrients availability in soil, binds the toxic elements (Khaled and Fawy, 2011), and reduces the mobility of heavy elements (Misra et al., 2009). The mineral composition of soil, however, may control the effect of humic substances on the formation of metal-humic acid complexes, because heavy metals and plant nutrients compete with each other for attachment on the humic surfaces (Spark et al., 1997b). Clemente and Bernal (2006) reported that humic acids reduced the mobility of heavy metals, e.g., zinc (Zn) and lead (Pb), however, showed a minor increase in the mobility of iron (Fe) and copper (Cu). Since, humic acids tend to form chelates with ferric ions (Fe^{3+}) via formation of ferrous-humate bridging which results in its transformation to bioavailable ferrous (Fe^{2+}) ions (Bocanegra et al., 2005). Despite the high Fe content in soil, the bioavailable portion of Fe is very negligible and is largely characterized by the very low soluble nature of Fe oxides (Colombo et al., 2014). As instance, the co-addition of humic acids and Fe-EDDHA complex to lemon trees enhanced Fe uptake and bioavailable Fe content (Sánchez-Sánchez et al., 2002). Humic acids are effective in the solubilization of very insoluble Fe phosphate minerals, which is attributed to the carboxylic and phenolic functional groups and high redox potential of humic acids (Yang et al., 2019b). Fuentes et al. (2013) evinced the involvement of carboxylic acids present in aliphatic domains in the Fe (III)-humic acid complexation. Humic acids can enhance both P and Fe nutrition due to the arrangement of humates with phosphates via Fe linkages, as well as by the complex formation between Fe molecule and ligands released from plant roots (Urrutia et al., 2014b).

Being redox reactive, humic acids induce chemical reduction of metal ions such as Fe^{3+} (Struyk and Sposito, 2001), favoured at acidic pH (< 4) while higher pH values limit this chemical reduction due to the formation of complexes between Fe^{3+} and humic molecules. The capability of humic materials to form chelates and enhance the bioavailable Fe pool depends on their solubility potential and molecular structure. Bocanegra et al. (2005) reported an increase in Fe nutrition and improved translocation from root to leaves of sunflower plants (*Helianthus annus* L.), under 15-days co-application of small-size humic acids (10,000; < 10 kDa) at the rate of 50–100 mg L^{-1} and EDTA. Water-soluble humic materials were reported to facilitate the rapid allocation of Fe into leaf veins and interveinal cells of cucumber plants (Zanin et al., 2015).

The metal ligands released from roots (rhizo-deposits) also tend to form a complex with Fe^{3+}, which may render it bioavailable for plants through various Fe uptake mechanisms. However, large-size humic substances can control Fe availability by the formation of stable amorphous Fe oxides (Schwertmann, 1991). These amorphous structures are formed after co-precipitation with insoluble humic material and constitutes a Fe pool, potentially available for ligand attachment (Colombo et al., 2012). The formation of a complex between Fe and insoluble humic matter is effective in alleviating the Fe-chlorosis disorder (under Fe-deficient conditions) in plants. Elena et al. (2009) investigated the effects of humic acids applied at variable rates (2, 5, 100, and 250 mg C L^{-1}) and Fe application (40 μM) as Fe-EDTA on the roots of cucumber (*Cucumis sativus* L.) and found upregulated activities of Fe (III)-chelate reductase and PM-ATPase. According to Tomasi et al. (2013), applying soluble humic materials with Fe (1 μM Fe; 5 mg C L^{-1}) improved the allocation of nutrients to leaf blade in Fe-deficient plants. This was demonstrated by the increased mobilization of potassium (K), copper (Cu), and zinc (Zn) from the vascular system as well as calcium (Ca) and manganese (Mn) from trichomes to the leaf blade, in addition to increasing the root NO_3-N uptake and CO_2 assimilation. The soluble Fe and humic matter complexes induced an upregulation of Fe(III)-chelate reductase and Fe transporter genes (LeIRT1 and LeIRT2) in roots of Fe-deficient tomato plants (Tomasi et al., 2013). Zamboni et al. (2016) reported that rapid response of Fe-deficient plants to Fe application was strongly dependent on the type of chelating agent. The applications of Fe-citrate and Fe-Phytosiderophore (PS) downregulated the genes and the molecular mechanisms induced by Fe deficiency, but the relevant changes in the root transcriptome were not determined by the co-application of Fe and soluble humic material (Zanin et al., 2019).

Cieschi and Lucena (2018) reported the accumulation of humic materials on soybean roots under co-application with Fe, which affected the Fe translocation from root to shoot. Humic matter may induce the different physiological mechanisms that modulated the Fe acquisition at the transcriptional and post-transcriptional levels, which signals the multifaceted role of humic matter in plant Fe nutrition (Zanin et al., 2019). Both solid and soluble humic materials were manifested to facilitate Fe (III)-oxide reduction in the sediments (Roden et al., 2010) as well as their bio-reduction in soils by the interchanging of electrons (e–) from bacteria to oxide surfaces (Zanin et al., 2019). The nature of the soil, the pH range (5–9 pH), and a low metal:humic substances (HS) ratio tends to favour the stability, while alkaline pH and a low metal:HS ratio extends the solubility. Thus, the plants grown under Fe-deficient calcareous soil conditions may gain other benefits from the stable and soluble complexation of Fe-humic complexes (Cieschi and Lucena, 2018).

8.4 EFFECT OF HUMIC ACIDS ON SOIL BIOLOGY

The rising trend towards the use of humic products in the agriculture sector has resultantly motivated the researchers to investigate their effects on soil biology. The effect of humic materials on soil micro-organisms is quite complex and highly debated among the scientific community. High-dose application of humic products to soil can induce peculiar shifts in soil microbial structure. Negative effects on the microbial community under humic acid additions can likely be imputed to its dose-toxicity and hydrophobicity aspects, with potential implications on environmental quality and safety (Guo et al., 2019). Humic substances can likely intensify the prolonged access to nutrients for soil microbes due to their slow-rate decomposition. Humic acids act as a growth stimulator for soil microbes and modulate cellular metabolic processes (Tikhonov et al., 2010). Peat-originated humic substances invigorated the population of hydrocarbon-oxidizing *Acinetobacter*-species bacteria (Tikhonov et al., 2010) and influenced C and N cycling processes, which may be attributed to the redox-active functional groups of humic materials that impart electrons (e–) for bacterial respiration due to rapid oxidation, leading to substantial variations in CO_2 and CH_4 emissions (Coates et al., 2002). Humic acids of low molecular weight tend to be relatively more bioactive than their high-molecular weight humic acids (Nardi et al., 2007). Humic substances can enhance microbial physiological processes and respiratory activities, stimulate soil microbes, and accelerate the nutrients' cycling in soil (Jillani and Ahmad, 1994). The humic products can also induce indirect effects on nutrients' availability after their subsequent transformation by soil microbes (Filip and Kubát, 2001).

Extracellular enzymes are critically involved in the degradation and mineralization of organic substrates (Nannipieri et al., 1996). Substrate attachment (humic materials) on the active sites of these enzyme entities, either via covalent or complex bondage, can explain such interactions (Verma et al., 1975). High-dose additions of humic acids resulted in a notable decline in extracellular enzyme activities in Hawaiian soil that was sterilized via irradiation, presumably by blocking the active sites mediated by humic acids and enzyme linkage (Allison, 2006). The effects of humic substances on microbial indicators like biodiversity and metabolic activities can be dose specific, as it was reported by Gorlenko et al. (2012) that a positive link between microbial indicators and humic acids at low rate was existed, while the CO_2 emissions were unaffected by the application rates during the 45-days. Humic particles are liable to adhere with soil minerals such as clay and calcium carbonates, which likely results in the development of small porous structure in soil, being considered ideal for microbial growth (Larcher, 2003).

8.5 SOIL QUALITY AND CROP YIELD

The addition of numerous organic amendments into soil supports crop growth especially at matured stages through stimulating the hormones involved in plant growth regulation (e.g., cytokinin, auxins, and gibberellins) (Quilty and Cattle, 2011), promotes the crop productivity of different field crops (Ulukan, 2008) and fertilizer-use efficiency (Pollhamer, 1993) and ameliorates soil health. Being heterogenous materials, humic materials include both hydrophilic and hydrophobic fractions in their macro-molecular structure. Hydrophilic groups act to attract water molecules, improving the water retention capacity of soil (Stevenson, 1994). Arancon et al. (2006) reported that vermicompost-derived humic acids increased the growth of marigold, pepper, and strawberry plants, and naturally derived humic acids enhanced the fruit yield of pepper plants compared with commercial humic acids. Tuefenkci et al. (2006) observed that humic acids addition at 500 to 2000 mg kg^{-1} positively enhanced the lettuce growth and nutrient concentrations. Fernandez-Escobar et al. (1996) reported that humic acid additions enhanced olive shoot growth and nutrients translocation in leaves (K, B, Mg, Ca, and Fe). A meta-analytic study by Verlinden et al. (2009) reported an overall effect on the dry matter yield and nutrient uptake (N, P, K, and Mg) with the application of humic acids, while there was no effect observed on the Na and Ca uptake.

8.6 HUMIC SUBSTANCES CONCERNED WITH CARBON SEQUESTRATION

Intensive farming practices contribute significantly to enhance the greenhouse gases emissions (notably CO_2 emissions) from the soil, which escalates the warming of earth's atmosphere (Paustian et al., 1997). This phenomenon of CO_2 transfer may be ceased with the enhancement of microbially-stable humic matter in soil. A major portion of C input to soil with plant litter additions is returned to the atmosphere when soil microbes degrade the added litter, while only a small fraction of this is integrated into soil humus (Stockmann et al., 2013). Humified organic matter indeed typifies the most stable and constant pool of soil organic carbon (SOC), with an average residing period of centuries (Piccolo, 1996). Humic substances tend to build up SOC content in soil (Song et al., 2014).

The research on SOC dynamics shows the gradual increase of alkyl organic groups into stable organic fractions, which are mainly derived from plant residue decomposition and microbial processing (Almendros et al., 1996). The plant-derived hydrophilic groups are usually sequestered in the hydrophobic domains of humus and got secured from further degradation (Spaccini et al., 2000). Spaccini et al. (2002) reported that humic matter may attenuate the SOM mineralization through promoting the SOC sequestration in its hydrophobic portions and limiting the rapid microbial processing of readily degradable organic materials. The addition of humic matter to soil enhanced the SOC sequestration by attenuating the decomposition of the labile aliphatic alcohol groups, indicating that hydrophobicity of humic matter is directly related to SOC sequestration in soil (Spaccini et al., 2002). It seems plausible that advanced soil management practices on enhanced SOC sequestration should focus on enhancing the hydrophobicity of SOM, which may likely be enhanced by soil humification through humic acid additions and the CO_2 release from soil may be mitigated and the soil be made a progressive sink of SOC (Spaccini et al., 2002). Song et al. (2014) reported that continuous manure additions into soil significantly enhanced the SOC sequestration in humic substances, notably in humin, which can be linked to the molecular changes in humic material structure. The addition of manure was found to enhance the aromaticity (recalcitrancy) of humic matter (Ferrari et al., 2011).

8.7 CONCLUSION

Humic acids are the complex heterogenous materials existing in soil, which are derived from a series of microbial processing on SOM and are resistant to further microbial decay. Due to their distinct nature, they can render numerous benefits to enhance soil fertility. Humic substances can induce hormonal changes and enhance root development as well as the outgrowth of root hairs. In addition to promoting root growth, these substances can promote PM-ATPase activity levels, which in turn enhances the nutrient uptake in response to higher H^+ release from the root cells. Root growth under humic acid additions can also be related to the development of soil acidity, which acidifies the root apoplast and promotes root elongation, which helps the plants to acquire nutrients from soil. Numerous benefits of SOM are related to the presence of humic substances, due in part to their highly chemical activity, and enhance their capability to bind nutrients and ensure their bioavailability throughout plant growth cycle. Humic acids also help to maintain the soil's structural stability and aggregation. These materials are also characterized by the ability to form specialized chelates with micronutrients, especially with Fe, and enhance its uptake by the plants. These characteristics of humic substances are associated with the presence of different functional groups and moieties in their structure. Being highly bioactive in nature, humic matter can induce substantial changes in the soil microbial community and functionalities, which in turn lead to enhanced nutrients mobility and availability in soil. To sum up, humic acids are the advanced biomaterials, and their utilization is gradually increasing in the agriculture sector. The usage of humic acids in the agriculture sector can be an eco-friendly way of advancing food security and enhancing the efficiency of different synthetic fertilizers.

REFERENCES

Albiach, R., Canet, R., Pomares, F., et al. (2001), 'Organic matter components and aggregate stability after the application of different amendments to a horticultural soil', *Bioresource Technology* 76(2), 125–129.

Allison, S.D. (2006), 'Soil minerals and humic acids alter enzyme stability: Implications for ecosystem processes', *Biogeochemistry* 81(3), 361–373.

Almendros, G., Guadalix, M.E., González-Vila, F.J., et al. (1996), 'Preservation of aliphatic macromolecules in soil humins', *Organic Geochemistry* 24(6), 651–659.

Arancon, N.Q., Edwards, C.A., Lee, S., et al. (2006), 'Effects of humic acids from vermicomposts on plant growth', *European Journal of Soil Biology* 42 S65-S69.

Atiyeh, R., Lee, S., Edwards, C., et al. (2002), 'The influence of humic acids derived from earthworm-processed organic wastes on plant growth', *Bioresource Technology* 84(1), 7–14.

Benedetti, M., Van Riemsdijk, W. and Koopal, L. (1996), 'Humic substances considered as a heterogeneous Donnan gel phase', *Environmental Science & Technology* 30(6), 1805–1813.

Benková, E., Michniewicz, M., Sauer, M., et al. (2003), 'Local, efflux-dependent auxin gradients as a common module for plant organ formation', *Cell* 115(5), 591–602.

Bocanegra, M., Lobartini, J. and Orioli, G. (2005), 'Iron-humate as a source of iron for plants', *Communications in Soil Science and Plant Analysis* 35(17–18), 2567–2576.

Brannon, C.A. and Sommers, L.E. (1985), 'Preparation and characterization of model humic polymers containing organic phosphorus', *Soil Biology and Biochemistry* 17(2), 213–219.

Calvo, P., Nelson, L. and Kloepper, J.W. (2014), 'Agricultural uses of plant biostimulants', *Plant and Soil* 383(1), 3–41.

Canellas, L., Teixeira Junior, L., Dobbss, L., et al. (2008), 'Humic acids crossinteractions with root and organic acids', *Annals of Applied Biology* 153(2), 157–166.

Canellas, L.P., da Silva, S.F., Olk, D.C., et al. (2015), 'Foliar application of plant growth-promoting bacteria and humic acid increase maize yields', *Journal of Food, Agriculture & Environment* 13(1), 131–138.
Canellas, L.P. and Olivares, F.L. (2014), 'Physiological responses to humic substances as plant growth promoter', *Chemical and Biological Technologies in Agriculture* 1(1), 1–11.
Chen, Y., Clapp, C. and Magen, H. (2004), 'Mechanisms of plant growth stimulation by humic substances: The role of organo-iron complexes', *Soil Science and Plant Nutrition* 50(7), 1089–1095.
Cieschi, M.T. and Lucena, J.J. (2018), 'Iron and humic acid accumulation on soybean roots fertilized with leonardite iron humates under calcareous conditions', *Journal of Agricultural and Food Chemistry* 66(51), 13386–13396.
Clemente, R. and Bernal, M.P. (2006), 'Fractionation of heavy metals and distribution of organic carbon in two contaminated soils amended with humic acids', *Chemosphere* 64(8), 1264–1273.
Coates, J.D., Cole, K.A., Chakraborty, R., et al. (2002), 'Diversity and ubiquity of bacteria capable of utilizing humic substances as electron donors for anaerobic respiration', *Applied and Environmental Microbiology* 68(5), 2445–2452.
Colombo, C., Palumbo, G., He, J.Z., et al. (2014), 'Review on iron availability in soil: Interaction of Fe minerals, plants, and microbes', *Journal of Soils and Sediments* 14(3), 538–548.
Colombo, C., Palumbo, G., Sellitto, V.M., et al. (2012), 'Characteristics of insoluble, high molecular weight iron-humic substances used as plant iron sources', *Soil Science Society of America Journal* 76(4), 1246–1256.
DiDonato, N., Chen, H., Waggoner, D., et al. (2016), 'Potential origin and formation for molecular components of humic acids in soils', *Geochimica et Cosmochimica Acta* 178 210–222.
Elena, A., Diane, L., Eva, B., et al. (2009), 'The root application of a purified leonardite humic acid modifies the transcriptional regulation of the main physiological root responses to Fe deficiency in Fe-sufficient cucumber plants', *Plant Physiology and Biochemistry* 47(3), 215–223.
Fahramand, M., Moradi, H., Noori, M., et al. (2014), 'Influence of humic acid on increase yield of plants and soil properties', *International Journal of Farming and Allied Sciences* 3(3), 339–341.
Fernandez-Escobar, R., Benlloch, M., Barranco, D., et al. (1996), 'Response of olive trees to foliar application of humic substances extracted from leonardite', *Scientia Horticulturae* 66(3–4), 191–200.
Ferrari, E., Francioso, O., Nardi, S., et al. (2011), 'DRIFT and HR MAS NMR characterization of humic substances from a soil treated with different organic and mineral fertilizers', *Journal of Molecular Structure* 998(1), 216–224.
Filip, Z. and Kubát, J. (2001), 'Microbial utilization and transformation of humic substances extracted from soils of long-term field experiments', *European Journal of Soil Biology* 37(3), 167–174.
Fuentes, M., Olaetxea, M., Baigorri, R., et al. (2013), 'Main binding sites involved in Fe (III) and Cu (II) complexation in humic-based structures', *Journal of Geochemical Exploration* 129, 14–17.
Glaser, B., Lehmann, J. and Zech, W. (2002), 'Ameliorating physical and chemical properties of highly weathered soils in the tropics with charcoal–A review', *Biology and Fertility of Soils* 35(4), 219–230.
Gorlenko, M., Yakimenko, O., Golichenkov, M., et al. (2012), 'Functional biodiversity of soil microbe colonies affected by organic substrates of different kinds', *Moscow University Soil Science Bulletin* 67(2), 71–78.
Greger, H. (1984), 'Alkamides: Structural Relationships, Distribution and Biological Activity1', *Planta Medica* 50(05), 366–375.
Guggenberger, G. (2005), 'Humification and mineralization in soils'. In *Microorganisms in soils: Roles in genesis and functions*, Springer, pp. 85–106.
Guo, C., Zhang, C., Sun, Z., et al. (2019), 'Synergistic impact of humic acid on the photo-reductive decomposition of perfluorooctanoic acid', *Chemical Engineering Journal* 360, 1101–1110.
Haynes, R. and Swift, R. (1990), 'Stability of soil aggregates in relation to organic constituents and soil water content', *Journal of Soil Science* 41(1), 73–83.
Hiemstra, T., Mia, S., Duhaut, P.-B., et al. (2013), 'Natural and pyrogenic humic acids at goethite and natural oxide surfaces interacting with phosphate', *Environmental Science & Technology* 47(16), 9182–9189.
Imbufe, A.U., Patti, A.F., Burrow, D., et al. (2005), 'Effects of potassium humate on aggregate stability of two soils from Victoria, Australia', *Geoderma* 125(3–4), 321–330.
Jillani, G. and Ahmad, R. (1994), *Substitution of organic manures with humic substances to enhance microbial activity*. National Congress on Soil Science NWFP, Agricultural University Peshawar.
Khaled, H. and Fawy, H.A. (2011), 'Effect of different levels of humic acids on the nutrient content, plant growth, and soil properties under conditions of salinity', *Soil and Water Research* 6(1), 21–29.

Larcher, W. (2003), *Physiological plant ecology: Ecophysiology and stress physiology of functional groups*, Springer Science & Business Media.
MacCarthy, P. (2001), 'The principles of humic substances', *Soil Science* 166(11), 738–751.
Misra, V., Tiwari, A., Shukla, B., et al. (2009), 'Effects of soil amendments on the bioavailability of heavy metals from zinc mine tailings', *Environmental Monitoring and Assessment* 155(1), 467–475.
Mora, V., Bacaicoa, E., Zamarreno, A.M., et al. (2010), 'Action of humic acid on promotion of cucumber shoot growth involves nitrate-related changes associated with the root-to-shoot distribution of cytokinins, polyamines and mineral nutrients', *Journal of Plant Physiology* 167(8), 633–642.
Mora, V., Baigorri, R., Bacaicoa, E., et al. (2012), 'The humic acid-induced changes in the root concentration of nitric oxide, IAA and ethylene do not explain the changes in root architecture caused by humic acid in cucumber', *Environmental and Experimental Botany* 76 24–32.
Morsomme, P. and Boutry, M. (2000), 'The plant plasma membrane H+-ATPase: Structure, function and regulation', *Biochimica et Biophysica Acta (BBA)-Biomembranes* 1465(1–2), 1–16.
Morsomme, P., Slayman, C.W. and Goffeau, A. (2000), 'Mutagenic study of the structure, function and biogenesis of the yeast plasma membrane H+-ATPase', *Biochimica et Biophysica Acta (BBA)-Reviews on Biomembranes* 1469(3), 133–157.
Mylonas, V. and McCants, C. (1980), 'Effects of humic and fulvic acids on growth of tobacco', *Plant and Soil* 54(3), 485–490.
Nannipieri, P., Sequi, P. and Fusi, P. (1996), 'Humus and enzyme activity'. In *Humic substances in terrestrial ecosystems*, Elsevier, pp. 293–328.
Nardi, S., Muscolo, A., Vaccaro, S., et al. (2007), 'Relationship between molecular characteristics of soil humic fractions and glycolytic pathway and Krebs cycle in maize seedlings', *Soil Biology and Biochemistry* 39(12), 3138–3146.
Nardi, S., Pizzeghello, D., Muscolo, A., et al. (2002), 'Physiological effects of humic substances on higher plants', *Soil Biology and Biochemistry* 34(11), 1527–1536.
Nisar, A. and Mir, S. (1989), 'Lignitic coal utilization in the form of HA as fertilizer and soil conditioner', *Sci. Tech. Develop* 8(1), 23–26.
Olk, D.C., Dinnes, D.L., Rene Scoresby, J., et al. (2018), 'Humic products in agriculture: potential benefits and research challenges—A review', *Journal of Soils and Sediments* 18(8), 2881–2891.
Omoarelojie, L.O., Kulkarni, M.G., Finnie, J.F., et al. (2021), 'Modes of action of biostimulants in plants'. In *Biostimulants for crops from seed germination to plant development*, Elsevier, pp. 445–459.
Palmgren, M.G. (2001), 'Plant plasma membrane H+-ATPases: Powerhouses for nutrient uptake', *Annual Review of Plant Biology* 52(1), 817–845.
Paustian, K., Andren, O., Janzen, H., et al. (1997), 'Agricultural soils as a sink to mitigate CO2 emissions', *Soil Use and Management* 13, 230–244.
Petronio, P., Vitorovic, D. and Jablanovic, M. (1982), 'Investigations of the biological effect of HA', *Acta Biologie et Medicinae Experimentalis* 7(1), 21–25.
Piccolo, A. (1996), 'Chapter 5 – Humus and soil conservation'. In A. Piccolo (Ed), *Humic substances in terrestrial ecosystems*, Elsevier Science B.V, pp. 225–264.
Piccolo, A., Conte, P., Trivellone, E., et al. (2002), 'Reduced heterogeneity of a lignite humic acid by preparative HPSEC following interaction with an organic acid. Characterization of size-separates by Pyr-GC-MS and 1H-NMR spectroscopy', *Environmental Science & Technology* 36(1), 76–84.
Pollhamer, E. (1993), 'Effect of humic acid, fulvic acid and NPK fertilizers on the quality of winter wheat varieties on chemical-free soil', *Novenytermeles (Hungary)*. https://agris.fao.org/agris-search-Pukalchik, M., Kydralieva, K., Yakimenko, O., et al. (2019), 'Outlining the potential role of humic products in modifying biological properties of the soil—A review', *Frontiers in Environmental Science* 7, 80.
Quan-Xian, H., Jian-Yun, L., Jian-Min, Z., et al. (2008), 'Enhancement of phosphorus solubility by humic substances in ferrosols', *Pedosphere* 18(4), 533–538.
Quilty, J. and Cattle, S. (2011), 'Use and understanding of organic amendments in Australian agriculture: A review', *Soil Research* 49(1), 1–26.
Ramírez-Chávez, E., López-Bucio, J., Herrera-Estrella, L., et al. (2004), 'Alkamides isolated from plants promote growth and alter root development in Arabidopsis', *Plant Physiology* 134(3), 1058–1068.
Ramos, A.C., Dobbss, L.B., Santos, L.A., et al. (2015), 'Humic matter elicits proton and calcium fluxes and signaling dependent on Ca2+-dependent protein kinase (CDPK) at early stages of lateral plant root development', *Chemical and Biological Technologies in Agriculture* 2(1), 1–12.
Roden, E.E., Kappler, A., Bauer, I., et al. (2010), 'Extracellular electron transfer through microbial reduction of solid-phase humic substances', *Nature Geoscience* 3(6), 417–421.

Rose, M.T., Patti, A.F., Little, K.R., et al. (2014), 'A meta-analysis and review of plant-growth response to humic substances: practical implications for agriculture'. In *Advances in agronomy*, Elsevier, pp. 37–89.

Sánchez-Sánchez, A., Sánchez-Andreu, J., Juárez, M., et al. (2002), 'Humic substances and amino acids improve effectiveness of chelate feeddha in lemon trees', *Journal of Plant Nutrition* 25(11), 2433–2442.

Sattelmacher, B., Gerendas, J., Thoms, K., et al. (1993), 'Interaction between root growth and mineral nutrition', *Environmental and Experimental Botany* 33(1), 63–73.

Schwertmann, U. (1991), 'Solubility and dissolution of iron oxides', *Plant and Soil* 130(1), 1–25.

Shah, Z.H., Rehman, H.M., Akhtar, T., et al. (2018), 'Humic substances: Determining potential molecular regulatory processes in plants', *Frontiers in Plant Science* 9 263.

Sharif, M., Khattak, R.A. and Sarir, M. (2002), 'Effect of different levels of lignitic coal derived humic acid on growth of maize plants', *Communications in Soil Science and Plant Analysis* 33(19–20), 3567–3580.

Sollins, P., Swanston, C. and Kramer, M. (2007), 'Stabilization and destabilization of soil organic matter—A new focus', *Biogeochemistry* 85(1), 1–7.

Song, X.Y., Liu, S.T., Liu, Q.H., et al. (2014), 'Carbon sequestration in soil humic substances under long-term fertilization in a wheat-maize system from North China', *Journal of Integrative Agriculture* 13(3), 562–569.

Spaccini, R., Piccolo, A., Conte, P., et al. (2002), 'Increased soil organic carbon sequestration through hydrophobic protection by humic substances', *Soil Biology and Biochemistry* 34(12), 1839–1851.

Spaccini, R., Piccolo, A., Haberhauer, G., et al. (2000), 'Transformation of organic matter from maize residues into labile and humic fractions of three European soils as revealed by 13C distribution and CPMAS-NMR spectra', *European Journal of Soil Science* 51(4), 583–594.

Spark, K., Wells, J. and Johnson, B. (1997a), 'Characteristics of the sorption of humic acid by soil minerals', *Soil Research* 35(1), 103–112.

Spark, K., Wells, J. and Johnson, B. (1997b), 'Sorption of heavy metals by mineral-humic acid substrates', *Soil Research* 35(1), 113–122.

Stevenson, F.J. (1994), *Humus chemistry: Genesis, composition, reactions*, John Wiley & Sons.

Stockmann, U., Adams, M.A., Crawford, J.W., et al. (2013), 'The knowns, known unknowns and unknowns of sequestration of soil organic carbon', *Agriculture, Ecosystems & Environment* 164, 80–99.

Struyk, Z. and Sposito, G. (2001), 'Redox properties of standard humic acids', *Geoderma* 102(3–4), 329–346.

Tikhonov, V., Yakushev, A., Zavgorodnyaya, Y.A., et al. (2010), 'Effects of humic acids on the growth of bacteria', *Eurasian Soil Science* 43(3), 305–313.

Tipping, E. (2002), *Cation binding by humic substances*, Cambridge University Press.

Tomasi, N., De Nobili, M., Gottardi, S., et al. (2013), 'Physiological and molecular characterization of Fe acquisition by tomato plants from natural Fe complexes', *Biology and Fertility of Soils* 49(2), 187–200.

Trevisan, S., Botton, A., Vaccaro, S., et al. (2011), 'Humic substances affect Arabidopsis physiology by altering the expression of genes involved in primary metabolism, growth and development', *Environmental and Experimental Botany* 74, 45–55.

Tuefenkci, S., Türkmen, Ö., Sönmez, F., et al. (2006), 'Effects of humic acid doses and application times on the plant growth, nutrient and heavy metal contents of lettuce grown on sewage sludge-applied soils', *Fresenius Environmental Bulletin* 15(4), 295–300.

Ulukan, H. (2008), 'Effect of soil applied humic acid at different sowing times on some yield components in wheat (Triticum spp.) hybrids', *International Journal of Botany* 4(2), 164–175.

Urrutia, O., Erro, J., Guardado, I., et al. (2014), 'Physico-chemical characterization of humic-metal-phosphate complexes and their potential application to the manufacture of new types of phosphate-based fertilizers', *Journal of Plant Nutrition and Soil Science* 177(2), 128–136.

Van Oosten, M.J., Pepe, O., De Pascale, S., et al. (2017), 'The role of biostimulants and bioeffectors as alleviators of abiotic stress in crop plants', *Chemical and Biological Technologies in Agriculture* 4(1), 1–12.

Verlinden, G., Pycke, B., Mertens, J., et al. (2009), 'Application of humic substances results in consistent increases in crop yield and nutrient uptake', *Journal of Plant Nutrition* 32(9), 1407–1426.

Verma, L., Martin, J. and Haider, K. (1975), 'Decomposition of carbon-14-labeled proteins, peptides, and amino acids; free and complexed with humic polymers', *Soil Science Society of America Journal* 39(2), 279–284.

Wang, Y., Qin, J., Zhou, S., et al. (2015), 'Identification of the function of extracellular polymeric substances (EPS) in denitrifying phosphorus removal sludge in the presence of copper ion', *Water Research* 73 252–264.

Yamaguchi, T., Takei, T., Yazawa, Y., et al. (2004), 'Effect of humic acid, sodium, and calcium additions on the formation of water-stable aggregates in Western Australian wheatbelt soils', *Soil Research* 42(4), 435–439.

Yang, F., Zhang, S., Song, J., et al. (2019a), 'Inside back cover: Synthetic humic acids solubilize otherwise insoluble phosphates to improve soil fertility (Angew. Chem. Int. Ed. 52/2019)', *Angewandte Chemie International Edition* 58(52), 19113–19113.

Yang, F., Zhang, S., Song, J., et al. (2019b), 'Synthetic humic acids solubilize otherwise insoluble phosphates to improve soil fertility', *Angewandte Chemie* 131(52), 18989–18992.

Yilmaz, E. (2011), 'Effects of different sources of organic matter on some soil fertility properties: A laboratory study on a Lithic Rhodoxeralf from Turkey', *Communications in Soil Science and Plant Analysis* 42(8), 962–970.

Zamboni, A., Zanin, L., Tomasi, N., et al. (2016), 'Early transcriptomic response to Fe supply in Fe-deficient tomato plants is strongly influenced by the nature of the chelating agent', *BMC Genomics* 17(1), 1–17.

Zandonadi, D.B., Matos, C.R.R., Castro, R.N., et al. (2019), 'Alkamides: a new class of plant growth regulators linked to humic acid bioactivity', *Chemical and Biological Technologies in Agriculture* 6(1), 1–12.

Zandonadi, D.B., Santos, M.P., Busato, J.G., et al. (2013), 'Plant physiology as affected by humified organic matter', *Theoretical and Experimental Plant Physiology* 25 13–25.

Zandonadi, D.B., Santos, M.P., Dobbss, L.B., et al. (2010), 'Nitric oxide mediates humic acids-induced root development and plasma membrane H+-ATPase activation', *Planta* 231(5), 1025–1036.

Zanin, L., Tomasi, N., Cesco, S., et al. (2019), 'Humic substances contribute to plant iron nutrition acting as chelators and biostimulants', *Frontiers in Plant Science* 10, 675.

Zanin, L., Tomasi, N., Rizzardo, C., et al. (2015), 'Iron allocation in leaves of Fe-deficient cucumber plants fed with natural Fe complexes', *Physiologia Plantarum* 154(1), 82–94.

Zanin, L., Tomasi, N., Zamboni, A., et al. (2018), 'Water-extractable humic substances speed up transcriptional response of maize roots to nitrate', *Environmental and Experimental Botany* 147 167–178.

Zhang, S., Du, Q., Cheng, K., et al. (2020), 'Efficient phosphorus recycling and heavy metal removal from wastewater sludge by a novel hydrothermal humification-technique', *Chemical Engineering Journal* 394, 124832.

9 Alkamides in Root Development

Atish Sardar

CONTENTS

9.0 INTRODUCTION

Plants are sessile organisms. The plant body of higher plants is differentiated into root, stem, and leaves. The stem and leaves, along with the reproductive structures, comprise the aerial portion, whereas the underground portion involves the root system. The aerial parts have been extensively studied; however, the root system are studied less due to the difficulty of observing them during the plant life cycle. Nowadays, with the emergence of modern imaging techniques, it has become possible to monitor real-time growth, development, and movement of roots inside the soil. Although the importance of the root system for nutrient and water uptake, anchorage, and mechanical support is well known, plant breeders have traditionally ignored the potential of the root system for increasing productivity, and it was not an important criterion of selection during crop development programs of the 1960s' green revolution (Waines and Ehdaie, 2007). Later, it was realized that the root system is essential for plant biomass and yield. Since then, researchers have focused on understanding the biology of the root system, and the term *root system architecture* was brought into prominence. Root system architecture (RSA) broadly comprises the primary root (PR), lateral root (LR), and accessory roots such as adventitious root (AR). But deep insights into the topic reveal that RSA also involves PR length and number, length and spread of LR, and the length and spread of root hairs (RH) present on the PR and LR (Smith and Smet, 2012). Based on extensive studies done on the model plant *Arabidopsis thaliana* and several other cereal plants belonging to dicots and monocots, RSA can be divided into two types based on their origin— roots emerging from embryonically derived meristematic tissue, such as the primary or tap root in dicots (Hochholdinger, 2009), and primary and seminal roots in monocots (Lynch and Brown, 2012): roots emerging post-embryonically either from the pericycle of PR, such as LR (Beeckman et. al., 2001), or from consecutive nodes of the stem, such as AR (Bellini et al., 2014). PR emerges from the seed after germination and grows into the soil. A transverse section through the PR shows that it consists of a layer of epidermis, followed by the cortex and then the endodermis. Next to the endodermis is the pericycle, followed by the central vascular strand, consisting of xylem and phloem; pith is absent in roots. The

DOI: 10.1201/9781003224525-12

growth of PR is caused by cell divisions at the root apical meristem, followed by subsequent cell elongation and dedifferentiation. A zone of undifferentiated cells is present within the root apical meristem, known as the quiescent centre. The quiescent cells are surrounded by initial cells, which may differentiate into specific cell types and give rise to the radial pattern of cell layer organization (Garay-Arroyo et al., 2012). The LR arises from lateral root primordium (LRP), formed in the pericycle cells of the PR due to the reactivation of cell cycle. This process is called priming. Actually, the pericycle cells adjacent to the xylem pole act as founder cells, and following priming, these founder cells undergo repeated divisions, giving rise to a dome-shaped LRP. At later stages of development, *de novo* creation of a new apical meristem within the emerging LR takes place (Dubrovsky et al., 2001; Smith and Smet, 2012). Interestingly, ARs emerge from AR founder cells, whose cellular origin is uncertain owing to their non-root origin and often arise from cells associated with stelar or cambial tissues (Bellini et al., 2014). Studies conducted in *Arabidopsis* reveal that ARs can either originate on etiolated hypocotyls upon transfer to light or from a cell layer resembling the pericycle founder cells of PR (Li et al., 2009, Sorin et al., 2005). Auxin plays important roles in the growth and maintenance of PR as well as in the formation and growth of LR and ARs. Endogenous auxin levels regulate cell division (at high concentration), cell elongation (at low concentration), positive geotropic growth, and LR and AR formation (Perrot-Rechenmann, 2010; De Smet, 2012). Other phytohormones like cytokinin, abscisic acid, gibberellin, ethylene, brassinosteroid, and jasmonic acid also regulate PR growth, LR and AR formation, and growth through crosstalk with auxin and transcriptional reprogramming (Liu et al., 2014, Nibau et al., 2008). Several small molecules of non-hormone origin, such as nitric oxide, strigolactones, karrikins, and alkamides, are reported to have regulatory effects on RSA.

Historically, alkamides are used as spices for their tingling pungency, distinct from those of phenolic compounds like capsaicin and piperine found in *Capsicum annuum* and *Piper nigrum*, respectively. Alkamides are a group of small, structurally diverse, bioactive molecules obtained naturally from fungi, animals, and several plant families. Presently, reports suggest that alkamides are found in more than 25 families in the plant kingdom. These families include Asteraceae, Brassicaceae, Rutaceae, Piperaceae, Menispermaceae, Aristolochiaceae, Convolvulaceae, Euphorbiaceae, Poaceae, and Solanaceae (Méndez-Bravo et al., 2011). Among these, alkamides are found mostly in the members of the genera *Acmella* (*Spilanthes*), *Heliopsis*, *Anacyclus* and *Echinacea* of Asteraceae, *Ruta*, *Dictamnus* and *Zanthoxylum* of Rutaceae, *Piper* of Piperaceae, *Capsicum* of Solanaceae, and *Lepidium* of Brassicaceae. Alkamides can be attributed to a wide variety of biological activities, such as high insecticidal toxicity, antifungal, antibacterial, antiprotozoal, molluscicidal, quorum-sensing blockage, cercaricidal, and acaricidal activities (Harald Greger, 2016). Additionally, they are important for pharmaceutical and cosmetics industries, owing to their anti-inflammatory, anesthetic, analgesic, wrinkle-smoothing, and anti-aging properties and are responsible for cannabimimetic and immunomodulatory effects. Alkamides are also reported to play an important role in various physiological processes that involve plant growth and development. Their effects can be observed on almost all organs of a plant, depending on the nature and dosage of the alkamide and the developmental stage of that particular plant. In this chapter, we discuss the effects of various alkamides on root development, at both the organ and cellular levels, with special emphasis on their interaction with different signaling molecules, already known to be involved during root development.

9.1 ALKAMIDES PROMOTE PLANT GROWTH AND ALTER ROOT DEVELOPMENT

N-Alkylamides or alkamides are low-molecular-weight secondary metabolites comprising straight-chain, N-substituted, α-unsaturated aliphatic acids linked to amines by amide linkages. Basically, alkamides are a group of compounds comprising more than 200 fatty acyl amides, derived from the condensation of α-unsaturated fatty acid and an amine. These compounds share structural similarities with N-acylethanolamines (NAEs) and N-acyl-L-homoserine lactones (AHLs)

(Buitimea-Cantúa et al., 2020). NAEs are bioactive molecules, comprising fatty acids linked to amino alcohols by amide linkages and are reported to regulate diverse physiological processes in plants and animals. AHLs, on the other hand, are small molecules containing a conserved HL-ring and an N-linked acyl side chain, which can readily diffuse through the bacterial plasma membrane and cell wall to participate in cell-to-cell signaling, also known as "quorum sensing", in Gram-negative bacteria (Faure et al., 2015). Researchers have found that alkamides, NAEs, and AHLs can act as small lipid signal molecules to modulate the genetic mechanisms involved in signal perception and to alter RSA in *Arabidopsis thaliana* in a dose-dependent manner (Morquecho-Contreras et al., 2010). The first report on the effect of alkamides on plant growth came in the early '90s, when amidenin (2E,4Z-decadienamide), an alkamide isolated from the actinomycete fungus *Amycolatopsis* sp., was able to stimulate the growth of rice seedlings at lower concentrations and to inhibit growth at higher concentrations (Figure 9.1). However, back then, very little information was available about the effect of plant-based alkamides on RSA (Kanbe et al., 1993). A decade later, it was found that plant-based alkamides, affinin (*N*-isobutyl-2*E*,6*Z*,8*E*-decatrienamide) and its reduced derivatives, *N*-isobutyl-2*E*-decenamide and *N*-isobutyl-decanamide, were also able to alter RSA in *Arabidopsis* in a dose-dependent manner (Ramírez-Chavez et al., 2004). Previously, it was reported that the natural source for affinin was the roots of *Heliopsis longipes* and for *N*-isobutyl-2*E*-decenamide and *N*-isobutyl-decanamide was the roots of *Acmella radicans* and *Cissampelos glaberrima* (Figure 9.1). These alkamides are rare or produced in minuscule quantities by aerial parts of the source plants (Laurerio-Rosario et al., 1996; Rios-Chavez et al., 2003). At lower concentrations (7×10^{-6} to 2.8×10^{-5} M), affinin promotes PR growth and LR formation and stimulates RH elongation. Moreover, affinin treatment increased the number of LRs, owing to its dual activity to induce cell division in pericycle founder cells, leading to formation of new LRPs, as well as to

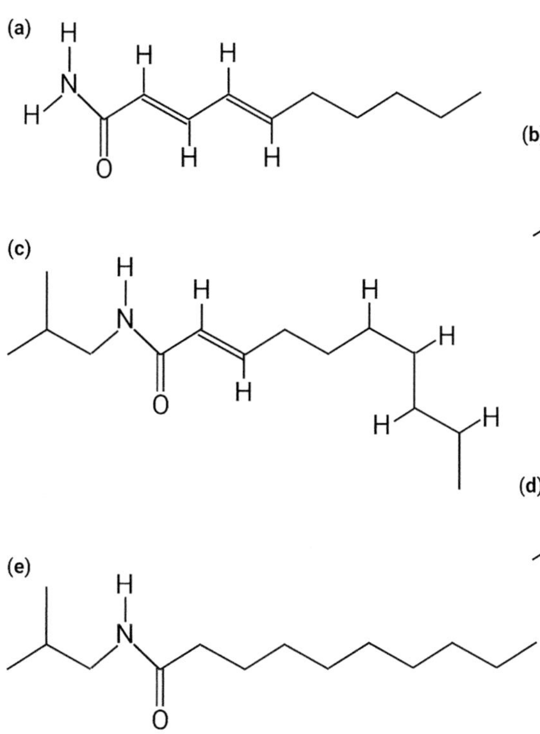

FIGURE 9.1 Chemical structures of various alkamides: (a) Amidenin (2E,4Z-decadienamide), (b) Affinin (*N*-isobutyl-2*E*,6*Z*,8*E*-decatrienamide), (c) *N*-isobutyl-2*E*-decenamide, (d) *N*-isobutyl-decanamide, (e) *N*-isopropyl-decanamide.

stimulate the emergence of existing LRPs. However, at higher concentrations (1.2×10^{-4} M), affinin inhibited PR growth and RH elongation. Similarly, *N*-isobutyl-decanamide inhibits PR growth and promotes LR formation. Also, *N*-isobutyl-decanamide showed the dual activity in terms of promoting LR emergence and increasing the number of total LRP in PR explants (Campos-Cuevas et al., 2008). In *in vitro* culture, *Arabidopsis* explants originated from stems and PR, when grown in varied concentrations of affinin and its derivative, *N*-isobutyl-decanamide, produced ARs and LRs, respectively. Another report revealed that *N*-isopropyl-decanamide, an alkamide detected in the unbound fraction of the humic supramolecular assembly isolated from vermicompost, was able to promote growth in the early and late stages of root development in maize seedlings in a dose-dependent fashion (Figure 9.1) (Zandonadi et al., 2019). Later, two new mutants, *drr1* (decanamide resistant root) and *dhm1* (decanamide hypersensitive mutant), were isolated from two different phenotypic screenings involving *Arabidopsis* seedlings grown under high and low concentrations of *N*-isobutyl-decanamide, respectively (Morquecho-Contreras et al., 2010; Pelagio-Flores et al., 2013). The *drr1* is a recessive mutant characterized by normal PR growth, reduced LR formation, late bolting and senescence, increased production of leaves, enhanced growth, and an overall extended longevity at the whole-plant level, suggesting that *drr1* plays a crucial role in LR formation and in the regulatory cascade of age-dependent plant senescence pathways. Moreover, *N*-isobutyl-decanamide-treated *drr1* mutants show resistance to PR growth inhibition and LR growth promotion. This phenotype might be due to the positive effect of the alkamide on pericycle cells to produce more LRP, rather than the byproduct of PR growth inhibition (Morquecho-Contreras et al., 2010). On the other hand, the *DHM1* mutant is also a recessive mutant and exhibits escalated PR growth with exacerbated LR and LRP formation, a constitutive tumourous shoot development, early bolting and senescence, decreased production of leaves, reduced growth, and an overall shortened longevity at the whole plant level, indicating that *DHM1* is a key component of regulation of cell division and differentiation, LR development, and age-dependent plant senescence (Pelagio-Flores et al., 2013). The *DHM1* mutant is highly sensitive to *N*-isobutyl-decanamide and manifests aggravated reduction and stimulation in PR growth and LR formation, respectively. The discovery of *drr1* and *DHM1* mutants, showing absolutely contrasting phenotypes, was able to shed some light on the effect of alkamides, including morphogenetic, hormonal modulation, and their crosstalk with other signaling molecules, on the growth and development of plants.

9.2 ALKAMIDE-INDUCED ALTERATIONS IN RSA INVOLVE HORMONAL MODULATION IN PLANTS

Auxin has been described as a key regulator of every phase of root development in both model and crop plants, including LR development, RH formation, and AR development (Bellini et al., 2014; Zandonadi et al., 2007). The concentration of auxin determines the phase of root development. For example, at lower concentrations (10^{-10} to 10^{-9} M), auxin stimulates PR growth, whereas at higher concentrations (10^{-8} to 10^{-6} M), auxin inhibits PR growth and stimulates LR formation and RH elongation (Ramírez-Chávez et al., 2004). During LR development, endogenous auxin activates cell division by enabling the CyclinB1-mediated G_2-to-M transition in founder cells of the pericycle region (Fukaki and Tasaka, 2009). In addition, LR organogenesis is controlled by fine-tuning the activators and repressors of the auxin response pathway through *Auxin/indole-3-acetic acid-auxin response factors* (AUX/IAA–ARF) modules. Different modules are responsible for different stages of LR development. For example, the IAA28-ARFs module regulates the specification of LR founder cells; *Solitary root/indole-3-acetic acid 14–auxin response factors 7* and *19* (SLR/IAA14-ARF7-ARF19) and *Bodenlos/indole-3-acetic acid 12/monopteros-auxin response factor 5* (BDL/IAA12/MP-ARF5) together produce a bimodal auxin response on LR initiation and organogenesis through activation of *Lateral organ boundaries/Asymmetric leaves-like* (LBD/ASL), depending on the endogenous auxin concentration; *Short hypocotyl 2/Auxin/indole-3-acetic acid-auxin response factors* (SHY2/IAA3-ARF) regulates LRP development (Khan et. al., 2016). Besides AUX/IAA–ARF, the strategic positioning of the auxin efflux carriers, *Pin-formed* (PIN), facilitates the generation of the targeted auxin maxima necessary for priming pericycle cells to form correctly patterned LRPs (Smith and Smet, 2012). Another phytohormone, cytokinin, promotes cytokinesis, root apical dominance, vascular cambium sensitivity, and vascular differentiation. In *Arabidopsis*, cytokinin is perceived by three histidine kinase receptors, *Arabidopsis histidine kinase 2* (AHK2), *Arabidopsis histidine kinase 3* (AHK3), and *Cytokinin response 1/Wooden leg/Arabidopsis histidine kinase 4* (CRE1/AHK4/WOL). All the single and double knock-out mutants of these genetic combinations produce apparently functional flowers with viable seeds, suggesting a functional redundancy among these proteins. However, the *cre1*-12 *ahk2*-2 *ahk3*-3 triple mutant exhibits a dwarf phenotype with nonfunctional flowers and inviable seeds, elaborating the importance of this mutant in studies involving cytokinin signaling. Reports suggest that cytokinin blocks G_2-to-M transition in founder cells of the pericycle region, suppresses LR initiation through *Arabidopsis crinkly 4* (ACR4) mediation, and inhibits LR development via downregulation of *PIN*-gene expression (Jing and Strader, 2019). Thus, cytokinin exerts an inhibitory effect on LR initiation and development and functions antagonistically to auxin (Schmulling et al., 2001). Intriguingly, the effect of jasmonic acid on root growth is intricate. Jasmonic acid inhibits PR elongation through repression of *Plethora 1* (PLT1) and *Plethora 2* (PLT2) by *Mouse-ear cress 2/Jasmonate insensitive 1* (MYC2/JAI1) but positively regulates the formation and distribution of LRs via activation of *Coronatine insensitive 1* (COI1). Furthermore, jasmonic acid signaling requires a functional auxin signaling pathway during LR formation (Xu et al., 2020). Primary observations reveal that alkamides alter RSA in an auxin-independent and cytokinin-dependent manner, and jasmonic acid also plays a key role. The analysis of mutants, defective in auxin, jasmonic acid, and cytokinin signaling, were used to study the effect of alkamides on RSA.

9.3 CROSSTALK BETWEEN ALKAMIDES AND CYTOKININ

N-isobutyl-decanamide inhibits PR growth and stimulates LR formation, accompanied by developmental transitions from LRs to callus-like structures. Evidence from the morphogenetic studies of *Arabidopsis* wild-type and cytokinin-signaling pathway mutants demonstrates that a functional cytokinin-signaling pathway is essential for alkamides to influence the meristematic activity and differentiation processes involving RSA (Figure 9.2) (López-Bucio et al., 2008).

9.3.1 Crosstalk between Alkamides and Auxin

At high concentrations (10^{-8} to 10^{-6} M), auxin inhibits PR growth and stimulates LR formation and RH elongation, whereas affinin promotes PR growth and LR formation and stimulates RH elongation at lower concentrations (7×10^{-6} to 2.8×10^{-5} M). Moreover, at a very high concentration, affinin and its derivatives, *N*-isobutyl-2*E*-decenamide and *N*-isobutyl-decanamide, failed to induce expression auxin-inducible gene markers, such as *DR5:uidA* and *BA3:uidA*. In addition, the auxin-resistant mutants *aux1-7*, *eir1-1*, and *axr4-2*, and the wild-type plants showed similar levels of sensitivity to affinin in terms of PR growth (Ramírez-Chávez et al., 2004). Mutant analysis showed that *drr1* mutants are not resistant to auxin-induced PR growth inhibition or LR formation (Morquecho-Contreras et al., 2010). Taken together, these reports demonstrate that although alkamides induce root developments similar to auxin, alkamides seem to operate by an auxin-independent signaling pathway. However, the *dhm1* mutants orchestrated increased *DR5:uidA* expression in PRs and LRs and diminution in LR formation upon treatment with the pharmacological inhibitor of auxin transport (Pelagio-Flores et al., 2013). The presence of normal *DRR1* and *DHM1* is essential for alkamide-induced phenotypes in *Arabidopsis*, and therefore, alkamides might alter RSA by modulating the auxin:cytokinin ratio (Figure 9.2).

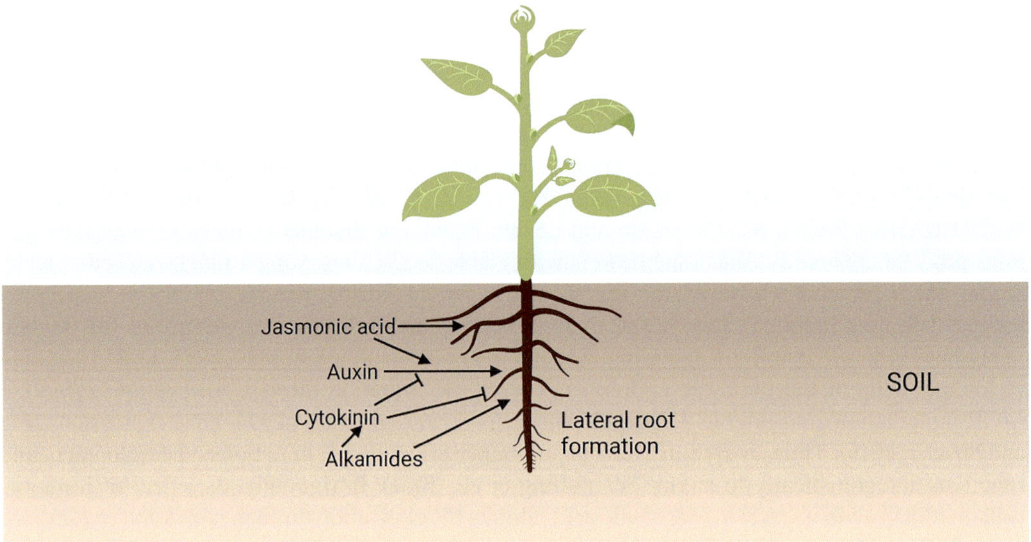

FIGURE 9.2 Crosstalk between alkamides and different hormones. During LR formation, auxin regulates various stages of development, starting from the priming of pericycle, reactivation of cell division in founder cells of the pericycle region, LRP formation to the patterning, and emergence of newly formed LRP. Cytokinin promotes root apical dominance and, hence, suppresses LR initiation as well as inhibiting LR development. Jasmonic acid promotes LR formation in an auxin-dependent fashion. Alkamides induce LR development by modulating the auxin:cytokinin ratio at the pericycle region of the PR.

9.3.2 Crosstalk between Alkamides and Jasmonic Acid

Very little information is available on the interaction between alkamides and jasmonic acid. In response to jasmonic acid treatment, the *drr1* mutants showed inhibition in PR growth as well as failure to active LR induction. This phenotype of *drr1* mutants is unique and different from that of the wild-type and *jasmonic acid-resistant 1* (JAR1) knock-out mutant, JAR1, upon jasmonic acid treatment. Hence, the activity of *DRR1* is partially modulated by jasmonic acid during root development (Morquecho-Contreras et al., 2010). Thus, alkamides might interact with jasmonic acid during PR development. However, as evidence suggests that jasmonic acid alters LR formation in an auxin-dependent process (Xu et al., 2020), alkamides may orchestrate indirect antagonism or operate independently of jasmonic acid during LR development (Figure 9.2).

9.3.3 Nitric Oxide Signaling Is Essential for Alkamide-Induced Alterations in RSA

Nitric oxide (NO) is a hydrophobic, highly active gaseous molecule and is produced as a bio-product in almost all types of organisms, including bacteria, plants, fungi, and animal. In plants, NO acts as a biological messenger and regulates various physiological processes, including leaf expansion, stomatal closure, LR and AR development, gravitropism, flowering, photomorphogenesis, programmed cell death, senescence, and xylem differentiation. Moreover, NO is strategically placed at the converging point in the signaling pathway of different plant hormones, such as auxin, cytokinin, abscisic acid, and jasmonic acid (Méndez-Bravo et al., 2010). In animals, NO synthase produces NO, but in plants, no such molecule exists. Instead, *NO-associated protein 1* (AtNOA), a member of the *Arabidopsis* circularly permuted GTPase family, was found to be structurally similar to NO synthase at the protein level (Moreau et al., 2008). In addition, NO is also produced by nitrate reductase via reduction of nitrite in plants. In *Arabidopsis*, two *NR* genes, *Nitrate reductase 1* (NIA1) and *Nitrate reductase 2* (NIA2), are present and encode for proteins that have an almost 83.5% similarity at the amino acid level. Pharmacological research involving NO led to the discovery of three types of molecules: i) NO donors, sodium nitropruside (SNP), and S-nitroso-N-acetylpenicillamine (SNAP) to mimic NO production; ii) NO scavengers, 2-phenyl-4,4,5,5-tetramethylimidazoline-1-oxyl-3-oxide (PTIO) or carboxy-PTIO (cPTIO) to remove the NO signal; and iii) detection of NO by using fluorescent dyes such as DAF-2DA (Goldstein et al., 2003; Wilson et al., 2008). NO is considered the central regulator of auxin-regulated signaling cascade determining growth and development of roots. Exogenous application of NO stimulated AR development in cucumber hypocotyls, emulating the effects of indole-3-acetic acid (IAA). However, depletion of nitric acid completely blocked auxin-promoted AR growth, though the inactivation was indirect and reversible (Pagnussat et al., 2002). Moreover, NO is considered to play a crucial role downstream of auxin-orchestrated LR formation. Like auxin, NO also inhibits PR growth and stimulates LR generation, whereas NO scavenging results in complete abolition of LR development and partial restoration of PR elongation, probably because NO regulates the transcription of genes involved in the cell cycle in pericycle founder cells (Correa-Aragunde et al., 2006). Also, accumulation of NO was reported during auxin-induced RH formation, and asymmetric accumulation of NO was observed in the root tip during the gravitropic response (Hu et al., 2005; Lombardo et al., 2006). Like auxin, *N*-isobutyl-decanamide was able to induce NO accumulation during AR formation in *Arabidopsis* shoot explants as well as in meristematic and proliferating root regions in *Arabidopsis* seedlings during LR formation (Campos-Cuevas et al., 2008; Méndez-Bravo et al., 2010). Although exogenous application of NO was found to mimic the effects of *N*-isobutyl-decanamide on RSA, depletion of nitric acid resulted in reduced LR initiation. Mutants compromised in endogenous NO production, *atnoa* and *nia1nia2*, also showed similar reduction in LR formation in response to alkamides, suggesting that genes representing different biosynthetic pathways of nitric acid with apparently redundant functions play a key role in the alkamide-regulated activation of the cell cycle in pericycle founder cells (Méndez-Bravo et al., 2010). Interestingly, *drr1* mutant showed reduction in both PR growth and LR development in the presence NO donor, indicating that *DRR1* is a key component of signaling pathway NO-induced LR formation. Furthermore, the NO donor was unable to induce auxin-responsive promoters, suggesting that NO function downstream of auxin signaling during LR formation either by acting as a second messenger regulating expression of cell cycle genes or by physically interacting with proteins responsible for cell cycle transition, affecting their function by nitration or nitrosylation. A recent report illustrated that *S*-nitrosylation of the auxin receptor protein, *Transport inhibitor response 1* (TIR1), resulted in degradation of AUX/IAA proteins and subsequent promotion in gene expression of the auxin-signaling pathway, depicting a direct influence of NO on auxin perception and signaling (Sun et al., 2021).

9.4 ALKAMIDES MODULATE THE ACTIVITY PROTON PUMPS TO INDUCE ALTERATIONS IN RSA

Humus is produced naturally by the biodegradation of tissues from dead organisms. Humic substances are organic compounds that are important components of humus. Humic substances can be divided into three fractions: humic acids (HA), fulvic acids, and humin (Baigorri et al., 2009). HA and fulvic acid are commonly used as soil supplements in agriculture. HA is a composite mixture of different acids containing carboxyl and phenolate groups (MacCarthy et al., 2001). Humic substances are biostimutants that render hormone-like activities to increase nutrient uptake and, in turn, plant growth and yield by altering plant physiology and metabolism. The increase in nutrient uptake by HA is attributed to its ability to control the synthesis and functionality of proton pumps present at the plasma membrane. HA increases the electrochemical proton gradient across the plasma membrane to enhance ion transport through the secondary transport systems (Morsomme and Boutry, 2000). This leads to the acidification of the apoplast and the unclasping of the cell wall, allowing cell elongation. The elongation-related differentiation zone of the root comprises a condensed meristematic region, which makes the zone more conducive for LR formation and RH proliferation (Frías et al., 1996). Thus, the dramatic increase in root surface area due to the formation of LRs and RHs is a testament to the increase in number and activity of the plasma membrane–situated proton pumps. Besides, HA treatment of maize plants resulted in upregulated expression in *Mha2*, a major isoform of H^+-ATPase mRNA, when compared to untreated plants. Interestingly, the synthesis of new proton pumps at the plasma membrane has been linked to auxin induction. Thus, the effect of HA on plasma membrane–situated proton pumps may rely on the auxin-dependent activation of the *Mha2* gene (Quaggiotti et al., 2004). In addition, NO mediates HA-induced root development and activation of proton pumps at the plasma membrane (Zandonadi et al., 2010). Zandonadi and his coworkers reported that N-isopropyl-decanamide promoted root growth and induced LR emergence in maize seedlings in a dose-dependent manner. They had isolated N-isopropyl-decanamide from the humic supramolecular assembly, a constituent of the unbound fraction of HA from cattle manure vermicompost. Treatment with varied concentrations of *N*-isopropyl-decanamide (10^{-4} to 10^{-8} M) and HA increased root length and superficial area significantly, stimulated plasma membrane–situated proton pump activity, and enhanced NO accumulation during LR generation in maize seedlings (Zandonadi et al., 2019). However, the introduction of auxin inhibitors could only partially compromise the bioactivity of HA, indicating that the bioactivity of HA is a cumulative effect of bioactivities of plant growth regulatory substances, including auxin and alkamides.

9.5 CONCLUSION

Evidence from the morphogenetic studies supports the hypothesis that during LR generation, alkamides exert dual modulation on the activity of nitric acid, which acts as an integrator of different signaling pathways that regulate various physiological processes (Figure 9.3). Firstly, alkamides utilize the cytokinin-signaling pathway to inhibit NO and auxin-induced root formation. Secondly, alkamides initiate cell proliferation and LR development by NO through the activation of DRR1. On the other hand, alkamides also activate DHM1, which inhibits LR formation, supposedly through feedback regulation of DRR1. Interestingly, alkamides also modulate the proton pumps of the plasma membrane to promote elongation of the cell wall, stimulating LR formation. Moreover, as the activation of proton pumps is involved with the formation of the electrochemical gradient across the plasma membrane, alkamides might interact with reactive oxygen species–signaling pathways, thereby activating the Ca^{2+} channel hyperpolarization and RH proliferation. Also, alkamides share structural similarities with animal endocannabinoids, and the animal endocannabinoid system requires the participation of G-protein-coupled receptors (GPCRs) (Silver, 2019).

Alkamides are a new class of plant growth regulators. Limited but interesting information is available on their role in root development. Based on the analyses of mutants, two antagonistically functioning recessive genes, *DRR1* and *DHM1*, have been identified as governing alkamide-induced responses during plant development. Alkamides seem to interact with different hormones, such as auxin, cytokinin, and jasmonic acid, and control root development, possibly by manipulating auxin and cytokinin levels in a tissue-specific fashion. Studies involving the interaction of alkamides and other hormones known to regulate root development, like abscisic acid, ethylene, gibberellin, and brassinosteroids, are required for a better understanding of the role of alkamides in PR growth and LR and LRP formation, as well as AR production. Moreover, future research on alkamides might consider the role of reactive oxygen species as well as GPCRs and heteromeric G-proteins in alkamide-induced root development. The modulatory roles of reactive oxygen species and heteromeric G-proteins are well documented (Chen et al., 2006; Zhou et al., 2020) Furthermore, the information on alkamide-induced root development in *Arabidopsis* is very promising and can be applied to agronomically important crops to enhance their root biomass and productivity.

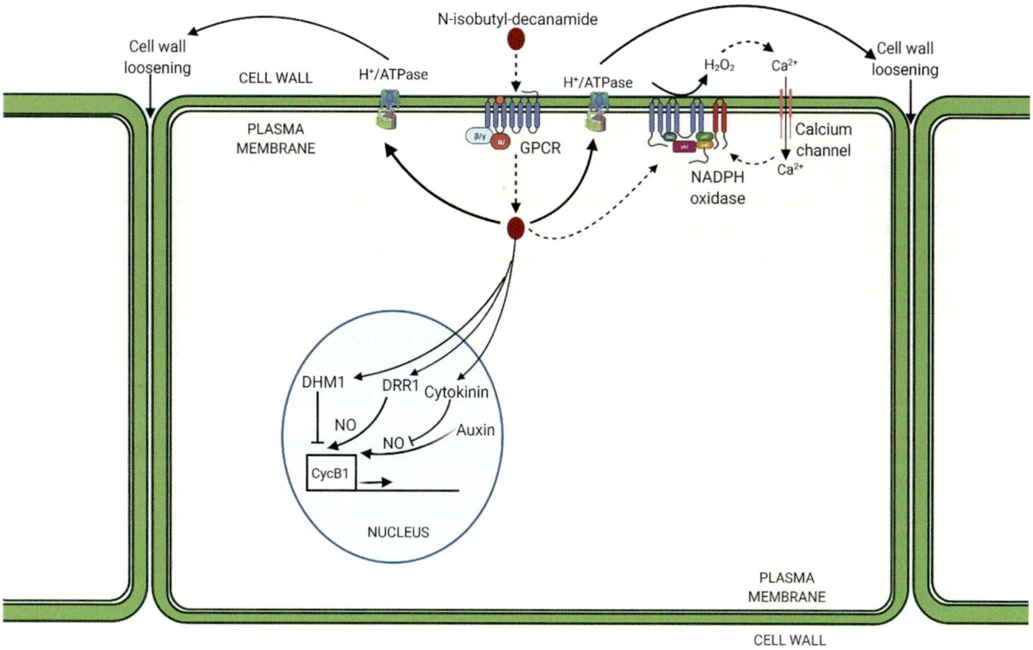

FIGURE 9.3 Effect of *N*-isobutyl-decanamide on pericycle founder cells during LR development. According to this model, *N*-isobutyl-decanamide is perceived by G-protein coupled receptors present at the plasma membrane of the pericycle founder cells. After entering the cell, *N*-isobutyl-decanamide activates the plasma membrane–situated proton pumps, which facilitate modification of extracellular pH, resulting in loosening of the cell wall, allowing cell elongation and, thereby, LR formation. Changes in extracellular pH might also result in the activation of the ROS signaling pathway by NADPH oxidase, inciting cell wall softening and LR generation as well as inducing Ca^{2+} channel and RH proliferation. *N*-isobutyl-decanamide also activates DRR1 and its downstream molecule, NO, which either modulates transcription of cell cycle genes (i.e., CycB1) or interacts directly with proteins responsible for cell cycle transition, initiating LRP formation. *N*-isobutyl-decanamide, on the other hand, inhibits auxin-induced LR formation through activation of the cytokinin-signaling pathway. Finally, *N*-isobutyl-decanamide activates DHM1, which inhibits LR formation, supposedly through feedback regulation.

ACKNOWLEDGMENTS

We acknowledge BioRender.com for generation of all the figures used in this chapter.

REFERENCES

Baigorri R, Fuentes M, González-Gaitano G, et al. Complementary multianalytical approach to study the distinctive structural features of the main humic fractions in solution: Gray humic acid, brown humic acid, and fulvic acid. J Agric Food Chem. 2009; 57(8): 3266–3272.

Beeckman T, Burssens S, Inze D. The peri-cell-cycle in *Arabidopsis*. J Exp Bot. 2001; 52: 403–411.

Bellini C, Pacurar DI, Perrone I. Adventitious roots and lateral roots: Similarities and differences. Annu Rev Plant Biol. 2014; 65: 639–666.

Buitimea-Cantúa GV, Marsch-Martinez N, Ríos-Chavez P, et al. Global gene expression analyses of the alkamide-producing plant heliopsis longipes supports a polyketide synthase-mediated biosynthesis pathway. Peer J. 2020; 8: e10074.

Campos-Cuevas JC, Pelagio-Flores R, Raya-González J, et al. Tissue culture of *Arabidopsis* thaliana explants reveals a stimulatory effect of alkamides on adventitious root formation and nitric oxide accumulation. Plant Sci. 2008; 174: 165–173.

Chen J-G, Gao Y, Jones AM. Differential Roles of *Arabidopsis* Heterotrimeric G-Protein Subunits in Modulating Cell Division in Roots. Plant Physiol. 2006; 141(3): 887–897.

Correa-Aragunde N, Graziano M, Chevalier C, et al. Nitric oxide modulates the expression of cell cycle regulatory genes during lateral root formation in tomato. J Exp Bot. 2006; 57: 581–588.

De Smet. Lateral root initiation: One step at a time. New Phytol. 2012; 193: 867–873.

Dubrovsky JG, Rost TL, Colon-Carmona A, et al. Early primordium morphogenesis during lateral root initiation in Arabidopsis thaliana. Planta. 2001; 214: 30–36.

Faure L, Cavazos R, Khan BR, et al. Effects of synthetic alkamides on Arabidopsis fatty acid amide hydrolase activity and plant development. Phytochem. 2015; 110: 58–71.

Frías I, Caldeira MT, Pérez-Castiñeira JR, et al. A major isoform of the maize plasma membrane H+-ATPase: Characterization and induction by auxin in coleoptiles. Plant Cell. 1996; 8: 1533.

Fukaki H, Tasaka M. Hormone interactions during lateral root formation. Plant Mol Biol. 2009; 69: 383–396.

Garay-Arroyo A, Sanchez MD, Garcia-Ponce B, et al. Hormone symphony during root growth and development. Dev Dyn. 2012; 241: 1867–1885.

Goldstein S, Russo A, Samuni A. Reactions of PTIO and carboxy-PTIO with NO, NO_2 and O_2. J Biol Chem. 2003; 278: 50949–50955.

Harald Greger. Alkamides: A critical reconsideration of a multifunctional class of unsaturated fatty acid amides. Phytochem Rev. 2016; 15: 729–770.

Hochholdinger F. "The maize root system: Morphology, anatomy, and genetics," in *Handbook of Maize: Its Biology*, eds J. L. Bennetzenand, S.C. Hake, New York: Springer, 2009, 145–160.

Hu X, Neill SJ, Tang Z, et al. Nitric oxide mediates gravitropic bending in soybean roots. Plant Physiol. 2005; 137: 663–670.

Jing HW, Strader LC. Interplay of auxin and cytokinin in lateral root development. Int J Mol Sci. 2019; 20(3): 486.

Kanbe K, Naganawa H, Okamura M, et al. Amidenin, a new plant growth-regulating substance isolated from Amycolatopsis sp. Biosci Biotechnol Biochem. 1993; 57: 1261–1263.

Khan MA, Gemenet DC, Villordon A. Root system architecture and abiotic stress tolerance: Current knowledge in root and tuber crops. Front Plant Sci. 2016; 7: 1584.

Laurerio-Rosario S, Silva A, Parente J. Alkamides from cissampelos glaberrima. Plant Med. 1996; 62: 376–377.

Li SW, Xue LG, Xu SJ, et al. Mediators, genes and signaling in adventitious rooting. Bot Rev. 2009; 75: 230–247.

Liu J, Rowe J, Lindsey K. Hormonal crosstalk for root development: A combined experimental and modeling perspective. Front Plant Sci. 2014; 5: 116.

Lombardo MC, Graziano M, Polacco J, et al. Nitric oxide functions as a positive regulator of root hair development. Plant Signal Behav. 2006; 1: 28–33.

López-Bucio J, Millán-Godínez M, Méndez-Bravo A, et al. Cytokinin receptors are involved in alkamide regulation of root and shoot development in Arabidopsis. Plant Physiol. 2008; 145(4): 1703–1713.

Lynch JP, Brown KM. New roots for agriculture: Exploiting the root phenome. Philos Trans R Soc Lond B Biol Sci. 2012; 367: 1598–1604.

MacCarthy P. The principles of humic substances. Soil Sci. 2001; 166(11): 738–751.

Méndez-Bravo A, Calderón-Vázquez C, Ibarra-Laclette E, et al. Alkamides activate jasmonic acid biosynthesis and signaling pathways and confer resistance to Botrytis cinerea in Arabidopsis thaliana. PLoS One. 2011; 6: e27251.

Méndez-Bravo A, Raya-González J, Herrera-Estrella L, et al. Nitric oxide is involved in alkamide-induced lateral root development in Arabidopsis. Plant Cell Physiol. 2010; 51: 1612–1626.

Moreau M, Lee GI, Wang Y, et al. AtNOS1/AtNOA1 is a functional *Arabidopsis thaliana* cGTPase and not a nitric-oxide synthase. J Biol Chem. 2008; 283: 32957–32967.

Morquecho-Contreras A, Méndez-Bravo A, Pelagio-Flores R, et al. Characterization of drr1, an alkamide resistant mutant of Arabidopsis reveals an important role for small lipid amides in lateral root development and plant senescence. Plant Physiol. 2010; 152: 1659–1673.

Morsomme P, Boutry M. The plant plasma membrane H+ – ATPase: Structure, function and regulation. Biochim Biophys Acta. 2000; 1465: 1–16.

Nibau C, Gibbs DJ, Coates JC. Branching out in new directions: The control of root architecture by lateral root formation. New Phytol. 2008; 179: 595–614.

Pagnussat GC, Simontacchi M, Puntarulo S, et al. Nitric oxide is required for root organogenesis. Plant Physiol. 2002; 129: 954–956.

Pelagio-Flores R, Ortiz-Castro R, López-Bucio J. dhm1, an Arabidopsis mutant with increased sensitivity to alkamides shows tumorous shoot development and enhanced lateral root formation. Plant Mol Biol. 2013; 81: 609–625.

Perrot-Rechenmann C. Cellular responses to auxin: Division versus expansion. Cold Spring Harb Perspect Biol. 2010; 2: a001446.

Quaggiotti S, Ruperti B, Pizzeghello D, et al. Effect of low molecular size humic substances on nitrate uptake and expression of genes involved in nitrate transport in maize (Zea mays L). J Exp Bot. 2004; 55: 803–813.

Ramírez-Chávez E, López-Bucio J, Herrrera-Estrella L, et al. Alkamides isolated from plants promote growth and alter root development in Arabidopsis. Plant Physiol. 2004; 134: 1058–1068.

Rios-Chavez P, Ramírez-Chávez E, Armenta-Salinas C, et al. Acmella radicans var. radicans: In vitro culture establishment and alkamide content. In Vitro Cell Dev Biol Plant. 2003; 39: 37–41.

Schmulling T, Werner T, Motyka V et al. Regulation of plant growth by cytokinin. Proc Natl Acad Sci USA. 2001; 98: 10487–10492.

Silver RJ. The endocannabinoid system of animals. Animals (Basel). 2019; 9(9): 686.

Smith S, Smet ID. Root system architecture: Insights from Arabidopsis and cereal crops. Phil Trans R Soc B. 2012; 367: 1441–1452.

Sorin C, Bussell JD, Camus I, et al. Auxin and light control of adventitious rooting in Arabidopsis require ARGONAUTE1. Plant Cell. 2005; 17: 1343–1359.

Sun C, Zhang Y, Liu L, et al. Molecular functions of nitric oxide and its potential applications in horticultural crops. Hortic Res. 2021; 8(1): 71.

Waines JG, Ehdaie B. Domestication and crop physiology: Roots of green-revolution wheat. Ann Bot. 2007; 100: 991–998.

Wilson, ID, Neill SJ, Hancock JT. Nitric oxide synthesis and signalling in plants. Plant Cell Environ. 2008; 31: 622–631.

Xu P, Zhao P, Cai X, et al. Integration of jasmonic acid and ethylene into Auxin signaling in root development. Front Plant Sci. 2020; 11: 271.

Zandonadi DB, Canellas LP, Façanha AR. Indoleacetic and humic acids induce lateral root development through a concerted plasmalemma and tonoplast H+ pumps activation. Planta. 2007; 225: 1583–1595.

Zandonadi DB, Matos CRR, Castro RN, et al. Alkamides: A new class of plant growth regulators linked to humic acid bioactivity. Chem Biol Technol Agric. 2019; 4: 1–12.

Zandonadi DB, Santos MP, Dobbss LB, et al. Nitric oxide mediates humic acids induced root development and plasma membrane H+-ATPase activation. Planta. 2010; 231: 1025–1036.

Zhou X, Xiang Y, Li C, et al. Modulatory role of reactive oxygen species in root development in model plant of Arabidopsis thaliana. Front Plant Sci. 2020; 11: 485932.

10 Alkamides and Plant Microbe Interactions in Rhizosphere

Current Status and Future Perspectives

Muhammad Fraz Ali, Irsa Ejaz, Muhammad Asad Naseer, and Zhencai Sun

CONTENTS

10.0 INTRODUCTION

Alkamides are secondary metabolites and lipophilic natural products consisting of over 250 correlated compounds found extensively in plant species. These compounds are composed of polyunsaturated aliphatic fatty acids and an amine substituent. Alkamides are composed of chains of varying lengths and acyl groups containing one or more double bonds (Faure et al., 2015). Besides their simple architecture, they are the most expanding group of natural and bioactive compounds and perform many biological activities. Alkamides have a strong and bitter taste and produce caustic and narcotic effects (Dubey et al., 2013) as well as an itching and salivating sensation in the mouth (Woelkart and Bauer, 2007).

10.1 TYPES, OCCURRENCE, AND IMPORTANCE OF ALKAMIDES

Generally, alkamides are novel plant secondary compounds and still not characterized properly, becoming a choice for scientists to study, along with their biosynthesis (Figure 10.1). They are divided into proto-alkaloids (Rios and Olivo, 2014) and pseudo-alkaloids (Percy et al., 2010). They are considered secondary metabolites, plant-growth regulators (PGRs), and natural compound complexes (Rios and Olivo, 2014) and are commonly distributed in more than 35 plant families (Joshi et al., 2020), mainly *Aristolochiaceae* (Holzbach and Lopes, 2010), *Asteraceae* (Althaus et al., 2017; Saadali et al., 2001), *Brassicaceae* (Zhao et al., 2005; Korkmaz, 2018), *Convolvulaceae* (Tofern et al., 1999), *Euphorbiaceae* (Kavit et al., 2013; Jiang et al., 2009), *Menispermaceae* (Rosario et al., 1996), *Piperaceae* (Do Nascimento et al., 2012; Dawid et al., 2012), *Poaceae* (Gamboa-Leon and William Scott Chilton, 2000), *Rutaceae* (Guetchueng et al., 2018; Saravanakumar et al., 2016), and *Solanaceae* (Kehie et al., 2015). The interest in studying the alkamides basically comes from the

DOI: 10.1201/9781003224525-13

tingling pungent smell of some compounds from spices such as capsaicin from chili peppers (Omolo et al., 2014) and peperino from black pepper (Srinivas and Madhusudana Rao, 1999). There are hundreds of species in the genuses *Acmella* (Nakatani and Nagashima, 1992), *Anacyclus* (Althaus et al., 2017), *Echinacea* (Hohmann et al., 2011), *Heliopsis* (Molinatorres et al., 1996), and *Zanthoxylum* (Guetchueng et al., 2018), all of which have been used for medicinal purposes for thousands of years. Biological activities involve immunomodulatory (Gertsch, 2008), antimicrobial (Joshi et al., 2020), antiviral (Vázquez-Martínez et al., 2020), larvicidal (Harald Greger, 2016), insecticidal (Faure et al., 2015), novel signaling (López-Bucio et al., 2006), and antioxidant activities (Fontenele et al., 2009). The common genera in which alkamides are present are listed in Table 10.1.

10.2 ANTIMICROBIAL PROPERTIES

The plant produces multiple intracellular and systemic responses against biotic and abiotic stresses. These pathways are dependent on plant hormones and lead to secondary metabolites with antimicrobial activity. Secondary metabolites from plants, including tannins, alkaloids, terpenoids, and flavonoids, have been shown *in vitro* to have antimicrobial properties. Alkamides also contains such secondary metabolites, which are used for pharmacological and medicinal purposes. The main purpose is to stop the growth of microorganisms such as bacteria, fungi, and viruses. The *Heliantheae* family contains many species, which have alkamides with antimicrobial characteristics (Molina-Torres et al., 1999). Afinin is the most common alkamide present in the *Heliantheae* family; olefinic alkamides are

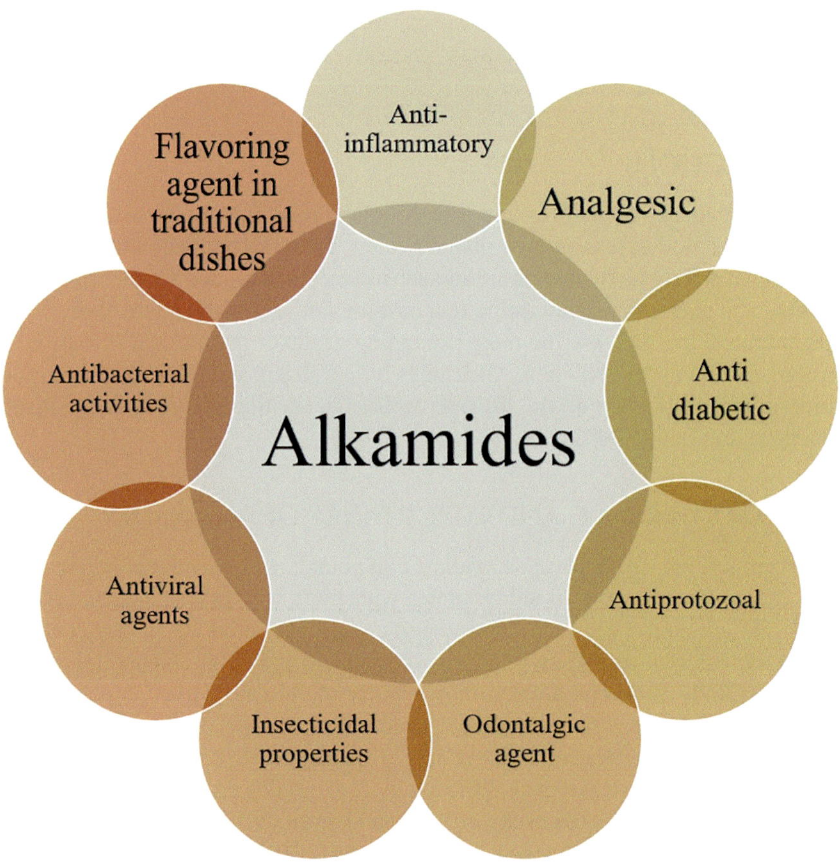

FIGURE 10.1 Common characteristics of alkamides.

TABLE 10.1
The Types and Common Species of Alkamides Found in the Form of Multiple Compounds and Their Inclusive Characteristics

Types	Families	Genera	Species	Common Alkamides	Characteristics	References
Aliphatic alkamides	Asteraceae	*Aaronsohnia*	*Aaronsohnia pubescens*	1. 6-thienylhexa-2*E* 2. 4*E*-dienoic acid 3. Isobutylamide	Anti-inflammatory purposes; used in folk medicine; toothache can be treated with the flowering parts of the plant	(Muller-Jakic et al., 1994)
		Achilea	*Achillea millefolium* *A. ptarmica* *A. wilhelmsii*	1. *p*-hydroxyphenethylamide *IV* 2. deca-2*E*,6*Z*,8*E*-trienoic acid 2-methylbutylamide *(homospilanthol)* 3. *deca-2E,4E*-dienoic acid *N*-methyl isobutylamide	Insecticide; anti-inflammation activity; antiprotozoal activity; and some immunological activities	(Saeidnia et al., 2011; Veryser et al., 2017; Si et al., 2006)
		Acmella	*Acmella decumbens*	1. (2*E*, 4*E*)-*N*-hydroxiphenylethyl-2,4-decadien-9-inamide 2. (4*E*, 6*E*)-*N*-isobutyl-4,6-undecen-10-inamide (2*Z*)-*N*-phenylethyl-2-nonane-6,8-diinamide 3. *N*-isobutyl 4. *N*-2-methylbutyl 5. *N*-phenethyl	Analgesic properties; odontalgic agent; insecticidal properties; immuno-stimulatory; antiviral agents	(Casado et al., 2009)
			Acmella radicans	1. *N*-(2-phenylethyl)-2*Z*,4*E*-octadienamide 2. 3-phenyl-*N*-(2-phenylethyl)-2 propenamide 3. (2*E*,6*Z*,8*E*)-*N*-isobutyl-2,6,8-decatrienamide 4. (2*Z*,4*E*)-*N*-(2-phenylethyl)-2,4-octadienamide 5. (2*E*)-*N*-(2-phenylethyl)-nona-2-en-6,8-diynamide 6. 3-phenyl-*N*-(2-phenylethyl)-2-propenamide	Famous for insecticide, fungicide, and antibacterial activities; used in traditional medicine for treatment of toothache; role in altering the growth and development of *Arabidopsis thaliana*	(Rios-Chavez et al., 2003; Cortez-Espinosa et al., 2011)

(Continued)

TABLE 10.1
(Continued)

Anacyclus	*Anacyclus pyrethrum*	1. Tetradeca-2*E*,4*E*-dien-8 2. 10-diynoic acid isobutylamide deca-2E,4E-dienoic acid 2-phenylethylamide	anti-plasmodial activity against malaria; folk medicine for treatment of inflammatory and painful diseases	(Althaus et al., 2017)
Artemisia	*Artemisia dracunculus*	1. *Neopellitorine* A 2. *Neopellitorine* B 3. *Pellitorine* 1 4. Coumarin 5. Herniarine	Insecticidal effect against the rice weevil and grain borer; used for flavoring foods; antibacterial and antifungal activity	(Mumivand et al., 2017; Saadali et al., 2001)
Echinaceae	*E. angustifolia*, *E. pallida* *E. purpurea*	1. Undeca-2*E*/*Z*-ene-8,10-diynoic acid isobutylamides 2. Dodeca-2*E*,4*Z*-diene-8,10-diynoic acid isobutylamide 3. Dodeca-2*E*-*en*-8,10-diynoic acid isobutylamide 4. Dodeca-2*E*,4*E*,8*Z*,10*E*/*Z*-tetraenoic acid isobutylamides 5. Dodeca-2*E*,4*E*,8*Z*-trienoic acid isobutylamide	Immunostimulant and prevents infections; prone to colds, respiratory diseases; protective against influenza	(Binns et al., 2001) (Woelkart et al., 2005)
Heliopsis	*Heliopsis longipes*	1. Affinin *(2E,6Z,8E)-N-isobutyl-2,6,8-decatrienamide*	Traditional medicine as a condiment; buccal anesthetic; analgesic in pain of toothache; antiparasitic and anti-inflammatory; fungistatic, and bacteriostatic activities; anti-arthritic activity	(Hernández et al., 2009; Rios et al., 2007)
Spilanthes	*S. acmella* *S. calva* *S. mauritiana* *S. paniculata* *S. oleraceae* *S. ocymifolia*	1. Spilanthol 2. Isobutylamide 3. 2*E*, 6*Z*, 8*E*-*N*-isobutylamide 4. 2, 6, 8-decatrienamide 5. Cinnamamidyl	Traditional medicine to treat toothaches, stammering, and stomatitis; immune stimulation; dietary supplements; powerful antiseptics	(Sharma and Arumugam, 2021; Dandin et al., 2014)

	Salmea	*Salmea scandens*	1. Isobutylamides From (2*E*,4*E*,8*Z*,10*Z* 2*E*,4*E*,8*Z*,10*E*-dodeca-2,4,8,10-tetraenoic acid)	Indigenous edible plant; flavoring agent in traditional dishes; anti-inflammatory and analgesic properties	(Molinatorres et al., 1996)
	Sanvitalia	*Sanvitalia ocymoides* DC.	1. Tetradeca-2*E*,4*E*,8*Z*,10*Z*,12*E*-pentaenoic acid isobutylamide	Indigenous plant; importance in medicinal industry; aesthetic importance: grown as horticultural plant	(Dominguez et al., 1987)
	Wedelia	*Wedelia parviceps* Blake	1. Afinin	Used in Chinese, Thai, Korean and South American folk medicine	(Johns et al., 1982)
Convolvulaceae	*Ipomoea*	*Ipomoea aquatic*	1. 1-(14-methylhexadecanoyl) 2. Pyrrolidine	Carminative agent useful in fever, jaundice, biliousness, bronchitis, liver complaints in Eastern medicine	(Manvar and Desai, 2013; Tofern et al., 1999)
	Merremia	*Merremia quinquefolia*	1. 1-hexadecanoylpyrrolidine 2. 1-octadecanoylpyrrolidine 3. 1-(14-methylhexadecanoyl) 4. Pyrrolidine 5. Ergolines 6. Lipophilic tropanes 7. Pyrrolizidine alkaloids	Used in the treatment of achy joints, rheumatoid arthritis, scurvy, loss of appetite, and upset stomach	(Tofern et al., 1999)
Euphorbiaceae	*Mallotus*	*Mallotus lianus*	1. *N*-isobutyl-2*E*,4*E*,12*Z*-octadecatrienamide (7*Z*,10*Z*,18*Z*) 2. Tricosa-7,10,18-trienamide	Anticarcinogenic, antiallergic, antibiotic, antioxidative, anti-inflammatory, anti-HIV, and antidiarrhoeal activities	(Jiang et al., 2008)
Menispermaceae	*Cissampelos*	*Cissampelos glaberrima* A.St.-Hil	1. Deca-2*E*,4*E*-dienoic acid isobutylamide 2. Octa-2*E*,4*E*-dienoic acid isobutylamide 3. Decen-2-oic acid isobutylamide 4. Decanoic acid isobutylamidel	Traditional medicine treatment of bronchial and respiratory diseases (asthma, bronchitis), urinary complaints, and bites of snakes and other poisonous animals	(Rosario et al., 1996)

(Continued)

TABLE 10.1
(Continued)

	Rutaceae	*Pilocarpus*	*P. trachylophus* Holmes *P. microphyllus* *P. jaborandi*	1. Epiisopiloturine 2. Pilocarpine 3. Isopilocarpine	Used for treating glaucoma as eye drops; scopolamine, atropine, and hyoscyamine poisonings can be treated	(Paula Santos et al., 2004)
		Zanthoxylum	*Z. ailanthoides* *Z. armatum* *Z. bungeanum* *Z. clava-herculis* L. *Z. integrifolium* (Merr,)	1. Lanyuamide I, II, III 2. Tetrahydrobungeanool 3. Hazaleamide 4. Zanthoamides G-I (1–3) 5. Araliopsine 6. *N*-methylplatydesminium cation 7. *N*-methylatanine 8. Sesamin 9. Hesperetin 10. Hesperidin	Often used in Asian cooking, like Sichuan pepper for its tingling sensation; typical of the culinary art of Northern Vietnam, spice forms the soul of that art; traditionally used to treat inflammatory-related ailments, such as toothaches	(Guetchueng et al., 2018; Nguyen et al., 2019; Chruma et al., 2018)
Aromatic alkamides	Solanaceae	Solanum	*S. khasianum* *S. lycopersicum* *S. citrullifolium* *S. paludosum*	1. *N-p*-coumaroyltyraminecapsaicin 2. Dihydrocapsaicin 3. Homocapsaicin-I & II 4. Homodihydrocapsaicin-I & II 5. Nonivamide 6. Nordihydrocapsaicin	Responsible for hot and pungent sensation; produces "numbing burn" in the throat; used as a topical analgesic and flavoring ingredient	(Mühlenbeck et al., 1996; Kozukue et al., 2005)

Piperaceae	*Piper*	*P. chaba* *P. nigrum* *P. chaba*	1. Guineensine 2. Retrofracamide 3. Pipernonaline 4. Piperrolein 5. Piperchabamide 6. Pellitorin 7. Dehydropipernonaline	Agricultural and pharmaceutical industries can benefit from piper plant amides; used as medicine in Korean, Thai, and Chinese traditional medicine; control of fungal growth and production	(Lee et al., 2008; Jeon et al., 2019)
	Ottonia/ Peperomia	*Ottonia propinqua*	1. Piperovatine *N*-isobutyl-6-(p-methoxyphenyl) 2E,4E-hexadieneamide	Intense drooling and salivation caused by the *Ottonia* genus; *Ottonia* is used as both a local anesthetic and hallucinogen; treatment of toothaches and sore throats	(Antunes et al., 2001)
Brassicaceae	*Lepidium*	*Lepidium meyenii*	1. *m*-methoxybenzyl, 2. Benzylalkamides, 3. Macamides	Boosts fertility and sexual behavior in humans; ancient herbal remedy for menopausal symptoms; immune stimulation and hormone secretion regulation	(Esparza et al., 2015)
Rutaceae	*Glycosmis*	*G. cyanocarpa* *G. mauritiana* *G. citirfolia*	1. Sinharine (cinnamic acid methylsulfidoethylamide) 2. *N*-methyl derivative methylsinharine skimmianine 3. Carbazole glycozolidol	Useful in treating cancer, anemia, rheumatism, fever, cough, and liver-related issues; useful in facial inflammation and skin ailments	(Harald Greger et al., 1992; Greger et al., 1996; Teja et al., 2021)

present in five species: *Wedelia parviceps* Blake (Cariño-Cortés et al., 2010), *Acmella ciliata* (Silveira et al., 2016), *Acmella oleracea* L. (Nakatani and Nagashima, 1992), *Acmella oppositifolia* (Lam.) Jansen (Cortez-Espinosa et al., 2011), and *Heliopsis longipes* (Gray) Blake (Molina-Torres et al., 2004).

Heliopsis longipes contains a greater amount of afinin and capsaicin alkamides than other species, which works contrary to *Escherichia coli* (E. coli), *Pseudomonas solanacearum* (Smith), *Bacillus subtilis* (grass bacillus), and *Saccharomyces cerevisiae* (brewer's yeast) as an antimicrobial agent (Molina-Torres et al., 1999). There is a lack of information on the antimicrobial properties of capsaicin. *Capsicum annuum* (sweet and chili pepper) and *Capsicum frutescens* (tabasco pepper), both bell peppers, exhibit antimicrobial and antibacterial activities and properties against food-borne bacteria (Koffi-Nevry et al., 2012).

10.3 ALKAMIDES IN RHIZOSPHERIC SIGNALING

Alkamides perform a wide range of biological activities in plants, so they are of growing interest as a study focus (Figure 10.2). In *Arabidopsis*, the naturally occurring alkamides of affinin affect root development and plant growth, as reported by Ramírez-Chávez et al. (2004). They also stimulate the lateral root formation and lengthening of root hairs in *Arabidopsis*. Alkamides can regulate plant growth activity and stimulate biomass production in tobacco and rice (Ortíz-Castro et al., 2009). The alkamide amidenin, which is isolated from an actinomycete fungus, can stimulate rice seedlings' growth (Kanbe et al., 1993). Cortez-Espinosa et al. (2011) reported *that Acmella radicans* and *Cissampelos glaberrima* contain the naturally occurring alkamides *N*-isobutyl-2E-decenamide and *N*-isobutyl-decanamide,

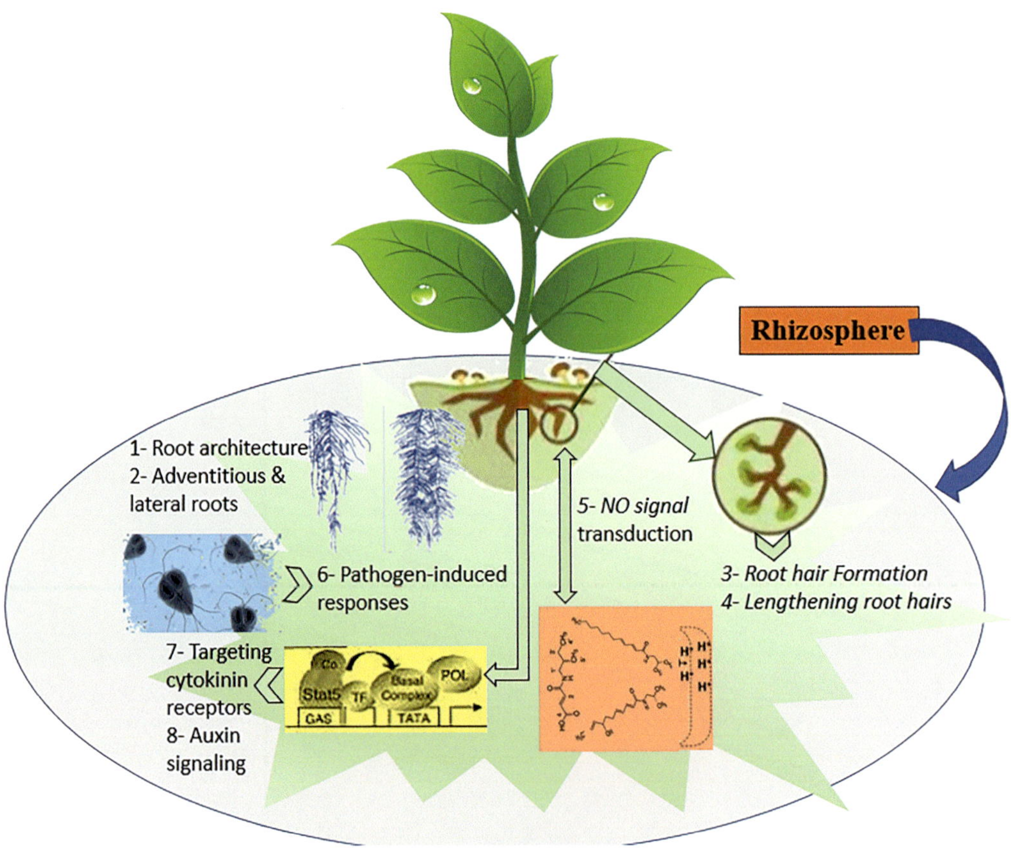

FIGURE 10.2 The roles of alkamides in rhizospheric interaction.

which stimulate root development. Campos-Cuevas et al. (2008) stated that external application of naturally occurring alkamides such as *N*-isobutyl decanamide, *N*-isobutyl-2*E*,6*Z*,8*E*-decatrienamide, and *N*-acylethanolamines promotes adventitious and lateral root creation from the stem and primary root in *Arabidopsis thaliana* seedlings *via* the NO signal transduction pathway. In response to the acylamide decanamide (*N*-isobutyl decanamide), nitric oxide (NO), which is formed by the catalytic reduction of affinin, plays a role in lateral root emergence, confirming that NO is an intermediate in the response to this acyl amide (Méndez-Bravo et al., 2010). In studies comparing the effects of *N*-decanoylethanolamide on root development, *NAE*s and alkamides were found to have biological activities that are similar to those that have been identified in higher plants that have more effects on root development.

As plants adjust their metabolism to abiotic and biotic stress, they also activate a variety of cellular and systemic responses. Numerous studies have demonstrated different interactions between alkamides and plants. They have been widely used in traditional medicines and also have the ability to develop chemical signals that can alter the plants' morphological, physiological, and adaptive processes. López-Bucio et al. (2007) reported that in plants that do not synthesize alkamides, alkamides can alter both developmental and stress response pathways. Méndez-Bravo et al. (2010) and Ramírez-Chávez et al. (2004) reported in their experiments that the alkamide affinin (*N*-isobutyl-2*E*,6*Z*,8*E*-decatrienamide) has altered *Arabidopsis thaliana* (Arabidopsis) root system architecture by increasing lateral root emergence and root hair elongation. A synthetic acyl amide formed by reducing affinin called *N*-isobutyl decanamide (decanamide) is shown to positively influence lateral root development and root establishment independent of auxin-responsive gene expression in *Arabidopsis* (Méndez-Bravo et al., 2011). Molinatorres et al. (1996) reported that alkamides from *E. purpurea*, *E. angustifolia*, and *H. longipes* plants are broadly used for management of the common cold and upper respiratory infections, Furthermore, they produce chemical signals and small lipophilic molecules that are potentially useful to plants. It was found that *N*-acylethanolamides (NAEs), which are a class of endogenous cannabinoids structurally related to both alkamides and anandamide, are involved in seed germination and pathogen-induced responses (Chapman, 2000; 2004).

Spilanthol (*Spilanthes acmella*) and its semi-synthetic derivatives were discovered to show a stimulatory effect on root development in *Arabidopsis* in a significant way. Among those investigated, *Spilanthes acmella* induced root growth and elongation of root hair, while its derivatives had greater potency in stimulating root growth. In *Arabidopsis*, it has been shown to stimulate adventitious root formation by targeting cytokinin receptors (López-Bucio et al., 2007; Campos-Cuevas et al., 2008). It has been demonstrated that *N*-alkylamides cause lateral root primordia to be formed

TABLE 10.2
Microbes/Microorganisms Associated with Alkamides' Signaling in Rhizosphere

Microbes	Microorganism-to-Plant Signaling	Plant-to-Microorganism Signaling	Reference
Nematodes	Acarosides	Caryophyllenes	(Berendsen et al., 2012; Venturi and Christoph Keel, 2016)
PGPR	QS molecules antimicrobials VOCs, phytohormones	LUxR-binding plant compounds QS mimics	
PGPF	QS molecules antimicrobials VOCs phytohormones	LUxR-binding plant compounds QS mimics	
Mycorrhiza	Myc-LCOs	Strigolactones cutin monomers	
Rhizobia	Nod-LCOs	Flavonoids	

PGPR: Plant Growth Promoting Rhizobacteria; PGPF: Plant Growth Promoting Fungi; VOCs: Volatile Organic Compounds; LCOs: Lipo-Chitooligosaccharides

independent of auxin signaling. Occurrence of *N*-alkylamides encourages the creation of nitric oxide, which plays a role in signal transduction.

10.4 IMPROVEMENT AND RESTORATION OF SOIL FERTILITY

Plant biostimulants in low concentrations can be used directly on plants to boost growth, development, and yield in order by increase nutrient uptake (Van Oosten et al., 2017). The phytochemical effects of humic compounds are believed to be a consequence of their putative hormone effects caused by gibberellins, cytokinins, NO, and ethylene, among other plant hormones. These hormones are brought into being in inconsequential amount in soils (Mora et al., 2010; 2014). Zandonadi et al. (2019) observed the conception of *N*-isopropyldecanamide and found that its effects on root growth were concentration dependent, and its high contents boosted root growth. However, low concentrations promoted root growth analogous to humic acid-(HA) prompted plasma membrane H^+-ATPase activity. *N*-isopropyldecanamide and HA are useful to enhance NO congregation during lateral root development. Overall, the special effects of biostimulants are formed via HA and plant growth–promoting bacteria (PGPB) on the physiology and metabolism of plants. The widespread dissemination of these lipids has the ability of unbound soil fractions and compost establishments.

10.5 FUTURE PERSPECTIVES

Based on previous scarce data availability about the biosynthesis and mechanisms in plants, Alkamides may affect a particular signal transduction that modifies root hair growth and differentiation. Further experimentation is needed to determine whether or not Alkamides alter the signaling pathways among plants' chemical behavior. The chemical structure of alkamides from previous studies shows that they are structurally related to the ceramide and glucosylceramide compounds (Ramírez-Chávez et al., 2004). The biological and chemical behavior of alkamides among these compounds through signaling pathways also need to be determined. The existence of new plant growth–regulating alkamide species needs be characterized, and it will be interesting to investigate the variation of signaling pathways and enzymatic activities among various species and compounds. More research is still vital to estimate and verify the molecular mechanisms to provide new insights into the development of Alkamides and their interaction with plants' development.

10.6 CONCLUSION

In the past few years, remarkable developments have been made in our understanding of alkamides and their interaction with plants. Alkamides show chemical signaling and alter metabolic pathways and gene regulation against biotic and abiotic stress. They also promote the growth and development of plants. Another important role of Alkamide has been recorded in traditional medicine: curing many diseases as an antimicrobial and antibacterial agent. It also has a potential role as a pesticide.

REFERENCES

Althaus, Julia B., Claudine Malyszek, Marcel Kaiser, Reto Brun, and Thomas J. Schmidt. 2017. "Alkamides from *Anacyclus Pyrethrum* L. and Their in Vitro Antiprotozoal Activity." *Molecules* 22 (5): 1–9. https://doi.org/10.3390/molecules22050796.

Antunes, P. A., G. O. Chierice, C. J.L. Constantino, and R. F. Aroca. 2001. "Spectroscopic Characterization of N-Isobutyl-6-(p-Methoxyphenyl) 2E,4E-Hexadieneamide Extracted from Ottonia Propinqua." *Vibrational Spectroscopy* 27 (2): 175–81. https://doi.org/10.1016/S0924-2031(01)00132-1.

Berendsen, Roeland L., Corné M.J. Pieterse, and Peter A.H.M. Bakker. 2012. "The Rhizosphere Microbiome and Plant Health." *Trends in Plant Science* 17 (8): 478–86. https://doi.org/10.1016/j.tplants.2012.04.001.

Binns, Shannon E., Iniyal Inparajah, Bernard R. Baum, and John T. Arnason. 2001. "Methyl Jasmonate Increases Reported Alkamides and Ketoalkene/Ynes in *Echinacea pallida* (Asteraceae)." *Phytochemistry* 57 (3): 417–20. https://doi.org/10.1016/S0031-9422(00)00444-1.
Campos-Cuevas, Juan Carlos, Ramón Pelagio-Flores, Javier Raya-González, Alfonso Méndez-Bravo, Randy Ortiz-Castro, and José López-Bucio. 2008. "Tissue Culture of *Arabidopsis thaliana* Explants Reveals a Stimulatory Effect of Alkamides on Adventitious Root Formation and Nitric Oxide Accumulation." *Plant Science* 174 (2): 165–73. https://doi.org/10.1016/j.plantsci.2007.11.003.
Cariño-Cortés, R., J. A. Gayosso-De-Lucio, M. I. Ortiz, M. Sánchez-Gutiérrez, P. B. García-Reyna, V. G. Cilia-López, N. Pérez-Hernández, E. Moreno, and H. Ponce-Monter. 2010. "Antinociceptive, Genotoxic and Histopathological Study of *Heliopsis Longipes* S.F. Blake in Mice." *Journal of Ethnopharmacology* 130 (2): 216–21. https://doi.org/10.1016/j.jep.2010.04.037.
Casado, Mariela, Maria Gabriela Ortega, Mariana Peralta, Alicia Mariel Agnese, and Jose Luis Cabrera. 2009. "Two New Alkamides from Roots of *Acmella Decumbens*." *Natural Product Research* 23 (14): 1298–1303. https://doi.org/10.1080/14786410802518201.
Chapman, Kent D. 2000. "Emerging Physiological Roles for N-Acylphosphatidylethanolamine Metabolism in Plants: Signal Transduction and Membrane Protection." *Chemistry and Physics of Lipids* 108 (1–2): 221–29.
Chapman, Kent D. 2004. "Occurrence, Metabolism, and Prospective Functions of N-Acylethanolamines in Plants." *Progress in Lipid Research* 43 (4): 302–27.
Chruma, Jason J., Douglas J. Cullen, Lydia Bowman, and Patrick H. Toy. 2018. "Polyunsaturated Fatty Acid Amides from the: Zanthoxylum Genus-from Culinary Curiosities to Probes for Chemical Biology." *Natural Product Reports* 35 (1): 54–74. https://doi.org/10.1039/c7np00044h.
Cortez-Espinosa, Nancy, Judit A. Aviña-Verduzco, Enrique Ramírez-Chávez, Jorge Molina-Torres, and Patricia Ríos-Chávez. 2011. "Valine and Phenylalanine as Precursors in the Biosynthesis of Alkamides in Acmella Radicans." *Natural Product Communications* 6 (6): 857–61. https://doi.org/10.1177/1934578x1100600625.
Dandin, V. S., P. M. Naik, H. N. Murthy, S. Y. Park, E. J. Lee, and K. Y. Paek. 2014. "Rapid Regeneration and Analysis of Genetic Fidelity and Scopoletin Contents of Micropropagated Plants of Spilanthes Oleracea L." *Journal of Horticultural Science and Biotechnology* 89 (1): 79–85. https://doi.org/10.1080/14620316.2014.11513052.
Dawid, Corinna, Andrea Henze, Oliver Frank, Anneke Glabasnia, Mathias Rupp, Kirsten Büning, Diana Orlikowski, Matthias Bader, and Thomas Hofmann. 2012. "Structural and Sensory Characterization of Key Pungent and Tingling Compounds from Black Pepper (*Piper Nigrum* L.)." *Journal of Agricultural and Food Chemistry* 60 (11): 2884–95. https://doi.org/10.1021/jf300036a.
Dominguez, X A, Sánchez H, Slim J S, J Jakupovic, L Lehmann, and F Bohlmann. 1987. "Highly Unsaturated Amides from *Sanvitalia Oxymoides*." *Revista Latinoamericana de Quimica* 18: 114–15.
Dubey, Suchita, Siddhartha Maity, Mahendra Singh, Shubhini A. Saraf, and Sudipta Saha. 2013. "Phytochemistry, Pharmacology and Toxicology of Spilanthes Acmella: A Review." *Advances in Pharmacological Sciences* 2013. https://doi.org/10.1155/2013/423750.
Esparza, Eliana, Antonella Hadzich, Waltraud Kofer, Axel Mithöfer, and Eric G. Cosio. 2015. "Bioactive Maca *(Lepidium Meyenii)* Alkamides Are a Result of Traditional Andean Postharvest Drying Practices." *Phytochemistry* 116 (1): 138–48. https://doi.org/10.1016/J.PHYTOCHEM.2015.02.030.
Faure, Lionel, Ronaldo Cavazos, Bibi Rafeiza Khan, Robby A. Petros, Peter Koulen, Elison B. Blancaflor, and Kent D. Chapman. 2015. "Effects of Synthetic Alkamides on Arabidopsis Fatty Acid Amide Hydrolase Activity and Plant Development." *Phytochemistry* 110: 58–71. https://doi.org/10.1016/j.phytochem.2014.11.011.
Fontenele, Juvenia B, L. Kalyne A.M Leal, Edilberto R Silveira, F. Helder Felix, Cícero F. Bezerra Felipe, and Glauce S.B Viana. 2009. "Antiplatelet Effects of Piplartine, an Alkamide Isolated from *Piper Tuberculatum*: Possible Involvement of Cyclooxygenase Blockade and Antioxidant Activity." *Journal of Pharmacy and Pharmacology* 61 (4): 511–15. https://doi.org/10.1211/jpp/61.04.0014.
Gamboa-Leon, Rubi, and William Scott Chilton. 2000. "Isobutylamide Numbing Agents of Toothache Grass, Ctenium Aromaticum." *Biochemical Systematics and Ecology* 28 (10): 1019–21. https://doi.org/10.1016/S0305-1978(00)00011-9.
Gertsch, Jürg. 2008. "Immunomodulatory Lipids in Plants: Plant Fatty Acid Amides and the Human Endocannabinoid System." *Planta Medica* 74 (6): 638–50. https://doi.org/10.1055/s-2008-1034302.
Greger, H., G. Zechner, O. Hofer, and S. Vajrodaya. 1996. "Bioactive Amides from *Glycosmis* Species." *Journal of Natural Products* 59 (12): 1163–68. https://doi.org/10.1021/NP9604238.

Greger, Harald. 2016. "Alkamides: A Critical Reconsideration of a Multifunctional Class of Unsaturated Fatty Acid Amides." *Phytochemistry Reviews* 15 (5): 729–70. https://doi.org/10.1007/s11101-015-9418-0.

Greger, Harald, Otmar Hofer, Hanspeter Kählig, and Gerald Wurz. 1992. "Sulfur Containing Cinnamides with Antifungal Activity from *Glycosmis Cyanocarpa*." *Tetrahedron* 48 (7): 1209–18. https://doi.org/10.1016/S0040-4020(01)90784-7.

Guetchueng, Stephanie T., Lutfun Nahar, Kenneth J. Ritchie, Fyaz M.D. Ismail, Andrew R. Evans, and Satyajit D. Sarker. 2018. "Zanthoamides G-I: Three New Alkamides from *Zanthoxylum Zanthoxyloides*." *Phytochemistry Letters* 26: 125–29. https://doi.org/10.1016/j.phytol.2018.05.031.

Hernández, Ivones, Lucía Márquez, Ioanna Martínez, Rodrigo Dieguez, Carla Delporte, Sylvia Prieto, Jorge Molina-Torres, and Gabino Garrido. 2009. "Anti-Inflammatory Effects of Ethanolic Extract and Alkamides-Derived from *Heliopsis Longipes* Roots." *Journal of Ethnopharmacology* 124 (3): 649–52. https://doi.org/10.1016/j.jep.2009.04.060.

Hohmann, Judit, Dóra Rédei, Peter Forgo, Pál Szabó, Tamás F. Freund, József Haller, Engin Bojnik, and Sándor Benyhe. 2011. "Alkamides and a Neolignan from *Echinacea Purpurea* Roots and the Interaction of Alkamides with G-Protein-Coupled Cannabinoid Receptors." *Phytochemistry* 72 (14–15): 1848–53. https://doi.org/10.1016/j.phytochem.2011.06.008.

Holzbach, Juliana C., and Lucia M.X. Lopes. 2010. "Aristolactams and Alkamides of *Aristolochia Gigantea*." *Molecules* 15 (12): 9462–72. https://doi.org/10.3390/molecules15129462.

Jeon, Hwang Ju, Kyeongnam Kim, Yong Deuk Kim, and Sung Eun Lee. 2019. "Naturally Occurring Piper Plant Amides Potential in Agricultural and Pharmaceutical Industries: Perspectives of Piperine and Piperlongumine." *Applied Biological Chemistry* 62 (1): 1–7. https://doi.org/10.1186/S13765-019-0471-Z/FIGURES/2.

Jiang, Liyan, Juanjuan Chen, Shan He, and Cuirong Sun. 2009. "High-Throughput Structural Elucidation of Amides in *Mallotus Lianus* Croiz Plant Materials by LC-ESI-MS-MS." *Chromatographia* 70 (3–4): 439–45. https://doi.org/10.1365/s10337-009-1160-6.

Jiang, Liyan, Yanbin Lu, Shan He, Yuanjiang Pan, Cuirong Sun, and Tianxing Wu. 2008. "Preparative Isolation and Purification of Two Amides from *Mallotus Lianus* Croiz by High-Speed Counter-Current Chromatography." *Journal of Separation Science* 31 (22): 3930–35. https://doi.org/10.1002/jssc.200800396.

Johns, Timothy, Kenneth Graham, and G. H.Neil Towers. 1982. "Molluscicidal Activity of Affinin and Other Isobutylamides from the Asteraceae." *Phytochemistry* 21 (11): 2737–38. https://doi.org/10.1016/0031-9422(82)83110-5.

Joshi, Veenu, G.D. Sharma, and S.K. Jadhav. 2020. "Alkamides: Multifunctional Bioactive Agents in Spilanthes Spp." *Journal of Scientific Research* 64 (01): 198–206. https://doi.org/10.37398/jsr.2020.640129.

Kanbe, Kenji, Hiroshi Naganawa, Mitiyasu Okamura, Toru Sasaki, Masa Hamada, Yoshiro Okami, and Tomio Takeuchi. 1993. "Amidenin, a New Plant Growth-Regulating Substance Isolated from *Amycolatopsis* Sp." *Bioscience, Biotechnology, and Biochemistry* 57 (8): 1261–63. https://doi.org/10.1271/bbb.57.1261.

Kavit, Mehta, B N Patel, and B K Jain. 2013. "Phytochemical Analysis of Leaf Extract of *Phyllanthus Fraternus*." *Research Journal of Recent Sciences* 2: 12–15.

Kehie, Mechuselie, Suman Kumaria, Pramod Tandon, and Nirala Ramchiary. 2015. "Biotechnological Advances on in Vitro Capsaicinoids Biosynthesis in Capsicum: A Review." *Phytochemistry Reviews* 14 (2): 189–201. https://doi.org/10.1007/s11101-014-9344-6.

Koffi-Nevry, Rose, Kouassi Clément Kouassi, Zinzerdof Yesse Nanga, Marina Koussémon, and Guillaume Yao Loukou. 2012. "Antibacterial Activity of Two Bell Pepper Extracts: *Capsicum Annuum* L. and *Capsicum Frutescens*." *Taylor & Francis* 15 (5): 961–71. https://doi.org/10.1080/10942912.2010.509896.

Korkmaz, Serol. 2018. "Antioxidants in Maca (*Lepidium Meyenii*) as a Supplement in Nutrition." *Antioxidants in Foods and Its Applications*. https://doi.org/10.5772/intechopen.75582.

Kozukue, Nobuyuki, Jae Sook Han, Etsuko Kozukue, Sin Jung Lee, Joung Ae Kim, Kap Rang Lee, Carol E. Levin, and Mendel Friedman. 2005. "Analysis of Eight Capsaicinoids in Peppers and Pepper-Containing Foods by High-Performance Liquid Chromatography and Liquid Chromatography – Mass Spectrometry." *Journal of Agricultural and Food Chemistry* 53 (23): 9172–81. https://doi.org/10.1021/JF050469J.

Lee, Seung Woong, Young Kook Kim, Koanhoi Kim, Hyun Sun Lee, Jung Ho Choi, Woo Song Lee, Chang Duk Jun, Jee Hun Park, Jeong Min Lee, and Mun Chual Rho. 2008. "Alkamides from the Fruits of *Piper Longum* and *Piper Nigrum* Displaying Potent Cell Adhesion Inhibition." *Bioorganic & Medicinal Chemistry Letters* 18 (16): 4544–46. https://doi.org/10.1016/J.BMCL.2008.07.045.

López-Bucio, José, Gustavo Acevedo-Hernández, Enrique Ramírez-Chávez, Jorge Molina-Torres, and Luis Herrera-Estrella. 2006. "Novel Signals for Plant Development." *Current Opinion in Plant Biology* 9 (5): 523–29. https://doi.org/10.1016/j.pbi.2006.07.002.

López-Bucio, José, Mayra Millán-Godínez, Alfonso Méndez-Bravo, Alina Morquecho-Contreras, Enrique Ramírez-Chávez, Jorge Molina-Torres, Anahí Pérez-Torres, Masayuki Higuchi, Tatsuo Kakimoto, and Luis Herrera-Estrella. 2007. "Cytokinin Receptors are Involved in Alkamide Regulation of Root and Shoot Development in Arabidopsis." *Plant Physiology* 145 (4): 1703–13. https://doi.org/10.1104/pp.107.107953.

Manvar, Mital N., and T. R. Desai. 2013. "Phytochemical and Pharmacological Profile of *Ipomoea Aquatica*." *Indian Journal of Medical Sciences* 67 (3): 49–60. https://doi.org/10.4103/0019-5359.121115.

Méndez-Bravo, Alfonso, Carlos Calderón-Vázquez, Enrique Ibarra-Laclette, Javier Raya-González, Enrique Ram\'\irez-Chávez, Jorge Molina-Torres, Angel A Guevara-Garcia, José López-Bucio, and Luis Herrera-Estrella. 2011. "Alkamides Activate Jasmonic Acid Biosynthesis and Signaling Pathways and Confer Resistance to Botrytis Cinerea in *Arabidopsis Thaliana*." *PloS One* 6 (11): e27251.

Méndez-Bravo, Alfonso, Javier Raya-González, Luis Herrera-Estrella, and José López-Bucio. 2010. "Nitric Oxide Is Involved in Alkamide-Induced Lateral Root Development in Arabidopsis." *Plant and Cell Physiology* 51 (10): 1612–26. https://doi.org/10.1093/pcp/pcq117.

Molina-Torres, J., A Garcia-Chavez, and E Ramirez-Chavez. 1999. "Antimicrobial Properties of Alkamides Present in Flavouring Plants Traditionally Used in Mesoamerica: Affinin and Capsaicin. PG – 241–8 AB – The Bioactive Amides Affinin and Capsaicin Isolated Respectively from *Heliopsis Longipes* Roots and Capsicum Spp Fr." *J Ethnopharmacol* 64 (3): 241–8.

Molina-Torres, Jorge, Carlos Salazar-Cabrera, Concepción Armenta-Salinas, and Enrique Ramírez-Chávez. 2004. "Fungistatic and Bacteriostatic Activities of Alkamides from *Heliopsis Longipes* Roots: Affinin and Reduced Amides." *Journal of Agricultural and Food Chemistry* 52 (15): 4700–4. https://doi.org/10.1021/JF034374Y.

Molina-Torres, Jorge, Rafael Salgado-Garciglia, Enrique Ramirez-Chavez, and Rosa E. Del Rio. 1996. "Purely Olefinic Alkamides in *Heliopsis Longipes* and *Acmella* (Spilanthes) Oppositifolia." *Biochemical Systematics and Ecology* 24 (1): 43–7. https://doi.org/10.1016/0305-1978(95)00099-2.

Mora, Verónica, Eva Bacaicoa, Angel Maria Zamarreño, Elena Aguirre, Maria Garnica, Marta Fuentes, and José Maria García-Mina. 2010. "Action of Humic Acid on Promotion of Cucumber Shoot Growth Involves Nitrate-Related Changes Associated with the Root-to-Shoot Distribution of Cytokinins, Polyamines and Mineral Nutrients." *Journal of Plant Physiology* 167 (8): 633–42. https://doi.org/10.1016/j.jplph.2009.11.018.

Mora, Verónica, Eva Bacaicoa, Roberto Baigorri, Angel M. Zamarreño, and José M. García-Mina. 2014. "NO and IAA Key Regulators in the Shoot Growth Promoting Action of Humic Acid in *Cucumis Sativus* L." *Journal of Plant Growth Regulation* 33 (2): 430–39. https://doi.org/10.1007/s00344-013-9394-9.

Mühlenbeck, Ute, Albertus Kortenbusch, and Wolfgang Barz. 1996. "Formation of Hydroxycinnamoylamides and α-Hydroxyacetovanillone in Cell Cultures of *Solanum Khasianum*." *Phytochemistry* 42 (6): 1573–79. https://doi.org/https://doi.org/10.1016/0031-9422(96)00173-2.

Muller-Jakic, B., W. Breu, A. Probstle, K. Redl, H. Greger, and R. Bauer. 1994. "In Vitro Inhibition of Cyclooxygenase and 5-Lipoxygenase by Alkamides from Echinaceae and Achillea Species." *Planta Medica* 60 (1): 37–40. https://doi.org/10.1055/s-2006-959404.

Mumivand, Hasan, Mesbah Babalar, Leila Tabrizi, Lyle E. Craker, Majid Shokrpour, and Javad Hadian. 2017. "Antioxidant Properties and Principal Phenolic Phytochemicals of Iranian Tarragon (*Artemisia Dracunculus* L) Accessions." *Horticulture, Environment, and Biotechnology* 58 (4): 414–22. https://doi.org/10.1007/S13580-017-0121-5.

Nakatani, Nobuji, and Mayumi Nagashima. 1992. "Pungent Alkamides from *Spilanthes Acmella* L. Var. Oleracea Clarke." *Bioscience, Biotechnology, and Biochemistry* 56 (5): 759–62. https://doi.org/10.1271/bbb.56.759.

Nascimento, Jeferson C. Do, Vanderlúcia F. De Paula, Jorge M. David, and Juceni P. David. 2012. "Occurrence, Biological Activities and 13C NMR Data of Amides from Piper (Piperaceae)." *Quimica Nova* 35 (11): 2288–311. https://doi.org/10.1590/S0100-40422012001100037.

Nguyen, DTX, TN Huynh, S Schwaiger, and H Stuppner. 2019. "The Alkamide Constituents of Zanthoxylum Rhetsa (Roxb.) DC. Fruits." *Planta Medica* 85 (18): 223. https://doi.org/10.1055/S-0039-3399933.

Omolo, Morrine A, Zen-Zi Wong, K Mergen, Jennifer C. Hastings, Nina C. Le, Holly A. Reil, Kyle A. Case, David J. Baumler, and others. 2014. "Antimicrobial Properties of Chili Peppers." *Journal of Infectious Diseases and Therapy* 2: 145. https://doi.org/10.4172/2332-0877.1000145

Oosten, Michael James Van, Olimpia Pepe, Stefania De Pascale, Silvia Silletti, and Albino Maggio. 2017. "The Role of Biostimulants and Bioeffectors as Alleviators of Abiotic Stress in Crop Plants." *Chemical and Biological Technologies in Agriculture* 4 (1): 1–12. https://doi.org/10.1186/s40538-017-0089-5.

Ortíz-Castro, Randy, Hexon Angel Contreras-Cornejo, Lourdes Macías-Rodríguez, and José López-Bucio. 2009. "The Role of Microbial Signals in Plant Growth and Development." *Plant Signaling and Behavior* 4 (8): 701–12. https://doi.org/10.4161/psb.4.8.9047.

Paula Santos, Ana, Paulo Roberto Hrihorowitsch Moreno, P R H Moreno, and Av Lineu Prestes. 2004. "Pilocarpus Spp: A Survey of Its Chemical Constituents and Biological Activities." *Revista Brasileira de Ciências Farmacêuticas* 40 (2): 116–37. https://doi.org/10.1590/S1516-93322004000200002.

Percy, Dane W., Jacqui L. Adcock, Xavier A. Conlan, Neil W. Barnett, Michelle E. Gange, Laura K. Noonan, Luke C. Henderson, and Paul S. Francis. 2010. "Determination of *Citrus Aurantium* Protoalkaloids Using HPLC with Acidic Potassium Permanganate Chemiluminescence Detection." *Talanta* 80 (5): 2191–95. https://doi.org/10.1016/j.talanta.2009.10.052.

Ramírez-Chávez, Enrique, José López-Bucio, Luis Herrera-Estrella, and Jorge Molina-Torres. 2004. "Alkamides Isolated from Plants Promote Growth and Alter Root Development in *Arabidopsis*." *Plant Physiology* 134 (3): 1058–68. https://doi.org/10.1104/pp.103.034553.

Rios-Chavez, P., E. Ramirez-Chavez, C. Armenta-Salinas, and J. Molina-Torres. 2003. "*Acmella Radicans* Var. Radicans: In Vitro Culture Establishment and Alkamide Content." *In Vitro Cellular and Developmental Biology – Plant* 39 (1): 37–41. https://doi.org/10.1079/IVP2002354.

Rios, María Yolanda, A. Berenice Aguilar-Guadarrama, and María del Carmen Gutiérrez. 2007. "Analgesic Activity of Affinin, an Alkamide from *Heliopsis Longipes* (Compositae)." *Journal of Ethnopharmacology* 110 (2): 364–67. https://doi.org/10.1016/j.jep.2006.09.041.

Rios, María Yolanda, and Horacio F. Olivo. 2014. *Natural and Synthetic Alkamides*. https://doi.org/10.1016/b978-0-444-63430-6.00003-5.

Rosario, S L, A J da Silva, and J P Parente. 1996. "Alkamides from *Cissampelos Glaberrima*." *Planta Medica* 62 (4): 376–7. https://doi.org/10.1055/s-2006-957913.

Saadali, Bouchra, Driss Boriky, Mohamed Blaghen, Maurice Vanhaelen, and Mohammed Talbi. 2001. "Alkamides from *Artemisia Dracunculus*." *Phytochemistry* 58 (7): 1083–86. https://doi.org/10.1016/S0031-9422(01)00347-8.

Saeidnia, S., A. R. Gohari, N. Mokhber-Dezfuli, and F. Kiuchi. 2011. "A Review on Phytochemistry and Medicinal Properties of the Genus Achillea." *DARU, Journal of Pharmaceutical Sciences* 19 (3): 173–86.

Saravanakumar, K., Adaikala Raj, and D. Umaiyambigai. 2016. "GC-MS and FT-IR Profiling of Leaves Methanol Extract from the *Pleiospermium Alatum* (Wall. Ex Wt. & Arn) Swingle Rutaceae Family." *The Journal of Phytopharmacology* 5 (5): 201–4. https://doi.org/10.31254/phyto.2016.5506.

Sharma, Rahul, and Neelakantan Arumugam. 2021. "N-Alkylamides of Spilanthes (Syn: Acmella): Structure, Purification, Characterization, Biological Activities and Applications – a Review." *Future Foods* 3: 100022. https://doi.org/10.1016/j.fufo.2021.100022.

Si, Xiao Tang, Man Li Zhang, Qing Wen Shi, and Hiromasa Kiyota. 2006. "Chemical Constituents of the Plants in the Genus Achillea." *Chemistry and Biodiversity* 3 (11): 1163–80. https://doi.org/10.1002/cbdv.200690119.

Silveira, Narjara, Julia Saar, Alan Diego C. Santos, Andersson Barison, Louis P. Sandjo, Marcel Kaiser, Thomas J. Schmidt, and Maique W. Biavatti. 2016. "A New Alkamide with an Endoperoxide Structure from *Acmella Ciliata* (Asteraceae) and Its in Vitro Antiplasmodial Activity." *Molecules* 21 (6). https://doi.org/10.3390/molecules21060765.

Srinivas, P V, and J Madhusudana Rao. 1999. "Isopiperolein B: An Alkamide from *Piper Nigrum*." *Phytochemistry* 52 (5): 957–58.

Teja, Parusu Kavya, Prachi Patel, Drashti Bhavsar, Chintakindi Bindusri, Kishori Jadhav, and Siddheshwar K. Chauthe. 2021. "Traditional Uses, Phytochemistry, Pharmacology, Toxicology and Formulation Aspects of Glycosmis Species: A Systematic Review." *Phytochemistry* 190: 112865. https://doi.org/10.1016/J.PHYTOCHEM.2021.112865.

Tofern, Britta, Petra Mann, MacKi Kaloga, Kristina Jenett-Siems, Ludger Witte, and Eckart Eich. 1999. "Aliphatic Pyrrolidine Amides from Two Tropical Convolvulaceous Species." *Phytochemistry* 52 (8): 1437–41. https://doi.org/10.1016/S0031-9422(99)00245-9.

Vázquez-Martínez, Juan, Juan Manuel Gutierrez-Villagomez, Tonatiu Campos-García, Jorge Molina-Torres, and Mercedes G. López. 2020. "Alkamides and Piperamides as Potential Antivirals against the Severe Acute Respiratory Syndrome Coronavirus 2 (SARS-CoV-2)." *Journal of Physical Chemistry Letters* 11 (19): 8008–16. https://doi.org/10.1021/acs.jpclett.0c01685.

Venturi, Vittorio, and Christoph Keel. 2016. "Signaling in the Rhizosphere." *Trends in Plant Science* 21 (3): 187–98. https://doi.org/10.1016/j.tplants.2016.01.005.

Veryser, Lieselotte, Lien Taevernier, Evelien Wynendaele, Yannick Verheust, Ann Dumoulin, and Bart De Spiegeleer. 2017. "N-Alkylamide Profiling of *Achillea Ptarmica* and *Achillea Millefolium* Extracts by Liquid and Gas Chromatography–Mass Spectrometry." *Journal of Pharmaceutical Analysis* 7 (1): 34–47. https://doi.org/10.1016/j.jpha.2016.09.005.

Woelkart, Karin, and Rudolf Bauer. 2007. "The Role of Alkamides as an Active Principle of Echinacea." *Planta Medica* 73 (7): 615–23. https://doi.org/10.1055/s-2007-981531.

Woelkart, Karin, Christoph Koidl, Andrea Grisold, J. David Gangemi, Ronald B. Turner, Egon Marth, and Rudolf Bauer. 2005. "Bioavailability and Pharmacokinetics of Alkamides From the Roots of *Echinacea Angustifolia* in Humans." *The Journal of Clinical Pharmacology* 45 (6): 683–89. https://doi.org/10.1177/0091270004273493.

Zandonadi, Daniel Basílio, Carlos Roberto Ribeiro Matos, Rosane Nora Castro, Riccardo Spaccini, Fábio Lopes Olivares, and Luciano Pasqualoto Canellas. 2019. "Alkamides: A New Class of Plant Growth Regulators Linked to Humic Acid Bioactivity." *Chemical and Biological Technologies in Agriculture* 6 (1): 1–12. https://doi.org/10.1186/s40538-019-0161-4.

Zhao, Jianping, Ilias Muhammad, D. Chuck Dunbar, Jamal Mustafa, and Ikhlas A. Khan. 2005. "New Alkamides from Maca (*Lepidium Meyenii*)." *Journal of Agricultural and Food Chemistry* 53 (3): 690–93. https://doi.org/10.1021/jf048529t.

11 Humic Acids and Fatty Acid Amides
Soil-Plant Interactions

Zandonadi DB, Santos MP, Olivares FL, and Canellas LP

CONTENTS

11.0 INTRODUCTION

Humic substances (HS) are the main components of organic carbon in waters, soils, and sediments (Stevenson, 1994), performing several ecological functions in both natural and agro-ecosystems, including the nutrients supply, by chelating minerals and, through their own degradation, shaping soil biotic communities and determining soil health and fertility (Nardi et al., 2018; Canellas and Olivares, 2017; Silva et al., 2021). Moreover, HS directly affects plant metabolism, resulting in mineral nutrient uptake and morphological changes (Nardi et al., 2018). Humic substances are the most active ingredients used for plant biostimulant formulation (du Jardin, 2015). Alkaline-extracted humic products are commercially available and are applied on farms at low concentrations, favouring both crop growth and economic yield (Olk et al., 2018).

Humic substances are a complex noncovalent supramolecular association of small heterogeneous molecules formed from the decomposition of organism tissues bound together by weak dispersive forces, hydrogen bonds, and metal-bridged intermolecular electrostatic interactions (Piccolo, 2001; 2002). Humic substance fractions have been defined by their solubility in water at different pH values: the fulvic acids (FA), which are soluble under all pH values; the humic acids (HA), which are insoluble at pH value < 2; and the humins, which are insoluble at any pH value (Stevenson, 1994).

The inherent chemical complexity of HS favours spontaneous interactions when in solution, resulting in the assembly of constituent molecules before contact with any surface. When many organic compounds are in mixture, the thermodynamic conditions drive to reduce the exposure of hydrophobic components to the aqueous medium, increasing the reciprocal attraction of each compound. Thus, the self-organization of hydrophobic and amphiphilic compounds leads to a progressive separation from the network of water structure. This increases the mixture entropy and the overall energy stabilization, forming a superstructure (Piccolo, 2002). This behavior is known as the hydrophobic effect: i.e., the entropy-driven tendency to exclude water molecules from humic aggregates, thus decreasing the total molecular energy in the solution (Piccolo, 2002). The system is dominated by non-molecular interactions and adopts a configuration in which the potential energy is the lowest possible, then leading supramolecular assemblies of biomolecules (e.g., proteins, lipids,

DOI: 10.1201/9781003224525-14

lignin, and nucleic acids) in solution through hydrogen bonds, hydrophobic forces, or electrostatic interactions.

Sequential oxidation of HA from peat released lipids with high unsaturated fatty acid composition (Lehtonen et al., 2001). The carbon distributions of the biomolecules from other HA sources also presented a relatively high content of lipids (Lamar et al., 2021). The role of these HA-derived hydrophobic molecules in plant growth are still poorly understood. Recent advances have been made toward a possible mechanism of action in which alkamides from HA stimulate maize root growth by inducing nitric oxide (NO) in root endodermis while simultaneously enhancing plasma membrane H^+-ATPase proton transport and ATP hydrolysis (Zandonadi et al., 2019). Interestingly, alkamides are molecules containing fatty acid amides with typical olefinic and acetylenic compounds with biological activity (Greger, 1984). Physiological effects of alkamides in plants have been reported, such as root morphological changes and NO production (Ramiréz-Chavez et al., 2004; Méndez-Bravo et al., 2010). Accordingly, plants uptake HA-derived moieties, which accumulate in root endodermis and partly translocate to shoots, modifying both polar and neutral lipids profiles in cucumber tissues (Kulikova et al., 2016). These findings depict the importance of HA at the soil-plant interface and the emergent role of alkamides. This chapter will focus on the chemistry of HS and how its hydrophobic properties could affect plant growth regulation.

11.1 THE HUMIC SUBSTANCES CHEMISTRY

For most practical purposes, a study of the chemical composition of HS can be limited to the broad groups of compounds: fats, waxes, and resins, or substances soluble in apolar solvents; carbohydrates and related compounds; proteins and their derivatives; and lignins and their derivatives. Nuclear magnetic resonance (NMR) spectroscopy is a potent tool for molecular investigation of complex chemical entities (Masoom et al., 2016). The chemical functional landscape of HS is readily provided by CP/MAS ^{13}C NMR, which allows access to the chemical nature at solid-state environmental samples. Humic acids isolated from vermicompost present typical distribution of functional groups (Figure 11.1). In solution, these groups showed mobility, and the main units of the building blocks of the supramolecular structure can be seen according to the specific diffusion coefficients assessed by diffusion-ordered nuclear magnetic resonance spectroscopy (DOSY-NMR) (Šmejkalová and Piccolo, 2008a, b). Humic acids treated with acetic acid change diffusivity due to the disruption of large assemblies in more rapidly diffusing smaller aggregates (Figure 11.2).

The typical HA spectrum (Figure 11.1) exhibited broad but resolved peaks for carbons in the alkyl (0–50 ppm), O-alkyl (50–108 ppm), aromatic (110–160 ppm), and carboxyl/carbonyl (160–220 ppm) regions. The alkyl area is dominated by a peak at 29 ppm and showed the presence of long-chain polymethylene structures. Other shoulders are also present at 15 and 20 ppm, attributed to the terminal methyl carbon from either short-chain hydrocarbons or long-chain n-alkanes. The O-alkyl region contains a resonance at 72 ppm, attributed to the presence of sugars, cellulose, alcohols, and/or ethers, and a peak in the 50–60 ppm region, characteristic of methoxy, N alkyl C from OCH_3, C-N, and complex aliphatic carbons. The presence of this shoulder is generally attributed to the presence of lignin. The aromatic carbon resonance region is dominated by a peak centred at 129 ppm and indicates the presence of unsubstituted and alkyl-substituted aromatic carbon. The presence of the phenolic carbons is indicated by a resonance in the chemical shift range of 145–162 ppm. The carboxyl-like carbon region presented a major peak at 172 ppm, indicating the presence of carboxyl, ester, or amide carbon.

The inventory of singular molecules in the HS assembly observed in each described region of bulk solid samples obtained by CP-MAS ^{13}C-NMR was provided by a sequential chemical fractionation designed by Nebbioso and Piccolo (2011) as humeomic in analogy with other omics sciences. The authors developed an analytical march capable of isolating step by step the different types of unassociated molecules weakly, moderately, and strongly associated with a humic matrix. Thousands of different types of molecules were isolated and identified by analytical methods from

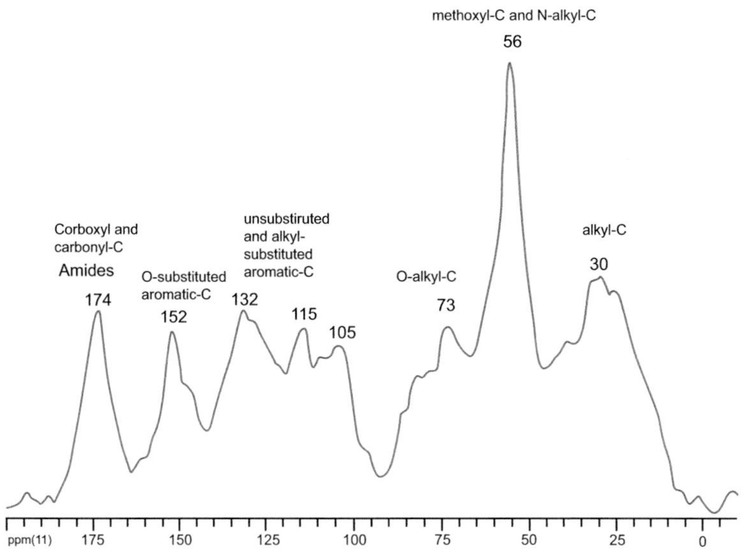

FIGURE 11.1 Solid-state NMR spectroscopy (^{13}C-NMR) cross-polarization magic angle spinning (CPMAS) spectrum of humic acid (HA) isolated from vermicompost. The ^{13}C CPMAS NMR spectrum was acquired with a Bruker AVANCE 300 NMR spectrometer, equipped with a 4-mm-wide bore MAS probe, operating at a ^{13}C resonating frequency of 75.475 MHz using 200 mg of HA. The spectrum was obtained by Professor Pier Luigi Mazzei at Centro Interdipartimentale di Ricerca per la Spettroscopia di Risonanza Magnetica Nucleare (CERMANU), Italy.

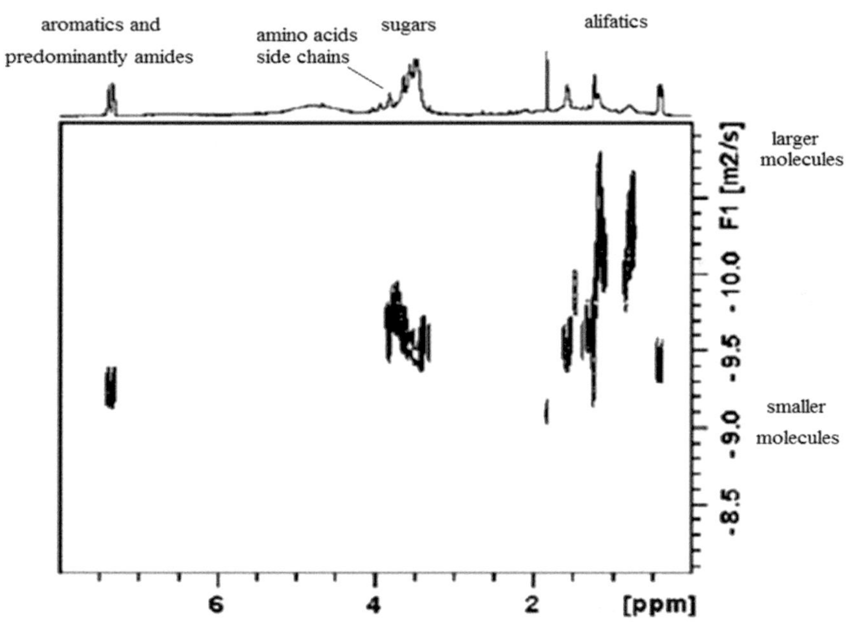

FIGURE 11.2 2-D DOSY projection of the HA isolated from cattle manure vermicompost. Solution-state DOSY-NMR spectra were obtained on a Bruker Avance 400 MHz instrument operating at a proton frequency of 400.13 MHz. 2-D-DOSY diffusion–ordered spectra were obtained using a stimulated echo pulse sequence with bipolar gradients (STEBPGP) provided by watergate 3-9-19 pulse train with gradients for presaturation of water signal. The spectrum was kindly obtained by Dr. Daniela Šmejkalová at Centro Interdipartimentale di Ricerca per la Spettroscopia di Risonanza Magnetica Nucleare (CERMANU), Italy.

different samples (HA, humins, whole-soil samples). Using ultra-high Fourier transform ion cyclotron resonance mass spectrometry in a high magnetic field (FT-ICR MS) and NMR spectroscopy, Monda et al. (2021) obtained unprecedented molecular-resolution data unveiling the presence of more than 10,000 compounds in the HA complex mixtures. The black box of humic matter was opened by humeomics, unveiling new possibilities to interpret the HS and develop sustainable technologies for soil stabilization, carbon sequestration, and remediation of organic and inorganic pollutants (Piccolo et al., 2019 and references therein), among others, including effects on plant growth (Canellas and Olivares, 2014). Advances in molecular-level analyses of complex nonrepetitive systems have been archived with the integrated use of high-performance separation, high-resolution organic structural spectroscopy, and mathematical data treatment (Hertkorn et al., 2007). The question now seems to be how these thousands of compounds are organized in the environment.

11.2 THE HUMIC SUBSTANCES MOLECULAR CHARACTERIZATION AND STRUCTURE-ACTIVITY RELATIONSHIP

Upon engagement in supramolecular association, an assembly's formation may induce changes to molecular properties, including charge, size, morphology, and function different from individual basic biomolecules (Tomaszewski et al., 2011). The HS complex mix of constituents may make the bioactivity hard to predict, although redox properties of HA quinones are related to biostimulant effect (Monda et al., 2021).

The systematic attempt to relate the structural aspect of HS with bioactivity considering macroscale chemical functionality was obtained using ^{13}C CPMAS NMR of HS and root traits, including lateral root emergence and activity of plasma membrane H^+-ATPase used as biological indicators (Canellas et al., 2012). Using partial least squares regression, it was possible to relate the hydrophobicity (ratio between hydrophobic-C/hydrophilic-C) of HS with their bioactivity. Afterward, using a similar approach, García et al. (2016) achieved similar results showing that hydrophobicity renamed by authors as ratio of lability (N, O aliphatic chains, and carboxyl group) and recalcitrance (unsubstituted aromatic and aliphatic structures) of HS were associated with the nature and strength of their biological activity on the root.

Some attempts at predicting the humic bioactivity was previously carried out, showing the ratio of hydrophilic to hydrophobic fragments in the structure of HA was one of the decisive factors of their physiological activity (Batuev et al., 2005). Using multivariate analysis, including principal component analysis (PCA), principal component regression, and Spearman correlation joined with NMR and DRIFT results Aguiar et al. (2013) showed that the main variables correlated with HA bioactivity were for NMR-positive loadings for lignin (56 ppm, 124 ppm, 148ppm, and 174 ppm) and COOH, ester and, amide groups (174 ppm) and for DRIFT were also aryl stretching and bending (1560, 1480, 860 and 780 cm^{-1}), probably from lignin. Further, such powerful analytical techniques as ultra-high-resolution FT-ICRMS, FT-ATR, and ^{13}C-NMR also found that lignin derivatives present in HA also played a leading part in the tomato yield, quality, and plant defense (Monda et al., 2021). Together, bioactivity can be interpreted as a feature that originates from the humic system in solution. Specifically, the final effect of HS applied directly to plants is not determined only by the presence of a single molecule but rather by a complex mixture of elements whose emergent features are also defined by plant-HS-environment interaction (Monda et al., 2021). Nevertheless, it could be interesting to study the organization pattern of the molecules most compatible with HS biostimulant effects, resulting in an agronomic efficiency increase. In this context, lipids' molecules are strong candidates both in the self-organization of the supramolecular arrangement and in the control of the HS bioactivity in solution.

Baigorri et al. (2007) found a molecular distribution pattern in the absence of discrete chemical structure in the HS; this pattern is a function of how the observation is made: i.e., a function of experimental design, instrumentation used, and the parameters to be evaluated and especially due to scale of (micro/macro) observation. Interestingly, the authors observed supramolecular assemblies, free small molecules, and a macro-molecular fraction. All proposed models of HA coexisted in the same suspension and each was observed with a different method. This is classical behavior of a dynamic complex system out of equilibrium (nonequilibrium state) characterized by the temperature-solvent condition. Wells and Stretz (2019) later proposed an interesting solution to this conundrum by considering a model of "supramolecular within supramolecular architecture originated by 'metachemical hydrogels' formed by near covalent strong hydrogen bonds that are encapsulated within dispersible physical hydrogel scaffolds produced by noncovalent weak intermolecular forces in a supramolecular within supramolecular self-assembling architecture" (Wells and Stretz, 2019). The core of the humic matrix is stabilized by the existence of a *continuum* of hydrogen bonding energies, ranging from <1 to 45 kcal mol^{-1}, assigned to weak, moderate, strong, and very strong classes of H-bonds, the latter very similar to near-covalent crosslinks according to Wells and Stretz (2019). The complex systems behavior understanding recommends that the recognition of the whole structure is essential to the behavior comprehension rather than isolated parts.

The physical and chemical heterogeneity of HS comprises a wide variety of sizes and shapes in solid and colloidal states. The occurrence of complex aggregation and dispersion phenomena in an aqueous medium with various degrees of roughness and irregularity of exposed surfaces is a typical behavior of complex chemical systems like lignins or HA (Shevchenko and Bailey, 1996). Numerous objects found in nature, such those with porous media, are disordered and irregular, which cannot be described by Euclidian geometry due to the scale-dependent measure of length, area, and volume. These objects are called fractals, and their dimensions are non-integral and defined as fractal dimensions. It is not surprising that both lignins and HA are fractal objects (Öesterberg and Mortensen, 1992; Rice and Lin, 1993). An essential principle of fractal geometry is that disorder exists at any characterization scale used to examine a substance (Rice and Lin, 1993). Different methods to assess of fractal dimension, like turbidimetry, small-angle neutron scattering, and X-ray diffraction, suggest that HA at pH 5 behave as random colloid aggregates with D-value varying from 2.3 to 2.5 and show properties very similar to kinetic growth processes in a state far from equilibrium. Östemberg and Mortensen (1992) showed that the HA aggregation process is pH dependent, and at pH 5.0, the aggregation proceeds until a steady state is reached. Humic acids exhibited fractal structures, which grow to large clusters with capacity to bind water, ions, and various organic molecules. The aggregation at low pH turns into a gel similar the metachemical hydrogels described by Wells and Stretz (2019). The gel formation has been observed in an HA system according to pH and temperature (Österberg and Mortensen, 1992). This structural conformation can be important in managing the physiological effect of HA. At low pH value, the carboxylate groups became strongly undissociated, and as short-range attraction might involve the formation of hydrogen bonds, protons are slowly consumed during the aggregation process and may also involves the association of hydrophobic weak and dispersive forces growing the self-organization process. The pH of HA suspension exposed to maize seedling roots drops along with the D-value (Figure 11.3).

The HA structure became soft and open and could further easily release bioactive fragments compared with denser and more enclosed structures. The zeta potential at pH 5.8 was −18 mV, and average particle size was 44 nm. The aggregation increased and became more stable at pH 3.1, showing zeta potential of −23 mV and 252 nm of average particle size. The chemically complex HA system can be described by assuming that the particles in solution consist of building units of a radial size, 40–50 nm, aggregated into clusters with an average radius of 252–350 nm. This result could be useful to associate the mechanism of aggregation of HA as well as their interactions with biological systems, including root seedlings. At a fixed range of pH value (5 to 7), HA appear as supramolecular particles with a conic shape and a hole in the centre, suggesting a pseudo-amphiphilic nature, with secluded hydrophobic domains and polar subunits in direct contact with the hydrophilic exterior, according to direct observation by atomic force microscopy (Colombo et al., 2015). Fischer (2017) also demonstrated that HA are spatially arranged in descending order of polarity, meaning that highly polar supramolecular subunits shield less

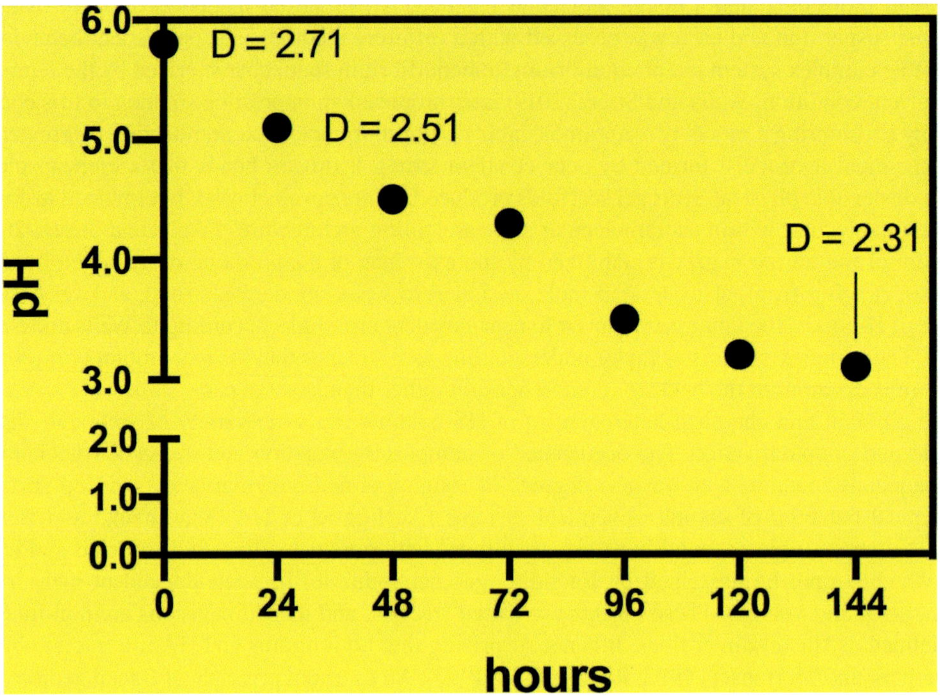

FIGURE 11.3 pH and D-value according the time of exposition of HA suspension to maize root seedlings. The initial pH value was adjusted to 5.8; HA concentration: 4 mM C L^{-1} of 2 mM CaCl2. The fractal dimension (D-value) was obtained according de Sá Mendonça et al. (2004) using turbidimetric measurement. The correction of carbon content was done after total organic carbon measurement using a TOC device from Shimadzu, Tokyo, Japan, and the pH was adjusted to the same value as when the aliquot was taken at each exposure time. The zeta potential was determined with Zetasizer apparatus (Malvern Inc., US).

polar subunits against the free solution and form layers of descending polarity of supramolecular subunits. The soft and permeable structure of HA (Duval et al., 2005) allows the penetration of hydrophobic organic molecules inside the aggregate. This property was evoked by Spaccini et al. (2002) and demonstrated the potential of the supramolecular structure of native HS to protect organic matter amended to soil against immediate mineralization. Using ^{13}C-labeled lipids, the authors stressed the role of HA hydrophobicity in carbon sequestration entrapping hydrophobic compounds like lipids and lignins derivatives. This soft and permeable structure, varying according to suspension conditions, is able to entrap new compounds, growing the aggregate structure. Piccolo et al. (2005) developed an HA biomimetics oxidative polymerization by porphyrin of iron (III) compound able to increase the apparent weight-average molecular weight of HA. The polymerization reaction modified the loosely bound humic supramolecular assemblies into more stable conformations, avoiding further disruption by the disaggregating effect of acetic acid. On the contrary, it is possible to design strategies to make the humic aggregate more sensitive and open. In fact, Canellas et al. (2008a) observed that root seedlings enhanced the organic acid exudation in the presence of HA, and the very small amount of citric acid in HA suspension enhanced their bioactivity significantly (Hottz-Rima et al., 2011).

Gibson (2017) showed, using X-ray scattering, that HA could show mass or surface fractal in accordance with its chemical composition. Mass fractals are soft and completely penetrable as they have no core interior while surface fractals have a solid core; therefore, they are completely impenetrable but may contain some material on the surface that will interact with the solvent. HS with low fractal dimension (D = 2.3) are found in a mass fractal and show chemical composition with a high content of aliphatic and carbohydrates and low aromatic carbon. Humic acids with surface fractals and high D values show limited anomeric carbon (Gibson, 2017). The hierarchical self-arrangement of HA was proposed with the interior of supramolecular aggregate formed by a core that would be composed of aromatic and aliphatic compound types, yielding a hydrophobic and hydrophilic region. The aromatic compounds are found in the hydrophobic region, separated from the solvent by a lipid layer, which polar head is interacting at the aqueous interface (Chilom and Rice, 2009). Here, we stress the role of polar lipids in the surface activity of humic assemble. Chilom and Rice (2009) describe that the hydrophobicity of HA is a fundamental property, and the self-organization of HA in solution can be viewed as an emergent characteristic resulting from the interaction of hydrophobic/hydrophilic domains with the aqueous medium.

Polar lipids have a key role in controlling the aggregation, softness, and roughness of fractal humic structure that may be the key to enhancing the bioactivity of HS in solution (Gibson, 2017). The control of molecular rigidity and diffusivity of HA as synonymous with mobility in suspension is exercised by the interaction of alkyl C moieties. Nebbioso and Piccolo (2009), using DOSY ^{1}H NMR, observed that the addition of Ca^{2+} (and Al^{3+}) decreased significantly the diffusivity of humic alkyl components and augmented that of aromatic and hydroxyalkyl components, therefore indicating that complexation caused a molecular-size increase in the former and a reduction in the latter. As previously reported, humic alkyl components were strongly associated with large hydrophobic-size aggregates and had great molecular rigidity, whereas oxidized carbons were more hydrated, less tightly aggregated, and less molecularly rigid (Piccolo et al., 2002; Conte et al., 2007; Šmejkalová and Piccolo, 2008). Another important aspect unveiled by Nebbioso and Piccolo (2009) was that that molecular rigidity increase was generally in the following order: aliphatic C > aromatic/double bonds C > carboxyl C. Acidic functions control the interaction with ions in solution, assume a mobile behaviour in the supramolecular arrangement, and may be weakly adhered or released.

Roots are active surfaces, and changes in ion fluxes were observed after HA treatment including Ca^{+2}, H^{+} and anions effluxes to rhizosphere (Ramos et al., 2015; de Azevedo et al., 2019). In addition, between the plasma membrane and soil solution, there is the cell wall, which is ten times thicker than the membrane and has a high cation exchange capacity, showing surface Donnan behaviour (Fernandes et al., 1984). Humic acids are rapidly adsorbed to the root surface (Figure 11.4), which becomes a surface for anchoring and stabilizing HA aggregate formation. During this process, it is possible that, depending on the concentration used, the adsorbed layer may clog some pores, reducing the root hydraulic conductivity (defined as the intrinsic ability of roots to conduct water across a water potential gradient between the root surface and the xylem in the stem) (Olaetxea et al., 2018). In fact, using high HA concentration, Asli and Neumann (2010) observed a decrease in root hydraulic conductivity. Olaetxea et al. (2015), using more realistic HA concentration, demonstrated that HA in solution can affect root hydraulic conductivity and water root uptake through ABA signaling pathways (Olaetxea et al., 2015, 2021). Furthermore, Olaetxea et al. (2021) hypothesized that HA promotes a transitory clogging of root pores that disappears gradually due to conformational changes as a result of root surface and HS interaction. Moreover, aquaporin activity is regulated by protein phosphorylation and cytoplasmic pH (Törnroth-Horsefield et al., 2006), and both are modified by HA (Ramos et al., 2015; Baía et al., 2020).

It is possible that HA aggregates formed around the root surface with mass fractal dimension and soft architecture are penetrable by mass and ions fluxes provide by cell membranes. The irregular and heterogeneous layer of HA deposited at rhizoplane (Figure 11.5) becomes a favourable niche for anchoring microorganisms. As showed by Jones et al. (2009), around 20% of net photosynthesis can be lost by rhizodeposition: mainly short-chain organic acids that have a notorious capacity to open the humic supramolecular arrangement (Piccolo et al., 2002). In addition, the exudation yield is significantly larger in plants treated with HA (Canellas et al., 2019a). The HS arrangement has also been suggested as an emergent, self-assembling system that follows a hierarchical or "structure-within-a-structure" architecture (Chilom and Rice, 2009). The first hierarchical level results from the lipids' self-assembly with strongly amphiphilic components into a nanostructured, composite

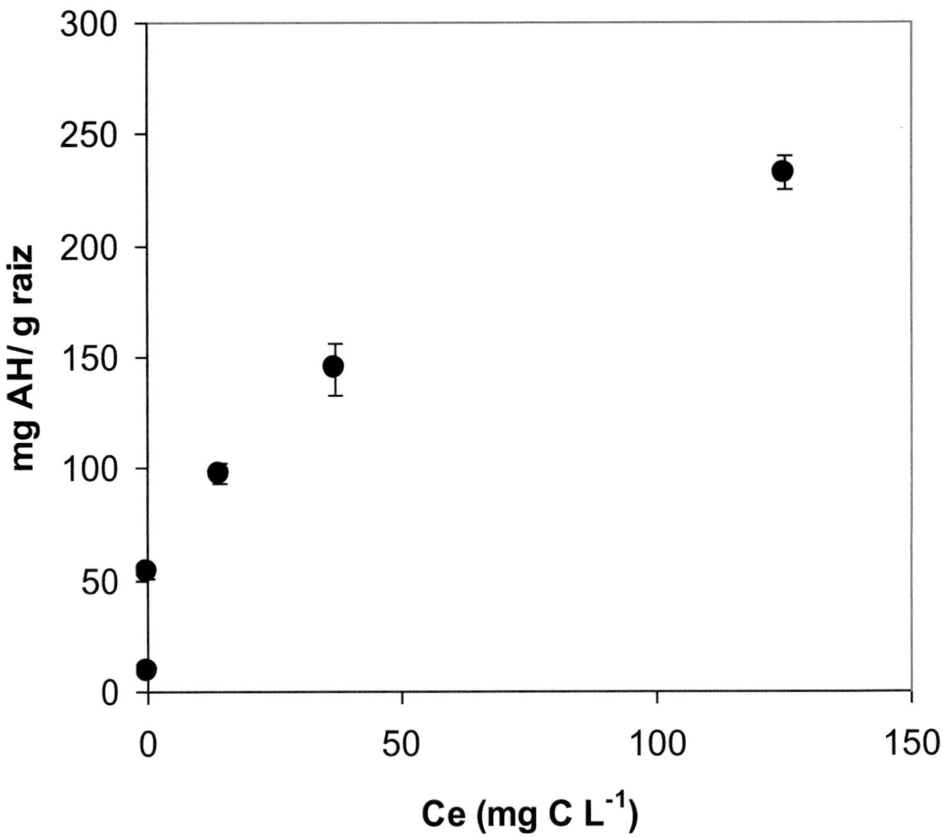

FIGURE 11.4 Typical L-curve of adsorption isotherm of HA isolated from vermicompost using maize root mass as adsorbent. The HA adsorption to the root surface was characterized by an initial slope that did not increase with the concentration of adsorptive in the soil solution (Sposito, 2008). This type of isotherm is the effect of a relatively high affinity of root surface for the HA at low surface coverage, mitigated by a decreasing amount of adsorbing surface remaining available as the surface excess increases.

material while the second level is an association of this composite material with nonamphiphilic components (Chilom et al., 2013). In accordance with this view, HS are organized in a hierarchical "supramolecular within supramolecular" self-assembling architecture (Wells and Stretz, 2019). In both interpretations, it is possible to indicate that polar lipids including alkylamides (also known as alkamides) are essential candidates to control humic conformation and its bioactivity.

11.3 PLANT PHYSIOLOGY AS AFFECTED BY HUMIC SUBSTANCES

After a century of the first discoveries on soil organic extracts' effects on plant growth (Bottomley, 1914), we start to heed both natural organic matter and HS "growth promoting" properties, but we are relatively far from understanding their mechanism of action on plants as compared to the profusion of plant hormone knowledge (Zandonadi et al., 2013; Scheres and van Der Putten, 2017; Nardi et al., 2018). Much of this difficulty may be for the reasons Bottomley pointed out earlier: the composition of organic matter in the soil varies not only with environmental aspects as both source of original materials and climate but also with the methods of organic matter extraction.

Most plant hormones have acidic functions such as auxins, abscisic acid, and gibberellins, and their presence on humic assembles can indicate the possibility of decoupling from the supramolecular

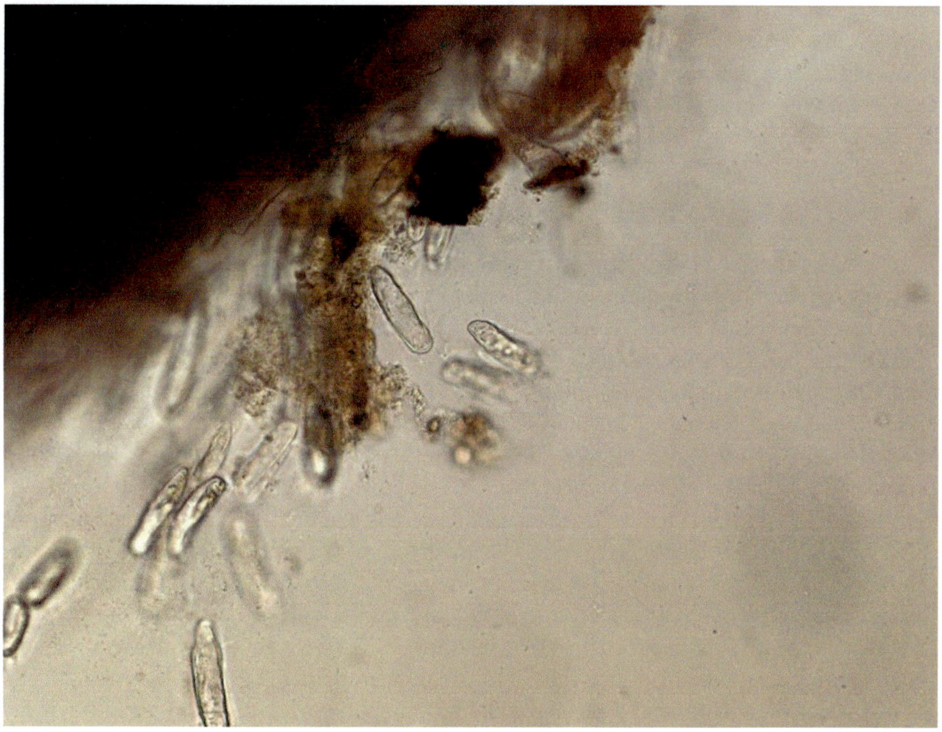

FIGURE 11.5 Optical microphotograph of (A) HA adsorbed to root surfaces; white star shows soft aggregation disrupted by root exudation (border cells) indicated by arrows; (B) aggregates formed in solution previous to root surface deposition. Note the growing aggregation. Humic acids concentration: 40 mg C L–1; Ca^{2+} concentration: 2 mM; initial pH: 5.8; C and D: Electronic microscopy of root maize surface of control plants (C) and (D) treated with HA. Note the irregular, rough, and soft surface of maize root recovered by HA.

arrangement depending on the suspension conditions. Plants modify the rhizosphere through various rhizodeposition mechanisms directly affecting the production of border cells (Canellas and Olivares, 2017), and the ion fluxes (de Azevedo et al., 2019; Ramos et al., 2015) to the solution. The study of the hydrophobic-hydrophilic properties of HS is key to the control of its bioactivity, which dictates solubility, morphological peculiarities, surface properties, and the nature of its interaction with water, nutrients, and organic molecules in the interface of roots-soil solution (Piccolo et al., 2019).

During plant growth, many environmental changes induce roots to adapt to dynamic conditions. For example, plants need to respond to both mineral nutrients and organic substances in substrate, resulting in root morphology adaptation through intermediation of hormones (Dobrenel et al., 2011; Giehl and Wirén, 2014; Jia et al., 2022). The levels of nitrogen as well as assimilated carbon are related to the regulation of target of rapamycin (TOR) (Dobrenel et al., 2011). This protein operates between nutritional status and growth regulation, balancing these two factors, transiently turning growth off or on depending on the availability of nutrients (Dobrenel et al., 2011). Moreover, TOR is also regulated by the presence of glucose in the roots or both glucose and light in the shoot (Li et al., 2017). The induction of TOR expression by HA has been shown but occurred at low sugar and amino acids levels in leaves and roots of plants treated with HA (Canellas et al., 2019b). Possibly, the detection of both nitrogen and carbon levels are over the perception of amino acids and sugars. The HA-simulated growth of both shoots and roots may consume the majority of the available sugars and amino acids. It has been reported that HA influence energy/respiratory metabolism (Merlo et al., 1991; Nardi et al., 2007; Roomi et al., 2018). It's noteworthy that HS enhanced the activity of enzymes of the glycolytic pathway (i.e., phosphofructokinase and pyruvate kinase) and of the

Krebs cycle (i.e., citrate synthase and malate dehydrogenase). The reactions of both glycolysis and the Krebs cycle are central and, together with oxidative phosphorylation, constitutes a key process delivering carbon skeletons, reduction equivalents, and energy for most biochemical pathways (Meyer et al., 2007). These authors addressed the metabolic signature associated with high plant-growth rates and reported a close link between biomass and a specific combination of metabolites.

11.4 BIOACTIVE FAT AND HUMIC ACIDS

Lipids are a water-insoluble, structurally diverse group that includes an extensive range of fatty acid–derived compounds, many pigments, and secondary metabolites that are not essentially related to fatty acid metabolism. A diversity of signaling lipids has been related to plant growth, development, and environmental responses, such as sterols (Nolan et al., 2020), fatty acids (Lim et al., 2017; Yang and Christoph Benning, 2018), sphingolipids (Jiang et al., 2019), and phospholipids (Colin and Jaillais, 2020). For instance, plant sterols are the precursors of compounds with physiological activity such as brassinosteroids, which induce an extensive spectrum of responses, including both cell division and elongation, proton pump activation, and ethylene-related genes (Nolan et al., 2020). Plant fatty acids, such as both oleic and linoleic acids, are relevant to plant defense mechanisms (Lim et al., 2017) and play key roles in physiology as signals for cell differentiation, auxin transport, and both the flowering and germination processes (Yang and Christoph Benning, 2018). Sphingolipid signaling is linked to plant-stress response to both biotic and abiotic stress of pathogens (Lung and Chye, 2019) and salinity (Jiang et al., 2019). Phospholipids (also known as glycerophospholipids) are essential to plant developmental cues, either in localized membranes (e.g., physicochemical parameters) or in a broad organism scale (e.g., phosphatidic acid and circadian clock), and both time and localization in the plant may influence phospholipids effects (Colin and Jaillais, 2020).

Plants possess many secondary compounds known as metabolic products without a known function in the primary metabolism. Alkamides are plant secondary (or specialized) metabolites containing fatty acid amides with typical olefinic and acetylenic compounds (Greger, 1984) and a wide range of biological activities in nature (Rizhsky et al., 2016). Within the group of fatty acid amides, there are also two other classes of related molecules: the N-acyl-l-homoserine lactones (N-AHLs or AHLs) from gram-negative bacteria and the N-acylethanolamines (NAEs) present in plants and mammals (Méndez-Bravo et al., 2011).

Although structurally different, alkamides and NAEs have similar biological effects. The NAEs are a family of small and diverse lipids derived from membrane phospholipids, the N-acyl phosphatidylethanolamine (NAPE). They are present in several tissues participating in important functions, such as the coordination of metabolic processes in several organisms. In plants, NAE participate in several processes such as seed germination, early seedling development, inhibition of ABA action on stomatal closure, flowering induction (reducing flower senescence), and response to pathogens (Wang et al., 2006; Keereetaweep et al., 2013; Austin-Brown and Chapman, 2002; Tripathy et al., 1999; Teaster et al., 2012; Kang et al., 2008). In cell culture (Chapman et al., 1998) and tobacco leaves (Tripathy et al., 1999) exposed to fungal elicitors, an increase in the amount of NAE was observed, which is possibly associated with the induction of defense responses to the pathogen, such as increased expression of the phenylalanine ammonia lyase gene (Tripathy et al., 1999). Both type and amount of NAE vary from species to species, tissue to tissue, and stage of plant development (Kang et al., 2008; Kilaru et al., 20112012). Concentrations of NAE are high in desiccated seeds and tend to decrease with seed swelling and plant development (Chapman et al., 1998; Blancaflor et al., 2003). N-acylethanolamines act in the regulation of plant growth and development in the ABA-induced signaling pathway. The effect of NAE is very similar, inducing genes of the ABA signaling pathway, which comprises the transcription factor ABI3 that regulates ABA responsive genes. The presence of NAE provokes plant sensitivity in ABA, thus acting as an early growth inhibitor (Teaster et al., 2007; Cotter et al., 2011).

There are few reports on alkamides and plant growth compared to plant hormones (Figure 11.6 and references cited therein). One of the seminal works using exogenous alkamides in plants showed that amidenin (isolated from the actinomycete *Amycolatopsis* sp.) regulated rice growth, inhibiting both shoot and root growth at higher concentrations tested (60 μmol.L^{-1}) or stimulating at intermediary concentration (6 to 18 μmol.L^{-1}) (Kanbe et al., 1993). Alkamides were first isolated from the plants *Echinacea pallida* and *Heliopsis longipes* and caused hormone-like effects in plants species that do not produce this metabolite (Ramírez-Chávez et al., 2004; López-Bucio et al., 2007; Méndez-Bravo et al., 2010; Méndez-Bravo et al., 2011). For instance, the alkamide affinin (*N*-isobutyl-2*E*,6*Z*,8*E*-decatrienamide) is a low–molecular weight α-unsaturated acyl-chain amide able to stimulate shoot and root biomass production in *Arabidopsis* (Ramírez-Chávez et al., 2004) and alter plant metabolic processes such as sugar, amino acids, fat acids, and organic acids metabolism in tomato (Campos-García and Molina-Torres, 2021). Alkamides participate in several plant-signaling pathways, possibly through chemical interactions involving hormones such as cytokinin (López-Bucio et al., 2007) and jasmonite (Romero et al., 2009). The saturated alkamide N-isobutyl decanamide changed both root and shoot in *Arabidopsis*, suggesting a proliferative effect due to the appearance of callus-like structures in these plant organs. The emergence of these structures was related to the genes of the cytokinin pathway, such as CRE1, AHK2, and AHK3 receptors, suggesting that alkamides act in this hormonal pathway by controlling meristematic activity (López-Bucio et al., 2007). Nevertheless, the effects of the alkamide affinin on the root architecture of *Arabidopsis* appear to be independent of auxin signaling since both auxin-resistant mutants and auxin-inducible gene markers were not modified by affinin (Ramírez-Chávez et al., 2004). Both morphologic and metabolic changes are also expected throughout plants treated with organic matter extracts or HS (Trevisan et al., 2011; Sleighter et al., 2015; Canellas et al., 2019b; Vujinović et al., 2020; Lamar et al., 2021).

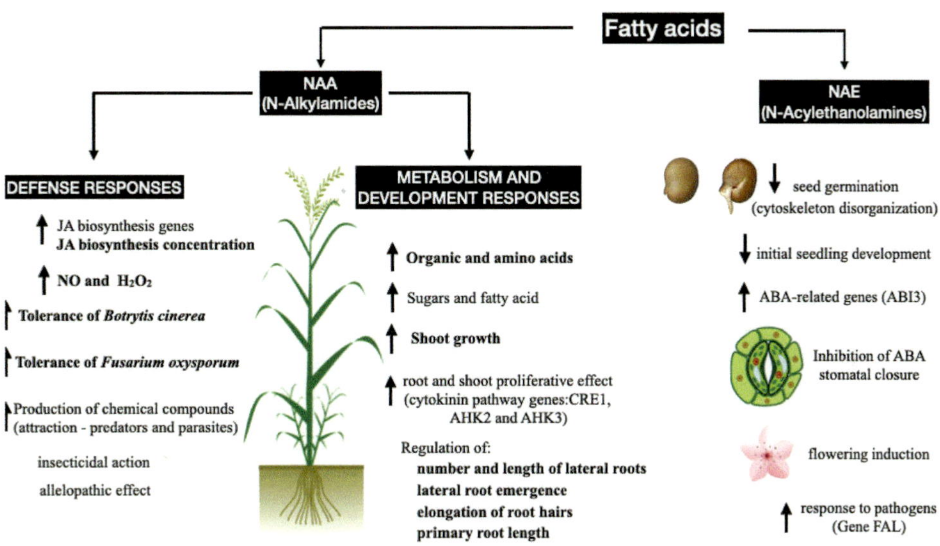

FIGURE 11.6 Plant response to N-Alkylamines (NAA). Bold text in the figure indicates effects reported for both NAA and HS. *Defense Responses:* increased transcriptions of genes related to jasmonic acid (JA) biosynthesis; increased nitric oxide (NO) and hydrogen peroxide (H_2O_2) concentration; increased tolerance of *Fusarium oxyporum* and *Botytis cinerea*; induced production of chemical compounds of predators and parasites attraction; insecticidal and allelopathic action. *Metabolism and development responses:* enhancement of organic acids, amino acids, sugar and fatty acids; induction of shoot growth at low concentration; root and shoot proliferative effect (affecting cytokinin pathway genes: CRE1, AHK2 and AHK3); increased number, emergence, and length of lateral roots; elongation of root hairs; reduced primary root length.

Source: Kanbe et al. (1993); Lait et al. (2003); Ramírez-Chávez, et al. (2004); Molina-Torres et al. (2004); Anaya et al. (2005); Tsao et al. (2003); López-Bucio et al. (2007); Méndez-Bravo et al. (2011); González-Morales et al. (2015); Da Luz et al. (2017); Campos-García and Molina-Torres (2021)

Both organic matter extracts and HS isolated from different sources directly affect plant growth and development (Reviewed by Zandonadi et al., 2013 and by Canellas and Olivares, 2014). Humic substances alter plant metabolic pathways such as the glycolytic and Krebs cycles, the photosynthetic process, and nutrient assimilation (Quaggiotti et al., 2004; Nardi et al., 2007; Conselvan et al., 2018; Souza et al., 2021) (Figure 11.7 and references cited therein). The variety of molecules detected in soil organic matter and HS includes auxin-like substances (Sarwar et al., 1992; Canellas et al., 2002). A portion of the observed humic effects such as root growth are auxin dependent while the remainder are auxin independent (Schmidt et al., 2007; Zandonadi et al., 2010; Trevisan et al., 2011; Mora et al., 2012; Zandonadi et al., 2019). Additionally, other plant hormones and signals appear to be related to the mode of action of HS (Mora et al., 2010; Zandonadi et al., 2010; Zandonadi et al., 2019). Humic acids endowed with indol-3-acetic acid (IAA) alter maize root growth and activate plasma membrane H^+-ATPase activity, recruiting nitric oxide (NO) as a messenger (Zandonadi et al., 2010). Moreover, auxin-free HA enhanced shoot growth and root plasma membrane H^+-ATPase activity, with simultaneous increases in both cytokinin and polyamine contents in plant shoot (Mora et al., 2010).

The presence of lipids on soil organic matter has long been recognized (Morrison, 1969, Mueller et al., 2013; Ding et al., 2020). Soil lipids are derived from plant and animal residues and microorganisms. Plant lipid properties influence some aliphatic lipids concentrations in soil (Mueller et al., 2013), and soil organic matter presents complex mixed sources of intact

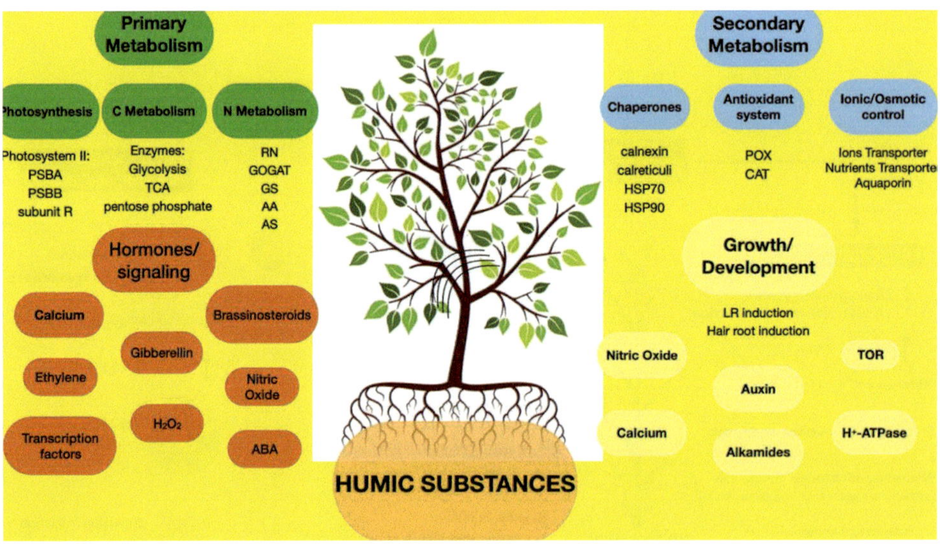

FIGURE 11.7 Plant response to humic substances (HS). *Primary metabolism:* carbon metabolism: photosynthesis: induce transcription of genes: chlorophyll binding protein D1 (PSBA); photosystem II 47 kDa protein (PSBB) and photosystem II subunit R; induced enzymes of gycolysis, tricarboxylic acid cycle (TCA), and the pentose phosphate pathway; nitrogen (N) metabolism: induced enzyme activity/gene expression: nitrate reductase (RN), glutamine synthetase (GS), glutamate oxoglutarate aminotransferase (GOGAT), aspartate aminotransferase (AA) and asparagine synthetase (AS). *Secondary metabolism:* stimulates secondary metabolism; chaperones: calnexin, calreticuli, HSP70 e HSP90; antioxidant system, peroxidase (POX); catalase (CAT). Ionic and osmotic control: ion transporter, nutrients transporter, aquaporin. *Growth/Development:* LR induced, hair root induction, signal molecules induction, nitric oxide, calcium, auxin, alkamides, H^+-ATPase, Target of Rapamycin (TOR). *Hormone and Signaling*: calcium, ethylene, gibberellin, brassinosteroid, H_2O_2, nitric oxide, ABA, transcriptions factors.

Source: Merlo et al. (1991); Nardi et al. (2007); Trevisan et al. (2010); Zandonadi et al. (2010); Roomi et al. (2018); Canellas et al. (2019b); Canellas et al. (2020); Souza et al. (2021); Silva et al. (2021)

polar lipids (Ding et al., 2020). Notably, despite plant residues' contribution to soil lipids, plant *in vivo* rhizodeposition of fatty acids has been shown (Sasse et al., 2018). Humic acids isolated from organic compost of fish beneficiation wastes contained lipids and N-alkyl groups (Busato et al., 2018). These lipid-rich HA enhanced both the lateral root average length and the root H^+ extrusion activity of lettuce. Along with both sugar and organic acids, lipids are delivered from plants, resulting in interaction with microbial communities essential as sources of lipids in soil organic matter. Interesting, exogenously applied HA in maize roots altered both biofilm production and colonization of *Herbaspirillum seropedicae*, an endophytic diazotrophic bacteria that promotes plant growth (Canellas and Olivares, 2017). In fact, HA can alter the composition of a bacterial community simultaneously with the stimulation of root development in rice (Silva et al., 2021). Humic acids can alter plant lipid content (Kulikova et al., 2016). This emerging scenario depicts an essential role of HA on soil lipids composition and their effects on plant growth.

There are several examples of analogous effects on plants of both HS and alkamides. It's noteworthy that HS increased secondary maize metabolites contents such as phenolic acids and flavonoids using both molecular and biochemical mechanisms (Schiavon et al., 2010). Similar to HA, alkamides regulate root and shoot development with cytokinin involvement (López-Bucio et al., 2007). The authors reported the expression of cell division markers and of the cytokinin-inducible marker *ARR5:uidA* both in roots and in shoots. Auxin and jasmonic acid (JA) crosstalk has been studied during the *Arabidopsis* lateral root formation induced by JA (Cai et al., 2014). Humic acids isolated from leonardite increased IAA content to nearly 20% in cucumber roots and increased nearly twofold the JA in roots (De Hita et al., 2020). Alkamide *N*-isobutyldecanamide also activates JA biosynthesis and signaling pathways, increasing endogenous levels of JA and their corresponding transcripts (Méndez-Bravo et al., 2011). Recently, the alkamide *N*-isopropyldecanamide was detected in the humic supramolecular assembly of earthworm compost–derived HA (Zandonadi et al., 2019). Maize plants were treated either with HA or alkamide, and the authors provided compelling evidence of early root development (cell division and root initiation) and increased root growth and mass (root area and dry mass). Alkamide and HA activated root plasma membrane H^+-ATPase concomitantly to root NO production on lateral root sites (Figure 11.8), and alkamide-induced lateral root development was related to NO production in *Arabidopsis* (Méndez-Bravo et al., 2010). As observed for HS, *N*-isobutyldecanamide affect the root similarly to auxin but independent of auxin signaling, as shown by *Arabidopsis* lines carrying auxin response elements *BA3:uidA* and *DR5:uidA* (Ramírez-Chávez et al., 2004). Analyzing the same elements, no evidence of auxin signaling was reported for "water-extractable" HS in *Arabidopsis*, although an increase in root hair density of about 22% was observed (Schmidt et al., 2007). In contrast, other HS-enhanced auxin synthetic reporter *DR5::GUS* in *Arabidopsis* (Trevisan et al., 2010) and in tomato (Dobbss et al., 2010). Interestingly, HA can partially overcome auxin transport and signaling inhibitors while NO scavenger abolishes HA effects on root growth (Zandonadi et al., 2010). Despite a direct hormone-like effect, depending on both source and age of HS, the electron-accepting (EACs) and donating capacities' (EDCs) moieties content may contrast. For instance, terrestrial HS showed smaller EDCs and larger EACs than aquatic HS (Aeschbacher et al., 2012). Noteworthy is the fact that HA isolated from the surface horizon of six different sites from a typical sequence of soil weathering presented EACs positively correlated to both enhanced root growth and plasma membrane H^+-ATPase activity (Canellas et al., 2008b). Accordingly, the anti- and pro-oxidant properties of HS should be considered part of the biostimulant response elicited in plants (Lamar, 2020) as observed in tomato growth and yield (Monda et al., 2021) and in maize morphological parameters such as total root length and root surface area (Lamar et al., 2021). These authors proposed a mechanism of action for HA based on the interplay of pro-oxidant activity from quinones and semiquinones (related to EACs) and antioxidant activity from polyphenols (related to EDCs).

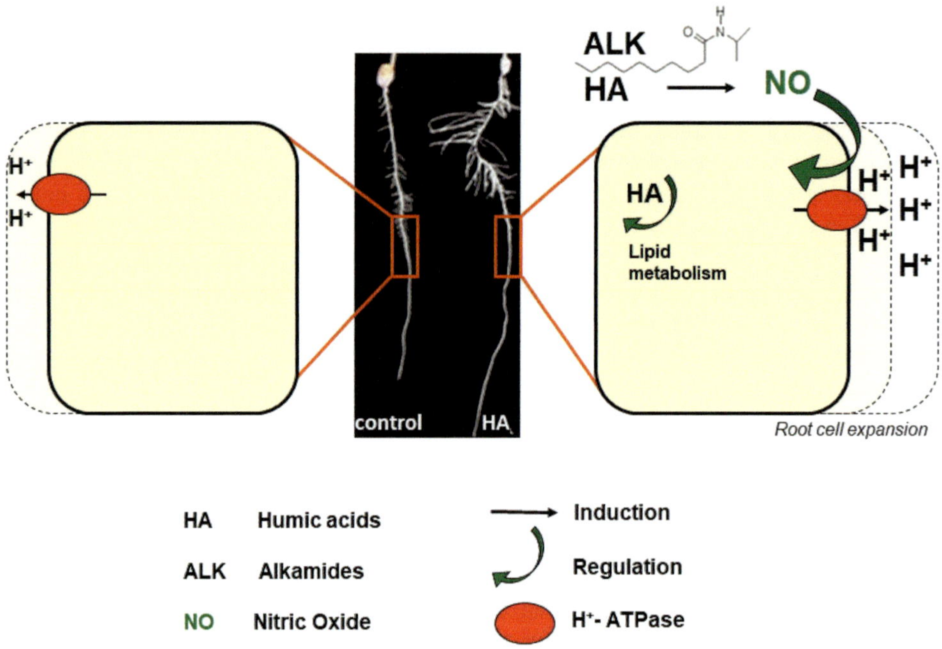

FIGURE 11.8 Effects of both alkamide (N-isopropyldecanamide) and humic acids (HA) on plasma membrane (PM) H^+-ATPase activity and on maize root development. N-isopropyldecanamide was detected in the unbound fraction of the humic supramolecular assembly. Humic acids uptaken by plant roots trigger endogenous lipid profile change (Kulikova et al., 2016).

Source: Adapted or modified with permission (copyright: author's own work) from Zandonadi et al. *Chem. Biol. Technol. Agric.* (2019) 6:23.

11.5 CONCLUDING REMARKS AND FUTURE PERSPECTIVES

The carbon distribution in the biomolecules in HA may account for its relatively high content of lipids. The uptake of HA by roots alters both polar and neutral endogenous lipids profiles, including free fatty acids. The direct effect of HA on plant endogenous lipids and the plant response to HA-derived hydrophobic structures highlight the significance of alkamides in HS mechanisms of action. Moreover, since P-type ATPase H^+-ATPase is linked to both HA and alkamide mechanisms of action during plant growth, the P-type P4 ATPases, which transports phospholipids across biological membranes (López-Marqués et al., 2021), should be of interest in further investigations. Given the complexity of HS-plant interaction, which involves signaling molecules, secondary metabolites, and hormones, additional studies will shed light on the physiological functions of HA-derived lipids and the endogenous lipid signaling triggered by HA.

REFERENCES

Aeschbacher, Michael, Cornelia Graf, René P. Schwarzenbach, and Michael Sander. "Antioxidant properties of humic substances." *Environmental Science & Technology* 46, no. 9 (2012): 4916–4925.

Aguiar, N. O., E. H. Novotny, A. L. Oliveira, V. M. Rumjanek, F. L. Olivares, and L. P. Canellas. "Prediction of humic acids bioactivity using spectroscopy and multivariate analysis." *Journal of Geochemical Exploration* 129 (2013): 95–102.

Anaya, Ana Luisa, Martha Macías-Rubalcava, Rocío Cruz-Ortega, Clara García-Santana, Perla N. Sánchez-Monterrubio, Blanca E. Hernández-Bautista, and Rachel Mata. "Allelochemicals from stauranthus perforatus, a rutaceous tree of the Yucatan Peninsula, Mexico." *Phytochemistry* 66, no. 4 (2005): 487–494.

Asli, Sare, and Peter M. Neumann. "Rhizosphere humic acid interacts with root cell walls to reduce hydraulic conductivity and plant development." *Plant and Soil* 336, no. 1 (2010): 313–322.

Austin-Brown, Shea L., and Kent D. Chapman. "Inhibition of phospholipase Dα by N-acylethanolamines." *Plant Physiology* 129, no. 4 (2002): 1892–1898.

Baía, Daiane Carvalho, Fábio L. Olivares, Daniel B. Zandonadi, Cleiton de Paula Soares, Riccardo Spaccini, and Luciano P. Canellas. "Humic acids trigger the weak acids stress response in maize seedlings." *Chemical and Biological Technologies in Agriculture* 7, no. 1 (2020): 1–13.

Baigorri, Roberto, Marta Fuentes, Gustavo González-Gaitano, and Jose Ma García-Mina. "Analysis of molecular aggregation in humic substances in solution." *Colloids and Surfaces A: Physicochemical and Engineering Aspects* 302, no. 1–3 (2007): 301–306.

Batuev, B. Ts, E. V. Atuev, N. V. Zoltoev, I. P. Bykov, and A. D. Dashitsyrenova. "Estimation of physiological activity of humic substances of oxidized coal (Buryatia)." *Chemistry for Sustainable Development* 13 (2005): 501–505.

Blancaflor, Elison B., Guichuan Hou, and Kent D. Chapman. "Elevated levels of N-lauroylethanolamine, an endogenous constituent of desiccated seeds, disrupt normal root development in Arabidopsis thaliana seedlings." *Planta* 217, no. 2 (2003): 206–217.

Bottomley, W. B. "Some accessory factors in plant growth and nutrition." *Proceedings of the Royal Society of London. Series B, Containing Papers of a Biological Character* 88, no. 602 (1914): 237–247.

Busato, Jader Galba, Caroline Moreira de Carvalho, Daniel Basilio Zandonadi, Fernando Fabriz Sodré, Alan Ribeiro Mol, Aline Lima de Oliveira, and Rodrigo Diana Navarro. "Recycling of wastes from fish beneficiation by composting: chemical characteristics of the compost and efficiency of their humic acids in stimulating the growth of lettuce." *Environmental Science and Pollution Research* 25, no. 36 (2018): 35811–35820.

Cai, XT., Xu P, Zhao PX., Liu R, Yu LH., and Xiang CB. "Arabidopsis ERF109 mediates cross-talk between jasmonic acid and auxin biosynthesis during lateral root formation." *Nature Communications* 5, no. 1 (2014): 1–13.

Campos-García, Tonatiu, and Jorge Molina-Torres. "Solanum lycopersicum Seedlings. Metabolic Responses Induced by the Alkamide Affinin." *Metabolites* 11, no. 3 (2021): 143.

Canellas, L. P., L. B. Dobbss, A. L. Oliveira, J. G. Chagas, N. O. Aguiar, V. M. Rumjanek, E. H. Novotny, F. L. Olivares, R. Spaccini, and A. Piccolo. "Chemical properties of humic matter as related to induction of plant lateral roots." *European Journal of Soil Science* 63, no. 3 (2012): 315–324.

Canellas, L. P., L. R. L. Teixeira Junior, L. B. Dobbss, C. A. Silva, L. O. Medici, D. B. Zandonadi, and A. R. Façanha. "Humic acids crossinteractions with root and organic acids." *Annals of Applied Biology* 153, no. 2 (2008a): 157–166.

Canellas, Luciano P., Daniel B. Zandonadi, Jader G. Busato, Marihus A. Baldotto, Marcelo L. Simões, Ladislau Martin-Neto, Arnoldo R. Façanha, Riccardo Spaccini, and Alessandro Piccolo. "Bioactivity and chemical characteristics of humic acids from tropical soils sequence." *Soil Science* 173, no. 9 (2008b): 624–637.

Canellas, Luciano P., and Fábio L. Olivares. "Physiological responses to humic substances as plant growth promoter." *Chemical and Biological Technologies in Agriculture* 1, no. 1 (2014): 1–11.

Canellas, Luciano P., and Fabio Lopes Olivares. "Production of border cells and colonization of maize root tips by herbaspirillum seropedicae are modulated by humic acid." *Plant and Soil* 417, no. 1 (2017): 403–413.

Canellas, Luciano P., Fabio Lopes Olivares, Anna L. Okorokova-Façanha, and Arnoldo Rocha Façanha. "Humic acids isolated from earthworm compost enhance root elongation, lateral root emergence, and plasma membrane H+-ATPase activity in maize roots." *Plant Physiology* 130, no. 4 (2002): 1951–1957.

Canellas, Luciano P., Fábio L. Olivares, Natália OA Canellas, Pierluigi Mazzei, and Alessandro Piccolo. "Humic acids increase the maize seedlings exudation yield." *Chemical and Biological Technologies in Agriculture* 6, no. 1 (2019a): 1–14.

Canellas, Luciano P., Natália OA Canellas, Luiz Eduardo Souza da S. Irineu, Fábio L. Olivares, and Alessandro Piccolo. "Plant chemical priming by humic acids." *Chemical and Biological Technologies in Agriculture* 7, no. 1 (2020): 1–17.

Canellas, Luciano P., Natália OA Canellas, Tatiane S. Soares, and Fabio L. Olivares. "Humic acids interfere with nutrient sensing in plants owing to the differential expression of TOR." *Journal of Plant Growth Regulation* 38, no. 1 (2019b): 216–224.

Chapman, Kent D., Swati Tripathy, Barney Venables, and Arland D. Desouza. "N-Acylethanolamines: formation and molecular composition of a new class of plant lipids." *Plant Physiology* 116, no. 3 (1998): 1163–1168.

Chilom, Gabriela, Andrea Baglieri, Cyndey A. Johnson-Edler, and James A. Rice. "Hierarchical self-assembling properties of natural organic matter's components." *Organic Geochemistry* 57 (2013): 119–126.

Chilom, Gabriela, and James A. Rice. "Structural organization of humic acid in the solid state." *Langmuir* 25, no. 16 (2009): 9012–9015.

Colin, Leia Axelle, and Yvon Jaillais. "Phospholipids across scales: Lipid patterns and plant development." *Current opinion in plant biology* 53 (2020): 1–9.

Colombo, Claudio, Giuseppe Palumbo, Ruggero Angelico, Hyen Goo Cho, Ornella Francioso, Andrea Ertani, and Serenella Nardi. "Spontaneous aggregation of humic acid observed with AFM at different pH." *Chemosphere* 138 (2015): 821–828.

Conselvan, Giovanni Battista, David Fuentes, Andrew Merchant, Cristina Peggion, Ornella Francioso, and Paolo Carletti. "Effects of humic substances and indole-3-acetic acid on Arabidopsis sugar and amino acid metabolic profile." *Plant and Soil* 426, no. 1 (2018): 17–32.

Conte, Pellegrino, Riccardo Spaccini, Daniela Šmejkalová, Antonio Nebbioso, and Alessandro Piccolo. "Spectroscopic and conformational properties of size-fractions separated from a lignite humic acid." *Chemosphere* 69, no. 7 (2007): 1032–1039.

Cotter, Matthew Q., Neal D. Teaster, Elison Blancaflor, and Kent D. Chapman. "N-acylethanolamine (NAE) inhibits growth in Arabidopsis thaliana seedlings via ABI3-dependent and-independent pathways." *Plant Signaling & Behavior* 6, no. 5 (2011): 671–679.

Da Luz, Shirlley FM, Lydia F. Yamaguchi, Massuo J. Kato, Oriel F. De Lemos, Luciana P. Xavier, José Guilherme S. Maia, Alessandra de R. Ramos, William N. Setzer, and Joyce Kelly do R. Da Silva. "Secondary metabolic profiles of two cultivars of piper nigrum (black pepper) resulting from infection by Fusarium solani f. sp. piperis." *International Journal of Molecular Sciences* 18, no. 12 (2017): 2434.

de Azevedo, Inga Gonçalves, Fábio Lopes Olivares, Alessandro Coutinho Ramos, Amanda Azevedo Bertolazi, and Luciano Pasqualoto Canellas. "Humic acids and Herbaspirillum seropedicae change the extracellular H+ flux and gene expression in maize roots seedlings." *Chemical and Biological Technologies in Agriculture* 6, no. 1 (2019): 1–10.

De Hita, David, Marta Fuentes, Victoria Fernández, Angel M. Zamarreño, Maite Olaetxea, and Jose M. García-Mina. "Discriminating the short-term action of root and foliar application of humic acids on plant growth: Emerging role of jasmonic acid." *Frontiers in Plant Science* 11 (2020): 493.

de Sá Mendonça, Eduardo, Alessandro Costa da Silva, Maurício Paulo Ferreira Fontes, and Efraim Lázaro Reis. "Fractal dimension of humic acids as influenced by complex formation." *Communications in Soil Science and Plant Analysis* 35, no. 19–20 (2004): 2747–2755.

Ding, S., Lange M, Lipp J, Schwab VF, Chowdhury S, Pollierer MM., . . . and Gleixner G. "Characteristics and origin of intact polar lipids in soil organic matter." *Soil Biology and Biochemistry* 151 (2020): 108045.

Dobbss, Leonardo Barros, Luciano Pasqualoto Canellas, Fábio Lopes Olivares, Natália Oliveira Aguiar, Lázaro Eustáquio Pereira Peres, Mariana Azevedo, Riccardo Spaccini, Alessandro Piccolo, and Arnoldo R. Facanha. "Bioactivity of chemically transformed humic matter from vermicompost on plant root growth." *Journal of Agricultural and Food Chemistry* 58, no. 6 (2010): 3681–3688.

Dobrenel, Thomas, Chloé Marchive, Rodnay Sormani, Manon Moreau, Milena Mozzo, Marie-Hélène Montané, Benoît Menand, Christophe Robaglia, and Christian Meyer. "Regulation of plant growth and metabolism by the TOR kinase." *Biochemical Society Transactions* 39, no. 2 (2011): 477–481.

Du Jardin, Patrick. "Plant biostimulants: definition, concept, main categories and regulation." *Scientia Horticulturae* 196 (2015): 3–14.

Duval, Jérôme FL, Kevin J. Wilkinson, Herman P. Van Leeuwen, and Jacques Buffle. "Humic substances are soft and permeable: Evidence from their electrophoretic mobilities." *Environmental Science & Technology* 39, no. 17 (2005): 6435–6445.

Fernandes, M. S., R. Rossiello, and M. Arruda. "Root cation-exchange capacity and aluminum toxicity relationships in 2 pasture grasses." *Pesquisa Agropecuaria Brasileira* 19, no. 5 (1984): 631–637.

Fischer, Thomas. "Humic supramolecular structures have polar surfaces and unpolar cores in native soil." *Chemosphere* 183 (2017): 437–443.

García, Andrés Calderín, Luiz Gilberto Ambrosio de Souza, Marcos Gervasio Pereira, Rosane Nora Castro, José María García-Mina, Everaldo Zonta, Francy Junior Gonçalves Lisboa, and Ricardo Luis Louro Berbara. "Structure-property-function relationship in humic substances to explain the biological activity in plants." *Scientific Reports* 6, no. 1 (2016): 1–10.

García, Andrés Calderín, Orlando Carlos Huertas Tavares, Dariellys Martínez Balmori, Vitor dos Santos Almeida, Luciano Pasqualoto Canellas, José María García-Mina, and Ricardo Luis Louro Berbara. "Structure-function relationship of vermicompost humic fractions for use in agriculture." *Journal of Soils and Sediments* 18, no. 4 (2018): 1365–1375.

Gibson, Dennis K. "Relationship of the fractal and chemical characteristics of self-assembled natural organic matter." PhD diss., South Dakota State University (2017).

Giehl, Ricardo FH., and Nicolaus von Wirén. "Root nutrient foraging." *Plant Physiology* 166, no. 2 (2014): 509–517.

González-Morales, Susana, Adalberto Benavides-Mendoza, Ema Laura García-Enciso, Edmundo Mario Rodríguez-Campos, and Alberto Flores-Olivas. "Effect of alkamides as tolerance inductors to biotic stress in tomato." *Revista Mexicana De Ciencias Agrícolas* 6, no. SPE12 (2015): 2371–2382.

Greger, Harald. "Alkamides: Structural relationships, distribution and biological activity1." *Planta Medica* 50, no. 5 (1984): 366–375.

Hertkorn, N., C. Ruecker, Markus Meringer, R. Gugisch, M. Frommberger, E. M. Perdue, M. Witt, and Ph Schmitt-Kopplin. "High-precision frequency measurements: Indispensable tools at the core of the molecular-level analysis of complex systems." *Analytical and Bioanalytical Chemistry* 389, no. 5 (2007): 1311–1327.

Hottz-Rima, Janana Aparecida, Silvia Aparecida Martim, Leonardo Barros Dobbss, Joseph Albert Medeiros Evaristo, Claudio Andres Retamal, Arnoldo Rocha Facanha, and Luciano Pasqualoto Canellas. "Citric acid addition improve humic acids action and change proteins profile from plasma membrane of maize roots." *Ciencia Rural* 41, no. 4 (2011): 614–620.

Jia, Zhongtao, Ricardo FH Giehl, and Nicolaus von Wirén. "Nutrient-hormone relations: Driving root plasticity in plants." *Molecular Plant* 15 (2022): 86–103.

Jiang, Zhonghao, Xiaoping Zhou, Ming Tao, Fang Yuan, Lulu Liu, Feihua Wu, Xiaomei Wu et al. "Plant cell-surface GIPC sphingolipids sense salt to trigger Ca2+ influx." *Nature* 572, no. 7769 (2019): 341–346.

Jones, Davey L., Christopher Nguyen, and Roger D. Finlay. "Carbon flow in the rhizosphere: Carbon trading at the soil–Root interface." *Plant and Soil* 321, no. 1 (2009): 5–33.

Kanbe, Kenji, Hiroshi Naganawa, Mitiyasu Okamura et al. "Amidenin, a new plant growth-regulating substance isolated from Amycolatopsis sp." *Bioscience, Biotechnology, and Biochemistry* 57, no. 8 (1993): 1261–1263.

Kang, Li, Yuh-Shuh Wang, Srinivasa Rao Uppalapati, Keri Wang, Yuhong Tang, Vatsala Vadapalli, Barney J. Venables, Kent D. Chapman, Elison B. Blancaflor, and Kirankumar S. Mysore. "Overexpression of a fatty acid amide hydrolase compromises innate immunity in Arabidopsis." *The Plant Journal* 56, no. 2 (2008): 336–349.

Keereetaweep, Jantana, Elison B. Blancaflor, Ellen Hornung, Ivo Feussner, and Kent D. Chapman. "Ethanolamide oxylipins of linolenic acid can negatively regulate Arabidopsis seedling development." *The Plant Cell* 25, no. 10 (2013): 3824–3840.

Kilaru, Aruna, Cornelia Herrfurth, Jantana Keereetaweep, Ellen Hornung, Barney J. Venables, Ivo Feussner, and Kent D. Chapman. "Lipoxygenase-mediated oxidation of polyunsaturated N-acylethanolamines in Arabidopsis." *Journal of Biological Chemistry* 286, no. 17 (2011): 15205–15214.

Kilaru, Aruna, Pamela Tamura, Giorgis Isaac, Ruth Welti, Barney J. Venables, Edith Seier, and Kent D. Chapman. "Lipidomic analysis of N-acylphosphatidylethanolamine molecular species in Arabidopsis suggests feedback regulation by N-acylethanolamines." *Planta* 236, no. 3 (2012): 809–824.

Kulikova, Natalia A., Dmitry P. Abroskin, Gennady A. Badun, Maria G. Chernysheva, Viktor I. Korobkov, Anton S. Beer, Eugenia A. Tsvetkova, Svetlana V. Senik, Olga I. Klein, and Irina V. Perminova. "Label distribution in tissues of wheat seedlings cultivated with tritium-labeled leonardite humic acid." *Scientific Reports* 6, no. 1 (2016): 1–10.

Lait, Cameron G., Hans T. Alborn, Peter EA Teal, and James H. Tumlinson. "Rapid biosynthesis of N-linolenoyl-L-glutamine, an elicitor of plant volatiles, by membrane-associated enzyme (s) in Manduca sexta." *Proceedings of the National Academy of Sciences* 100, no. 12 (2003): 7027–7032.

Lamar, Richard T. "Possible role for electron shuttling capacity in elicitation of PB activity of humic substances on plant growth enhancement." *The Chemical Biology of Plant Biostimulants* (2020): 97–121.

Lamar, Richard T., Hiarhi Monda, and Rachel Sleighter. "Use of ore-derived humic acids with diverse chemistries to elucidate structure-activity relationships (SAR) of humic acids in plant phenotypic expression." *Frontiers in Plant Science* 12 (2021): 758424–758424.

Lehtonen, Keijo, Kari Hänninen, and Martti Ketola. "Structurally bound lipids in peat humic acids." *Organic Geochemistry* 32, no. 1 (2001): 33–43.

Li, Xiaojuan, Wenguo Cai, Yanlin Liu, Hui Li, Liwen Fu, Zengyu Liu, Lin Xu, Hongtao Liu, Tongda Xu, and Yan Xiong. "Differential TOR activation and cell proliferation in Arabidopsis root and shoot apexes." *Proceedings of the National Academy of Sciences* 114, no. 10 (2017): 2765–2770.

Lim, Gah-Hyun, Richa Singhal, Aardra Kachroo, and Pradeep Kachroo. "Fatty acid–And lipid-mediated signaling in plant defense." *Annual review of Phytopathology* 55 (2017): 505–536.

López-Bucio, José, Mayra Millán-Godínez, Alfonso Méndez-Bravo, Alina Morquecho-Contreras, Enrique Ramírez-Chávez, Jorge Molina-Torres, Anahí Pérez-Torres, Masayuki Higuchi, Tatsuo Kakimoto, and Luis Herrera-Estrella. "Cytokinin receptors are involved in alkamide regulation of root and shoot development in Arabidopsis." *Plant Physiology* 145, no. 4 (2007): 1703–1713.

López-Marqués, Rosa L., James A. Davis, Jeffrey F. Harper, and Michael Palmgren. "Dynamic membranes: the multiple roles of P4 and P5 ATPases." *Plant Physiology* 185, no. 3 (2021): 619–631.

Lung, Shiu-Cheung, and Mee-Len Chye. "Arabidopsis acyl-CoA-binding proteins regulate the synthesis of lipid signals." *New Phytologist* 223, no. 1 (2019): 113–117.

Masoom, Hussain, Denis Courtier-Murias, Hashim Farooq, Ronald Soong, Brian P. Kelleher, Chao Zhang, Werner E. Maas et al. "Soil organic matter in its native state: Unravelling the most complex biomaterial on earth." *Environmental Science & Technology* 50, no. 4 (2016): 1670–1680.

Méndez-Bravo, Alfonso, Carlos Calderón-Vázquez, Enrique Ibarra-Laclette, Javier Raya-González, Enrique Ramírez-Chávez, Jorge Molina-Torres, Angel A. Guevara-García, José López-Bucio, and Luis Herrera-Estrella. "Alkamides activate jasmonic acid biosynthesis and signaling pathways and confer resistance to *Botrytis cinerea* in *Arabidopsis thaliana*." *PloS One* 6, no. 11 (2011): e27251.

Méndez-Bravo, Alfonso, Javier Raya-González, Luis Herrera-Estrella, and José López-Bucio. "Nitric oxide is involved in alkamide-induced lateral root development in Arabidopsis." *Plant and Cell Physiology* 51, no. 10 (2010): 1612–1626.

Merlo, L., R. Ghisi, C. Passera, and N. Rascio. "Effects of humic substances on carbohydrate metabolism of maize leaves." *Canadian Journal of Plant Science* 71, no. 2 (1991): 419–425.

Meyer, Rhonda C., Matthias Steinfath, Jan Lisec, Martina Becher, Hanna Witucka-Wall, Ottó Törjék, Oliver Fiehn et al. "The metabolic signature related to high plant growth rate in Arabidopsis thaliana." *Proceedings of the National Academy of Sciences* 104, no. 11 (2007): 4759–4764.

Molina-Torres, Jorge, Carlos Junior Salazar-Cabrera, Concepción Armenta-Salinas, and Enrique Ramírez-Chávez. "Fungistatic and bacteriostatic activities of alkamides from Heliopsis longipes roots: Affinin and reduced amides." *Journal of Agricultural and Food Chemistry* 52, no. 15 (2004): 4700–4704.

Monda, Hiarhi, Amy M. McKenna, Ryan Fountain, and Richard T. Lamar. "Bioactivity of humic acids extracted from shale ore: molecular characterization and structure-activity relationship with tomato plant yield under nutritional stress." *Frontiers in Plant Science* 12 (2021): 958.

Mora, Verónica, Eva Bacaicoa, Angel-Maria Zamarreno, Elena Aguirre, Maria Garnica, Marta Fuentes, and José-Maria García-Mina. "Action of humic acid on promotion of cucumber shoot growth involves nitrate-related changes associated with the root-to-shoot distribution of cytokinins, polyamines and mineral nutrients." *Journal of Plant Physiology* 167, no. 8 (2010): 633–642.

Mora, Verónica, Roberto Baigorri, Eva Bacaicoa, Angel M. Zamarreno, and José M. García-Mina. "The humic acid-induced changes in the root concentration of nitric oxide, IAA and ethylene do not explain the changes in root architecture caused by humic acid in cucumber." *Environmental and Experimental Botany* 76 (2012): 24–32.

Morrison, RI. Soil lipids. In *Organic Geochemistry*, pp. 558–575. Springer, Berlin, 1969.

Mueller, Kevin E., David M. Eissenstat, Carsten W. Müller, Jacek Oleksyn, Peter B. Reich, and Katherine H. Freeman. "What controls the concentration of various aliphatic lipids in soil?." *Soil Biology and Biochemistry* 63 (2013): 14–17.

Nardi, Serenella, Adele Muscolo, Silvia Vaccaro, Salvatore Baiano, Riccardo Spaccini, and Alessandro Piccolo. "Relationship between molecular characteristics of soil humic fractions and glycolytic pathway and Krebs cycle in maize seedlings." *Soil Biology and Biochemistry* 39, no. 12 (2007): 3138–3146.

Nardi, Serenella, Diego Pizzeghello, and Andrea Ertani. "Hormone-like activity of the soil organic matter." *Applied Soil Ecology* 123 (2018): 517–520.

Nebbioso, Antonio, and Alessandro Piccolo. "Basis of a humeomics science: Chemical fractionation and molecular characterization of humic biosuprastructures." *Biomacromolecules* 12, no. 4 (2011): 1187–1199.

Nebbioso, Antonio, and Alessandro Piccolo. "Molecular rigidity and diffusivity of Al3+ and Ca2+ humates as revealed by NMR spectroscopy." *Environmental Science & Technology* 43, no. 7 (2009): 2417–2424.

Nolan, Trevor M., Nemanja Vukašinović, Derui Liu, Eugenia Russinova, and Yanhai Yin. "Brassinosteroids: multidimensional regulators of plant growth, development, and stress responses." *The Plant Cell* 32, no. 2 (2020): 295–318.

Olaetxea, Maite, David De Hita, C. Andrés Garcia, Marta Fuentes, Roberto Baigorri, Verónica Mora, Maria Garnica et al. "Hypothetical framework integrating the main mechanisms involved in the promoting action of rhizospheric humic substances on plant root-and shoot-growth." *Applied Soil Ecology* 123 (2018): 521–537.

Olaetxea, Maite, Verónica Mora, Eva Bacaicoa, Roberto Baigorri, Maria Garnica, Marta Fuentes, Esther Casanova et al. "ABA-regulation of root hydraulic conductivity and aquaporin gene-expression is crucial to the plant shoot rise caused by rhizosphere humic acids." *Plant Physiology* 169 (2015): 2587–2596.

Olaetxea, Maite, Veronica Mora, Roberto Baigorri, Angel M. Zamarreño, and Jose M. García-Mina. "The Singular molecular conformation of humic acids in solution influences their ability to enhance root hydraulic conductivity and plant growth." *Molecules* 26, no. 1 (2021): 3.

Olk, Daniel C., Dana L. Dinnes, J. Rene Scoresby, Chad R. Callaway, and Jerald W. Darlington. "Humic products in agriculture: Potential benefits and research challenges—A review." *Journal of Soils and Sediments* 18, no. 8 (2018): 2881–2891.

Österberg, R., and K. Mortensen. "Fractal dimension of humic acids." *European Biophysics Journal* 21, no. 3 (1992): 163–167.

Piccolo, Alessandro. "The supramolecular structure of humic substances." *Soil Science* 166, no. 11 (2001): 810–832.

Piccolo, Alessandro. "The supramolecular structure of humic substances: A novel understanding of humus chemistry and implications in soil science." *Advances in Agronomy* 75 (2002): 57–134.

Piccolo, Alessandro, Pellegrino Conte, Enrico Trivellone, Barend van Lagen, and Peter Buurman. "Reduced heterogeneity of a lignite humic acid by preparative HPSEC following interaction with an organic acid. Characterization of size-separates by Pyr-GC-MS and 1H-NMR spectroscopy." *Environmental Science & Technology* 36, no. 1 (2002): 76–84.

Piccolo, Alessandro, Pellegrino Conte, and Pietro Tagliatesta. "Increased conformational rigidity of humic substances by oxidative biomimetic catalysis." *Biomacromolecules* 6, no. 1 (2005): 351–358.

Piccolo, Alessandro, Riccardo Spaccini, Davide Savy, Marios Drosos, and Vincenza Cozzolino. "The soil humeome: Chemical structure, functions and technological perspectives." In *Sustainable Agrochemistry*, pp. 183–222. Springer, Cham, 2019.

Ph Benning. "Functions of triacylglycerols during plant development and stress." *Current Opinion in Biotechnology* 49 (2018): 191–198.

Quaggiotti, Silvia, Benedetto Ruperti, Diego Pizzeghello, Ornella Francioso, Vitaliano Tugnoli, and Serenella Nardi. "Effect of low molecular size humic substances on nitrate uptake and expression of genes involved in nitrate transport in maize (Zea mays L.)." *Journal of Experimental Botany* 55, no. 398 (2004): 803–813.

Ramírez-Chávez, Enrique, José López-Bucio, Luis Herrera-Estrella, and Jorge Molina-Torres. "Alkamides isolated from plants promote growth and alter root development in Arabidopsis." *Plant Physiology* 134, no. 3 (2004): 1058–1068.

Ramos, Alessandro C., Leonardo B. Dobbss, Leandro A. Santos, Mânlio S. Fernandes, Fábio L. Olivares, Natália O. Aguiar, and Luciano P. Canellas. "Humic matter elicits proton and calcium fluxes and signaling dependent on Ca 2+-dependent protein kinase (CDPK) at early stages of lateral plant root development." *Chemical and Biological Technologies in Agriculture* 2, no. 1 (2015): 1–12.

Rice, James A., and Jar Shyong Lin. "Fractal nature of humic materials." *Environmental Science & Technology* 27, no. 2 (1993): 413–414.

Romero, Fredy R., Kathleen Delate, George A. Kraus, Avery K. Solco, Patricia A. Murphy, and David J. Hannapel. "Alkamide production from hairy root cultures of Echinacea." *In Vitro Cellular & Developmental Biology-Plant* 45, no. 5 (2009): 599–609.

Roomi, Sohaib, Antonio Masi, Giovanni Battista Conselvan, Sara Trevisan, Silvia Quaggiotti, Micaela Pivato, Giorgio Arrigoni, Tayyaba Yasmin, and Paolo Carletti. "Protein profiling of Arabidopsis roots treated with humic substances: insights into the metabolic and interactome networks." *Frontiers in Plant Science* 9 (2018): 1812.

Sarwar, Muhammad, Muhammad Arshad, Dean A. Martens, and W. T. Frankenberger. "Tryptophan-dependent biosynthesis of auxins in soil." *Plant and Soil* 147, no. 2 (1992): 207–215.

Sasse, Joelle, Enrico Martinoia, and Trent Northen. "Feed your friends: Do plant exudates shape the root microbiome?." *Trends in Plant Science* 23, no. 1 (2018): 25–41.

Scheres, Ben, and Wim H. Van Der Putten. "The plant perceptron connects environment to development." *Nature* 543, no. 7645 (2017): 337–345.

Schiavon, M., PizzeghelloD, Muscolo A, Vaccaro S, Francioso O., and Nardi S. (2010). "High molecular size humic substances enhance phenylpropanoid metabolism in maize (Zea mays L)." *Journal of Chemical Ecology*, *36*(6), 662–669.

Schmidt, Wolfgang, Simonetta Santi, Roberto Pinton, and Zeno Varanini. "Water-extractable humic substances alter root development and epidermal cell pattern in Arabidopsis." *Plant and Soil* 300, no. 1 (2007): 259–267.

Shevchenko, Sergey M., and George W. Bailey. "Life after death: Lignin-humic relationships reexamined." *Critical Reviews in Environmental Science and Technology* 26, no. 2 (1996): 95–153.

Silva, Maura Santos Reis de Andrade, Tavares, Orlando Carlos Huertas, Thiago Gonçalves Ribeiro, José Maria García-Mina, Vera Lúcia Divan Baldani, Andrés Calderín García, Ricardo Luiz Louro Berbara, and Ederson da Conceição Jesus. "Humic acids enrich the plant microbiota with bacterial candidates for the suppression of pathogens." *Applied Soil Ecology* 168 (2021): 104146.

Sleighter, Rachel L., Paolo Caricasole, Kristen M. Richards, Terry Hanson, and Patrick G. Hatcher. "Characterization of terrestrial dissolved organic matter fractionated by pH and polarity and their biological effects on plant growth." *Chemical and Biological Technologies in Agriculture* 2, no. 1 (2015): 1–19.

Šmejkalová, Daniela, and Alessandro Piccolo. "Aggregation and disaggregation of humic supramolecular assemblies by NMR diffusion ordered spectroscopy (DOSY-NMR)." *Environmental Science & Technology* 42, no. 3 (2008a): 699–706.

Šmejkalová, Daniela, and Alessandro Piccolo. "Host-guest interactions between 2, 4-dichlorophenol and humic substances as evaluated by 1H NMR relaxation and diffusion ordered spectroscopy." *Environmental Science & Technology* 42, no. 22 (2008b): 8440–8445.

Souza, Aline Costa, Daniel Basílio Zandonadi, Mirella Pupo Santos, Natália Oliveira Aguiar Canellas, Cleiton de Paula Soares, Luiz Eduardo Souza da Silva Irineu, Carlos Eduardo de Rezende et al. "Acclimation with humic acids enhances maize and tomato tolerance to salinity." *Chemical and Biological Technologies in Agriculture* 8, no. 1 (2021): 1–13.

Spaccini, R., A. Piccolo, P. Conte, G. Haberhauer, and M. H. Gerzabek. "Increased soil organic carbon sequestration through hydrophobic protection by humic substances." *Soil Biology and Biochemistry* 34, no. 12 (2002): 1839–1851.

Sposito, Garrison. *The Chemistry of Soils*. Oxford University Press, Oxford, 2008.

Stevenson, FJ. *Humus Chemistry: Genesis, Composition, Reactions*, 2nd ed., John Wiley & Sons, New York, 1994.

Teaster, Neal D., Christy M. Motes, Yuhong Tang, William C. Wiant, Matthew Q. Cotter, Yuh-Shuh Wang, Aruna Kilaru et al. "N-Acylethanolamine metabolism interacts with abscisic acid signaling in Arabidopsis thaliana seedlings." *The Plant Cell* 19, no. 8 (2007): 2454–2469.

Teaster, Neal D., Jantana Keereetaweep, Aruna Kilaru, Yuh-Shuh Wang, Yuhong Tang, Christopher NQ Tran, Brian G. Ayre, Kent D. Chapman, and Elison B. Blancaflor. "Overexpression of fatty acid amide hydrolase induces early flowering in Arabidopsis thaliana." *Frontiers in Plant Science* 3 (2012): 32.

Tomaszewski, Jeanne E., René P. Schwarzenbach, and Michael Sander. "Protein encapsulation by humic substances." *Environmental Science & Technology* 45, no. 14 (2011): 6003–6010.

Törnroth-Horsefield, Susanna, Yi Wang, Kristina Hedfalk, Urban Johanson, Maria Karlsson, Emad Tajkhorshid, Richard Neutze, and Per Kjellbom. "Structural mechanism of plant aquaporin gating." *Nature* 439, no. 7077 (2006): 688–694.

Trevisan, S., Pizzeghello D, Ruperti B, Francioso O, Sassi A, Palme K., . . . and Nardi S. "Humic substances induce lateral root formation and expression of the early auxin-responsive IAA19 gene and DR5 synthetic element in Arabidopsis." *Plant Biology* 12, no. 4 (2010): 604–614.

Trevisan, Sara, Alessandro Botton, Silvia Vaccaro, Alice Vezzaro, Silvia Quaggiotti, and Serenella Nardi. "Humic substances affect Arabidopsis physiology by altering the expression of genes involved in primary metabolism, growth and development." *Environmental and Experimental Botany* 74 (2011): 45–55.

Tripathy, Swati, Barney J. Venables, and Kent D. Chapman. "N-Acylethanolamines in signal transduction of elicitor perception. Attenuation of alkalinization response and activation of defense gene expression." *Plant Physiology* 121, no. 4 (1999): 1299–1308.

Tsao, Rong, Athula B. Attygalle, Frank C. Schroeder, Chris H. Marvin, and Brian D. McGarvey. "Isobutylamides of unsaturated fatty acids from chrysanthemum m orifolium associated with host-plant resistance against the western flower thrips." *Journal of Natural Products* 66, no. 9 (2003): 1229–1231.

Vujinović, Tihana, Laura Zanin, Silvia Venuti, Marco Contin, Paolo Ceccon, Nicola Tomasi, Roberto Pinton, Stefano Cesco, and Maria De Nobili. "Biostimulant action of dissolved humic substances from a conventionally and an organically managed soil on nitrate acquisition in maize plants." *Frontiers in Plant Science* 10 (2020): 1652.

Wang, Yuh-Shuh, Rhidaya Shrestha, Aruna Kilaru, William Wiant, Barney J. Venables, Kent D. Chapman, and Elison B. Blancaflor. "Manipulation of Arabidopsis fatty acid amide hydrolase expression modifies plant growth and sensitivity to N-acylethanolamines." *Proceedings of the National Academy of Sciences* 103, no. 32 (2006): 12197–12202.

Wells, Martha JM., and Holly A. Stretz. "Supramolecular architectures of natural organic matter." *Science of the Total Environment* 671 (2019): 1125–1133.

Yang, Yang, and Christo Rizhsky, Ludmila, Huanan Jin, Michael R. Shepard, Harry W. Scott, Alicen M. Teitgen, M. Ann Perera, Vandana Mhaske et al. "Integrating metabolomics and transcriptomics data to discover a biocatalyst that can generate the amine precursors for alkamide biosynthesis." *The Plant Journal* 88, no. 5 (2016): 775–793.

Zandonadi, Daniel B., Mirella P. Santos, Leonardo B. Dobbss, Fábio L. Olivares, Luciano P. Canellas, Marla L. Binzel, Anna L. Okorokova-Façanha, and Arnoldo R. Façanha. "Nitric oxide mediates humic acids-induced root development and plasma membrane H+-ATPase activation." *Planta* 231, no. 5 (2010): 1025–1036.

Zandonadi, Daniel Basílio, Carlos Roberto Ribeiro Matos, Rosane Nora Castro, Riccardo Spaccini, Fábio Lopes Olivares, and Luciano Pasqualoto Canellas. "Alkamides: A new class of plant growth regulators linked to humic acid bioactivity." *Chemical and Biological Technologies in Agriculture* 6, no. 1 (2019): 1–12.

Zandonadi, Daniel Basílio, Mirella Pupo Santos, Jader Galba Busato, Lázaro Eustáquio Pereira Peres, and Arnoldo Rocha Façanha. "Plant physiology as affected by humified organic matter." *Theoretical and Experimental Plant Physiology* 25 (2013): 13–25.

12 Alkamides Symbiosis in Plants as Source of New Antibiotics

Juan Bueno MD, MSc

CONTENTS

> Life is not found in atoms or molecules or genes as such, but in organization; not in symbiosis but in synthesis.
>
> **—Edwin Conklin (1863–1952)**

> I suspect that some people also dislike the idea that natural selection has no foresight. The process itself, in effect, does not know where to go. It is the "environment" that provides the direction, and over the long run its effects are largely unpredictable in detail.
>
> **—Francis Crick (1916–2004)**

> From the paramecium to the human race, all life forms are meticulously organized, sophisticated aggregates of evolving microbial life. Far from leaving microorganisms behind on an evolutionary "ladder", we are both surrounded by them and composed of them.
>
> **—Lynn Margullis (1938–2011)**

> Above all, a plant, an oak for example, is an animal. An enormous animal in which live parasites or rather symbionts, an infinite multitude of small microscopic green organisms, of the species of unicellular "algae" cyano-phyceae.
>
> **—Konstantín Merezhkovsky (1855–1921)**

12.0 INTRODUCTION

Symbiosis is the most successful path of adaptive evolution in plants from the origin of vegetable eukaryotic cells to be the product of the binding of three prokaryotic cells, between them a cyanobacteria that would give rise to what is known as chloroplast (Mereschkowsky, 1905; Coba de la

DOI: 10.1201/9781003224525-15

Peña et al., 2018; Stadnichuk and Kusnetsov, 2021). In this way, secondary metabolites from plants are products of symbiogenesis, and the need to optimize energy flows before environmental changes (Akula and Ravishankar, 2011; Yang et al., 2018). And it is possible to affirm this because the secondary metabolism evolved to optimize primary metabolism in order to keep cells with sufficient energy for adaptation (Maeda, 2019; Yamagishi et al., 2021). Thus, secondary metabolites have as a primary function using the mechanisms of molecular communication to increase the survival of the living organism that produces them (Chevrette et al., 2020). In this order of ideas, phytochemicals come from the interaction and communication that allow plants to establish symbiosis with their ecosystem (Figure 12.1) (Babalola et al., 2020). This constitutes the role of metabolites in response to interaction with microorganisms that integrate endophytes, epiphytes, and rhizobia (Uroz et al., 2019; Noman et al., 2021). For this reason, in plants, the processes of secondary metabolism such as the shikimic acid and mevalonic acid pathways are linked to the primary metabolism of carbon (Bhat et al., 2020; Kandar, 2021). Among the mediators of secondary metabolism that optimize the primary metabolism for the maintenance of the homeostasis of plants in the search for survival are the alkamides (Pang et al., 2021; Teklić et al., 2021). Thus, alkamides are regulatory compounds of plant growth, as well as part of the defense molecules produced by plant organisms in the presence of phytopathogenic microorganisms (Zandonadi et al., 2019; Köhl et al., 2019). In this order of ideas, this kind of molecule, composed of an unsaturated fatty acid and an amine, has the ability to mediate molecular communication in the quorum-sensing mechanism, which is the signal system employing bacteria to form microbial associations such as biofilms (Williams, 2007; Trebino et al., 2021). That is to say, when modulating the quorum sensing, the alkamides mediate in the interaction between microorganisms and plants. This means that they are integrative of the holobiont, conformed by the host (vegetable eukaryotic cells) and its symbiont microorganisms (microbiome) (Chagas et al., 2018; Nobori et al., 2018). In that order, the interactions between the plants the

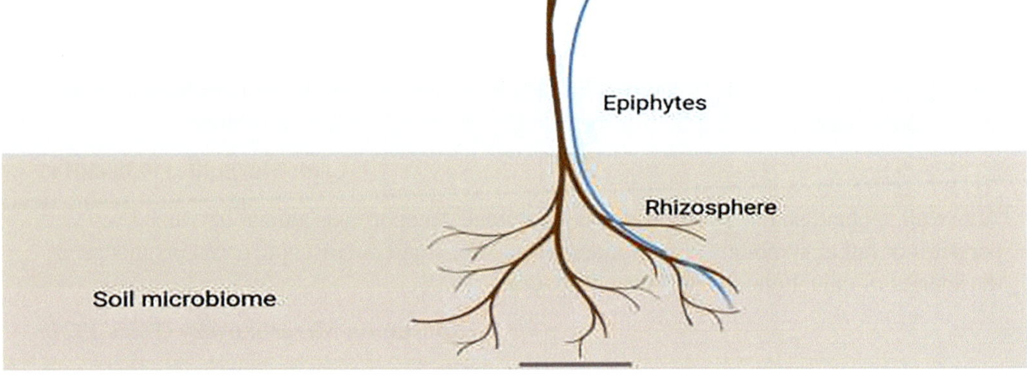

FIGURE 12.1 Plant symbiotic holobiont. The plant holobiont is made up of the host and its microbiome (endophytes, epiphytes, and rhizosphere). This interaction induces the production of secondary metabolites for communication and defense.

Source: Created using BioRender (https://biorender.com/)

alkamides are part of, as well as the phytohormones and the virulence networks of the microbiome where the quorum sensing participates, allow the establishment of the symbiosis that will boost the plant as holobiont (Gupta and Sharma, 2021). Therefore, the objective of this chapter is to explore the biological activity of the alkamides in the molecular communication of plants as a symbiotic holobiont in order to evaluate its potential as a future antimicrobial agent for use in medicine and agriculture (Maggini et al., 2017; Castronovo et al., 2021).

12.1 PLANT HOLOBIONT

By definition, each macroorganism this plant or animal is associated with has its microbiota formed by symbiotic microorganisms, in which a sharing of genetic and molecular information occurs in the hologenome (horizontal genetic transfer) (Lyu et al., 2021; Rosenberg, 2021; Zilber-Rosenberg and Rosenberg, 2021). Thus, it has been demonstrated that the interaction of microbiome plants promotes the adaptive evolution of plant species, as well as their chemical diversity (Compant et al., 2019; Dastogeer et al., 2020). In this order of ideas, there are the internal microbiome or endosphere and the external or episphere, among which fungi and bacteria predominate, which allow the development of a system of molecular communication and genetic exchange, as well as epigenetic adaptation. That is, they promote gene expression in response to environmental changes (Vannier et al., 2015; Ashapkin et al., 2020; Tsaballa et al., 2021). In this sense, the plant as a holobiont has molecular communication systems capable of establishing information networks and exchange with bacterial and fungal biofilms (Deveau et al., 2018; Kumawat et al., 2022). (It should be remembered that the environmental form of sociomicrobiology is biofilm.) (See Figure 12.2.) In that way, there is molecular signaling throughout the plant that defines nutrition, the transformation of energy, and the response to phytopathogenic diseases (Phour et al., 2020; Gorshkov and Tsers, 2022). These molecular communication pathways that are activated in the holobiont before the microbial invasion

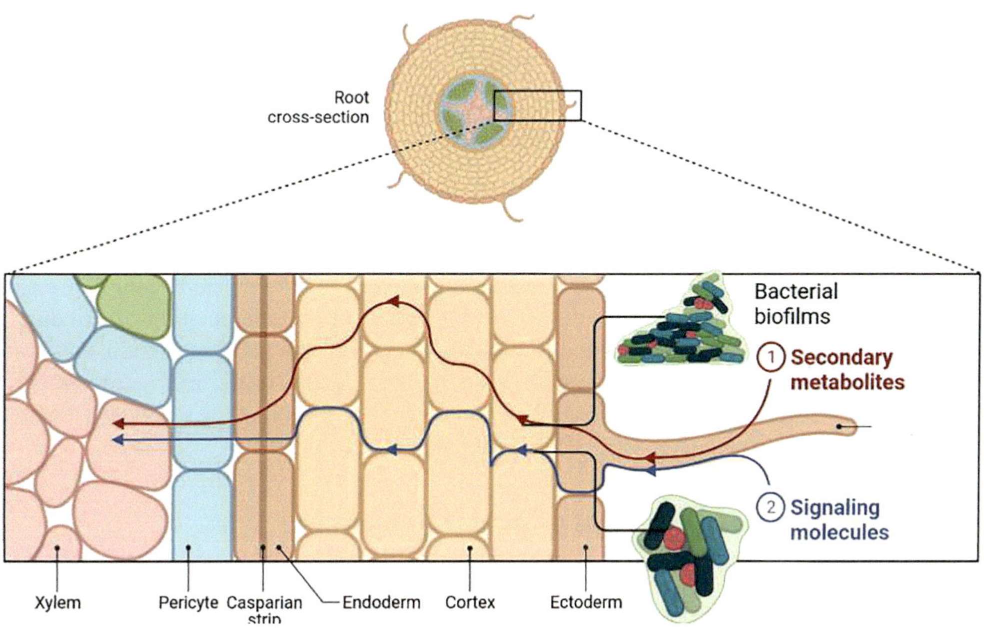

FIGURE 12.2 Biofilm in the holobiont. Biofilms are microbial groupings that allow the persistence and survival of bacterial microorganisms; communication between microbiome biofilms and host constitutes the holobiont and symbiotic exchange.

Source: Created using BioRender (https://biorender.com/)

are the ones that, in the synthesis of alkamides, are present by epigenetic expression (Dudler and Eberl, 2006; Mullard, 2009).

12.2 PLANT HOLOBIONT IN HEALTH AND DISEASE

The current concept of the symbiotic holobiont assumes that the alteration of the microbiome in contact with the host, the phenomenon known as dysbiosis, is the one that causes disease in both plants and animals (O'Malley, 2017; Tipton et al., 2019; Rosenberg, 2021). In this order of ideas, dysbiosis as alteration of microbial populations both in rhizobium and in endophytes and epiphytes causes the inadequate immune response of the plant, allowing the colonization and invasion of phytopathogens, as well as deficiencies in growth from poor absorption and metabolism of nutrients (Chakraborty et al., 2021; Zhang et al., 2021). Likewise, the production of various metabolites in both symbiosis and dysbiosis determines the chemotype of the plant, as well as the accumulation of compounds with biological activity or toxicity (Sangwan et al., 2001; Félix et al., 2018). This production of metabolites in response to pathobionts and symbionts will determine the characteristics of the plant and its subsequent uses in both medicine and agroindustry (Alves et al., 2022). In this way, the growth of the plant under different environmental conditions will be epigenetically modulated by the interactions with the microbiome. This means there are beneficial microorganisms that promote metabolites that induce growth and harmful microorganisms that cause nutritional deficits in the plant and its consequent death (Chen et al., 2018). Thus, the conformation and establishment of a symbiotic plant-microbiota consortium will allow the expression of certain secondary metabolites that are health or disease biomarkers (Mercado-Blanco et al., 2018; Sergaki et al., 2018). These consortiums between the microorganisms and the host plant allow the acquisition of nutrition through the rhizome; they also control possible pathogens by means of the endophytes of the root that communicate with the rhizosphere (Singha et al., 2021; Vandana et al., 2021). In this way, the secondary metabolites that mediate the molecular signaling in the symbiosis are fundamental to establishing the chemical diversity of plants in adaptation with their environment (Hiruma, 2019). To understand this, the metabolomics of symbiosis acquire important characteristics that guide an analysis of the biosynthetic mechanisms of the plants (Sharma et al., 2021).

12.2.1 Metabolomics of Plant-Microbe Symbiosis

The metabolome of a plant changes depending on the interaction with its microbiome. This determines the adaptive capacity of the plant cell by virtue of the epigenetic control of the clusters of biosynthetic genes that express the enzymes that synthesize the secondary metabolites (Adeniji et al., 2020; Aghdam et al., 2021). Thus, the biosynthesis of the alkamides is stimulated by the expression of these gene clusters in interaction with the molecular signals of the microbiota, which induce a response by chloroplasts and a transduction toward the nucleus of the plant cell (Vidhyasekaran, 2015; Peres et al., 2019). In that way, chloroplasts are stimulated against stress, and microbiome signals and induced hormetic responses are transduced into gene expression of the secondary metabolites that produce plant growth (Abbasi et al., 2021; Jalal et al., 2021).

12.2.2 Genetic Clusters for Synthesis of Bioactive Molecules in the Symbiosis Microbiome-Plant

The natural history of the biosphere is the product of communication and interaction to obtain information and cognition. In that way, the molecular signaling is the mechanism that initiates adaptation by virtue of genetic expression (Cohen and Marron, 2020; Slijepcevic, 2021). Thus, in plants, the epigenetic control of enzymes that synthesize the molecules of secondary metabolism through DNA methylation and histone acetylation allows expression of the environmental change in biosynthetic gene clusters (BGCs) (Scherlach et al., 2021). Among these molecules are hormones such as jasmonic acid, abscisic acid, brassinosteroids, auxin, gibberellins, strigolactones, ethylene, and salicylic acids (Bhatt et al., 2020; Salvi et al., 2021). The production of phytohormones is regulated by the epigenetic expression of BCGs in order to respond to interactions with the environment through symbiosis with the microbiome of the plant (Krishna et al., 2019). Likewise, this induction of phytohormones by the microbiome represented in endophytes, epiphytes, and rhizobium stimulates the growth of plants as well as the acquisition of nutrients and photosynthesis (Egamberdieva et al., 2017; Kudoyarova et al., 2019). Likewise, these phytohormones, such as jasmonic acid, can also be activated by alkamides in response to phytopathogens invasion, which establishes a signaling cascade between alkamides and microbiome to activate phytohormones in adaptive response (Ramírez-Chávez et al., 2004; Hentrich et al., 2013). That is to say, the alkamides as metabolites of the defense of the plants respond against the changes of microbiomes in phytopathogens by perceiving the dysbiosis signals of the microbial invasion (Ravanbakhsh et al., 2018; Pascale et al., 2020). In this order of ideas, the immune system of plants requires signaling molecules that induce the expression of BCGs in interaction with the environment and microbiota; in this, alkamides have an important function (Tóth and Stacey, 2015).

12.3 ALKAMIDES AS A MOLECULAR DEFENSE IN PLANTS

The molecular defense mechanisms in plants are formed by both signaling compounds and antimicrobials capable of inhibiting invasion by phytopathogens and stimulating growth before environmental changes (Ponce de León and Montesano, 2013; Enebe et al., 2019). In this way, the plants interact with plant growth–promoting microbes (PGPM), which stimulate the plant immune system to repel phytopathogens (Backer et al., 2018). Thus, alkamides have the ability to induce an immune response in collaboration with the beneficial microbiota of the plant in a eubiotic stress process (González-Morales et al., 2015; Hou et al., 2016). In this order of ideas, a eubiotic stress is one that induces an adaptive response in the host organisms without producing dysbiosis in the entire symbiotic holobiont (Duperron et al., 2020). The alkamides are a group of compounds that, in addition to mediating in molecular communication, present an antimicrobial biological activity due to their lipophilicity, which allows them to penetrate the cell wall of the microorganisms (Cruz et al., 2014; Wink, 2015). They can also induce the production of antimicrobial compounds by the microbe-plant interaction (Tewari and Arora, 2013). By virtue of this phenomenon, we are facing a set of biological activities that allow alkamides to participate in signaling processes and inhibition of microorganisms, so we can conclude that they are regulation molecules between the growth and control of microbiota (Ortíz-Castro et al., 2009; Al-Ani et al., 2020).

12.4 ALKAMIDES BETWEEN EUBIOSIS AND ANTIBIOSIS

So, by definition, eubiosis is a microbiome in balance with its host on the holobiont, which allows maintaining the health of the organism as a whole, be it plant or animal (Berg et al., 2021). In that way, the restoration of the microbiome is the opposite of dysbiosis in the communities that make up endophytes, epiphytes, and rhizobium (Müller et al., 2016). Thus, alkamides are produced in response to phytopathogens' invasion, as well as signal mediators and antimicrobial molecules (Mohan et al., 2018; Ortiz-Castro and López-Bucio, 2019). This double function has been observed in the different antibiotics where its role is reported in molecular communication inducing stimulation or microbial inhibition (Romero et al., 2011; Sengupta et al., 2013). Likewise, the role of alkamides has been found in the regulation of biofilm formation, which means that they can signal themselves in the quorum-sensing system, molecular communicators involved in establishing this microbial formation (Mohan et al., 2018; Gutierrez-Villagomez et al., 2020). In this way, the regulation of the quorum sensing allows alkamides to interact with the microbiota of the plant to maintain microbial populations in equilibrium, avoiding with this biological activity the emergence of pathobionts (symbionts that acquire virulence factors) (Baltenneck et al., 2021). These interactions between alkamides and plant symbionts make this type of molecule an interesting scaffold for the development of new antimicrobials that can regulate the quorum-sensing signals in bacteria, inhibiting the invasion and subsequent formation of infectious biofilms (Zhang, 2003; Aziz et al., 2020).

12.5 ALKAMIDES AS PROMISING ANTIMICROBIAL COMPOUNDS

The antibiotic activity of alkamides is due to their ability to cause the disruption of microbial walls, causing cell lysis of phytopathogenic microorganisms (Mir-Rashed et al., 2010; Silveira et al., 2018). Likewise, alkamides can regulate microbial biofilms in the signaling process known as quorum sensing, which is an important therapeutic target in the inhibition of the invasion process (Zhao et al., 2015; Barrera-Ortiz et al., 2021). Alkamides especially interact with the quorum-sensing regulatory molecules known as N-Acyl-L-Homoserine lactones, which influence both biofilm development and the stimulation of plant growth (Rankl et al., 2016; García et al., 2021). Thus, the regulation of molecular signals in the biofilms represented in quorum sensing is one of the promising therapeutic targets for the development of new antibiofilm drugs, which are the form of association, environmental persistence, and invasion of bacterial microorganisms in infectious diseases (Fleitas Martínez et al., 2019; Nadar et al., 2022). In this way, the role of alkamides in antibiofilm activity – modulating molecular signaling – is one of the results of the interaction of microbiota-plant symbiosis (Korenblum et al., 2020; Vesty et al., 2020). In this order of ideas, molecular communication through the secondary metabolites of plants can modulate microbial pathogenesis, avoiding invasion and tissue destruction in infectious diseases (Pusztahelyi et al., 2015; Isah, 2019).

12.6 ALKAMIDES AS ANTIBIOFILM AGENTS

Biofilms, the microbial formations that facilitate the communication, survival, and persistence of bacterial species, have also been linked to the pathogenesis and invasion of tissues during the infection process (Vestby et al., 2020; Schulze et al., 2021). In this order of ideas, biofilms are made up of molecular signs mediated by acyl homoserine lactones in a process known as quorum sensing (Verbeke et al., 2017; Muras et al., 2020). For this reason, one of the antimicrobial therapeutic targets is the inhibition or regulation of molecular communication in quorum sensing; this can be carried out using the molecules present in the symbiosis within the host microbiome-interactions (Whiteley et al., 2017; White et al., 2020). Thus, the alkamides are constituted in one of these secondary metabolites with the ability to interact as molecular communicators in the biofilms with the possibility of inhibiting invasion and pathogenesis (Netzker et al., 2015). This is possible because the alkamides produced by plants are compounds similar to the N-Acyl-L-Homoserine lactones that can mimic the quorum-sensing signals (Ortíz-Castro et al., 2009; Singh et al., 2022). Finally, this indicates that alkamides can modulate the formation of biofilms and inhibit the establishment of this mechanism of microbial survival, which becomes a promising mediator in the search for and development of new antibiotics (Dusane et al., 2014; Chaudhari et al., 2015).

12.7 PERSPECTIVES

In this order of ideas, the holobiont symbiosis in plants with their microbiome will allow the development of new secondary metabolites, produced in the defense of phytopathogens (Vandenkoornhuyse et al., 2015). Thus, the determination of the biosynthetic gene clusters that are expressed during these molecular interactions will be the starting point for a modern biotechnology of natural products in order to obtain bioactive compounds such as alkamides in the industrial quantities required for medicine and agriculture (Chen et al., 2019). Finally, the chemical diversity of symbiosis will provide genetic information for the development of green synthetic models with which to establish modern synthetic industries (Chevrette and Currie, 2019).

12.8 CONCLUSIONS

The search for new anti-infective agents is a priority due to the increase in antimicrobial resistance enhanced by the recent coronavirus (COVID-19) pandemic (Hirabayashi et al., 2021; Nieuwlaat et al., 2021). In this order of ideas, it is necessary to develop medications capable of inhibiting the mechanisms of bacterial resistance such as biofilms (Das and Mehta, 2018; Varela et al., 2021). Thus, the symbiosis between plants and microbiome can be an important approximation to find signaling molecules capable of inhibiting the establishment of biofilms and preventing mechanisms that favour microbial invasion (Mendes et al., 2013; Leach et al., 2017). For this reason, the molecules that activate phytohormones, such as the alkamides capable of performing molecular signaling in biofilms, are alternatives to study in order to obtain antimicrobial medications with eubiotic activity: that is, that do not alter the microbiome of the host during treatment (Spisni et al., 2021). Finally, the development of eubiotic antibiotics based on plant-signaling molecules will be of great importance in preventing dysbiosis predisposed to acquiring a greater number of both chronic and infectious diseases (Isibor et al., 2021).

ACKNOWLEDGMENT

The author thanks the School of Medicine, University of Applied and Environmental Sciences, as well as the Research Center of Bioprospecting and Biotechnology for Biodiversity Foundation (BIOLABB) for their collaboration and invaluable support during the writing of this chapter.

REFERENCES

Abbasi, Sakineh, et al. "The stimulators and responsive genes to induce systemic resistance against pathogens: An exclusive focus on tomato as a model plant." Biocatalysis and Agricultural Biotechnology 33 (2021): 101993.

Adeniji, Adetomiwa Ayodele, and Olubukola Oluranti Babalola. "Metabolomic applications for understanding complex tripartite plant-microbes interactions: Strategies and perspectives." Biotechnology Reports 25 (2020): e00425.

Aghdam, Shiva Abdollahi, and Amanda May Vivian Brown. "Deep learning approaches for natural product discovery from plant endophytic microbiomes." Environmental Microbiome 16.1 (2021): 1–20.

Akula, Ramakrishna, and Gokare Aswathanarayana Ravishankar. "Influence of abiotic stress signals on secondary metabolites in plants." Plant Signaling & Behavior 6.11 (2011): 1720–1731.

Al-Ani, Laith Khalil Tawfeeq, et al. "The role of microbial signals in plant growth and development: Current status and future prospects." in *New and future developments in microbial biotechnology and bioengineering*. Elsevier, 2020, 225–242.

Alves, Nayara Sabrina Freitas, et al. "Variation in Peperomia pellucida growth and secondary metabolism after rhizobacteria inoculation." Plos One 17.1 (2022): e0262794.

Ashapkin, Vasily V., et al. "Epigenetic mechanisms of plant adaptation to biotic and abiotic stresses." International Journal of Molecular Sciences 21.20 (2020): 7457.

Aziz, Mina, and Kent D. Chapman. "Fatty acid amide hydrolases: An expanded capacity for chemical communication?." Trends in Plant Science 25.3 (2020): 236–249.

Babalola, Olubukola Oluranti, et al. "The nexus between plant and plant microbiome: Revelation of the networking strategies." Frontiers in Microbiology 11 (2020): 2128.

Backer, Rachel, et al. "Plant growth-promoting rhizobacteria: context, mechanisms of action, and roadmap to commercialization of biostimulants for sustainable agriculture." Frontiers in Plant Science (2018): 1473.

Baltenneck, Julie, Sylvie Reverchon, and Florence Hommais. "Quorum sensing regulation in phytopathogenic bacteria." Microorganisms 9.2 (2021): 239.

Barrera-Ortiz, Salvador, et al. "Bacterial quorum-sensing signaling-related drr1 mutant influences abscisic acid responsiveness in Arabidopsis thaliana L." Journal of Plant Growth Regulation (2021): 1–15.

Berg, Gabriele, et al. "Microbiome modulation—Toward a better understanding of plant microbiome response to microbial inoculants." Frontiers in Microbiology 12 (2021): 803.

Bhat, Basharat Ahmad, et al. "Role of micronutrients in secondary metabolism of plants." in *Plant micronutrients*. Springer, Cham, 2020, 311–329.

Bhatt, Deepesh, et al. "Role of growth regulators and phytohormones in overcoming environmental stress." *Protective Chemical Agents in the Amelioration of Plant Abiotic Stress: Biochemical and Molecular Perspectives* (2020): 254–279.

Castronovo, Lara Mitia, et al. "Medicinal plants and their bacterial microbiota: A review on antimicrobial compounds production for plant and human health." Pathogens 10.2 (2021): 106.

Chagas, Fernanda Oliveira, et al. "Chemical signaling involved in plant–microbe interactions." Chemical Society Reviews 47.5 (2018): 1652–1704.

Chakraborty, Arindam, Jatinder Singh, and Hina Upadhyay. "Fungal diversity and ecology: Major challenges for crop production in present climate scenario." in *Recent trends in mycological research*. Springer, Cham, 2021, 89–107.

Chaudhari, Gunjan M., and Raghunath T. Mahajan. "Bioprospecting of Terminalia arjuna stem bark and its flavonoids for antimicrobial and anti-biofilm potential." Journal of Chemical and Pharmaceutical Research 7.12 (2015): 834–841.

Chen, Feng, et al. "Terpenoid secondary metabolites in bryophytes: Chemical diversity, biosynthesis and biological functions." Critical Reviews in Plant Sciences 37.2–3 (2018): 210–231.

Chen, Ray, Hon Lun Wong, and Brendan Paul Burns. "New approaches to detect biosynthetic gene clusters in the environment." Medicines 6.1 (2019): 32.

Chevrette, Marc G., and Cameron R. Currie. "Emerging evolutionary paradigms in antibiotic discovery." Journal of Industrial Microbiology and Biotechnology 46.3–4 (2019): 257–271.

Chevrette, Marc G., et al. "Evolutionary dynamics of natural product biosynthesis in bacteria." Natural product reports 37.4 (2020): 566–599.

Coba de la Peña, Teodoro, et al. "The symbiosome: legume and rhizobia co-evolution toward a nitrogen-fixing organelle?." Frontiers in Plant Science 8 (2018): 2229.

Cohen, Irun R., and Assaf Marron. "The evolution of universal adaptations of life is driven by universal properties of matter: Energy, entropy, and interaction." F1000Research 9 (2020).

Compant, Stéphane, et al. "A review on the plant microbiome: Ecology, functions, and emerging trends in microbial application." Journal of advanced research 19 (2019): 29–37.

Cruz, I., et al. "Alkamides from Echinacea disrupt the fungal cell wall-membrane complex." Phytomedicine 21.4 (2014): 435–442.

Das, Rina, and Dinesh K. Mehta. "Microbial biofilm and quorum sensing inhibition: Endowment of medicinal plants to combat multidrug-resistant bacteria." Current Drug Targets 19.16 (2018): 1916–1932.

Dastogeer, Khondoker MG, et al. "Plant microbiome–An account of the factors that shape community composition and diversity." Current Plant Biology (2020): 100161.

Deveau, Aurélie, et al. "Bacterial–fungal interactions: Ecology, mechanisms and challenges." FEMS Microbiology Reviews 42.3 (2018): 335–352.

Duperron, Sébastien, et al. "Microbiome-aware ecotoxicology of organisms: Relevance, pitfalls, and challenges." Frontiers in Public Health (2020): 407.

Dusane, Devendra H., et al. "Alkaloids modulate motility, biofilm formation and antibiotic susceptibility of uropathogenic Escherichia coli." PLoS One 9.11 (2014): e112093.

Egamberdieva, Dilfuza, et al. "Phytohormones and beneficial microbes: Essential components for plants to balance stress and fitness." Frontiers in Microbiology 8 (2017): 2104.

Enebe, Matthew Chekwube, and Olubukola Oluranti Babalola. "The impact of microbes in the orchestration of plants' resistance to biotic stress: a disease management approach." Applied Microbiology and Biotechnology 103.1 (2019): 9–25.

Félix, Carina, et al. "Production of toxic metabolites by two strains of Lasiodiplodia theobromae, isolated from a coconut tree and a human patient." Mycologia 110.4 (2018): 642–653.

Fleitas Martínez, Osmel, et al. "Recent advances in anti-virulence therapeutic strategies with a focus on dismantling bacterial membrane microdomains, toxin neutralization, quorum-sensing interference and biofilm inhibition." *Frontiers in Cellular and Infection Microbiology* (2019): 74.

GaPlrcía, Tonatiu Campos, Jorge Molina-Torres, and Kirk L. Overmyer. "Reactive oxygen species signalling is involved in alkamide-induced altered root development." Biorxiv (2021).

González-Morales, Susana, et al. "Effect of alkamides as tolerance inductors to biotic stress in tomato." Revista Mexicana De Ciencias Agrícolas 6.SPE12 (2015): 2371–2382.

Gorshkov, Vladimir, and Ivan Tsers. "Plant susceptible responses: The underestimated side of plant–pathogen interactions." Biological Reviews 97.1 (2022): 45–66.

Gupta, Vadakattu VSR, and Anil K. Sharma, eds. *Rhizosphere biology: Interactions between microbes and plants*. Springer, Singapore, 2021.

Gutierrez-Villagomez, Juan Manuel, et al. "From natural to synthetic quorum sensing active compounds: Insights to develop specific quorum sensing modulators for microbe-plant interaction." in *Quorum sensing: Microbial rules of life. American chemical society*. American Chemical Society, 2020, 87–113.

Hentrich, Mathias, et al. "The jasmonic acid signaling pathway is linked to auxin homeostasis through the modulation of YUCCA 8 and YUCCA 9 gene expression." The Plant Journal 74.4 (2013): 626–637.

Hirabayashi, Aki, et al. "Impact of the COVID-19 pandemic on the surveillance of antimicrobial resistance." Journal of Hospital Infection 117 (2021): 147–156.

Hiruma, Kei. "Roles of plant-derived secondary metabolites during interactions with pathogenic and beneficial microbes under conditions of environmental stress." Microorganisms 7.9 (2019): 362.

Hou, Quancan, Guido Ufer, and Dorothea Bartels. "Lipid signalling in plant responses to abiotic stress." Plant, Cell & Environment 39.5 (2016): 1029–1048.

Isah, Tasiu. "Stress and defense responses in plant secondary metabolites production." Biological Research 52 (2019).

Isibor, Patrick Omoregie, et al. "Significance of African Diets in Biotherapeutic Modulation of the Gut Microbiome." Bioinformatics and Biology Insights 15 (2021): 11779322211012697.

Jalal, Arshad, et al. "Hormesis in plants: Physiological and biochemical responses." Ecotoxicology and Environmental Safety 207 (2021): 111225.

Kandar, Chandi Charan. "Secondary metabolites from plant sources." in *Bioactive natural products for pharmaceutical applications*. Springer, Cham, 2021, 329–377.

Köhl, Jürgen, Rogier Kolnaar, and Willem J. Ravensberg. "Mode of action of microbial biological control agents against plant diseases: relevance beyond efficacy." Frontiers in Plant Science 10 (2019): 845.

Korenblum, Elisa, et al. "Rhizosphere microbiome mediates systemic root metabolite exudation by root-to-root signaling." Proceedings of the National Academy of Sciences 117.7 (2020): 3874–3883.

Krishna, Suresh BN, et al. "Integrating microbiome network: establishing linkages between plants, microbes and human health." The Open Microbiology Journal 13.1 (2019).

Kudoyarova, Guzel, et al. "Phytohormone mediation of interactions between plants and non-symbiotic growth promoting bacteria under edaphic stresses." Frontiers in Plant Science (2019): 1368.

Kumawat, Kailash Chand, Nadia Razdan, and Krishna Saharan. "Rhizospheric microbiome: Bio-based emerging strategies for sustainable agriculture development and future perspectives." Microbiological Research 254 (2022): 126901.

Leach, Jan E., et al. "Communication in the phytobiome." Cell 169.4 (2017): 587–596.

Lyu, Dongmei, et al. "Plant holobiont theory: The phytomicrobiome plays a central role in evolution and success." Microorganisms 9.4 (2021): 675.

Maeda, Hiroshi A. "Harnessing evolutionary diversification of primary metabolism for plant synthetic biology." Journal of Biological Chemistry 294.45 (2019): 16549–16566.

Maggini, Valentina, et al. "Plant-endophytes interaction influences the secondary metabolism in Echinacea purpurea (L) Moench: An in vitro model." Scientific Reports 7.1 (2017): 1–8.

Mendes, Rodrigo, Paolina Garbeva, and Jos M. Raaijmakers. "The rhizosphere microbiome: Significance of plant beneficial, plant pathogenic, and human pathogenic microorganisms." FEMS Microbiology Reviews 37.5 (2013): 634–663.

Mercado-Blanco, Jesús, et al. "Belowground microbiota and the health of tree crops." Frontiers in Microbiology 9 (2018): 1006.

Mereschkowsky, Constantin. "Uber natur und ursprung der chromatophoren im pflanzenreiche." Biologisches Centralblatt 25 (1905): 293–604.

Mir-Rashed, Nadereh, et al. "Disruption of fungal cell wall by antifungal Echinacea extracts." Medical Mycology 48.7 (2010): 949–958.

Mohan, Rajinikanth, et al. "Quorum sensing and biofilm formation in pathogenic and mutualistic plant-bacterial interactions." in *Implication of quorum sensing system in biofilm formation and virulence*. Springer, Singapore, 2018, 133–160.

Mullard, Asher. "Microbiology: Tinker, bacteria, eukaryote, spy." Nature News 459.7244 (2009): 159–161.

Müller, Daniel B., et al. "The plant microbiota: Systems-level insights and perspectives." Annual Review of Genetics 50 (2016): 211–234.

Muras, Andrea, et al. "Acyl homoserine lactone-mediated quorum sensing in the oral cavity: A paradigm revisited." Scientific Reports 10.1 (2020): 1–14.

Nadar, Sahaya, et al. "Development of antibiofilm therapeutics strategies to overcome antimicrobial drug resistance." Microorganisms 10.2 (2022): 303.
Netzker, Tina, et al. "Microbial communication leading to the activation of silent fungal secondary metabolite gene clusters." Frontiers in Microbiology 6 (2015): 299.
Nieuwlaat, Robby, et al. "Coronavirus disease 2019 and antimicrobial resistance: Parallel and interacting health emergencies." Clinical Infectious Diseases 72.9 (2021): 1657–1659.
Nobori, Tatsuya, Akira Mine, and Kenichi Tsuda. "Molecular networks in plant–pathogen holobiont." FEBS Letters 592.12 (2018): 1937–1953.
Noman, Muhammad, et al. "Plant–Microbiome crosstalk: Dawning from composition and assembly of microbial community to improvement of disease resilience in plants." International Journal of Molecular Sciences 22.13 (2021): 6852.
O'Malley, Maureen A. "From endosymbiosis to holobionts: Evaluating a conceptual legacy." Journal of Theoretical Biology 434 (2017): 34–41.
Ortiz-Castro, Randy, and José López-Bucio. "Phytostimulation and root architectural responses to quorum-sensing signals and related molecules from rhizobacteria." Plant Science 284 (2019): 135–142.
Pang, Zhiqiang, et al. "Linking Plant Secondary Metabolites and Plant Microbiomes: A Review." Frontiers in Plant Science 12 (2021): 300.
Pascale, Alberto, et al. "Modulation of the root microbiome by plant molecules: the basis for targeted disease suppression and plant growth promotion." Frontiers in Plant Science 10 (2020): 1741.
Peres, Ana Laura GL., et al. "Brassinosteroids, the sixth class of phytohormones: A molecular view from the discovery to hormonal interactions in plant development and stress adaptation." International Journal of Molecular Sciences 20.2 (2019): 331.
Phour, Manisha, et al. "Interkingdom signaling in plant-rhizomicrobiome interactions for sustainable agriculture." Microbiological Research (2020): 126589.
Ponce de León, Inés, and Marcos Montesano. "Activation of defense mechanisms against pathogens in mosses and flowering plants." International Journal of Molecular Sciences 14.2 (2013): 3178–3200.
Pusztahelyi, Tünde, Imre J. Holb, and István Pócsi. "Secondary metabolites in fungus-plant interactions." Frontiers in Plant Science 6 (2015): 573.
Ramírez-Chávez, Enrique, et al. "Alkamides isolated from plants promote growth and alter root development in Arabidopsis." Plant Physiology 134.3 (2004): 1058–1068.
Rankl, Simone, et al. "Microbial homoserine lactones (AHLs) are effectors of root morphological changes in barley." Plant Science 253 (2016): 130–140.
Ravanbakhsh, Mohammadhossein, et al. "Microbial modulation of plant ethylene signaling: ecological and evolutionary consequences." Microbiome 6.1 (2018): 1–10.
Romero, Diego, et al. "Antibiotics as signal molecules." Chemical Reviews 111.9 (2011): 5492–5505.
Rosenberg, Eugene. "Evolution of holobionts: The hologenome concept." in *Microbiomes*. Springer, Cham, 2021, 317–352.
Salvi, Prafull, et al. "Phytohormone signaling and crosstalk in regulating drought stress response in plants." Plant Cell Reports 40.8 (2021): 1305–1329.
Sangwan, N. S., et al. "Regulation of essential oil production in plants." Plant Growth Regulation 34.1 (2001): 3–21.
Scherlach, Kirstin, and Christian Hertweck. "Mining and unearthing hidden biosynthetic potential." Nature Communications 12.1 (2021): 1–12.
Schulze, Adina, et al. "Biofilms by bacterial human pathogens: Clinical relevance-development, composition and regulation-therapeutical strategies." Microbial Cell 8.2 (2021): 28.
Sengupta, Saswati, Madhab Kumar Chattopadhyay, and Hans-Peter Grossart. "The multifaceted roles of antibiotics and antibiotic resistance in nature." Frontiers in Microbiology 4 (2013): 47.
Sergaki, Chrysi, et al. "Challenges and approaches in microbiome research: from fundamental to applied." Frontiers in Plant Science 9 (2018): 1205.
Sharma, Vinay, et al. "Metabolomics intervention towards better understanding of plant traits." Cells 10.2 (2021): 346.
Silveira, Narjara, Louis Pergaud Sandjo, and Maique Weber Biavatti. "Spilanthol-containing products: A patent review (1996–2016)." Trends in Food Science & Technology 74 (2018): 107–111.
Singh, Kshitij, Ram Chandra, and Diane Purchase. "Unraveling the secrets of rhizobacteria signaling in rhizosphere." Rhizosphere (2022): 100484.
Singha, K. Malabika, Brahmanand Singh, and Piyush Pandey. "Host specific endophytic microbiome diversity and associated functions in three varieties of scented black rice are dependent on growth stage." Scientific Reports 11.1 (2021): 1–17.

Slijepcevic, Predrag. "Principles of information processing and natural learning in biological systems." Journal for General Philosophy of Science 52.2 (2021): 227–245.
Spisni, Enzo, et al. "Natural compounds in the modulation of the intestinal microbiota: Implications in human physiology and pathology." Alternative Medicine: Update (2021): 175.
Stadnichuk, I. N., and Kusnetsov V. V. "Endosymbiotic origin of chloroplasts in plant cells' evolution." Russian Journal of Plant Physiology 68.1 (2021): 1–16.
Teklić, Tihana, et al. "Linking abiotic stress, plant metabolites, biostimulants and functional food." Annals of Applied Biology 178.2 (2021): 169–191.
Tewari, Sakshi, and Naveen Kumar Arora. "Transactions among microorganisms and plant in the composite rhizosphere habitat." Plant Microbe Symbiosis: Fundamentals and Advances (2013): 1–50.
Tipton, Laura, John L. Darcy, and Nicole A. Hynson. "A developing symbiosis: enabling cross-talk between ecologists and microbiome scientists." Frontiers in Microbiology 10 (2019): 292.
Trebino, Michael A., et al. "Strategies and approaches for discovery of small molecule disruptors of biofilm physiology." Molecules 26.15 (2021): 4582.
Tsaballa, Aphrodite, et al. "Vegetable grafting from a molecular point of view: The involvement of epigenetics in rootstock-scion interactions." Frontiers in Plant Science (2021): 2129.
Tóth, Katalin, and Gary Stacey. "Does plant immunity play a critical role during initiation of the legume-rhizobium symbiosis?." Frontiers in Plant Science 6 (2015): 401.
Uroz, Stephane, Pierre Emmanuel Courty, and Phil Oger. "Plant symbionts are engineers of the plant-associated microbiome." Trends in Plant Science 24.10 (2019): 905–916.
Vandana, Udaya Kumar, et al. "The endophytic microbiome as a hotspot of synergistic interactions, with prospects of plant growth promotion." Biology 10.2 (2021): 101.
Vandenkoornhuyse, Philippe, et al. "The importance of the microbiome of the plant holobiont." New Phytologist 206.4 (2015): 1196–1206.
Vannier, Nathan, et al. "Epigenetic mechanisms and microbiota as a toolbox for plant phenotypic adjustment to environment." Frontiers in Plant Science 6 (2015): 1159.
Varela, Manuel F., et al. "Bacterial resistance to antimicrobial agents." Antibiotics 10.5 (2021): 593.
Verbeke, Frederick, et al. "Peptides as quorum sensing molecules: measurement techniques and obtained levels in vitro and in vivo." Frontiers in Neuroscience 11 (2017): 183.
Vestby, Lene K., et al. "Bacterial biofilm and its role in the pathogenesis of disease." Antibiotics 9.2 (2020): 59.
Vesty, Eleanor F., et al. "Cross-kingdom signalling regulates spore germination in the moss Physcomitrella patens." Scientific Reports 10.1 (2020): 1–13.
Vidhyasekaran, P. *Plant hormone signaling systems in plant innate immunity*. Springer, Dordrecht, 2015, 27–122.
White, Joni Renee, et al. "The role of host molecules in communication with the resident and pathogenic microbiota: a review." Medicine in Microecology 4 (2020): 100005.
Whiteley, Marvin, Stephen P. Diggle, and E. Peter Greenberg. "Bacterial quorum sensing: The progress and promise of an emerging research area." Nature 551.7680 (2017): 313.
Williams, Paul. "Quorum sensing, communication and cross-kingdom signalling in the bacterial world." Microbiology 153.12 (2007): 3923–3938.
Wink, Michael. "Modes of action of herbal medicines and plant secondary metabolites." Medicines 2.3 (2015): 251–286.
Yamagishi, Jumpei F., Nen Saito, and Kunihiko Kaneko. "Adaptation of metabolite leakiness leads to symbiotic chemical exchange and to a resilient microbial ecosystem." PLoS Computational Biology 17.6 (2021): e1009143.
Yang, Li, et al. "Response of plant secondary metabolites to environmental factors." Molecules 23.4 (2018): 762.
Zandonadi, Daniel Basílio, et al. "Alkamides: A new class of plant growth regulators linked to humic acid bioactivity." Chemical and Biological Technologies in Agriculture 6.1 (2019): 1–12.
Zhang, Jing, et al. "Plant community and soil properties drive arbuscular mycorrhizal fungal diversity: A case study in tropical forests." Soil Ecology Letters 3.1 (2021): 52–62.
Zhang, Lian-Hui. "Quorum quenching and proactive host defense." Trends in Plant Science 8.5 (2003): 238–244.
Zhao, Qian, et al. "Involvement of calmodulin in regulation of primary root elongation by N-3-oxo-hexanoyl homoserine lactone in Arabidopsis thaliana." Frontiers in Plant Science 5 (2015): 807.
Zilber-Rosenberg, Ilana, and Eugene Rosenberg. "Microbial-driven genetic variation in holobionts." FEMS Microbiology Reviews 45.6 (2021): fuab022.

13 Alkamides as Metabolites in Plants

Hitesh Chopra, Shabana Bibi, Santosh Kumar Gupta, Mohammad Mehedi Hasan, Muhammad Aurang Zeb, Muhammad Saad Khan, and Qudsia Yousafi

CONTENTS

13.0 INTRODUCTION

Alkamides are natural products that show broad variability in their distribution patterns, functions, and structures and are now recommended as medicinal extracts for the treatment of many diseases (Samykutty et al., 2013). These are secondary metabolites comprising more than 200 different compounds (Rios-Chavez et al., 2003). This bioactive natural compound has been used as traditional medicine since long ago by different populations throughout the world for the treatment of local health problems toothaches and sore throats (Ramírez-Chávez et al., 2004). These are α unsaturated N-acyl amides thar are structurally related to the amide and lactone derivatives (N-acyl ethanolamides, N-acyl-L-homoserine lactones) known as NAEs and AHLs, respectively (Aziz and Chapman, 2020). These compounds are reported in more than 33 plant families and different parts like roots, shoots, leaves, flowers, and seeds. These natural compounds have an important and broad range of biological activities, such as antimicrobial, anti-inflammatory, antiobesity, antidiabetic, insecticidal, antiviral, larvicidal, analgesic, diuretic, antioxidant, and immunomodulatory activities (Figure 13.1). Many plant extracts and their metabolites are recognized worldwide for their native activities against several fatal diseases, such as bacterial and viral infections (Singla et al., 2021a; Tufail et al., 2022; Mustafa et al., 2022; Bhattacharya, Das et al., 2022; Kour et al., 2022; Walia et al., 2022), and different body organ–related diseases such as heart, liver, and kidney diseases associated with inflammatory mechanisms in the human body (Johnstone and Laster, 2021). Alkamides play a very important role in plant growth and development (Ramírez-Chávez et al., 2004). These secondary metabolites are linked to the cytokinin-signaling system to regulate meristematic activity and cell differentiation, which finally results in their role in root development (Zandonadi et al., 2019). Various reports are available about the role of alkamides in plant defense activities (Johnstone and Laster, 2021). Alkamides decreased the area of necrosis in various tissues of plants caused by fungal infections and prevented the growth of these pathogens (Johnstone, 2021). It was established that alkamides in biotic stress conditions play an active role via the jasmonic acid pathway and MAP kinase-regulated signaling to stop the expansion of fungal necrotic lesions (Méndez-Bravo et al., 2011).

DOI: 10.1201/9781003224525-16

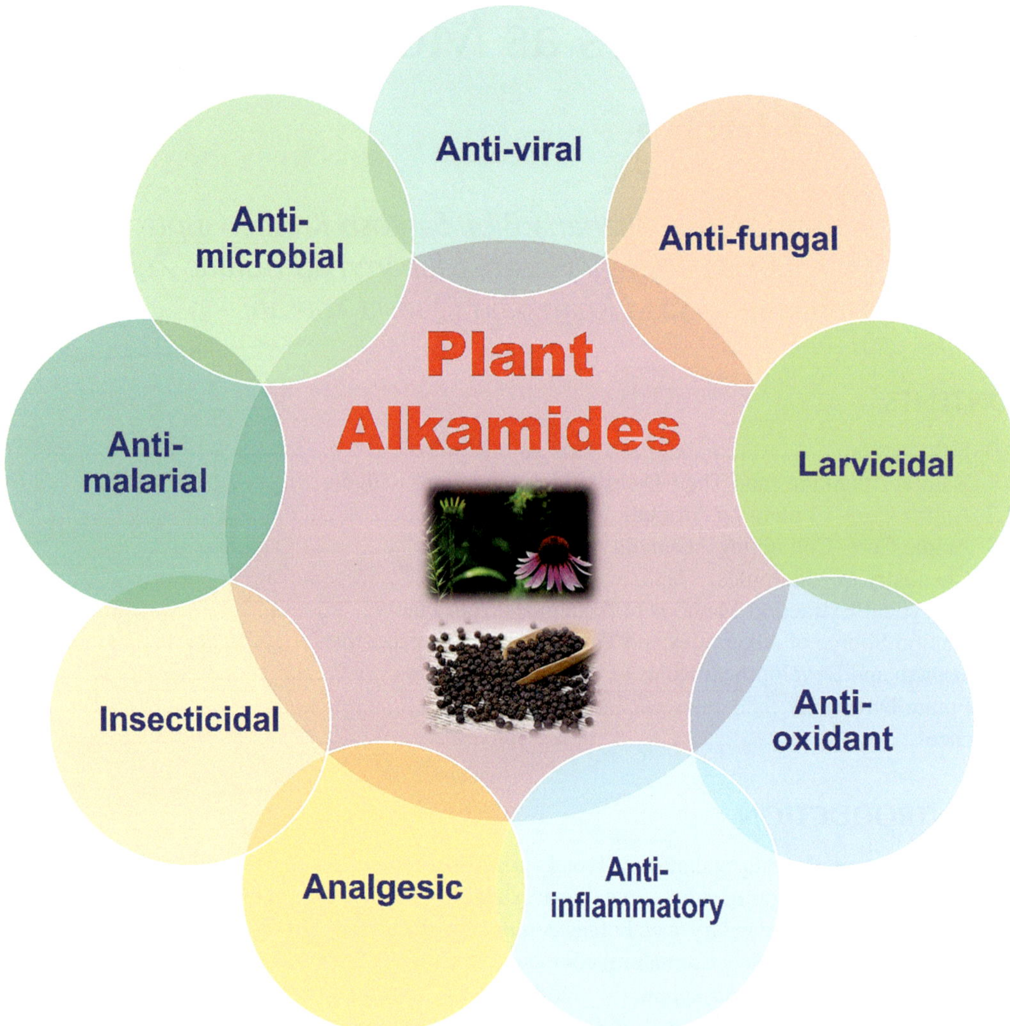

FIGURE 13.1 Plant alkamides and pharmacological importance in multiple diseases.

13.1 NATURAL ALKAMIDES AND THEIR DISTRIBUTION IN THE PLANT KINGDOM

Biosynthesis of both amine and acid aliphatic residues containing natural alkamides in plants is distributed in different parts. A total of six plant families are reported to contain aliphatic alkamides in a different genus. These are Asteraceae, Convolvulaceae, Euphorbiaceae, Menispermaceae, and Rutaceae. These families further contain different genera of plants with different types of aliphatic alkamides. The different genera of plants are further categorized into species and sub-species containing different aliphatic and aromatic alkamides. In the family Asteraceae, there are 11 genera with different alkamides distributed through different species of plants. These are *Aaronsohnia*, *Achilea*, *Acmella*, *Anacyclus*, *Artemisia*, *Echinacea*, *Heliopsis*, *Spilanthes*, *Salmea*, *Sanvitalia*, and *Wedelia*. Aliphatic alkamides extracted from plant species fall within the genera of *Echinacea*, *Achilea*, *Acmella*, and *Spilanthes* contain 93, 71, 31, and 17, respectively. Other genera of this family contain few natural metabolites as aliphatic alkamides. A total of 38 plant species were reported to contain 230 aliphatic alkamides distributed in 11 genera of the family Asteraceae (Figures 13.2A

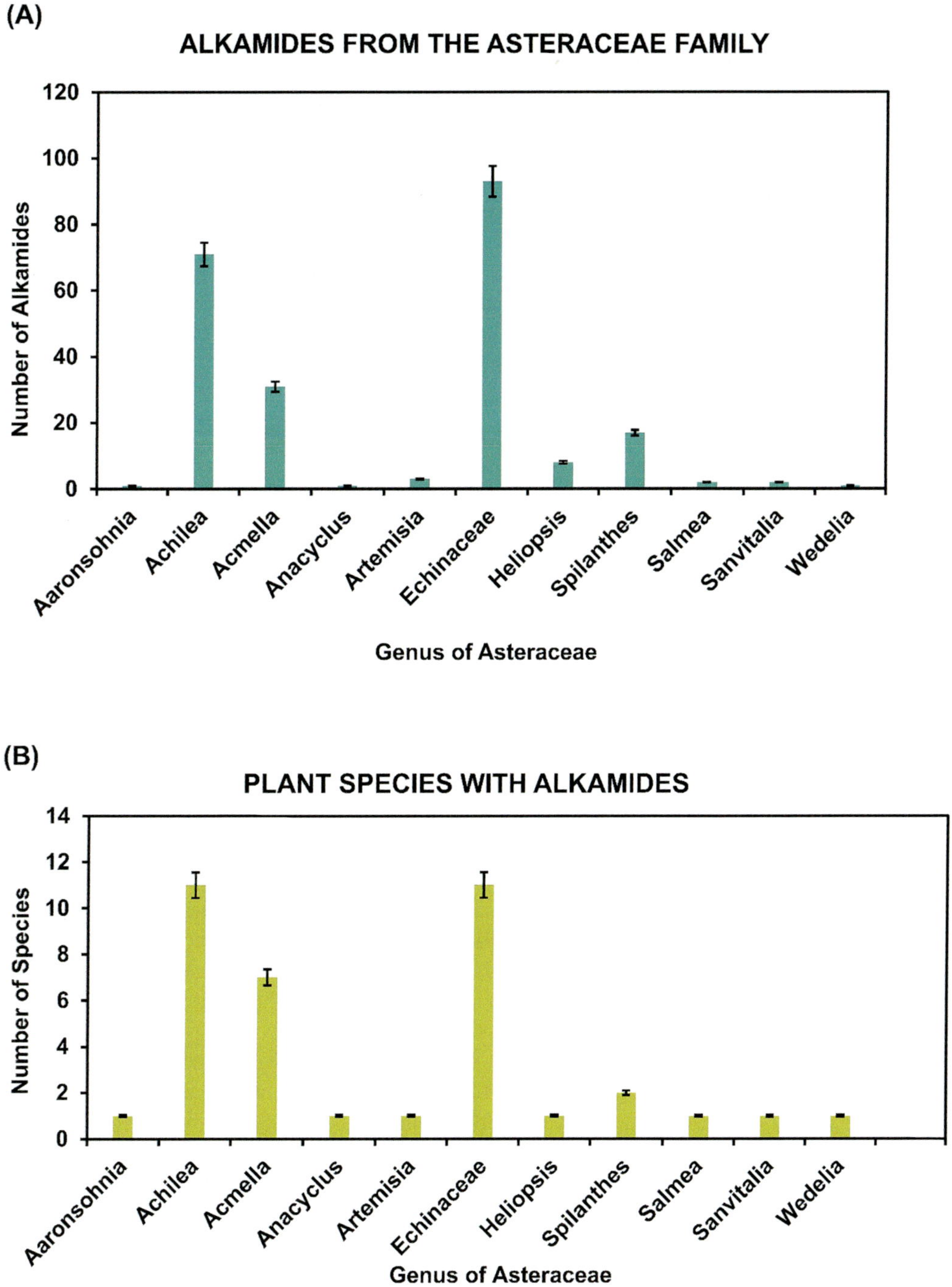

FIGURE 13.2 The Asteraceaeplant family genusand species with the distribution pattern of aliphatic natural alkamides. (A) The number of alkamides is distributed in the different genera of Asteraceae. (B) Different plant species of different genera in Asteraceae with aliphatic alkamides distribution pattern.

and B). These metabolites have a biogenetic capacity that helps to unite them with extensively dispersed functional groups like piperidinyl (piperidine), which is a cyclic amine, and N-isobutyl, N-2-methylbutyl with C8 to C18 olefinic and acetylenic acid residues. C17 is the exception in this category. Few minor amides were also reported in this category. These are N-4-methylbutyl, N-tyramidal, and O-methyl-tyramidyl residues. Approximately 72% of aliphatic alkamides are extracted from Asteraceaeplants containing C10, C11, and C12 long side-chain residues. Sidechains representing C14 to C18 residues were represented by 13% of the total alkamides isolated from this family. The alkamides isolated from this family have various medicinal properties (Rios, 2012).

Aliphatic and aromatic alkamides were isolated from a group of various plant families. A total of 90% of alkamides were isolated from Asteraceae. The rest of the aliphatic alkamides were distributed as follows:4% Convolvulaceae, 3% Rutaceae, 2% Menispermaceae, and 1% Euphorbeace (Rios, 2012). Aromatic alkamides were categorized in Piperaceae (46%), Rutaceae (33%), Solanaceae (14%), and Brassicaceae (7%) (Figures 13.3 A and B). Aromatic plant alkamides isolated from different plant species are categorized into two groups. These are amine core–containing aromatic alkamides and acidic core–containing aromatic alkamides.

13.2 AMINE CORE–CONTAINING AROMATIC ALKAMIDES

These alkamides were isolated from the plant families *Solanaceae*, *Brassicaceae*, and *Rutaceae*. In the *Solanaceae* family, the *Capsicum* genus is rich in these alkamides. These alkamides are known as capsaicinoids and are isolated from chili. The *Capsicum* extracts containing these alkamides are used in various medicinal therapies: toothache, muscle pain, and aerosols used as body massagers after exercises to tone muscles. Two important alkamides isolated from this genus are capsaicin and dihydrocapsaicin, which constitute approximately 90% of the total capsaicinoids in many varieties of peppers (Kozukue et al., 2005). Other alkamides from *Brassicaceae* and *Rutaceae* also contain various medicinal properties. The very rare sulphur-containing alkamides were isolated from the *Rutaceae*. These alkamides are known as glycosmis after their genus name, *Glycosmis* (Rios, 2012).

13.3 ACIDIC CORE–CONTAINING AROMATIC ALKAMIDES

Piperine and its analogues were reported in this category of natural alkamides (Das et al., 1996). A carboxylic acid fragment is characteristic of this group of metabolites, which are isolated from plants of the *Piperaceae* family. Pharmacologically, this family of alkamides is very important (Woelkart et al., 2005). Natural alkamides containing both acidic and amine aromatic residues are extensively distributed among 28 plant families. The root extract of most of the plants of this family is used in treatment as an anesthetic and for its hallucinogenic effects in natural medicines used in the treatment of toothaches and sore throats (Rios, 2012).

Piperovatine alkamides are the best example in this group. The *Piper* species is a very good source of these natural, clinically important secondary metabolites that have been used in traditional medicine for thousands of years in Mexico, in China, and on the Indian sub-continent (Das et al., 1996). These are used for the treatment of diseases related to women, such as gonorrhea and menstrual and chronic intestinal pain; human respiratory diseases, such as tuberculosis, respiratory infections, bronchitis, asthma, cough, viral fevers, and malaria; and joint pain disorders, such as arthritis (Flórez-Álvarez et al., 2022). These compounds are used by tribes for the relaxation of muscles in hypertension situations and anxiety due to their analgesic and diuretic properties (Rios, 2012).

13.4 CHEMISTRY OF ALKAMIDES

N-acyl-l-homoserine lactones (AHLs) are a kind of fatty acid amide that is structurally similar to the N-acylethanolamines (NAEs) found in plants and mammals (Méndez-Bravo et al., 2011). Alkamides are fatty acid amides found in bacteria, plants, and mammals (Greger, 2016). A varied

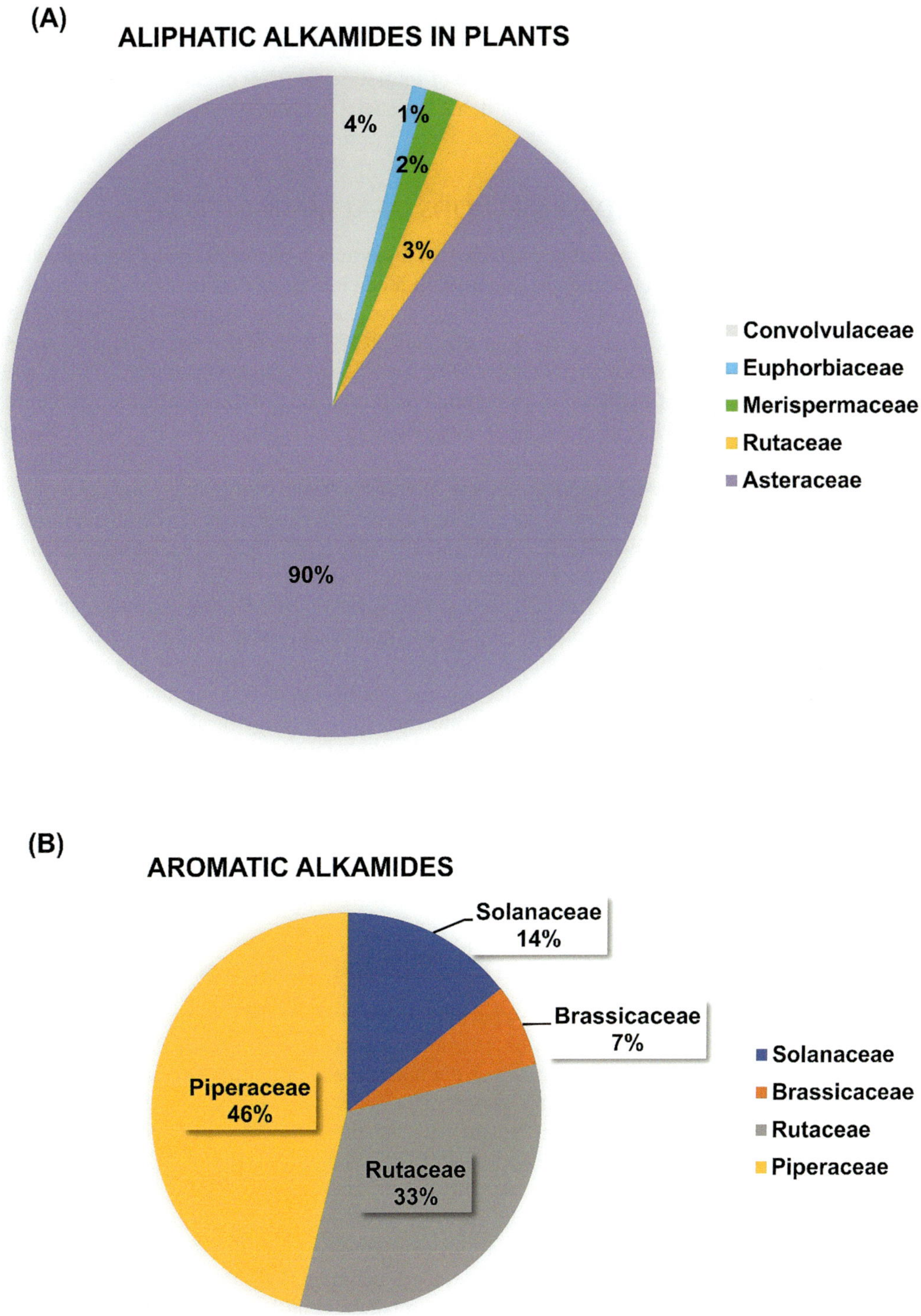

FIGURE 13.3 Aliphatic and aromatic alkamides and their distribution in the plant kingdom. (A) Aliphatic natural alkamides and their categorization in five major plant families Asteraceae, Convolvulaceae, Rutaceae, Menispermaceae, and Euphorbiaceae. (B) Aromatic natural alkamides are distributed in Piperaceae, Rutaceae, Solanaceae, and Brassicaceae.

range of structural diversity exists in alkamides, even though they have a relatively basic molecular composition. A saturated or unsaturated (including double- and triple-bond) fatty acid chain spanning in length from C8 to C18, as well as aliphatic, cyclic, or aromatic amine moiety, is found in each of these fatty acids. Acidic properties (degree of saturation, number of carbon atoms, and stereochemistry), as well as those of the amide part, are beneficial in chemotaxonomy because of their similarities (Zandonadi et al., 2019).

13.5 ANALYTICAL/EXTRACTION METHODS OF ALKAMIDES FROM PLANTS

To assure the quality of herbal medicines, a variety of analytical procedures, including both chromatographic and spectroscopic techniques, are used in medicinal plant research and development. Individual or hyphenated analytical procedures might solve the problem. Various techniques, such as thin-layer chromatography (TLC), gas chromatography (GC), high-performance liquid chromatography (HPLC), and electrophoretic methods (ECs), have been employed for plant analysis, both separately and in a variety of hyphenated configurations. Combining chromatographic procedures with spectroscopic detectors has been the method of choice in recent years for acquiring structural information on the compounds in herbal samples.

For the examination of herbal products, the use of column liquid chromatography (LC) or gas chromatography (GC) in conjunction with an ultraviolet-visible spectrum (UV-VIS) or a mass spectrometer (HPLC-DAD, CE-DAD, GC-MS, and LC-MS, respectively) is a useful technique (Kamboj, 2012). It is possible to use hyphenated methods such as LC-IR, LC-MS, LC-NMR, and GC-MS in addition to other techniques such as HPLC-DAD, HPLC-MS, or a hyphenation of EC. The methodology used will be dictated by the study aim and the availability of equipment. In general, liquid chromatography combined with mass spectrometry (LC-MS) is a more sensitive and selective analytical approach for detecting and identifying chemicals than other methods available today (Marston, 2007). However, despite the rising interest in alkylamides as medicines, only a small number of analytical techniques for their quantification have been published in the scientific literature. They developed a sensitive and specific method for the identification and quantification of alkylamides in human plasma using HPLC-electrospray ionization (ESI) ion trap mass spectrometry (MS). This method was applied to samples obtained after oral administration of *Echinacea angustifolia* DC, Asteraceae, and root extracts, and it was found to be effective (Woelkart et al., 2005).

13.6 NATURAL SOURCE OF ALKAMIDES AND THEIR MEDICINAL PERSPECTIVES

All peppers, especially chili peppers, have a particular pungency due to the capsaicinoid alkaloid capsaicin (8-methyl-N-vanillyl-6-nonenamide), which is produced in various concentrations by the fruits of the *Capsicum* genus (a member of the *Solanaceae* plant family). Capsaicin derivatives have a highly intense and blistering taste that protects the plants from herbivores, which is why they are so popular as a spice in cuisines around the world. But in addition, traditional medicine uses capsaicin for topical pain relief, weight loss, and cardioprotection (Bode and Dong, 2011; Fattori et al., 2016).*

However, the interaction of capsaicin with the TRPV1 receptor cannot be credited with other positive benefits, such as cardioprotective qualities (Rollyson et al., 2014). As an interesting side note, the neuron's refractory phase following receptor attachment has an exceptionally long duration, making it insensitive to additional (apparently unrelated) stimuli (Bode and Dong, 2011). This is known as desensitization, and it is commonly used in topical creams for the analgesic action of capsaicin (Fattori et al., 2016). In the stomach and intestine, capsaicin is passively absorbed, with absorption rates ranging from 50 to 90% (Rollyson et al., 2014) after oral treatment.

Pepsin is activated when gastric juice secretes hydrochloric acid, which maintains a pH of 2.0 in the stomach, allowing it to digest and destroy proteins (Xiang et al., 2021). Indigestion, gastric

TABLE 13.1
Chemical Structures of Capsaicinoids, Capsinoids, and Capsiconinoids

	Name	Molecular Formula	Chemical Structure
Capsaicinoids	Capsaicin	$C_{18}H_{27}NO_3$	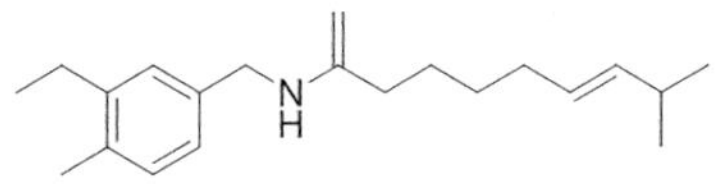
	Dihydrocapsaicin	$C_{18}H_{29}NO_3$	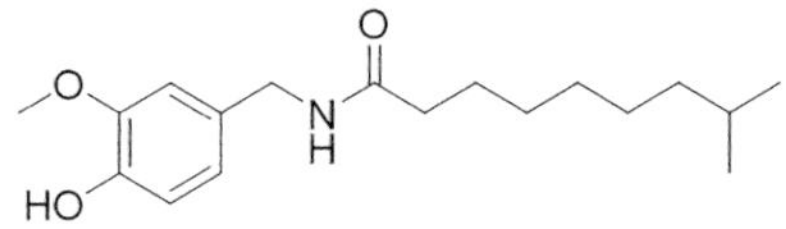
	Nordihydrocapsaicin	$C_{17}H_{27}NO_3$	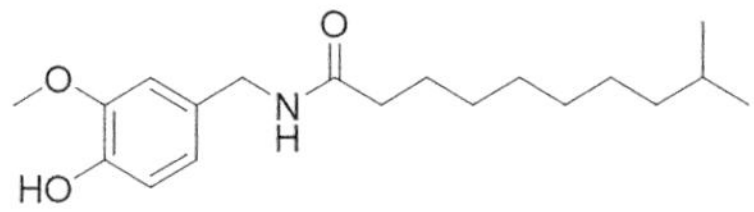
	Homocapsaicin	$C_{19}H_{29}NO_3$	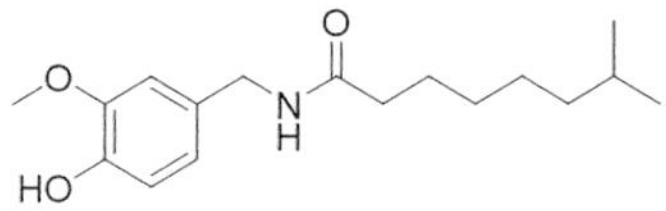
	Homodihydrocapsaicin	$C_{19}H_{31}NO_3$	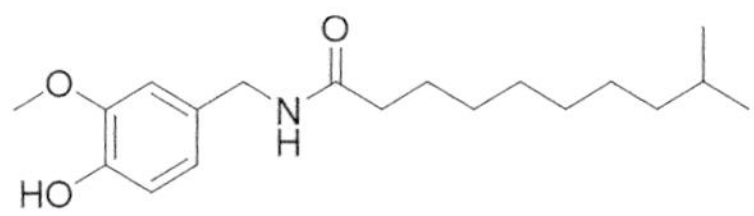
Capsinoids	Capsiate	$C_{18}H_{26}O_4$	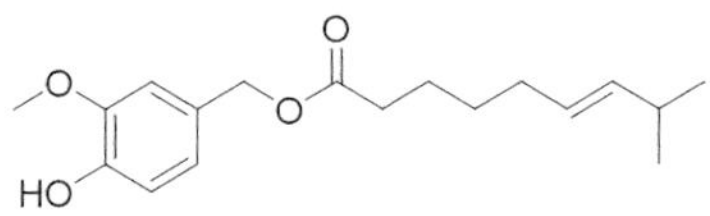
	Dihydrocapsiate	$C_{18}H_{28}O_4$	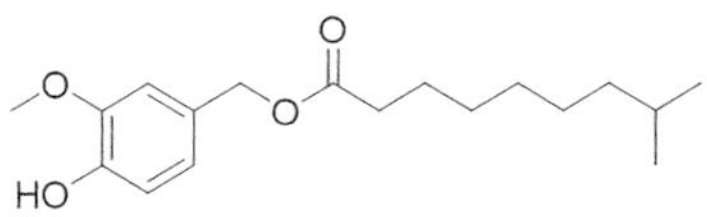
	Nordihydrocapsiate	$C_{18}H_{26}O_4$	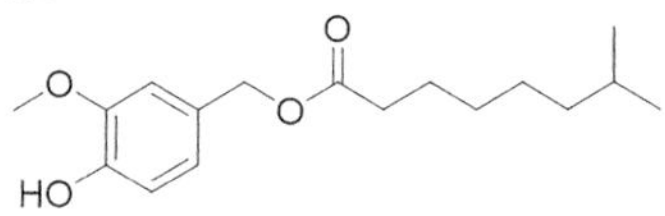
Capsiconinoids	Capsiconiate	$C_{20}H_{28}O_4$	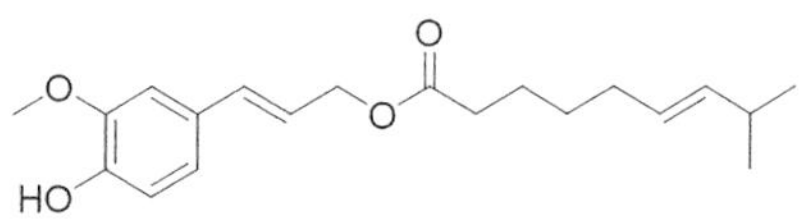
	Dihydrocapsiconiate	$C_{20}H_{30}O_4$	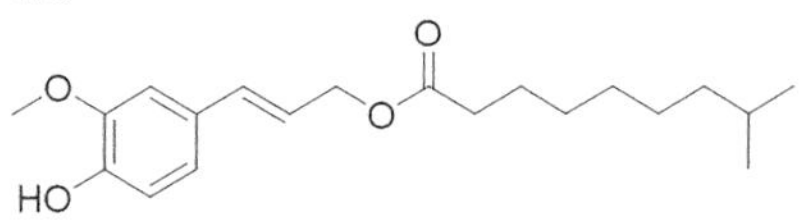

ulcers, and other stomach diseases can all be caused by an overabundance of gastric acid in the stomach (Dong et al., 2015). Low levels of gastric acid can impede digestion and nutrient absorption, which can lead to weight gain and other health problems (Xiang et al., 2021). Due to the sensitive main nerves, low dosages of capsaicin have been proven to block gastric acid output in animals that have had their pylori surgically removed. Large doses of capsaicin can cause hyperacidity, which increases the risk of gastric mucosal damage when administered intragastrically (Mózsik et al., 2001). Capsaicin, on the other hand, has no effect on stomach acid secretion at rest but can reduce gastric acid secretion induced by vagus nerve stimulation (Imatake et al., 2009).

When the vagal pathway is activated, acetylcholine and gastrin-releasing peptide are produced, which may lead to an increase in gastrin release from G cells. Gastrin increases digestion by delaying stomach emptying and stimulating gastric acid output directly (Xiang et al., 2021). Helicobacter pylori thrive in an acidic environment, which raises the risk of stomach cancer because of the effect of capsaicin on gastrin secretion (López-Carrillo et al., 2012). However, this finding was based on the consumption of capsaicin over an extended time. The release of gastrin is inhibited by a diet rich in capsicum (Ericson et al., 2009).

Patients with gastrointestinal malignancies and functional gastrointestinal disorders (FGIDs) are more sensitive than healthy people (Fikree and Byrne, 2021). Irritable bowel syndrome (IBS) and functional dyspepsia are examples of FGIDs, which are digestive diseases caused by the interactions of physiological, mental, and social factors (FD). When stomach and duodenum contents flow back into the esophagus, heartburn, acid reflux, and difficulty in swallowing are all symptoms of GERD (Fikree and Byrne, 2021).

IBS is a nonorganic gastrointestinal condition that is accompanied by changes in bowel habits, defecation, visceral hypersensitivity, and alterations in mucosal immune function (Platel and Srinivasan, 2001). Patients with IBS have higher levels of TRPV1 nerve fibres and a low-grade inflammatory response, which may contribute to their increased sensitivity to pain and discomfort in the abdomen (Akbar et al., 2008). There is an immediate increase in pain and an increase in visceral sensitivities as a result of capsaicin's ability to activate TRPV1. Patients with IBS who consume capsaicin will experience worsening symptoms of abdominal pain and burning, as well as an increase in anxiety (Gonlachanvit et al., 2009; Van Wanrooij et al., 2014).

Epigastric discomfort, burning, early satiety, postprandial fullness (or bloating), and nausea are all prevalent FGID symptoms associated with FD. In people with FD, the pathophysiology of stomach discomfort and other symptoms may be influenced by the consumption of chili peppers (Gonlachanvit et al., 2009), which have been shown to aggravate the symptoms of individuals with FD, which is substantially distinct from patients with other gastrointestinal disorders (Hammer, 2018). The use of capsaicin capsules for FD diagnosis has thus been the subject of technological studies (Führer et al., 2011).

The permeability of the intestines serves as a strong mechanical barrier and is a key sign of gut health. As a result of gut barrier malfunction, the gram-negative microbiota's Lipopolysaccharide (LPS) leaks into the bloodstream, resulting in higher LPS concentrations and chronic low-grade inflammation (CLGI). CLGI may play a role in the development of obesity, according to several lines of evidence (Gasmi et al., 2020; Scheithauer et al., 2020). LPS-induced endotoxemia is the primary cause of CLGI. Activation of the intestine's toll-like receptor 4 and an increase in LPS levels in mice on a high-fat diet have both been documented (Mohammad and Thiemermann, 2021). After that, an inflammatory response in the area causes an increase in intestinal permeability by enlarging the space between intestinal epithelial cells. Endotoxemia is caused when LPS enters the bloodstream through a damaged intestinal epithelium (Massier et al., 2021). Obesity may be caused by increases in intestinal permeability caused by LPS, according to these studies. An increase in LPS content in mice caused by a high-fat diet is caused by bacteria in the S24-7 family, according to Kang et al. (2017). To reduce LPS levels in the colon, feeding capsaicin reduces S24-7 production in the intestine, weakens local inflammation, and reduces intestinal permeation. Wang et al. (2020) came to the same conclusion, but their investigation revealed that proteobacteria cause CLGI and that capsaicin decreases intestinal LPS levels by reducing their population. Other research, on the other hand, found that proteobacteria, not S24-7, is responsible for CLGI in the gut and that this can have therapeutic benefits on diabetes mellitus and obesity (Wei et al., 2020). Capsaicin appears to lower LPS levels in the gut, but more research is needed to determine if this is due to a decrease in S24-7 levels or a decrease in proteobacteria numbers (as shown in Figure 13.4). LPS and inflammatory factors such as interleukin 1, interleukin 6, tumour necrosis factor-alpha, and the intestinal barrier impact were measured by Kang et al. (2017). The experimental design used by Wang et al. (2020) was not the same. Maintaining intestinal barrier function is dependent on preventing damage to tight junction proteins such as ZO-1 and occludin (Ghosh et al., 2021). Capsaicin, according to Kang et al. (2017), enhances ZO-1 and occludin mRNA and protein expression. It was also discovered that *bifidobacterium*, paeoniflorin, theaflavins, and other compounds that repair the intestinal

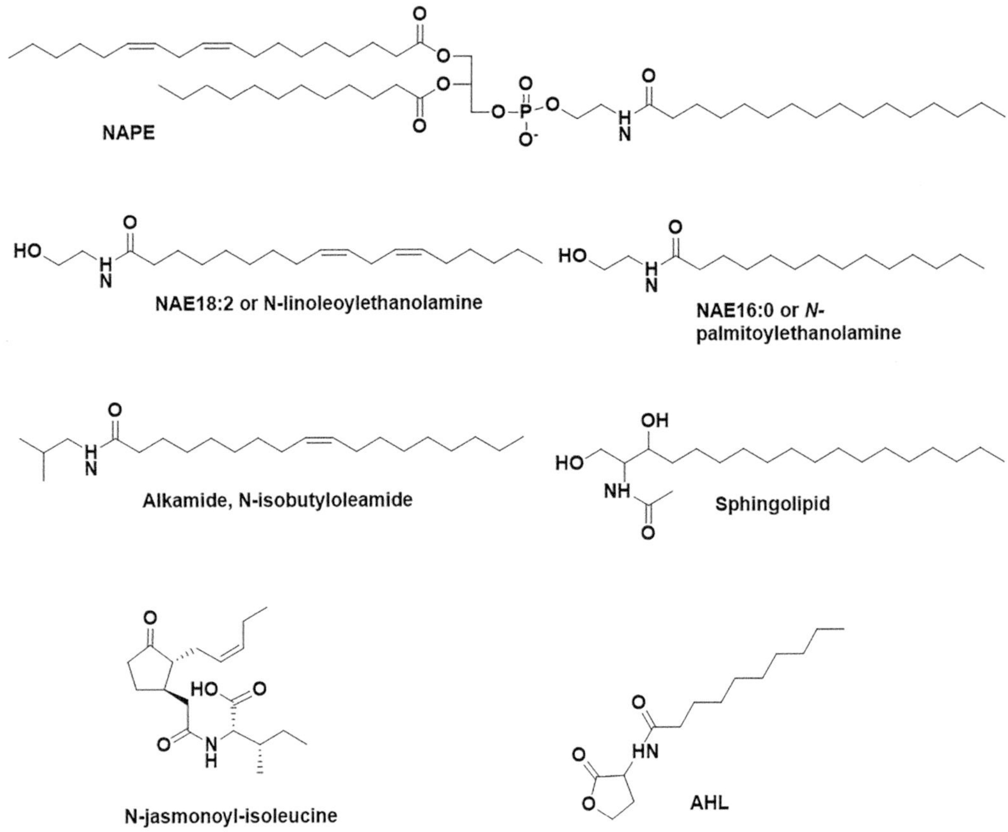

FIGURE 13.4 Chemical structural representation of alkamides containing acyl groups.

barrier function had similar results. Researchers now believe capsaicin may play a role in mediating intestinal barrier restoration. Capsaicin's role in intestinal barrier function, especially *in vitro* transmembrane resistance values, needs further investigation. Alkamides extracted from capsaicin are involved in the anti-obesity mechanism (Figure 13.5).

There are several mechanisms underlying the capsaicin's pain-killing properties, including its action on the TRPV1 receptors and the production of neuropeptides, such as substance P and calcitonin gene-related peptides (Kaneko and Szallasi, 2014; Moran and Szallasi, 2018) Capsaicin releases these substances into the body, causing the patient to experience greater discomfort. As a result, if the alkaloid is delivered multiple times, the vanilloid receptor gets desensitized, resulting in a reduction in pain sensation. Longer exposure to capsaicin results in a change in spatial conformation and inactivation of TRPV1, according to research. This indicates that even though the activating stimulus is still present, the receptor is no longer activated.

Spilanthes oleracea's capitula yielded crude spilanthol when it was originally isolated (Alonso et al., 2018). It was later determined to be N-isobutyl deca-2,6,8-trienamide from an insecticidal amide found in the roots of a Mexican plant called *Erigeron affinis* DC and given the name "Affinin". *Heliopsis longipes* proved to be the plant's true identification in succeeding years, and the term "Affinin" was requested to be dropped from any future deliberations. Affinin's toxicity against house flies was further proved to have been reduced to all trans-spilanthol isomers when it was exposed to UV or selenium (Acree et al., 1945). Even though they are considered synonyms, the generic names "Spilanthol" and "Affinin" are still employed differently in literary works (Molinatorres et al., 1996; Castro-Ruiz et al., 2017).

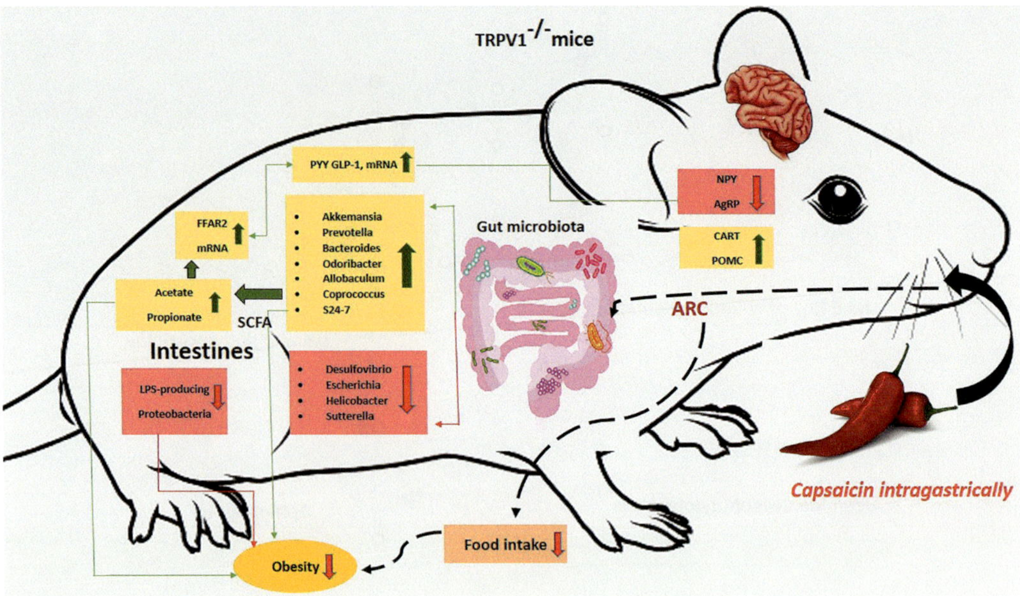

FIGURE 13.5 Alkamides extracted from capsaicin involved in the antiobesity mechanism.

Figure 13.6 depicts the chemical structure of an NAA. One nitrogen atom is bound to three variable R groups in almost all NAAs. A fatty acid chain is always R1; R2 is the residue of an amino acid, and R3 is almost always a hydrogen atom. Isotope labelling techniques and biochemical enzyme tests are used to find biosynthetic pathways (Rizhsky et al., 2016; Cortez-Espinosa et al., 2011). For example, studies show that NAAs are formed by the coupling of an unsaturated fatty acid chain moiety with an amine group through the O = C single bond NH linkage (Boonen et al., 2012). The combination of highly electronegative O and N atoms close to each other creates a partial double bond in this amide linkage, which is common in protein-peptide bonds. Fatty chains are linked to over 30 plant families including Asteraceae, Aristolochiaceae, and Aristolochioides and are present in over 200 amine-bonded fatty chains found in the Asteraceae, Aristolochiacae, and Aristolochioidea families (Martin and Becker, 1984; Boonen et al., 2012). Long fatty acid chains, high unsaturation, different geometric fatty chain orientations, and various sorts of amine moieties provide these compounds a great amount of variety (Boonen et al., 2012; Rizhsky et al., 2016). Plants' NAA biosynthesis pathways are shown in Figure 13.3. Unsaturated fatty acid precursors like oleic (18:1), linoleic (18:2), and linolenic (18:3) acids are used to make NAAs by shortening chains at both the carboxyl and methyl ends (terminal) with subsequent dehydrogenation, dehydration, and oxidative processes that result in the generation of characteristic alkene/alkyne products and chain-length modification, respectively (Greger, 2016).

Piperine has four isomeric forms: namely, trans-trans isomer, cis-trans isomer, cis-cis isomer, and trans-cis isomer (isochavicine) in black pepper. In contrast to piperine, the geometrical isomers of piperine exhibit no pungency. Figure 13.7 depicts that piperine has led to the discovery of additional alkaloids in black pepper extract, including piperanine, piperettine, piperylin A, piperylin B, and pipericine. Light intensity and duration of exposure enhance the likelihood of isomerization. Piperine, which may be transformed into isopiperine 2, chavicine 3, and isochavicine 4, can be shown to undergo light-induced isomerization. Piperine may be detected on storage as chavicine progressively and spontaneously loses its pungency.

Different drugs' absorption and bioavailability are improved by piperine (Khatri and Awasthi, 2016). As a result, data suggest that piperine has a variety of pharmacological actions that affect

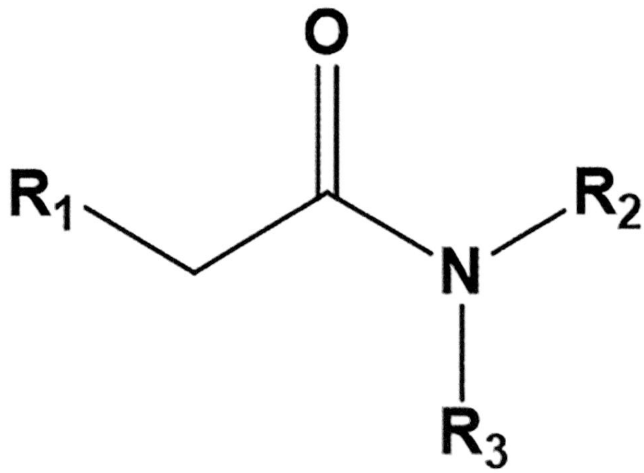

FIGURE 13.6 Basic skeleton structure of NAA.

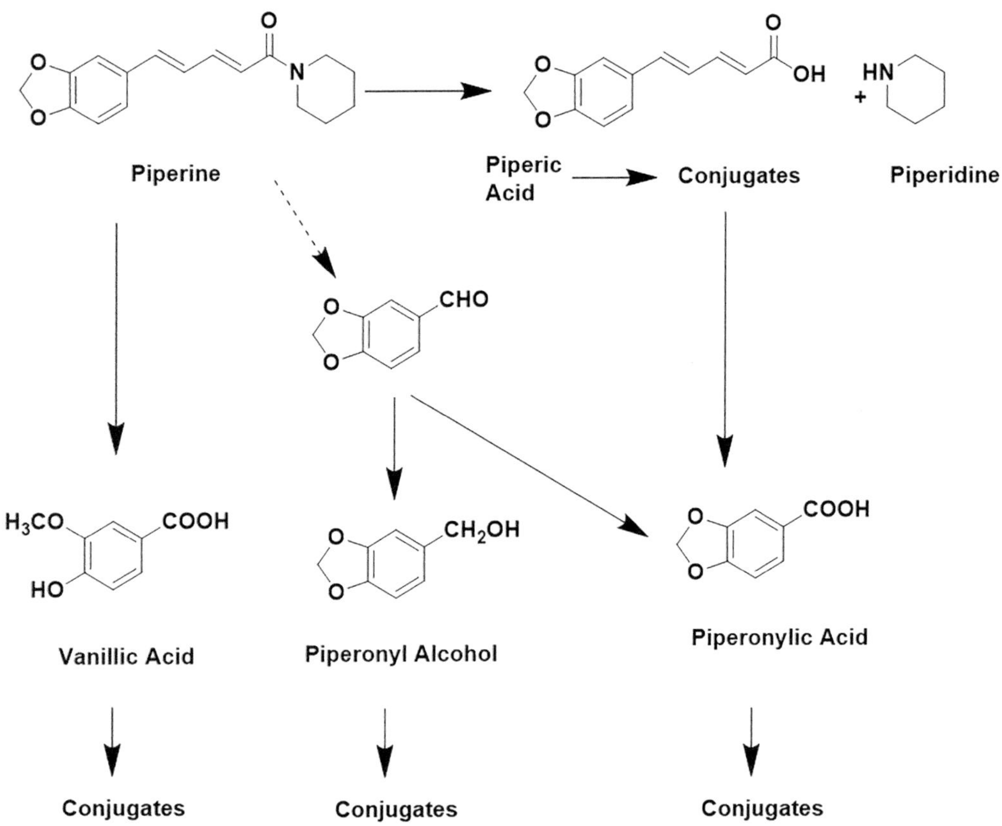

FIGURE 13.7 Alkamide metabolites extracted from piperine.

transporter and metabolic enzyme activity modulation. Piperine may also be used as a complementary medication. Park et al. found that piperine inhibited the expression of PPARin 3T3-L1 cell lines, which may be utilised to treat obesity-related disorders (Park et al., 2012). Piper nigrum has a wide range of biological activities, including its function in the digestive process, its antioxidant characteristics, and its ability to treat a variety of conditions (Goswami and Meghwal, 2012).

The anti-allergic properties of piperine were reported by Aswar et al. Mast cell stabilization is achieved by the suppression of secreted mediators such as interleukin-6, interleukin-1b, immunoglobulin-E, and histamine release. Xenobiotic agents, such as piperine, may be useful in the treatment of allergic rhinitis caused by ovalbumin (Aswar et al., 2015). They found that piperine inhibited airway inflammation in a mouse asthma model by decreasing the Th-2 cytokines interleukin-4, interleukin-5, and interleukin-13 and increasing the expression of the transforming growth factor gene and the CCR3 receptor on eosinophils in particular. As a potential downregulator of Th-2 cytokines, piperine may be useful (Kim et al., 2017). According to Dong et al., piperine has a dose-dependent impact on periodontitis, decreasing alveolar bone loss, and tissue breakdown. This action is attributed to piperine's suppression of interleukin-1 and metalloproteinase-8 and-13. (MMP-8 and MMP-13). That piperine has a significant anti-inflammatory effect in the periodontitis model (Dong et al., 2015) may thus be claimed. Piperine strongly inhibits the cyclooxygenase (COX) enzyme at a dosage of 25 g/ml by 30 to 80%, according to Liu et al. (2010). As an agonist, piperine has been shown to activate the pregnane X receptor (PXR), which, in turn, activates the enzyme CYP-3A4 gene expression. According to these findings, piperine may prevent and reduce colonic inflammation disease activity through PXR receptor activation (Hu et al., 2015).

Many spices and herbs, including black piper, are known to contain multiple active components, including flavonoids, terpenoids, phytoestrogens, and minerals. Piperine was shown to have antioxidant properties, which may help reduce cellular oxidative stress as a result of a high-fat diet (Vijayakumar et al., 2004). Piperine has also been demonstrated to reduce the levels of thiobarbituric acid–reactive chemicals by maintaining catalase, glutathione, glutathione peroxidase, Glutathione-S-transferase, and superoxide dismutase concentrations (Vijayakumar et al., 2004). Biotransformation enzyme activity may be improved in the liver by taking this drug at a higher concentration (Singh and Rao, 1993). Studies on piperine's antioxidant action have shown that it reduces lung metastatic incidence in B16F-10 melanoma cells via altering lipid peroxidation and stimulating antioxidant enzymes (Sunila and Kuttan, 2004; Pradeep and Kuttan, 2002).

13.7 GROWTH AND DEVELOPMENT ROLES OF ALKAMIDES IN PLANTS

When compared to NAE in plants, alkamides have several actions that seem to be overlapping. It has been shown that exogenous N-isobutyl-decanamide (affinin) affects seedling growth, particularly root development. When used at low doses (7 mM), affinin was shown to cause an increase in the length of main roots as well as the formation of lateral roots and the growth of root hairs in plants (López-Bucio et al., 2007). There are two possible explanations for the growth of lateral roots: either the encouragement of the development of pre-existing lateral root primordials or the synthesis of brand-new ones. Higher quantities of affinin (120 mM), on the other hand, had the opposite impact on growth (Ramírez-Chávez et al., 2004). It was shown that high concentrations of this alkamide prevented root hairs and root roots from growing longer than they should. In contrast to the NAE, the reduction in root development did not seem to be accompanied by a disturbance of cell/tissue organization in the test subjects. Despite this, both forms of lipids have the potential to significantly impact seedling growth. Alkamides have the potential to interact with the ethylene pathway by affecting the synthesis of ethylene or the signaling function of this hormone (Ramírez-Chávez et al., 2004), among other things. It is also possible that affinin interacts with the cytokinin-signaling system to regulate meristematic activity and cell differentiation (López-Bucio et al., 2007). In addition, alkamides, like NAEs, affect plants' defensive responses against pathogens (Méndez-Bravo et al., 2011). Affinin decreases the area of necrosis caused by infections and prevents the growth of fungal pathogens. The exact mechanism of action of this alkamide in terms of plant defense is still being investigated. It may interact with the body via the jasmonic acid pathway and MAP kinase-regulated signaling. In light of the structural similarity between alkamides and NAEs, it may be reasonable to speculate that some of the effects of alkamides on plants may be mediated through endogenous targets of NAEs. This is especially true given that NAEs appear to be ubiquitous throughout the plant kingdom, whereas alkamides are more limited in their distribution.

13.8 FUTURE DIRECTIONS

Currently, application of alkamide-based drug discovery has been restricted due to their weak pharmacokinetics such as low oral bioavailability, low stability in GIT, and high metabolism rate. Therefore, there is a strong need to develop advance computational drug design methods (Bibi and Sakata, 2016) and nanotechnology-based formulations that can be used for increasing its bioavailability and reduce the chances of adverse drug effects in late stages (Chopra, Shuvra Dey et al., 2021; Bhattacharya, Borges e Soares et al., 2022; Singla et al., 2021b; Chopra, Bibi, Inderbir Singh et al., 2022; Khalil et al., 2022). It is reported that the nanostructure may form a layer that can prevent its degradation by various proteolytic enzymes and reduce the burden of overdosing on patients. so these applications could be highly significant for alkamides research because nanoformulations will allow the drug moiety to target organs and reduce the side effects. Also, conjugation with biopolymers in case of a transdermal route could also be possible, such as in wound dressings (Singh Bakshi et al., 2022; Chopra, Gandhi et al., 2021; Chopra, Bibi, Kumar et al., 2022; Chopra, Kumar et al., 2021). Therefore, it is expected that, in future, the use of alkamide metabolites will reach to the maximum against fatal diseases.

REFERENCES

Acree, Fred, Martin Jacobson, and H.L. Haller. 1945. "An Amide Possessing Insecticidal Properties from the Roots of Erigeron Affinis DC." *Journal of Organic Chemistry* 10 (3). https://doi.org/10.1021/jo01179a014.

Akbar, A., Y. Yiangou, P. Facer, J.R.F. Walters, P. Anand, and S. Ghosh. 2008. "Increased Capsaicin Receptor TRPV1-Expressing Sensory Fibres in Irritable Bowel Syndrome and Their Correlation with Abdominal Pain." *Gut* 57 (7). https://doi.org/10.1136/gut.2007.138982.

Alonso, Isabella G., Lais T. Yamane, Verônica S. de Freitas-Blanco, Luiz F.T. Novaes, Michelle Franz-Montan, Eneida de Paula, Marili V.N. Rodrigues, Rodney A.F. Rodrigues, and Julio C. Pastre. 2018. "A New Approach for the Total Synthesis of Spilanthol and Analogue with Improved Anesthetic Activity." *Tetrahedron* 74 (38). https://doi.org/10.1016/j.tet.2018.06.034.

Aswar, Urmila, Sumit Shintre, Shilpa Chepurwar, and Manoj Aswar. 2015. "Antiallergic Effect of Piperine on Ovalbumin-Induced Allergic Rhinitis in Mice." *Pharmaceutical Biology* 53 (9). https://doi.org/10.3109/13880209.2014.982299.

Aziz, Mina, and Kent D. Chapman. 2020. "Fatty Acid Amide Hydrolases: An Expanded Capacity for Chemical Communication?" *Trends in Plant Science* 25 (3): 236–49.

Bhattacharya, Tanima, Giselle Amanda Borges e Soares, Hitesh Chopra, Md. Mominur Rahman, Ziaul Hasan, Shasank S. Swain, and Simona Cavalu. 2022. "Applications of Phyto-Nanotechnology for the Treatment of Neurodegenerative Disorders." *Materials 2022, Vol. 15, Page 804* 15 (3): 804. https://doi.org/10.3390/MA15030804.

Bhattacharya, Tanima, Debashrita Das, Giselle A. Borges E Soares, Prasun Chakrabarti, Zhaoquan Ai, Hitesh Chopra, Madalin Alexandru Hasan, and Simona Cavalu. 2022. "Novel Green Approaches for the Preparation of Gold Nanoparticles and Their Promising Potential in Oncology." *Processes*. https://doi.org/10.3390/pr10020426.

Bibi, Shabana, and Katsumi Sakata. 2016. "Current Status of Computer-Aided Drug Design for Type 2 Diabetes." *Current Computer Aided-Drug Design* 12 (2). https://doi.org/10.2174/157340991266616042 6120709.

Bode, Ann M., and Zigang Dong. 2011. "The Two Faces of Capsaicin." *Cancer Research*. https://doi.org/10.1158/0008-5472.CAN-10-3756.

Boonen, Jente, Antoon Bronselaer, Joachim Nielandt, Lieselotte Veryser, Guy De Tré, and Bart De Spiegeleer. 2012. "Alkamid Database: Chemistry, Occurrence and Functionality of Plant N-Alkylamides." *Journal of Ethnopharmacology*. https://doi.org/10.1016/j.jep.2012.05.038.

Castro-Ruiz, Jesús Eduardo, Alejandra Rojas-Molina, Francisco J. Luna-Vázquez, Fausto Rivero-Cruz, Teresa García-Gasca, and César Ibarra-Alvarado. 2017. "Affinin (Spilanthol), Isolated from Heliopsis Longipes, Induces Vasodilation via Activation of Gasotransmitters and Prostacyclin Signaling Pathways." *International Journal of Molecular Sciences* 18 (1). https://doi.org/10.3390/ijms18010218.

Chopra, Hitesh, Shabana Bibi, Sandeep Kumar, Muhammad Saad Khan, Pradeep Kumar, and Inderbir Singh. 2022. "Preparation and Evaluation of Chitosan/PVA Based Hydrogel Films Loaded with Honey for Wound Healing Application." *Gels* 8 (2): 111. https://doi.org/10.3390/GELS8020111.

Chopra, Hitesh, Shabana Bibi, Inderbir Singh, Mohammad Mehedi Hasan, Muhammad Saad Khan, Qudsia Yousafi, Atif Amin Baig, et al. 2022. "Green Metallic Nanoparticles: Biosynthesis to Applications." *Frontiers in Bioengineering and Biotechnology* 0 (April): 548. https://doi.org/10.3389/FBIOE.2022.874742.

Chopra, Hitesh, Protity Shuvra Dey, Debashrita Das, Tanima Bhattacharya, Muddaser Shah, Sidra Mubin, Samka Peregrine Maishu, et al. 2021. "Curcumin Nanoparticles as Promising Therapeutic Agents for Drug Targets." *Molecules* 26 (16): 4998. https://doi.org/10.3390/MOLECULES26164998.

Chopra, Hitesh, Shweta Gandhi, Rupesh Kumar Gautam, and Mohammad Amjad Kamal. 2021. "Bacterial Nanocellulose Based Wound Dressings: Current and Future Prospects." *Current Pharmaceutical Design* 28 (7): 570–80. https://doi.org/10.2174/1381612827666211021162828.

Chopra, Hitesh, Sandeep Kumar, and Inderbir Singh. 2021. "Strategies and Therapies for Wound Healing: A Review." *Current Drug Targets*. https://doi.org/10.2174/1389450122666210415101218.

Cortez-Espinosa, Nancy, Judit A. Aviña-Verduzco, Enrique Ramírez-Chávez, Jorge Molina-Torres, and Patricia Ríos-Chávez. 2011. "Valine and Phenylalanine as Precursors in the Biosynthesis of Alkamides in Acmella Radicans." *Natural Product Communications* 6 (6). https://doi.org/10.1177/1934578x1100600625.

Das, Biswanath, A Kashinatham, and KVNS Srinivas. 1996. "Alkamides and Other Constituents of Piper Longum1." *Planta Medica* 62 (06): 582.

Dong, Y., Z. Huihui, and C. Li. 2015. "Piperine Inhibit Inflammation, Alveolar Bone Loss and Collagen Fibers Breakdown in a Rat Periodontitis Model." *Journal of Periodontal Research* 50 (6). https://doi.org/10.1111/jre.12262.

Ericson, A., E. Mohammed Nur, F. Petersson, and S. Kechagias. 2009. "The Effects of Capsaicin on Gastrin Secretion in Isolated Human Antral Glands: Before and after Ingestion of Red Chilli." *Digestive Diseases and Sciences* 54 (3): 491–98. https://doi.org/10.1007/S10620-008-0400-1.

Fattori, Victor, Miriam S.N. Hohmann, Ana C. Rossaneis, Felipe A. Pinho-Ribeiro, and Waldiceu A. Verri. 2016. "Capsaicin: Current Understanding of Its Mechanisms and Therapy of Pain and Other Pre-Clinical and Clinical Uses." *Molecules*. https://doi.org/10.3390/molecules21070844.

Fikree, Asma, and Peter Byrne. 2021. "Management of Functional Gastrointestinal Disorders." *Clinical Medicine* 21 (1): 44.

Flórez-Álvarez, Lizdany, Jahnnyer Mart\'\inez-Moreno, Mar\'\ia I Zapata-Cardona, Elkin Galeano, Fernando Alzate-Guarin, and Wildeman Zapata. 2022. "In Vitro Antiviral Activity against SARS-CoV-2 of Plant Extracts Used in Colombian Traditional Medicine." *Vitae* 29 (1).

Führer, M., H. Vogelsang, and J. Hammer. 2011. "A Placebo-Controlled Trial of an Oral Capsaicin Load in Patients with Functional Dyspepsia." *Neurogastroenterology and Motility* 23 (10). https://doi.org/10.1111/j.1365-2982.2011.01766.x.

Gasmi, Amin, Sadaf Noor, Alain Menzel, Alexandru Doşa, Lyudmila Pivina, and Geir Bjørklund. 2020. "Obesity and Insulin Resistance: Associations with Chronic Inflammation, Genetic and Epigenetic Factors." *Current Medicinal Chemistry* 28 (4). https://doi.org/10.2174/0929867327666200824112056.

Ghosh, Sweta, Caleb Samuel Whitley, Bodduluri Haribabu, and Venkatakrishna Rao Jala. 2021. "Regulation of Intestinal Barrier Function by Microbial Metabolites." *CMGH*. https://doi.org/10.1016/j.jcmgh.2021.02.007.

Gonlachanvit, S., A. Mahayosnond, and P. Kullavanijaya. 2009. "Effects of Chili on Postprandial Gastrointestinal Symptoms in Diarrhoea Predominant Irritable Bowel Syndrome: Evidence for Capsaicin-Sensitive Visceral Nociception Hypersensitivity." *Neurogastroenterology and Motility* 21 (1). https://doi.org/10.1111/j.1365-2982.2008.01167.x.

Goswami, T.K., and Murlidhar Meghwal. 2012. "Chemical Composition, Nutritional, Medicinal And Functional Properties of Black Pepper: A Review." *Journal of Nutrition & Food Sciences* 01 (S1). https://doi.org/10.4172/scientificreports.172.

Greger, Harald. 2016. "Alkamides: A Critical Reconsideration of a Multifunctional Class of Unsaturated Fatty Acid Amides." *Phytochemistry Reviews*. https://doi.org/10.1007/s11101-015-9418-0.

Hammer, Johann. 2018. "Identification of Individuals with Functional Dyspepsia With a Simple, Minimally Invasive Test: A Single Center Cohort Study of the Oral Capsaicin Test." *American Journal of Gastroenterology* 113 (4). https://doi.org/10.1038/ajg.2018.16.

Hu, Donghua, Yuguang Wang, Zhiwu Chen, Zengchun Ma, Qing You, Xianxie Zhang, Qiande Liang, et al. 2015. "The Protective Effect of Piperine on Dextran Sulfate Sodium Induced Inflammatory Bowel Disease and Its Relation with Pregnane X Receptor Activation." *Journal of Ethnopharmacology* 169. https://doi.org/10.1016/j.jep.2015.04.006.

Imatake, Kazuhiro, Teruaki Matsui, and Mitsuhiko Moriyama. 2009. "The Effect and Mechanism of Action of Capsaicin on Gastric Acid Output." *Journal of Gastroenterology* 44 (5): 396–404. https://doi.org/10.1007/S00535-009-0018-X.

Johnstone, Stephanie E., and Scott M. Laster. 2021. "The Structure and Function of Alkamides in Mammalian Systems." In *Pharmacognosy-Medicinal Plants*. IntechOpen.

Kamboj, Anjoo. 2012. "Analytical Evaluation of Herbal Drugs." In *Drug Discovery Research in Pharmacognosy*. https://doi.org/10.5772/26109.

Kaneko, Yosuke, and Arpad Szallasi. 2014. "Transient Receptor Potential (TRP) Channels: A Clinical Perspective." *British Journal of Pharmacology*. https://doi.org/10.1111/bph.12414.

Kang, Chao, Bin Wang, Kanakaraju Kaliannan, Xiaolan Wang, Hedong Lang, Suocheng Hui, Li Huang, et al. 2017. "Gut Microbiota Mediates the Protective Effects of Dietary Capsaicin against Chronic Low-Grade Inflammation and Associated Obesity Induced by High-Fat Diet." *MBio* 8 (3). https://doi.org/10.1128/mBio.00470-17.

Khalil, Muhammad Saqib, Muhammad Shakeel, Naila Gulfam, Syed Umair Ahmad, Aamir Aziz, Junaid Ahmad, Shabana Bibi, et al. 2022. "Fabrication of Silver Nanoparticles from Ziziphus Nummularia Fruit Extract: Effect on Hair Growth Rate and Activity against Selected Bacterial and Fungal Strains." *Journal of Nanomaterials* 2022: 1–14. https://doi.org/10.1155/2022/3164951.

Khatri, Smriti, and Rajendra Awasthi. 2016. "Piperine Containing Floating Microspheres: An Approach for Drug Targeting to the Upper Gastrointestinal Tract." *Drug Delivery and Translational Research* 6 (3). https://doi.org/10.1007/s13346-016-0285-z.

Kim, Jisu, Kang Pa Lee, Dae Won Lee, and Kiwon Lim. 2017. "Piperine Enhances Carbohydrate/Fat Metabolism in Skeletal Muscle during Acute Exercise in Mice." *Nutrition and Metabolism* 14 (1). https://doi.org/10.1186/s12986-017-0194-2.

Kour, Jasmeet, Monika Hans, Hitesh Chopra, Renu Sharma, Breetha Ramaiyan, and Bharti Mittu. 2022. "Phytohaemagglutinins." In *Handbook of Plant and Animal Toxins in Food*. https://doi.org/10.1201/9781003178446-11.

Kozukue, Nobuyuki, Jae-Sook Han, Etsuko Kozukue, Sin-Jung Lee, Joung-Ae Kim, Kap-Rang Lee, Carol E Levin, and Mendel Friedman. 2005. "Analysis of Eight Capsaicinoids in Peppers and Pepper-Containing Foods by High-Performance Liquid Chromatography and Liquid Chromatography-Mass Spectrometry." *Journal of Agricultural and Food Chemistry* 53 (23): 9172–81.

Liu, Yunbao, Vivek R. Yadev, Bharat B. Aggarwal, and Muraleedharan G. Nair. 2010. "Inhibitory Effects of Black Pepper (Piper Nigrum) Extracts and Compounds on Human Tumor Cell Proliferation, Cyclooxygenase Enzymes, Lipid Peroxidation and Nuclear Transcription Factor-Kappa-B." *Natural Product Communications* 5 (8). https://doi.org/10.1177/1934578x1000500822.

López-Bucio, José, Mayra Millán-Godínez, Alfonso Méndez-Bravo, Alina Morquecho-Contreras, Enrique Ramírez-Chávez, Jorge Molina-Torres, Anahí Pérez-Torres, Masayuki Higuchi, Tatsuo Kakimoto, and Luis Herrera-Estrella. 2007. "Cytokinin Receptors Are Involved in Alkamide Regulation of Root and Shoot Development in Arabidopsis." *Plant Physiology* 145 (4). https://doi.org/10.1104/pp.107.107953.

López-Carrillo, Lizbeth, M. Constanza Camargo, Barbara G. Schneider, Liviu A. Sicinschi, Raúl U. Hernández-Ramírez, Pelayo Correa, and Mariano E. Cebrian. 2012. "Capsaicin Consumption, Helicobacter Pylori CagA Status and IL1B-31C>T Genotypes: A Host and Environment Interaction in Gastric Cancer." *Food and Chemical Toxicology : An International Journal Published for the British Industrial Biological Research Association* 50 (6): 2118–22. https://doi.org/10.1016/J.FCT.2012.02.043.

Marston, Andrew. 2007. "Role of Advances in Chromatographic Techniques in Phytochemistry." *Phytochemistry*. https://doi.org/10.1016/j.phytochem.2007.08.004.

Martin, Rainer, and Hans Becker. 1984. "Spilanthol-Related Amides from Acmella Ciliata." *Phytochemistry* 23 (8). https://doi.org/10.1016/S0031-9422(00)83490-1.

Massier, Lucas, Matthias Blüher, Peter Kovacs, and Rima M. Chakaroun. 2021. "Impaired Intestinal Barrier and Tissue Bacteria: Pathomechanisms for Metabolic Diseases." *Frontiers in Endocrinology*. https://doi.org/10.3389/fendo.2021.616506.

Méndez-Bravo, Alfonso, Carlos Calderón-Vázquez, Enrique Ibarra-Laclette, Javier Raya-González, Enrique Ramírez-Chávez, Jorge Molina-Torres, Angel A. Guevara-García, José López-Bucio, and Luis Herrera-Estrella. 2011. "Alkamides Activate Jasmonic Acid Biosynthesis and Signaling Pathways and Confer Resistance to Botrytis Cinerea in Arabidopsis Thaliana." *PLoS ONE* 6 (11). https://doi.org/10.1371/journal.pone.0027251.

Mohammad, Shireen, and Christoph Thiemermann. 2021. "Role of Metabolic Endotoxemia in Systemic Inflammation and Potential Interventions." *Frontiers in Immunology*. https://doi.org/10.3389/fimmu.2020.594150.

Molinatorres, Jorge, Rafael Salgado-Garciglia, Enrique Ramirez-Chavez, and Rosa E. Del Rio. 1996. "Purely Olefinic Alkamides in Heliopsis Longipes and Acmella (Spilanthes) Oppositifolia." *Biochemical Systematics and Ecology* 24 (1). https://doi.org/10.1016/0305-1978(95)00099-2.

Moran, Magdalene M., and Arpad Szallasi. 2018. "Targeting Nociceptive Transient Receptor Potential Channels to Treat Chronic Pain: Current State of the Field." *British Journal of Pharmacology*. https://doi.org/10.1111/bph.14044.

Mózsik, Gyula, Áron Vincze, and János Szolcsányi. 2001. "Four Response Stages of Capsaicin-Sensitive Primary Afferent Neurons to Capsaicin and Its Analog: Gastric Acid Secretion, Gastric Mucosal Damage and Protection." *Journal of Gastroenterology and Hepatology* 16 (10): 1093–97. https://doi.org/10.1046/J.1440-1746.2001.02598.X.

Mustafa, Faheem, Hitesh Chopra, Atif Amin Baig, Satya Kumar Avula, Sony Kumari, Tapan Kumar Mohanta, Muthupandian Saravanan, Awdhesh Kumar Mishra, Nanaocha Sharma, and Yugal Kishore Mohanta. 2022. "Edible Mushrooms as Novel Myco-Therapeutics: Effects on Lipid Level, Obesity, and BMI." *Journal of Fungi*. https://doi.org/10.3390/jof8020211.

Park, Ui Hyun, Hong Suk Jeong, Eun Young Jo, Taesun Park, Seung Kew Yoon, Eun Joo Kim, Ji Cheon Jeong, and Soo Jong Um. 2012. "Piperine, a Component of Black Pepper, Inhibits Adipogenesis by Antagonizing PPARγ Activity in 3T3-L1 Cells." *Journal of Agricultural and Food Chemistry* 60 (15). https://doi.org/10.1021/jf204514a.

Platel, K., and K. Srinivasan. 2001. "Studies on the Influence of Dietary Spices on Food Transit Time in Experimental Rats." *Nutrition Research* 21 (9). https://doi.org/10.1016/S0271-5317(01)00331-1.

Pradeep, C.R., and G. Kuttan. 2002. "Effect of Piperine on the Inhibition of Lung Metastasis Induced B16F-10 Melanoma Cells in Mice." *Clinical and Experimental Metastasis* 19 (8). https://doi.org/10.1023/A:1021398601388.

Ramírez-Chávez, Enrique, José López-Bucio, Luis Herrera-Estrella, and Jorge Molina-Torres. 2004. "Alkamides Isolated from Plants Promote Growth and Alter Root Development in Arabidopsis." *Plant Physiology* 134 (3). https://doi.org/10.1104/pp.103.034553.

Rios-Chavez, P., E. Ramirez-Chavez, C. Armenta-Salinas, and J. Molina-Torres. 2003. "Acmella Radicans Var. Radicans: In Vitro Culture Establishment and Alkamide Content." *In Vitro Cellular \& Developmental Biology-Plant* 39 (1): 37–41.

Rios, Mar\'\ia Yolanda. 2012. *Natural Alkamides: Pharmacology, Chemistry and Distribution*. Vol. 1. InTech Rijeka.

Rizhsky, Ludmila, Huanan Jin, Michael R. Shepard, Harry W. Scott, Alicen M. Teitgen, M. Ann Perera, Vandana Mhaske, et al. 2016. "Integrating Metabolomics and Transcriptomics Data to Discover a Biocatalyst That Can Generate the Amine Precursors for Alkamide Biosynthesis." *Plant Journal* 88 (5). https://doi.org/10.1111/tpj.13295.

Rollyson, William D., Cody A. Stover, Kathleen C. Brown, Haley E. Perry, Cathryn D. Stevenson, Christopher A. McNees, John G. Ball, Monica A. Valentovic, and Piyali Dasgupta. 2014. "Bioavailability of Capsaicin and Its Implications for Drug Delivery." *Journal of Controlled Release : Official Journal of the Controlled Release Society* 196 (December): 96–105. https://doi.org/10.1016/J.JCONREL.2014.09.027.

Samykutty, Abhilash, Aditya Vittal Shetty, Gajalakshmi Dakshinamoorthy, Mary Margaret Bartik, Gary Leon Johnson, Brian Webb, Guoxing Zheng, Aoshuang Chen, Ramaswamy Kalyanasundaram, and Gnanasekar Munirathinam. 2013. "Piperine, a Bioactive Component of Pepper Spice Exerts Therapeutic Effects on Androgen Dependent and Androgen Independent Prostate Cancer Cells." *PLoS ONE* 8 (6). https://doi.org/10.1371/journal.pone.0065889.

Scheithauer, Torsten P.M., Elena Rampanelli, Max Nieuwdorp, Bruce A. Vallance, C. Bruce Verchere, Daniël H. van Raalte, and Hilde Herrema. 2020. "Gut Microbiota as a Trigger for Metabolic Inflammation in Obesity and Type 2 Diabetes." *Frontiers in Immunology*. https://doi.org/10.3389/fimmu.2020.571731.

Singh, A., and A.R. Rao. 1993. "Evaluation of the Modulatory Influence of Black Pepper (Piper Nigrum, L.) on the Hepatic Detoxication System." *Cancer Letters* 72 (1–2). https://doi.org/10.1016/0304-3835(93)90003-R.

Singh Bakshi, Inderbir, Hitesh Chopra, Madhu Sharma, Deepak Kaushik, Rakesh Pahwa, and Haryanto. 2022. "Herbal Bioactives for Wound Healing Application." *Herbal Bioactive-Based Drug Delivery Systems*, 259–82. https://doi.org/10.1016/b978-0-12-824385-5.00003-0.

Singla, Rajeev K., Xuefei He, Hitesh Chopra, Christos Tsagkaris, Li Shen, Mohammad Amjad Kamal, and Bairong Shen. 2021a. "Natural Products for the Prevention and Control of the COVID-19 Pandemic: Sustainable Bioresources." *Frontiers in Pharmacology* 12 (December): 3215. https://doi.org/10.3389/FPHAR.2021.758159/BIBTEX.

Singla, Rajeev K., Chandragiri Siva Sai, Hitesh Chopra, Sahar Behzad, Himangini Bansal, Rajat Goyal, Rupesh K. Gautam, et al. 2021b. "Natural Products for the Management of Castration-Resistant Prostate Cancer: Special Focus on Nanoparticles Based Studies." *Frontiers in Cell and Developmental Biology* 9 (November). https://doi.org/10.3389/FCELL.2021.745177.

Sunila, E.S., and G. Kuttan. 2004. "Immunomodulatory and Antitumor Activity of Piper Longum Linn. and Piperine." *Journal of Ethnopharmacology* 90 (2–3). https://doi.org/10.1016/j.jep.2003.10.016.

Tufail, Tabussam, Jasmeet Kour, Farhan Saeed, Huma Bader Ul Ain, Ali Ikram, Waseem Khalid, Muzzamal Hussain, and Hitesh Chopra. 2022. "An Overview of Food Toxins." In *Handbook of Plant and Animal Toxins in Food*. https://doi.org/10.1201/9781003178446-1.

Vijayakumar, R.S., D. Surya, and N. Nalini. 2004. "Antioxidant Efficacy of Black Pepper (Piper Nigrum L.) and Piperine in Rats with High Fat Diet Induced Oxidative Stress." *Redox Report* 9 (2). https://doi.org/10.1179/135100004225004742.

Walia, Vaibhav, Deepak Kaushik, Vineet Mittal, Kuldeep Kumar, Ravinder Verma, Jatin Parashar, Rokeya Akter, et al. 2022. "Correction to: Delineation of Neuroprotective Effects and Possible Benefits of AntioxidantsTherapy for the Treatment of Alzheimer's Diseases by Targeting Mitochondrial-Derived Reactive Oxygen Species: Bench to Bedside." *Molecular Neurobiology* 59 (1), 657–80. https://doi.org/10.1007/s12035-021-02665-7.

Wang, Yuanwei, Cheng Tang, Yong Tang, Haiyan Yin, and Xiong Liu. 2020. "Capsaicin Has an Anti-Obesity Effect through Alterations in Gut Microbiota Populations and Short-Chain Fatty Acid Concentrations." *Food and Nutrition Research* 64. https://doi.org/10.29219/fnr.v64.3525.

Wanrooij, Sander J.M. Van, Mira M. Wouters, Lukas Van Oudenhove, Winde Vanbrabant, Stéphanie Mondelaers, Patrick Kollmann, Florian Kreutz, Michael Schemann, and Guy E. Boeckxstaens. 2014. "Sensitivity Testing in Irritable Bowel Syndrome with Rectal Capsaicin Stimulations: Role of Trpv1 Upregulation and Sensitization in Visceral Hypersensitivity?" *American Journal of Gastroenterology* 109 (1). https://doi.org/10.1038/ajg.2013.371.

Wei, Guohua, Yong Ye, Xiang Yan, Xingyu Chao, Fan Yang, Mengyang Wang, Wencheng Zhang, Chuanxun Yuan, and Qingmei Zeng. 2020. "Effect of Banana Pulp Dietary Fibers on Metabolic Syndrome and Gut Microbiota Diversity in High-Fat Diet Mice." *Journal of Food Biochemistry* 44 (9). https://doi.org/10.1111/jfbc.13362.

Woelkart, Karin, Christoph Koidl, Andrea Grisold, J. David Gangemi, Ronald B. Turner, Egon Marth, and Rudolf Bauer. 2005. "Bioavailability and Pharmacokinetics of Alkamides from the Roots of Echinacea Angustifolia in Humans." *Journal of Clinical Pharmacology* 45 (6). https://doi.org/10.1177/0091270004273493.

Xiang, Qunran, Weiling Guo, Xin Tang, Shumao Cui, Feng Zhang, Xiaoming Liu, Jianxin Zhao, Hao Zhang, Bingyong Mao, and Wei Chen. 2021. "Capsaicin—the Spicy Ingredient of Chili Peppers: A Review of the Gastrointestinal Effects and Mechanisms." *Trends in Food Science & Technology* 116: 755–65.

Zandonadi, Daniel Bas\'\ilio, Carlos Roberto Ribeiro Matos, Rosane Nora Castro, Riccardo Spaccini, Fábio Lopes Olivares, and Luciano Pasqualoto Canellas. 2019. "Alkamides: A New Class of Plant Growth Regulators Linked to Humic Acid Bioactivity." *Chemical and Biological Technologies in Agriculture* 6 (1): 1–12.

14 Pharmacological Potential of Plant-Derived Alkylamides

Anju Jose, Chenmala Karthika, Athira KV, Md. Habibur Rahman, and Shrouq H. Sweilam

CONTENTS

14.0 INTRODUCTION

Alkylamides are bioactive natural amides that have an aromatic or aliphatic unsaturated fatty-acid residue linked to an aliphatic or aromatic amine residue. They are also known as N-alkylamides, alkamides, alkenamides, or alkenylamides (Akter et al., 2020). They originate from abundant natural plant families such the Euphorbiaceae, Rutaceae, Aristolochiaceae, Asteraceae, Convolvulaceae, Menispermaceae, Brassicaceae, Piperaceae, Poaceae, and Solanaceae. Several Echinacea species accumulate unsaturated alkamides with 12 to 18 carbon atoms (Méndez-Bravo et al., 2011). These are present in different organs of the plant, such as leaves and stems of *Aristolochia gehrtii*, *Phyllanthus fraternus*, *Amaranthus hypochondriacus*, *Achyranthes ferruginea*, etc.; the pericarpium of *Zanthoxylum piperitum* and *Piper spp*; the placenta of *Capsicum spp*; the fruits of *Piper longum*; the freduces of *Spilanthes acmella*; the seeds of the *Piper* species and tubers of *Lepidium meyenii*; and the roots of *Heliopsis longipes*, *Echinaceae purpurea*, *Achillea wilhelmsii*, *Acmella oppositifolia*, *Asarum heterotropoides*, and *Cissampelos glaberrima* (Yolanda Rios Gomez, 2012). Due to their structural diversity and functionality, alkylamides have a wide range of pharmacological effects. Alkamid, an online database with further information on alkylamide categorization, is presently accessible for public use (Boonen et al., 2012). As natural compounds, more than 300 alkamides have been identified. Plants from the Asteraceae, Solanaceae, Rutaceae, and Piperaceae families are the most prevalent sources of these natural alkaloids (Rios and Olivo, 2014). Alkylamides have numerous potential activities such as antimalarial, antibacterial, antiviral, antifungal, insecticidal, immunostimulant, analgesic, anti-inflammatory, antioxidant, diuretic, pungent, and analgesic (Veryser, 2016). Other properties include the alkylamide-comprising plants, which

DOI: 10.1201/9781003224525-17

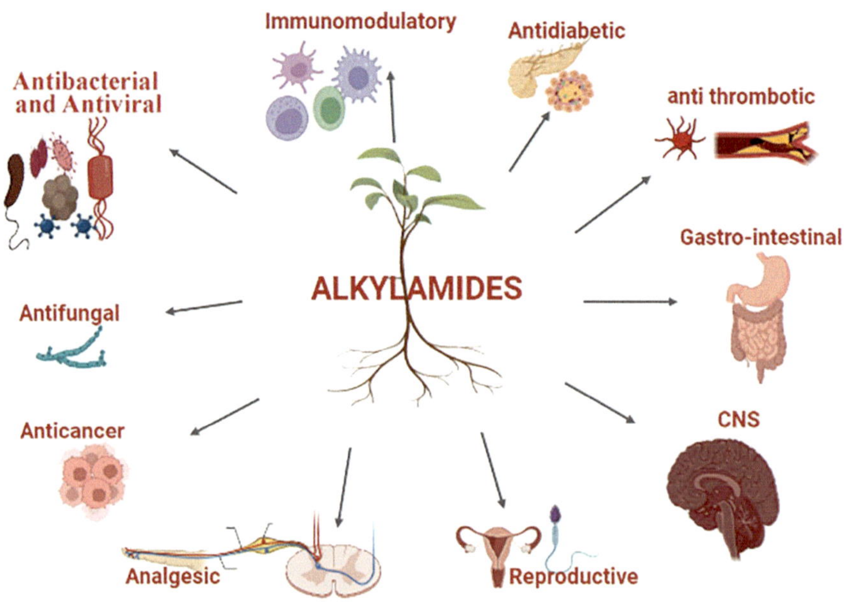

FIGURE 14.1 Effects of alkylamides.

are employed as spices because of their pungent and tingling flavors; they're also used in topical cosmetics because of their wrinkle-softening and anti-aging effects. They are also employed for toothaches, dermal and stomach illnesses, sexual dysfunctions, and viral infections (Veryser, 2016). Figure 14.1 represents the action of plant-derived alkylamides in various diseases.

14.0.1 Alkylamides in Various Plant Species

Plants containing alkamides are frequently utilized as seasonings in various regions of the world during the preparation of meals. Alkylamides may be found in a variety of plant organs, including roots, leaves, stems, fruits, flowers, seeds, and tubers (Evidente and Masi, 2021). The plant that contains N-alkylamides is made up of an aliphatic chain of polyunsaturated fatty acids connected to a short-chain amine (Veryser, 2016). N-Alkylamides are a class of bioactive compounds discovered in plants of the genera *Echinacea*, *Xanthoxylum*, and *Spilanthe* (Boonen et al., 2010). Chloroform is the ideal solvent for alkamide extraction, while methanol and ethanol have also been utilized. When alkylamide comes in contact with the mouth, cheeks, tongue, gums, palate, teeth, and lips, it generates acute ptyalism (salivation) and tingling. (Rios and Olivo, 2014),(Yolanda Rios Gomez, 2012). Alkylamides from *Echinacea angustifolia* have also been successfully extracted using the SC-CO2 method (Sun et al., 2002). Pure alkamides of the plants are susceptible to oxidation and polymerization of double and triple bonds that occur during drying, handling, and storage (Yolanda Rios Gomez, 2012). The stems of *Zanthoxylum nitidum* belonging to the *Rutaceae* family yield a novel alkylamide, (2E,6E,8E)-N-(2-methyl propyl)-10-oxo-2,6,8-decatrienamide, along with 22 recognized compounds (2–23) with mild antioxidant activities (Chakthong et al., 2019). The alkylamides are a key component class in Echinacea species. Alkylamide with a carboxylic acid functional group has recently been identified from E. purpurea (Bruni et al., 2018). Alkylamides, phenylpropanoids (caffeic acid derivatives), polysaccharides, and glycoproteins are now hypothesized to be the source of action in the echinacea species (Spelman et al., 2009). E. purpurae and E. angustifolia contain higher concentrations of alkylamide content in the root than in the arial parts (Woelkart and Bauer, 2007). E. purpurae has also shown mosquitocidal property activity with alkamides with isobutylamide moiety (Manayi et al., 2015).

Capsaicinoids are unique alkamides from the family *Solanaceae*, which can only be found in plants of the *Capsicum* genus. They are the metabolite responsible for the spicy and pungent sensation felt while eating capsicum. Capsaicinoids possess antimicrobial properties (Choi et al., 2006) and also have various other medical uses (Hernández-Pérez et al., 2020). The presence of aromatic ring, often a 3,4-methylenedioxyphenyl group attached to the end of an unsaturated C3–C16 fatty-acid chain, distinguishes alkamides isolated from the *Piperaceae* family (Rios and Olivo, 2014). Chili peppers contain capsaicinoids, which cause them to burn. Between C8 and C11 on the fatty-acid NAA, they cause a strong burning sensation. This action is attributed to the existence of an amide bond connecting the an acyl chain to the vanillyl ring (Boonen et al., 2012). The root of bell pepper (Capsicum annuum var. grossum) has various newly described phenolic amides N-cis-feruloyltyramine (NCFT), grossamide, and N-trans-feruloyltyramine, N-trans-p-coumaroyl-octopamine, N-trans-p-coumaroyltyramine and N-trans-feruloyl octopamine (Evidente and Masi, 2021). The alkylamide, as well as ethyl caffeate, N-trans feruloyl octopamine, isoscopoletin, N-trans-p-coumaroyl octopamine, vanillin, p-aminobenzaldehyde, and ferulic acid were isolated for the first time from eggplant roots. Crinum biflorum Rottb. (Crinum distichum), an Amaryllidaceae species, has been detected with N-p-coumaroyltyramine, an alkylamide, and four other compounds that are pharmacologically significant (Masi et al., 2021). Novel alkylamides, (2E,5Z)-N-isobutyl undeca-2,5-diene-8,10-diynamide, and four known analogues, were isolated from Acmella oleracea flowers belonging to the family Asteraceae that have been used as a folk medicine to cure pain such as toothache, throat pain, stomatitis, and malaria (Cheng et al., 2015). In the *Acmella* genus, Wedelia acapulcensis var. parviceps (S.F. Blake) Strother (Asteraceae) and Heliopsis longipes (A. Gray) S.F. Blake (Asteraceae) N-alkylamides have been identified as the primary bioactive chemicals that have various pharmacological properties including antioxidant, analgesic, anticancer, neuroprotective, anti-inflammatory, antimutagenic, antimicrobial (Lalthanpuii and Lalchhandama, 2020), and insecticidal. The primary ingredients of A. oleracea were identified as alkylamides, vanillic acid, 3-acetyl aleuritolic acid, scopoletin, β-sitostenone, trans-ferulic acid, and trans-isoferulic acid (Prachayasittikul et al., 2009). Acmella oleracea, belonging to Asteraceae, is an ornamental or medicinal plant that is also referred as the toothache plant (jambu), which has the most ample alkylamide, spilanthol (Sut et al., 2020). The presence of lipophilic N-alkylamides in A. oleracea is largely responsible for the biological activities reported (Stein et al., 2021). Two previously unknown NAAs in *Achillea ptarmica* were identified as deca-2E,6Z,8E-trienoic acid 2-methylbutylamine (homospilanthol) or a similar isomeric molecule and deca-2E,4E-dienoic acid N-methyl isobutylamide. Homospilanthol and other compounds were also identified in *Achillea millefolium* species. The primary component of *Achillea millefolium* is found to be 2E,4E,6Z-octadecatrienoic acid piperidine, with only trace quantities of the corresponding all trans-isomer. NAAs found by GC-MS in the roots of A. distans Willd. include N-isobutyl-2,4-decadiene amide (pellitorine), 2,4-decadienoic acid piperidide, 2,4-decadienoic acid piperidide, and 2,4,6-decatrienoic acid piperidide (Veryser et al., 2017). Spilanthol is an olefinic alkylamide with an isobutyl side chain that is responsible for the compound's sensory action (Savic et al., 2021). Spilanthol is also an analgesic, anti-inflammatory, and antimutagenic that is abundantly found in *Spilanthes acmella* (Dias et al., 2012). When applied to the human tongue, α-HO-sanshool from Zanthoxylum produces a powerful tingling sensation, equivalent to a moderate electric shock, as well as cooling (Epple et al., 2001). *S. acmella* Murr. were evaluated for antimicrobial, antioxidant, and cytotoxic activities (Prachayasittikul et al., 2009). *Zanthoxylum bungeanum* (Szechuan pepper) have various alkylamidesm such as hydroxy-γ-sanshool, Hydroxy-ε-sanshool, hydroxy-α-sanshool, hydroxy-β-sanshool, isobungeanool, and bungeanool. These were found to be the key pungent compounds. The molecules hydroxy-α and hydroxyl-γ-sanshool with one cis and three or four trans bonds are strongly pungent, but hydroxyl-β-sanshool with all trans double bonds is significantly less pungent. The amount and composition of alkylamides in them determine their pungency (Zhang et al., 2019). Pellitorine alkylamide is found in a variety of plants and is most prevalent in Anacyclus pyrethrum. Pellitorine is antithrombotic, antiseptic, antiprotozoal, and anticancer (Veryser, 2016). Since alkylamides are lipophilic in nature, they are easily soluble in the organic solvents and can be extracted using compounds such as n-hexane, chloroform, ethanol, or ethanol. Plants are extracted using a variety of processes, including maceration, hydrodistillation, supercritical fluid extraction, and Soxhlet extraction (Veryser, 2016). P. longum has also been used to treat lung infections, cough, spleen illnesses, viral hepatitis, malaria, bronchitis, asthma, upset stomach, and gonorrhea (Kumar et al., 2011). Spilanthol possess inhibitory action on NF-1κB by inhibiting the phosphorylation of cytoplasmic inhibitor-B and the DNA binding activity of NF-1κB (R. Sharma and Arumugam, 2021). On reaching systemic circulation, alkylamides are processed by cytochromes P 450 in liver. The breakdown of NAAs in the liver microsomal fractions was demonstrated to be time dependent and dependent on NADPH, suggesting that NAA metabolism occurs in the human liver (R. Sharma and Arumugam, 2021).

14.1 ANTIMICROBIAL ACTIVITY

In pathological situations such as infections, M1 macrophages have been shown to have an improved bactericidal capability to remove intracellular microorganisms. The phosphorylation of transcription factors by MAPKs and NF-κB regulates the production of inflammatory cytokines, and the phosphorylation of MAPKs can activate NF-κB through a JNK-dependent mechanism. *Echinacea purpurea* promotes bactericidal macrophages M1 functional polarization (Fu et al., 2017). Echinacea has the capacity to destroy a variety of bacteria, including *Staphylococcus aureus* (Akhavan et al., 2016). Higher levels of viral infection inhibitors have been observed in both hydrophilic and lipophilic extract of Echinacea. Different test techniques, such as dilution, disc diffusion, and even TLC-hyphenated bioassays, were employed to investigate the antimicrobial activity of NAAs (Boonen et al., 2012).

14.2 ANTIMALARIAL ACTIVITY

By blocking erythrocytic schizogony in the ring stage, (2E,6Z,8E)-N-isobutyl-2,6,8 decatrienamide (spilanthol) and (2E,4Z)-N-isobutyl-2,4-undecadiene-8,10-diynamide (UDA), from *Spilanthes paniculata* inhibits the malaria parasite leading to the death of the parasite (Rajendran et al., 2017). N-alkylamides such as Spilanthol and Undeca-2E-ene-8,10-diynoic acid isobutylamide present in *S. acmella* were found to produce antimalarial activity with a decrease in the presence of parasites in the blood (Spelman et al., 2011).

14.2.1 Antifungal Activity

Oral and injectable formulations of *E. purpurea* showed a considerable clinical decrease in recurrent candidiasis. Antifungal activity was found in root extracts from the eight *Echinacea taxa* against the majority of pathogenic fungi by the significant inhibition of LOX (Merali et al., 2003). The *E. purpurea* extract has hindered the growth of *Candida albicans* and *Saccharomyces cerevisiae* (Manayi et al., 2015).

14.2.2 Antithrombotic Activity

N-p-coumaroyltyramine was extracted from *Annona montana* (Annonaceae) stem portions and demonstrated considerable suppression of rabbit platelet aggregation generated by thrombin, arachidonic acid, collagen, and platelet-activating factor (PAF), as well as specific cytotoxicity against the P-388 cell line (Wu et al., 1995);(Masi et al., 2021).

14.2.3 Analgesic Activity

The interaction of affinin with many signaling pathways, including nitric oxide (NO), the gamma-aminobutyric acid pathway, serotonin, opioids, and the activation of possible transient receptor (TRP) channels, notably TRPV1, may provide orofacial pain relief (García-Badillo et al., 2021). Sanshool major alkylamide in extracts from *Zanthoxylum* inhibits the activity of numerous voltage-gated sodium channel subtypes, Nav1.7 being the most substantially affected. It also suppresses the activity of Aδmechanonociceptors, which mediate both acute and inflammatory pain. Antinociception action is demonstrated by a cold water extract of *A. oleracea* freduces that inhibits prostaglandin production, interrupts nociception transmission, and acts as an antihistamine (Ratnasooriya and Pieris, 2005). The N-alkylamides pilanthol in the *acmella oleracea* exerts analgesic activity. The nociceptic effect of Spilanthes is expected due to inhibition of prostaglandin production; activation of opioid, serotoninergic, and GABAergic systems; and anaesthetic action via voltage-gated Na channels blockade (Nomura et al., 2013).

14.2.4 Antidiabetic Activity

Diabetes is the most prevalent condition in both males and females (Refardt et al., 2020). TRPV1 and CB1 receptors are critical in the progress of diabetes; TRPV1 activator, in particular, enhances insulin secretion and glucose homeostasis. Obesity promote the death of β-cells, which can be prevented by inhibiting the CB1 receptor (Gram et al., 2007). Alkylamide also upregulates the expression of pancreatic-duodenal homeobox 1 (PDX-1), glucose transporter-2 (GLUT2), glucokinase (GK), and TRPV1 in diabetic rats' pancreases (Ren et al., 2017). A dose-dependent response of alkylamide in Zanthoxylum shows significant decrease in the fasting blood glucose, fructosamine levels, and the weight of the organs such as the liver and kidney. There is a prominent decline in the enzyme expression levels including phosphoenolpyruvate carboxykinase and glucose-6-phosphatase in the liver. In the pancreas, increased expression of pancreatic duodenal homeobox-1, glucokinase, and glucose transporter (You et al., 2015). Alkylamides can reduce glycogen production, restore islet function, and improve insulin secretion by inhibiting gluconeogenesis (Oh et al., 2013). The anti-α-glucosidase and anti-α-amylase activity of N-p-coumaroyltyramine was identified in the Crinum species (Masi et al., 2021). Ethanolic extract of *Echinacea purpurea* ethanol extract has shown significant effect on hyperglycemia and insulin resistance (Mao et al., 2021).

14.2.5 Anticancer Activity

Echinacea reduces the symptoms of xenobiotic-induced hepatotoxicity by reducing oxidative stress, inducing apoptosis, and enhancing intracellular enzyme functioning (Xu et al., 2021). This is the only component that has been shown to pass across the intestinal barrier. By reducing the expression of triggering receptors expressed on myeloid cells 2 (TREM2) and the PI3K/AKT signaling pathway, echinacoside (100 g/ml) can limit cell growth and cause cell death in HepG2 cells (Ye et al., 2019). The N-p-coumaroyltyramine from *C. biflorum* was found to be harmful to A431 and HeLa cancer cells, but it protected immortalized HaCaT cells from oxidative stress employed by hydrogen peroxide. New alkylamide derivatives of racemocin B, indolo[3,2-*a*]carbazole from the from the green alga *Caulerpa racemose* have shown that in aggressive triple-negative breast cancer cell line MDA-MB-23, they triggered G2/M cell cycle arrest and apoptosis. It also inhibited autophagic flux and accumulation of autophagy protein 1 light chain 3, LC3II, and p62 (Xiao et al., 2018). It is found that after oral dosing, spilanthol and pellitorine may pass the Caco-2 cell monolayer *in vitro* and the intestinal membrane *in vivo* in rats (Veryser et al., 2016; Matthias et al., 2005; Veryser et al., 2014). The anti-proliferative properties of n alkylamide (H-10) compound were observed in the B16, HeLa, HepG2, and HCT116 cell lines. In

2D and 3D culture, it has been found to inhibit cancer cell proliferation. In 2D-cultured HCT116 cells, compound H-10 promotes G2 phase arrest, whereas in 3D-cultured HCT116 cells, compound H-10 induces G1 phase arrest. thus exhibiting a significant cancer potential (Jiang et al., 2020). Long pepper (*Piper longum L.* Piperaceae) comprises an active alkaloid/amide called piplartine, which is employed in the mitigation of cancer. Piplartine has been found to have cytotoxic effect by inducing oxidative stress in the cancer cells, genotoxic, and anticancer properties (Bezerra et al., 2013).

14.2.6 Effects of Alkylamides in CNS and Neuro Conditions

There are numerous CNS conditions that causes significant burdens to normal, healthy life, including depression, which is prevalent and has comorbidities (Jose et al., 2021). Since Alkamides are structurally similar to endocannabinoids, the N-arachidonoylethanolamine (anandamide), which is an endogenous cannabinoid brain neurotransmitter with striking structural similarities to plant alkamides, is most active in the central nervous system to produce remarkable effect (García-Badillo et al., 2021). CB1 and CB2 receptor ligands provide many mitigation options for a variety of CNS-related illnesses. Alkylamides such as dodeca-2E,4E-dienoic acid isobutylamide and dodeca-2E,4E,8Z,10Z-tetraenoic acid isobutylamide bind to human CB 2 receptor, which has agonist activity (Dimmito et al., 2021). *Otanthus Maritimus L.*, which comprises two alkylamides, is an effective binder of CB1 and CB2 receptors with a Ki value of 0.8 μM and 0.16 μM, respectively (Dimmito et al., 2021). Similarly, only a few pharmacokinetic data on the capability of N-alkylamides to penetrate the BBB is currently available. Dodeca-2E,4E,8Z,10E/Z-tetraenoic acid isobutylamides (Echinacea-derived tetraenes) have been found to be bioavailable and can penetrate the blood-brain barrier quickly when administered orally to rats (Woelkart et al., 2009).

14.3 GASTROINTESTINAL ACTIVITY

Gastrointestinal conditions are common and may vary among individuals in symptoms and physiology (Black et al., 2020). *Achillea millefolium* L (A. millefolium) and *Achillea ptarmica* L. (*A. ptarmica*) plants are employed to cure stomach diseases and stomach discomfort, respectively (Veryser et al., 2017). *Clausena lansium*, from the family Rutaceae, is a traditional medicine whose leaves and roots are employed to cure gastric ailments and whose seeds are employed for ulcers and gastrointestinal inflammation (Evidente and Masi, 2021). An *in vivo* study in mice has shown that intestinal mucositis induced by the administration of 5-fluorouracil (5-FU) is suggestively decreased by the administration of spilanthol (30mg/kg), which confirms the therapeutic effect of the alkylamides present in the *Acmella oleracea* (De Freitas-Blanco et al., 2019).

14.4 ANTIPROTOZOAL ACTIVITY AND ANTI-HELMINTHIC ACTIVITY

Echinaceae has been found to reduce or inhibit the multiplication of parasites (Canlas et al., 2010). E. purpurea aerial portions have been shown to have antitrypanosomal and antileishmanial properties (Zimmermann et al., 2012). *Spilanthes acmella* also have shown that antiplasmodial bioactivity may be enhanced by N-alkylamides (Mbeunkui et al., 2011). Antiprotozoal activity was observed *in vitro* (Althaus et al., 2014). *Acmella oleracea* includes significant bioactive chemicals that are responsible for broad-spectrum anthelmintic action by causing detrimental anthelmintic effects by directly attacking the cuticle (Lalthanpuii and Lalchhandama, 2020). *N*-isobutyl-(2*E*,4*Z*,8*Z*,10*E*)-dodecatetraenamide, from *Spilanthes acmella Murr*, which belongs to the family Asteraceae, has reported remarkable antihelmentic activity with significant shrinkage and folding on the main body; the spines and suckers are also destroyed (Lalthanpuii et al., 2020).

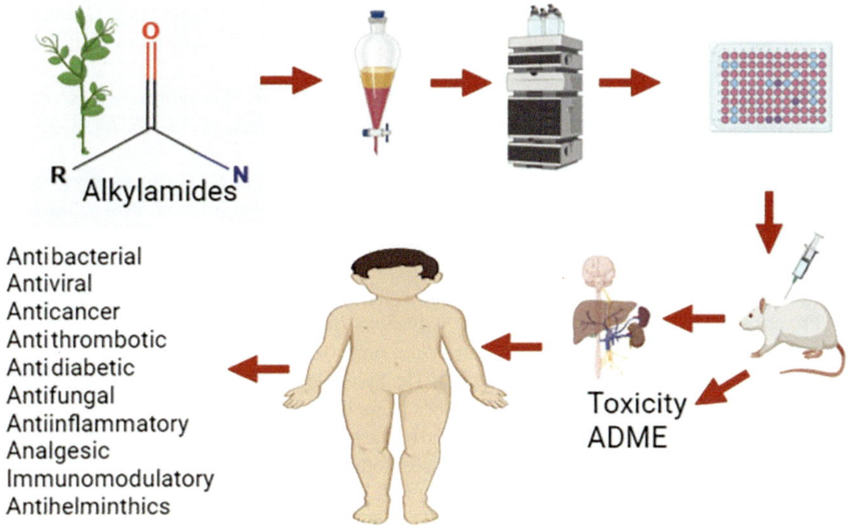

FIGURE 14.2 Development of alkylamides as a drug from the plant.

14.5 OTHER ACTIVITIES

Affinin causes vasodilation by interacting with endothelial TRV1 and TRPA1 channels, which causes the NO/cGMP, H2S/KATP, and HNO/TRPA1/CGRP signaling pathways to be activated, where the association of affine with TRP channel affect angiogenesis (García-Badillo et al., 2021). Through the activation of GTs (NO, H2S, and CO), as well as potassium channels in a endothelium-dependent action, affinin might cause vasodilation in the rat aorta, which provides significant mitigation possibilities in cardiovascular disease and hypertension (Castro-Ruiz et al., 2017). Figure 14.2 represents the isolation, characterization, *in vitro* preclinical toxicity, and clinical studies of alkylamides and their therapeutic actions.

14.6 CONCLUSION

Alkylamides are metabolites that are present in a variety of medicinal plants from different families. According to research on the pharmacological properties,Alkylamides have important roles in the pharmaceutical, cosmetic, and food industries. Dodeca-2E,4E,8Z,10E/Z-tetraenoic acid isobutylamide and dodeca-2E,4E-dienoic acid isobutylamide, both chemically pure alkylamides, inhibited CYP 2C19, 2D6, and 3A4 but did not induce CYP 1A2 inhibition. Szechuan pepper (*Xanthoxylum piperitum*) and jambu fruit (*Acmella oleracea*) comprise two natural alkylamides, a-hydroxy-sanshool and spilanthol, respectively. They are commonly contained in ethnic foods to give distinct mouth experiences when eating. In numerous other fields, alkylamide has activity that shows that osteoclast-specific gene expression was reduced by echinalkamide therapy. The inhibitory action of echinalkamide was mediated via disrupting MAPK phosphorylation and activation of the osteoclast transcription factors c-Fos and NFATc1 (Chang et al., 2020). Due to the significant pharmacological potential of alkylamides, it has a lot of potential for new product innovations.

REFERENCES

Akhavan, Saeedeh, Reza Mohammadhassan, Aida Mahmoudi, Atieh Khalkhali, and Roya Barzin. 2016. "Antiviral Activity of Echinacea (Echinacea Purpurea)." *IJBPAS* 5: 999–1005. www.researchgate.net/publication/325324332.

Akter, Rokeya, Md Arifur Rahman Chowdhury, and Md Habib Ur Rahman. 2020. "Flavonoids and Polyphenolic Compounds as Potential Talented Agents for the Treatment of Alzheimer's Disease with Their Antioxidant Activities." *Current Pharmaceutical Design* 27 (3): 345–56. doi: 10.2174/1381612826666201102102810. PMID: 33138754.

Althaus, Julia B., Marcel Kaiser, Reto Brun, and Thomas J. Schmidt. 2014. "Antiprotozoal Activity of Achillea Ptarmica (Asteraceae) and Its Main Alkamide Constituents." *Molecules* 19 (5): 6428–38. https://doi.org/10.3390/molecules19056428.

Bezerra, Daniel P., Claudia Pessoa, Manoel O. De Moraes, Nicolau Saker-Neto, Edilberto R. Silveira, and Leticia V. Costa-Lotufo. 2013. "Overview of the Therapeutic Potential of Piplartine (Piperlongumine)." *European Journal of Pharmaceutical Sciences* 48 (3): 453–63. https://doi.org/10.1016/j.ejps.2012.12.003.

Black, Christopher J., Douglas A. Drossman, Nicholas J. Talley, Johannah Ruddy, and Alexander C. Ford. 2020. "Functional Gastrointestinal Disorders: Advances in Understanding and Management." *Lancet (London, England)* 396 (10263): 1664–74. https://doi.org/10.1016/S0140-6736(20)32115-2.

Boonen, J., B. Baert, N. Roche, C. Burvenich, and B. De Spiegeleer. 2010. "Transdermal Behaviour of the N-Alkylamide Spilanthol (Affinin) from Spilanthes Acmella (Compositae) Extracts." *Journal of Ethnopharmacology* 127 (1): 77–84. https://doi.org/10.1016/j.jep.2009.09.046.

Boonen, Jente, Antoon Bronselaer, Joachim Nielandt, Lieselotte Veryser, Guy De Tré, and Bart De Spiegeleer. 2012. "Alkamid Database: Chemistry, Occurrence and Functionality of Plant N-Alkylamides." *Journal of Ethnopharmacology* 142 (3): 563–90. https://doi.org/10.1016/j.jep.2012.05.038.

Bruni, Renato, Virginia Brighenti, Lindsay K. Caesar, Davide Bertelli, Nadja B. Cech, and Federica Pellati. 2018. "Analytical Methods for the Study of Bioactive Compounds from Medicinally Used Echinacea Species." *Journal of Pharmaceutical and Biomedical Analysis* 160 (October): 443–77. https://doi.org/10.1016/j.jpba.2018.07.044.

Canlas, Judith, James B. Hudson, Manju Sharma, and Devki Nandan. 2010. "Echinacea and Trypanasomatid Parasite Interactions: Growth-Inhibitory and Anti-Inflammatory Effects of Echinacea." *Pharmaceutical Biology* 48 (9): 1047–52. https://doi.org/10.3109/13880200903483468.

Castro-Ruiz, Jesús Eduardo, Alejandra Rojas-Molina, Francisco J. Luna-Vázquez, Fausto Rivero-Cruz, Teresa García-Gasca, and César Ibarra-Alvarado. 2017. "Affinin (Spilanthol), Isolated from Heliopsis Longipes, Induces Vasodilation via Activation of Gasotransmitters and Prostacyclin Signaling Pathways." *International Journal of Molecular Sciences* 18 (1). https://doi.org/10.3390/ijms18010218.

Chakthong, Suda, Rangsima Ampaprom, Sakawduen Inparn, Uraiwan Phetkul, Sasitorn Chusri, Surasak Limsuwan, and Supayang Piyawan Voravuthikunchai. 2019. "New Alkylamide from the Stems of Zanthoxylum Nitidum." *Natural Product Research* 33 (2): 153–61. https://doi.org/10.1080/14786419.2018.1440218.

Chang, Bo Yoon, Seul Ki Lee, Da Eun Kim, Jin Hye Bae, Thanh Tam Ho, So-Young Park, Mi Kyeong Lee, and Sung Yeon Kim. 2020. "Effect of Echinalkamide Identified from Echinacea Purpurea (L.) Moench on the Inhibition of Osteoclastogenesis and Bone Resorption." *Scientific Reports* 10 (1): 10914. https://doi.org/10.1038/s41598-020-67890-x.

Cheng, Yuan-Bin, Rosa Huang Liu, Meng-Chi Ho, Tung-Ying Wu, Ching-Yeu Chen, I-Wen Lo, Ming-Feng Hou, Shyng-Shiou Yuan, Yang-Chang Wu, and Fang-Rong Chang. 2015. "Alkylamides of Acmella Oleracea." *Molecules (Basel, Switzerland)* 20 (4): 6970–77. https://doi.org/10.3390/molecules20046970.

Choi, Suk-Hyun, Bong-Soon Suh, Etsuko Kozukue, Nobuyuki Kozukue, Carol E Levin, and Mendel Friedman. 2006. "Analysis of the Contents of Pungent Compounds in Fresh Korean Red Peppers and in Pepper-Containing Foods." *Journal of Agricultural and Food Chemistry* 54 (24): 9024–31. https://doi.org/10.1021/jf061157z.

Dias, A.M.A., P. Santos, I.J. Seabra, R.N.C. Júnior, M.E.M. Braga, and H.C. De Sousa. 2012. "Spilanthol from Spilanthes Acmella Flowers, Leaves and Stems Obtained by Selective Supercritical Carbon Dioxide Extraction." *Journal of Supercritical Fluids* 61: 62–70. https://doi.org/10.1016/j.supflu.2011.09.020.

Dimmito, Marilisa Pia, Azzurra Stefanucci, Alice Della Valle, Giuseppe Scioli, Angelo Cichelli, and Adriano Mollica. 2021. "An Overview on Plants Cannabinoids Endorsed with Cardiovascular Effects." *Biomedicine & Pharmacotherapy = Biomedecine & Pharmacotherapie* 142: 111963. https://doi.org/10.1016/j.biopha.2021.111963.

Epple, G., B.P. Bryant, I. Mezine, and S. Lewis. 2001. "Zanthoxylum Piperitum, an Asian Spice, Inhibits Food Intake in Rats." *Journal of Chemical Ecology* 27 (8): 1627–40. https://doi.org/10.1023/a:1010462309244.

Evidente, Antonio, and Marco Masi. 2021. "Natural Bioactive Cinnamoyltyramine Alkylamides and Co-Metabolites." *Biomolecules* 11 (12). https://doi.org/10.3390/biom11121765.

Freitas-Blanco, Veronica Santana De, Karin Maia Monteiro, Patrícia Rosa De Oliveira, Ellen Cristina Souza De Oliveira, Lucia Elaine De Oliveira Braga, João Ernesto De Carvalho, and Rodney Alexandre Ferreira Rodrigues. 2019. "Spilanthol, the Principal Alkylamide from Acmella Oleracea, Attenuates 5-Fluorouracil-Induced Intestinal Mucositis in Mice." *Planta Medica* 85 (3): 203–9. https://doi.org/10.1055/a-0715-2002.

Fu, Aikun, Yang Wang, Yanping Wu, Hongliang Chen, Shasha Zheng, Yali Li, Xin Xu, and Weifen Li. 2017. "Echinacea Purpurea Extract Polarizes M1 Macrophages in Murine Bone Marrow-Derived Macrophages Through the Activation of JNK." *Journal of Cellular Biochemistry* 118 (9): 2664–71. https://doi.org/10.1002/jcb.25875.

García-Badillo, Paola Estefanía, Anaguiven Avalos-Soriano, Josué López-Martínez, Teresa García-Gasca, and Jesús Eduardo Castro-Ruiz. 2021. "Proangiogenic Effect of Affinin and an Ethanolic Extract from Heliopsis Longipes Roots: Ex Vivo and in Vivo Evidence." *Molecules* 26 (24): 1–14. https://doi.org/10.3390/molecules26247670.

Gram, Dorte X., Bo Ahrén, Istvan Nagy, Uffe B Olsen, Christian L Brand, Frank Sundler, René Tabanera, et al. 2007. "Capsaicin-Sensitive Sensory Fibers in the Islets of Langerhans Contribute to Defective Insulin Secretion in Zucker Diabetic Rat, an Animal Model for Some Aspects of Human Type 2 Diabetes." *The European Journal of Neuroscience* 25 (1): 213–23. https://doi.org/10.1111/j.1460-9568.2006.05261.x.

Hernández-Pérez, Talía, María Del Rocío Gómez-García, María Elena Valverde, and Octavio Paredes-López. 2020. "Capsicum Annuum (Hot Pepper): An Ancient Latin-American Crop with Outstanding Bioactive Compounds and Nutraceutical Potential. A Review." *Comprehensive Reviews in Food Science and Food Safety* 19 (6): 2972–93. https://doi.org/10.1111/1541-4337.12634.

Jiang, Kunxiu, Yantao Xing, Qinghua Quan, Qianqian Sun, Jingyun Tian, Cen Liu, Xingzhuo Song, Xirui Wang, and Yonggang Liu. 2020. "Synthesis and Biological Evaluation of N-Alkylamide Derivatives as Anti-Tumor Agents." *Journal of Traditional Chinese Medical Sciences* 7 (4): 393–403. https://doi.org/10.1016/j.jtcms.2020.09.004.

Jose, Anju, Anju Elizabath Babu, S. Prashant, and K.V. Athira. 2021. "Cardiovascular Disease Comorbidity in Major Depressive Disorder." In *Cardiovascular Disease Comorbidity in Major Depressive Disorder*, 225–36. Recent Advancements and Research in Biological Sciences. India: Thanuj International Publishers.

Kumar, Suresh, Jitpal Kamboj, Suman, and Sunil Sharma. 2011. "Overview for Various Aspects of the Health Benefits of Piper Longum Linn. Fruit." *Journal of Acupuncture and Meridian Studies* 4 (2): 134–40. https://doi.org/10.1016/S2005-2901(11)60020-4.

Lalthanpuii, Pawi Bawitlung, and Kholhring Lalchhandama. 2020. "Chemical Composition and Broad-Spectrum Anthelmintic Activity of a Cultivar of Toothache Plant, Acmella Oleracea, from Mizoram, India." *Pharmaceutical Biology* 58 (1): 393–99. https://doi.org/10.1080/13880209.2020.1760316.

Lalthanpuii, Pawi Bawitlung, Zar Zokimi, and Kholhring Lalchhandama. 2020. "Anthelmintic Activity of Praziquantel and Spilanthes Acmella Extract on an Intestinal Cestode Parasite." *Acta Pharmaceutica (Zagreb, Croatia)* 70 (4): 551–60. https://doi.org/10.2478/acph-2020-0039.

Manayi, Azadeh, Mahdi Vazirian, and Soodabeh Saeidnia. 2015. "Echinacea Purpurea: Pharmacology, Phytochemistry and Analysis Methods." *Pharmacognosy Reviews* 9 (17): 63–72. https://doi.org/10.4103/0973-7847.156353.

Mao, Chien Feng, Sabri Sudirman, Chi Chih Lee, David Tsou, and Zwe Ling Kong. 2021. "Echinacea Purpurea Ethanol Extract Improves Male Reproductive Dysfunction With Streptozotocin–Nicotinamide-Induced Diabetic Rats." *Frontiers in Veterinary Science* 8: 297. https://doi.org/10.3389/fvets.2021.651286.

Masi, Marco, Manoj Koirala, Antonella Delicato, Roberta Di Lecce, Natacha Merindol, Seydou Ka, Matar Seck, et al., 2021. "Isolation and Biological Characterization of Homoisoflavanoids and the Alkylamide N-p-Coumaroyltyramine from Crinum Biflorum Rottb., an Amaryllidaceae Species Collected in Senegal." *Biomolecules* 11 (9). https://doi.org/10.3390/biom11091298.

Matthias, Anita, Kerry G. Penman, Nick J. Matovic, Kerry M. Bone, James J. De Voss, and Reg P. Lehmann. 2005. "Bioavailability of Echinacea Constituents: Caco-2 Monolayers and Pharmacokinetics of the Alkylamides and Caffeic Acid Conjugates." *Molecules (Basel, Switzerland)* 10 (10): 1242–51. https://doi.org/10.3390/10101242.

Mbeunkui, Flaubert, Mary H. Grace, Carmen Lategan, Peter J. Smith, Ilya Raskin, and Mary Ann Lila. 2011. "Isolation and Identification of Antiplasmodial N-Alkylamides from Spilanthes Acmella Flowers Using Centrifugal Partition Chromatography and ESI-IT-TOF-MS." *Journal of Chromatography. B, Analytical Technologies in the Biomedical and Life Sciences* 879 (21): 1886–92. https://doi.org/10.1016/j.jchromb.2011.05.013.

Méndez-Bravo, Alfonso, Carlos Calderón-Vázquez, Enrique Ibarra-Laclette, Javier Raya-González, Enrique Ramírez-Chávez, Jorge Molina-Torres, Angel A Guevara-García, José López-Bucio, and Luis Herrera-Estrella. 2011. "Alkamides Activate Jasmonic Acid Biosynthesis and Signaling Pathways and Confer Resistance to Botrytis Cinerea in Arabidopsis Thaliana." *PloS One* 6 (11): e27251. https://doi.org/10.1371/journal.pone.0027251.

Merali, S., S. Binns, M. Paulin-Levasseur, C. Ficker, M. Smith, B. Baum, E. Brovelli, and J.T. Arnason. 2003. "Antifungal and Anti-Inflammatory Activity of the Genus Echinacea." *Pharmaceutical Biology* 41 (6): 412–20. https://doi.org/10.1076/phbi.41.6.412.17828.

Nomura, Ellen Cristine Ogata, Melissa Raboni Alves Rodrigues, Carla Francielle da Silva, Letícia Alencar Hamm, Adamara Machado Nascimento, Lauro Mera de Souza, Thales Ricardo Cipriani, Cristiane Hatsuko Baggio, and Maria Fernanda de Paula Werner. 2013. "Antinociceptive Effects of Ethanolic Extract from the Flowers of Acmella Oleracea (L.) R.K. Jansen in Mice." *Journal of Ethnopharmacology* 150 (2): 583–9. https://doi.org/10.1016/j.jep.2013.09.007.

Oh, Kyoung-Jin, Hye-Sook Han, Min-Jung Kim, and Seung-Hoi Koo. 2013. "Transcriptional Regulators of Hepatic Gluconeogenesis." *Archives of Pharmacal Research* 36 (2): 189–200. https://doi.org/10.1007/s12272-013-0018-5.

Prachayasittikul, Supaluk, Saowapa Suphapong, Apilak Worachartcheewan, Ratana Lawung, Somsak Ruchirawat, and Virapong Prachayasittikul. 2009a. "Bioactive Metabolites from Spilanthes Acmella Murr." *Molecules* 14 (2): 850–67. https://doi.org/10.3390/molecules14020850.

Ratnasooriya, W.D., and K.P.P. Pieris. 2005. "Attenuation of Persistent Pain and Hyperalgesia by Spilanthus Acmella Flowers in Rats." *Pharmaceutical Biology* 43 (7): 614–19. https://doi.org/10.1080/13880200500301944.

Rajendran, Radhika, Balaji Sitthu Narashimman, Vishal Trivedi, and Rakhi Chaturvedi. 2017. "Isolation and Quantification of Antimalarial N-Alkylamides from Flower-Head Derived in Vitro Callus Cultures of Spilanthes Paniculata." *Journal of Bioscience and Bioengineering* 124 (1): 99–107. https://doi.org/10.1016/j.jbiosc.2017.02.001.

Refardt, Julie, Bettina Winzeler, and Mirjam Christ-Crain. 2020. "Diabetes Insipidus: An Update." *Endocrinology and Metabolism Clinics of North America* 49 (3): 517–31. https://doi.org/10.1016/j.ecl.2020.05.012.

Ren, Tingyuan, Yuping Zhu, Xuejuan Xia, Yongbo Ding, Jing Guo, and Jianquan Kan. 2017. "Zanthoxylum Alkylamides Ameliorate Protein Metabolism Disorder in STZ-Induced Diabetic Rats." *Journal of Molecular Endocrinology* 58 (3): 113–25. https://doi.org/10.1530/JME-16-0218.

Rios, María Yolanda, and Horacio F. Olivo. 2014. "Natural and Synthetic Alkamides: Applications in Pain Therapy." *Studies in Natural Products Chemistry* 43: 79–121. (Elsevier). https://doi.org/10.1016/B978-0-444-63430-6.00003-5.

Savic, Sasa, Sanja Petrovic, Sanela Savic, and Nebojsa Cekic. 2021. "Identification and Photostability of N-Alkylamides from Acmella Oleracea Extract." *Journal of Pharmaceutical and Biomedical Analysis* 195 (February): 113819. https://doi.org/10.1016/j.jpba.2020.113819.

Sharma, Rahul, and Neelakantan Arumugam. 2021. "N-Alkylamides of Spilanthes (Syn: Acmella): Structure, Purification, Characterization, Biological Activities and Applications – a Review." *Future Foods* 3 (February): 100022. https://doi.org/10.1016/j.fufo.2021.100022.

Spelman, Kevin, Delphine Depoix, Megan McCray, Elisabeth Mouray, and Philippe Grellier. 2011. "The Traditional Medicine Spilanthes Acmella, and the Alkylamides Spilanthol and Undeca-2E-Ene-8,10-Diynoic Acid Isobutylamide, Demonstrate in Vitro and in Vivo Antimalarial Activity." *Phytotherapy Research : PTR* 25 (7): 1098–101. https://doi.org/10.1002/ptr.3395.

Spelman, Kevin, Matthew H. Wetschler, and Nadja B. Cech. 2009. "Comparison of Alkylamide Yield in Ethanolic Extracts Prepared from Fresh versus Dry Echinacea Purpurea Utilizing HPLC-ESI-MS." *Journal of Pharmaceutical and Biomedical Analysis* 49 (5): 1141–49. https://doi.org/10.1016/j.jpba.2009.02.011.

Stein, Renan, Markus Berger, Bianca Santana de Cecco, Luana Peixoto Mallmann, Paula Barros Terraciano, David Driemeier, Eliseu Rodrigues, Walter Orlando Beys-da-Silva, and Eduardo Luis Konrath. 2021. "Chymase Inhibition: A Key Factor in the Anti-Inflammatory Activity of Ethanolic Extracts and Spilanthol Isolated from Acmella Oleracea." *Journal of Ethnopharmacology* 270 (April): 113610. https://doi.org/10.1016/j.jep.2020.113610.

Sun, Li, Karamatollah A. Rezaei, Feral Temelli, and Buncha Ooraikul. 2002. "Supercritical Fluid Extraction of Alkylamides from Echinacea Angustifolia." *Journal of Agricultural and Food Chemistry* 50 (14): 3947–53. https://doi.org/10.1021/jf0200265.

Sut, Stefania, Irene Ferrarese, Shyam Sharan Shrestha, Gourav Kumar, Antonio Slaviero, Simone Sello, Adriano Altissimo, Luca Pagni, Francesco Gattesco, and Stefano Dall'Acqua. 2020. "Comparison of Biostimulant Treatments in Acmella Oleracea Cultivation for Alkylamides Production." *Plants (Basel, Switzerland)* 9 (7). https://doi.org/10.3390/plants9070818.

Veryser, Lieselotte. 2016. *Analytical, Pharmacokinetic and Regulatory Characterisation of Selected Plant N-Alkylamides. Public Defense*. Ghent University, 2016. https://lib.ugent.be/catalog/rug01:002323987.

Veryser, Lieselotte, Lien Taevernier, Tanmayee Joshi, Pratima Tatke, Evelien Wynendaele, Nathalie Bracke, Sofie Stalmans, et al., 2016. "Mucosal and Blood-Brain Barrier Transport Kinetics of the Plant N-Alkylamide Spilanthol Using in Vitro and in Vivo Models." *BMC Complementary and Alternative Medicine* 16: 177. https://doi.org/10.1186/s12906-016-1159-0.

Veryser, Lieselotte, Lien Taevernier, Evelien Wynendaele, Yannick Verheust, Ann Dumoulin, and Bart De Spiegeleer. 2017. "N-Alkylamide Profiling of Achillea Ptarmica and Achillea Millefolium Extracts by Liquid and Gas Chromatography-Mass Spectrometry." *Journal of Pharmaceutical Analysis* 7 (1): 34–47. https://doi.org/10.1016/j.jpha.2016.09.005.

Veryser, Lieselotte, Evelien Wynendaele, Lien Taevernier, Frederick Verbeke, Tanmayee Joshi, Pratima Tatke, and Bart de Spiegeleer. 2014. "N-Alkylamides: From Plant to Brain." *Functional Foods in Health and Disease* 4 (6): 264–75. https://doi.org/10.31989/ffhd.v4i6.6.

Woelkart, Karin, and Rudolf Bauer. 2007. "The Role of Alkamides as an Active Principle of Echinacea." *Planta Medica* 73 (7): 615–23. https://doi.org/10.1055/s-2007-981531.

Woelkart, Karin, Reginald F Frye, Hartmut Derendorf, Rudolf Bauer, and Veronika Butterweck. 2009. "Pharmacokinetics and Tissue Distribution of Dodeca-2E,4E,8E,10E/Z-Tetraenoic Acid Isobutylamides after Oral Administration in Rats." *Planta Medica* 75 (12): 1306–13. https://doi.org/10.1055/s-0029-1185631.

Wu, Y.C., G.Y. Chang, F.N. Ko, and C.M. Teng. 1995. "Bioactive Constitutents from the Stems of Annona Montana." *Planta Medica* 61 (2): 146–49. https://doi.org/10.1055/s-2006-958035.

Xiao, Xiao, Mei Xu, Chao Yang, Yao Yao, Li-Na Liang, Philip Ed Chung, Qun Long, et al., 2018. "Novel Racemosin B Derivatives as New Therapeutic Agents for Aggressive Breast Cancer." *Bioorganic & Medicinal Chemistry* 26 (23–24): 6096–104. https://doi.org/10.1016/j.bmc.2018.11.014.

Xu, Wenqian, Hongkang Zhu, Bin Hu, Yuliang Cheng, Yahui Guo, Weirong Yao, and He Qian. 2021. "Echinacea in Hepatopathy: A Review of Its Phytochemistry, Pharmacology, and Safety." *Phytomedicine : International Journal of Phytotherapy and Phytopharmacology* 87: 153572. https://doi.org/10.1016/j.phymed.2021.153572.

Ye, Ying, Yanan Song, Juhua Zhuang, Guoyu Wang, Jing Ni, and Wei Xia. 2019. "Anticancer Effects of Echinacoside in Hepatocellular Carcinoma Mouse Model and HepG2 Cells." *Journal of Cellular Physiology* 234 (2): 1880–88. https://doi.org/10.1002/jcp.27063.

Yolanda, Rios Gomez. 2012. "Natural Alkamides: Pharmacology, Chemistry and Distribution." In *Drug Discovery Research in Pharmacognosy*, 107–44. I-Tech Education and Publishing.

You, Yuming, Ting Ren, Shiqi Zhang, Gerald Gasper Shirima, Ya Jiao Cheng, and Xiong Liu. 2015. "Hypoglycemic Effects of Zanthoxylum Alkylamides by Enhancing Glucose Metabolism and Ameliorating Pancreatic Dysfunction in Streptozotocin-Induced Diabetic Rats." *Food and Function* 6 (9): 3144–54. https://doi.org/10.1039/c5fo00432b.

Zhang, Lu-Lu, Lei Zhao, Hou-Yin Wang, Bo-Lin Shi, Long-Yun Liu, and Zhong-Xiu Chen. 2019. "The Relationship between Alkylamide Compound Content and Pungency Intensity of Zanthoxylum Bungeanum Based on Sensory Evaluation and Ultra-Performance Liquid Chromatography-Mass Spectrometry/Mass Spectrometry (UPLC-MS/MS) Analysis." *Journal of the Science of Food and Agriculture* 99 (4): 1475–83. https://doi.org/10.1002/jsfa.9319.

Zimmermann, Stefanie, Semira Thomi, Marcel Kaiser, Matthias Hamburger, and Michael Adams. 2012. "Screening and HPLC-Based Activity Profiling for New Antiprotozoal Leads from European Plants." *Scientia Pharmaceutica* 80 (1): 205–13. https://doi.org/10.3797/scipharm.1111–13.

Section III

Karrikins in Plant Biology

15 Karrikins
Molecular Association with Strigolactones: Role in Plant Growth Regulation and Quorum Sensing Modulation

Atif Khurshid Wani, Nahid Akhtar, Ajit Prakash, Sourbh Suren Garg, Tahir ul Gani Mir, and Rattandeep Singh

CONTENTS

15.0 INTRODUCTION

Angiosperms start their new phase of plant growth and development with the germination of seeds. Since dormancy of seeds prevents germination, it is considered essential for the survival of plants, particularly during unfavourable conditions. Dormancy is broken by several endogenous and environmental signals inducing SG. Phytohormones are key in SG and plant development. Gibberellins (GA) induce SG by breaking seed dormancy, whereas abscisic acid (ABA) promotes dormancy and hampers germination (Oracz and Stawska, 2016; Shu et al., 2013). Other than phytohormones, other compounds like reactive oxygen species (ROS) and nitrogen oxide also act as plant growth regulators (PGRs) and control development (Bethke et al., 2006; Oracz et al., 2009). Kks are PGRs found in burnt plant material (BPM). Kks as PGRs stimulate plant development and SG by mimicking SLs (Flematti et al., 2015). SL phytohormones increase the growth of symbiotic arbuscular mycorrhizal (AM) fungi, which results in the enhancement of plant growth and plant branching (Aroca et al., 2013). Smoke emerging from bushfires or wildfires had been associated with SG until butanolide karrikinolide was found to be responsible in 2004 (Nelson et al., 2012). After that, multiple compounds have been characterized in smoke, collectively termed Kks. These compounds were extracted by liquid fractionation and subsequently tested for their role in SG. Bioassays identified Kks designated Kk1, Kk2, Kk6 – Kk1 being the first discovered Kk and also the most active, along with Kk4 (Nelson et al., 2012; Hrdlička et al., 2019). These are formed by the combustion of carbohydrates, particularly cellulose. Upon the burning of plants, the carbohydrates present in BPM convert into Kks. It is believed

DOI: 10.1201/9781003224525-19

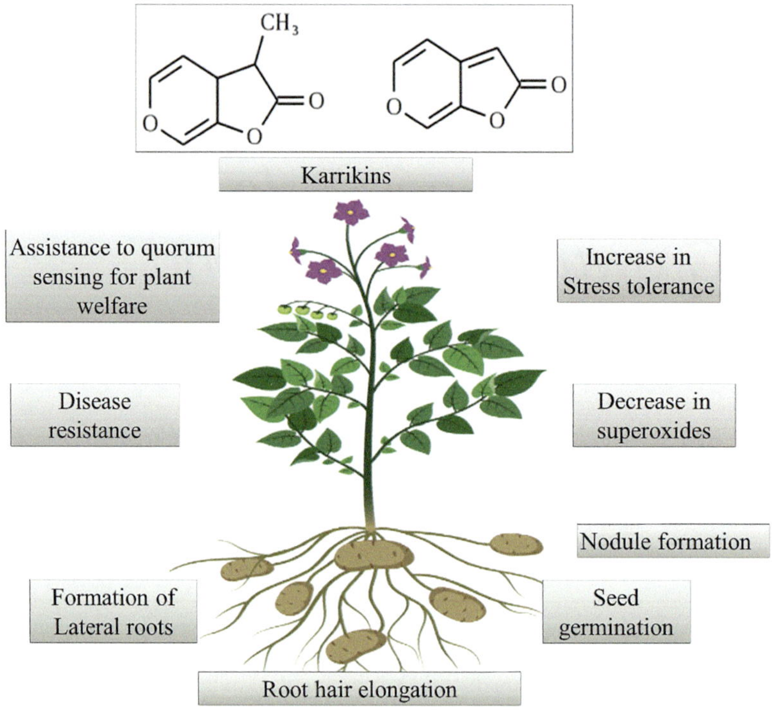

FIGURE 15.1 Summary Diagram

that Kk pyran (C5H6O) moiety originates from pyranose sugar. Upon plant burning, Kks diffuses into the air and then deposit on the soil surface after rainfall to induce SG (Flematti et al., 2011). Importantly, plantlike foothill deervetch (*Acmispon bracycarpus*), goldback fern (*Pentagramma triangularis*), California broom (*Acmispon glaber*), and coffee fern (*Pellaea andromedifolia*), collectively known as fire followers, are not able to germinate in the absence of Kks. They require rain to germinate after fires, thus they may remain dormant for a long time until a fire occurs followed by rain at a suitable time. Kk action has largely been studied through genetic reservoirs of *Arabidopsis thaliana* (Flematti et al., 2004; Lamont et al., 2019). Kk recognition by Arabidopsis is dependent on α/β-fold hydrolase known as Kk-INSENSITIVE-2 (KAI2) (De Cuyper et al., 2017). It also depends on the F-box protein in Arabidopsis called MORE AXILLARY GROWTH-2 (MAX2), useful in SL signaling. Kk-signaling requires SUPPRESSOR OF MORE AXILLARY GROWTH-1 (SMAX1), a DWARF53 protein homolog essential for SL signaling in rice. The response of Arabidopsis to Kk1 and Kk2 signals has been shown by various studies. MAX2 and KAI2 are vital in understanding the activity of Kks. Kks influence the seedling photomorphogenesis, giving rise to short hypocotyls and large cotyledons (Nelson et al., 2011; Waters et al., 2012; Zhang et al., 2020). Generally, KAI2 is found in algae and mosses, whereas DWARF14 is believed to have evolved with seeds due to KA12 duplication (He et al., 2012). This chapter digs deep into the signaling network of Kks in several plant parts for their potential role in plant embryogenesis and morphogenesis, besides highlighting their role in modulating crosstalk between plants and the associated bacterial population (Figure 15.1).

15.1 KKS AND SLS: A MOLECULAR TIE-UP

Kks and SLs as signaling molecules have a high degree of similarity. Both consist of a butanolide ring, which is essential for its activity, and both take part in SG (Cook et al., 1966; Umehara et al., 2008). Since Kk synthesis in living cells has not been reported, they aren't true hormones.

They act like hormones in signal transduction and reception, so they are sometimes described as hormones (Janssen and Snowden, 2012). Since Kks are active SG stimulants, their activity is to revive plant growth after fires (Antala et al., 2019). Kk-insensitive mutants like MAX2 and KAI2 in Arabidopsis have been reported to have longer hypocotyls, hooked epinastic cotyledons, and reduced germination efficiency (Nelson et al., 2011; Waters et al., 2012). It is imperative to note that, unlike MAX2, the activity of KAI2 is confined to the early seedling development (SD) stage. Taking the significance of Kks in SG into account, the signaling system has always been a focal point in this field. SMXL family members mediate the signaling pathways between SLs and Kks (Stanga et al., 2013). Upon treatment with SLs, SMXL2 and D14 show an interaction. And then D14 also shows interaction with MAX2. On the other hand, in treatment with Kks, SMXL2 interacts with KAI2 instead of showing interaction with D14 (Lei Wang et al., 2020). Thus, KI2 shows the interactions toward SMXL2. As soon as the interactions are complete, they all form a complex among themselves: i.e., SMXL2, KAI2, MAX2 and SMXL2, D14, MAX2. As of now, who mediates the interaction between MAX2 and KAI2 through Kks is not known (Li et al., 2017). But it has been found that the addition of GR24*ent*-5DS enhances the association between MAX2 and KAI2 (Conn and Nelson, 2015). Thereafter, SMXL2 undergoes ubiquitination followed by degradation (Lei Wang et al., 2015). It leads to suppression in the functioning of a certain gene such as DLK2 and KUF1. Repression in the functioning of these genes results in the inhibition of elongation of hypocotyls and shoots branching (Waters and Smith, 2013). KAI2, a Kk signaling receptor, when bound by Kks undergoes a series of conformational changes (Guo et al., 2013). This results in the formation of the Skp1-cullin1-F-box (SCF) E3 ligase complex with MAX2. The SCF complex promotes the SMAX1 degradation, which acts as a repressor in Kks' signaling pathway (Kagiyama et al., 2013). One of the studies has also reported SMAX1-LIKEs as repressors in Kk signaling (Stanga et al., 2013). The involvement of repressors in Kk signaling pathways most likely diversifies its biological functioning. Both SL and Kk molecules depend on α/β hydrolase (KAI2 or D14/DAD2) and LRR (leucine-rich repeat) type F-box for signal transduction. Figure 15.2 illustrates the molecular tie-up between SL and Kk signaling (Hamiaux et al., 2012).

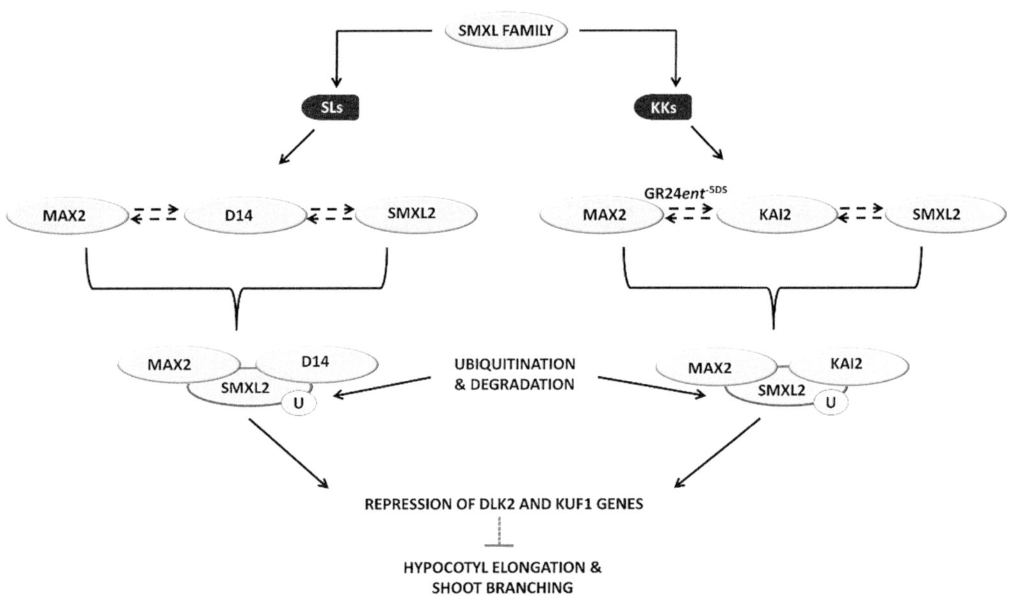

FIGURE 15.2 Molecular tie-up between strigolactones (SLs) and karrikins (Kks).

15.2 ROLE OF KARRIKINS IN PLANT GROWTH AND DEVELOPMENT

Smoke stimulates SG in several plant species through Kk signaling (Table 15.1). Many studies have reported an increase in SG of monocots and dicots, including perennials, annuals, weeds, horticulture crops, and woody plants. A study of the smoke derived from more than 25 plant species showed varying results in the SG of *Themeda triandra*. The Kk quantification through liquid

TABLE 15.1
Role of Karrikins (Kks) on the Growth and Development of Plants

Plant	Effect of Kks	References
Tomato (*Lycopersicon esculentum*)	Increase in the SV indices and seedling weight (SW) during extreme temperatures.	Neeru Jain et al. (2006)
Maize (*Zea mays*)	Improvement in shoot height, number of leaves, and survival percentage.	van Staden et al. (2006)
Okra (*Abelmoschus esculentus*)	Significant increase in absolute growth rate, shoot/root length, total leaf area, and stem thickness.	Kulkarni et al. (2007)
Thale cress (*Arabidopsis thaliana*)	SG stimulation under appropriate conditions. Hypocotyl elongation repression in low light.	Waters et al. (2017)
Arabidopsis thaliana	Elicits proteolysis and ubiquitination of SMXL2 for hypocotyl elongation regulation.	Lei Wang et al. (2020)
Amaranth (*Amaranthus hybridus*)	Increase in chlorophyll, protein, and carbohydrate content using various biostimulants, including Kks.	Ngoroyemoto et al. (2019)
Arabidopsis thaliana	Enhances light responses during SG and SD.	Nelson et al. (2010)
Witchweed (*Striga hermonthica*)	Initiation of germination of the dormant seeds by Kk interaction with ShKAI2ib receptor.	Zhang et al. (2020)
Rice (Oryza sativa)	Regulation of mesocotyl elongation in darkness through *OsSMAX1* overexpression.	Zheng et al. (2020)
Kangaroo paw (*Anigozanthos flavidus*), Corkybark (*Gyrostemon ramulosus*) and Net veined gyrostemon (*Gyrostemon racemiger*).	All three plants were stimulated to germinate by smoke water/seed starter, but none of these responded to Kk1.	Downes et al. (2013)
Teff (*Eragrostis tef*)	Improved seedling length (SL) and vigor index. Increased percentage imbibition from higher osmotic potential to lower.	Ghebrehiwot et al. (2008)
Cabbage (*Brassica oleracea*)	Alleviates cadmium stress by improving antioxidant defense mechanism, photosynthetic parameters, and plant growth.	Shah et al. (2020)
Kale (*Brassica oleracea*)	Kk1 positively augmented the growth and some physiochemical properties in post-germination process. Reduced the negative effect of Cd and temperature stress.	A. Ahmad et al. (2021)
Arabidopsis thaliana	Stimulates dormant SG and SD through post-fire chemical cues.	Kochanek et al. (2016)

Plant	Effect of Kks	References
Soybean (*Glycine max*)	Kk and SL signaling mediate soyabean-rhizobia interaction.	M. Z. Ahmad et al. (2020)
Red sage (*Salvia miltiorrhiza*)	Stimulates tanshinones biosynthesis.	Zhou et al. (2019)
Arabidopsis	Regulation of carbohydrate metabolism, protein transport, redox homeostasis, and photosynthesis.	Baldrianová et al. (2015)
Chinese broccoli *(Brassica alboglabra)*	Alleviates Cd and heat stressor by modulating biochemical attributes and osmoregulators.	A. Ahmad et al. (2021)
Wheat (*Triticum aestivum*)	Ameliorates saline stress and improves biochemical, physiological, and growth responses.	Shabir et al. (2021)
Carrot (*Daucus carota*)	Promotes SG and reduces auxin stress.	Asaf et al. (2014)
Green needlegrass (*Nassella viridula*), Altai wildrye (Leymus angustus), Kura clover (Trifolium ambiguum), and Orchard grass (Dactylis glomerata)	SG improvement and increase in biomass production.	Abu et al. (2016)

chromatography and mass spectrometry revealed significant differences in the Kk levels in total as well as relative (Baxter et al., 1995; Hrdlička et al., 2019). Kks are also useful through biochar by improving the nutrient- and water-holding capacity of the soil (Kochanek et al., 2016). Arabidopsis seeds are Kk sensitive, which helps them quickly overcome the dormancy (Lu Wang et al., 2018). Kk2 is effective in hypocotyl elongation inhibition and germination stimulation in Arabidopsis (Nelson et al., 2010). SG of *Brassica tournefortii*, a weed, is also stimulated by Kk1 (Sun et al., 2020). Wang et al. investigated KAI2's role in Arabidopsis seeds in response to abiotic stress and reported that Kk-KAI2 signaling provides stress tolerance and inhibits germination under unfavourable seedling conditions (Wang et al., 2018). Shah *et al.* identified a higher mortality rate of Arabidopsis mutants and concluded that the overexpression of the Kk-receptor gene *Sapium sebiferum* (bioenergy plant) KAI2 increases the cold stress tolerance through the regulation of redox homeostasis in *A. thaliana* (Shah et al., 2021). In another study carried out by Carbonnel *et al.*, the Kk-signaling regulator SMAX1 in *Lotus japonicus* controls root hair development by inhibiting ethylene biosynthesis (Carbonnel et al., 2020). Besides improving plant height, stem thickness, weight, and stress tolerance, Kks also affect secondary metabolism (Li et al., 2017). KAI2 Arabidopsis mutants generally have less anthocyanin content due to transcription misregulation. The beta-carotene and ascorbic acid content were found to increase due to Kk-primed seeds of carrot roots. The roots of carrots grown from Kk-soaked seeds were found to be heavier and bigger with more pigmentation than plants used as control (Akeel et al., 2019).

15.2.1 Seed Germination (SG), Seedling Development (SD), and Seedling Vigor (SV)

SG and early SDs are crucial during the life cycle of a plant. These processes are highly regulated by several factors like phytohormones, light, nutrient availability, etc. Kks promote SG under slightly fewer light conditions and induce seed response to light. However, the Kk-acceleration effect on germination fades during darkness (Nelson et al., 2010). Fresh seeds of *Arabidopsis* are not sensitive to Kks, but once they are treated with ripening agents, the sensitivity toward Kks increases (Waters et al., 2014). The effect of Kks on SG has been reported to be dependent on ecotype and seed dormancy depth in *Arabidopsis* (Nelson et al., 2009). Keeley and Fotheringham reported the

TABLE 15.2
Role of Quorum-Sensing (QS) Molecules Like C6 (N-hexanoyl-L-HSL), C8 (N-octanoyl-L-HSL), C10 (N-decanoyl-L-HSL), C12 (N-dodecanoyl-HSL), and C14 (N-tetradecanoyl-HSL) in Plant Growth and Protection

Plant/Plant Part	QS Molecule	Role	References
Barley (*Hordeum vulgare*)	Oxo-C8 and Oxo-C10	Root development and systemic resistance	Sieper et al. (2014)
H. vulgare	Oxo-C6 and Oxo-C8	Modulation of plant growth, root architecture, accumulation of nitric oxide (NO), and membrane hyperpolarization leading to K^+ uptake	Rankl et al. (2016))
Yam bean (*Pachyrhizus erosus*), and *H. vulgare*	Oxo-C6, Oxo-C8, and Oxo-C10	Decrease in superoxide dismutase in roots of barley; plant growth and pigmentation showed lesser response to applied AHLs	Götz-Rösch et al. (2015)
Arabidopsis thaliana	Oxo-C14	Confers pathogenic resistance, enhances callose deposition, cell wall lignification, and accumulation of phenolic compounds	Schenk et al. (2014)
Cucumber (*Cucumis sativus*)	Oxo-C14	Defense against *Pseudoperonospora cubensis* and modulation of plant growth	Pazarlar et al. (2020)
A. Thaliana and P. erosus	Oxo-C14	Offers resistance toward hemibiotrophic and biotrophic pathogens through AtMPK activation	Schikora et al. (2011)
A. Thaliana	Oxo-C8	Resistance against *Pseudomonas syringae*	Liu et al. (2020)
A. thalina	Oxo-C6 and Oxo-C8	Root growth through GPCR candidates such as *Cand2* and *Cand7*.	Jin et al. (2012)
T. aestivum and A. thaliana	Oxo-C8, Oxo-C10, and Oxo-C12	N-AHL-depleted rhizobium radiobacter mutant shows reduced growth and reduced resistance activities	Alabid et al. (2020)
H. vulgare	Oxo-C10	Improves plant-bacteria interaction for root growth and conferring resistance	Han et al. (2016)
O. sativa	HSLs	Growth promotion	Cai et al. (2020)
G.max	Isovaletyl-HSL	Induces Bjal (LuxI homolog) expression	Lindemann et al. (2011)
T. aestivum	Oxo-C6	Priming wheat seed with C6-HSL improves plant growth and seed yield	Moshynets et al. (2019)
A. thaliana	Oxo-C6	Root elongation	von Rad et al. (2008)

effectiveness of smoke in inducing *California chaparral*'s SG (Keeley and Fotheringham, 1998). Gupta *et al.* studied the role of smoke on the SG of *Lactuca sativa* and demonstrated the promotion of lettuce SG by Kk1 by alleviating the ABA levels and enhancing the hydrolytic enzyme activity (Gupta et al., 2019). Elsadek and Yousef studied the effect of smoke water on horticultural crops and reported the enhancement of SG and SD (Elsadek and Yousef, 2019). Several other studies have focused on the role of Kks on seeds and associated networks through phytohormone crosstalk. Meng *et al.* reported a delay in soybean SG by regulating GA and ABA biogenesis during shaded conditions. They reported that during seed imbibition, the Kk promoted ABA biogenesis, whereas GA biosynthesis was impaired (Meng et al., 2016). Sparg *et al.* highlighted the application of Kks and smoke solutions on SV and SG on *Merwilla natalensis*, *Tulbaghia violacea*, and *Albuca pachychlamys*. *T. violacea* and *A. pachychlamys* exhibited higher SV when exposed to smoke rich in Kks than non-smoked seeds. However, a negligible effect was observed in the case of *M. natalensis* seedlings (Sparg et al., 2005).

15.2.2 Abiotic Stress Tolerance

Abiotic stresses are the main hindrances in crop productivity. The stresses, such as salinity and drought, could cause several morphological, physiological, and anatomical complications in plants (Danquah et al., 2014). To maintain growth during stress conditions, plants generally depend on hormones like ABA, and GA, which regulate downstream stress responses (Egamberdieva et al., 2017). Besides playing roles in SG, SD, and SV, Kks have been reported to reduce the impact of abiotic stresses through redox homeostasis (Table 15.1). Shah *et al.* reported improvement in salt and osmotic stress tolerance through redox homeostasis in *Sapium sebeiferum* by Kk1 supplementation. They observed that one nanomolar of Kk1 increased lateral roots and taproot length and improved seedling biomass under stress conditions (Shah et al., 2020). In one of the recent studies, the overexpression of KAI2, a Kk receptor gene, was reported to promote cold stress tolerance in *Arabidopsis thaliana* through the redox homeostasis regulation (Shah et al., 2021). In tomatoes, a Kk-primed seed produces more vigorous seedlings than non-primed or water-primed seeds under salinity, osmoticum, or temperature stresses (N. Jain and Van, 2007). KAI2 has been reported to be involved in anthocyanin biosynthesis, stomatal closure, membrane integrity, and regulation of cuticle development, all contributing to an increase in stress tolerance (Li et al., 2017). Wang et al. reported that the Kk-KAI2 signaling network provides stress tolerance in Arabidopsis by germination inhibition under drastic conditions (Lu Wang et al., 2018).

15.3 KARRIKINS AND QUORUM SENSING (QS)

QS molecules ensure communication among bacterial populations (Abisado et al., 2018). The plant has also evolved to recognize and respond to these molecules (Fuller et al., 2017). N-acyl homoserine lactones (HSLs), the QS molecules, have been widely examined for their impact on different plants. Molecules of AHLs consist of an acyl chain and a lactone ring, ranging from 8 to 18 carbons (Schuster et al., 2013). Even though the exact mechanism involved in perceiving these signaling molecules is not clear, the role of AHL in root development, gene expression, and proteomes has been reported. Mathesius *et al.* were the first to report the impact of bacterial AHL on plants (Mathesius et al., 2003). *Arabidopsis* seedlings have shown variations in protein accumulation involved in the metabolism of carbon in response to the AHL (Miao et al., 2012). The proteomic analysis has also shown large variations in *Medicago truncatula*'s physiological activities, including hormone metabolism, oxidative stress, and flavonoid synthesis in the rhizosphere when subjected to oxo-C16-HSL and oxo-C12-HSL (members of AHL family) (Ding et al., 2016; Veliz-Vallejos et al., 2014; Wani, Akhtar, Singh, et al. 2022). Shrestha *et al.* studied the impact of QS molecules on PG and the immune system and reported that triple, quadruple, and double combinations with long-chained AHLs improve the plant's resistance to a pathogen *Pseudomonas syringae* (Shrestha et al., 2020). AHLs with shorter acyl chains such as C4 or C6 have been observed to improve root elongation and growth rate (Bai et al., 2012). Ortiz-Castro et al. evaluated the growth response of *Arabidopsis thaliana* to different AHL molecules in post-embryonic root development. They reported that the compounds affected root formation, root growth, and development of root hair with C10-HL as the most active AHL in modulating root architecture (Ortíz-Castro et al., 2009). The ubiquitous association between plants and bacteria and structural similarities between Kks and QS molecules have led to the evaluation of their role in plant development. Some studies have suggested that Kks are utilized by plants to manipulate some members of the phytomicrobiome to their usefulness in a post-fire environment to provide them protection and nutrition. The plants recognize QS molecules such as oxo-C12-HSL of *Pseudomonas aeruginosa*, known to affect the gene expression in plants (Shiner et al., 2005; Zhu et al., 2000). Mandabi *et al.* also reported modulation of bacterial QS by Kks from plant smoke. The modulation was studied in *Agrobacterium tumefaciens, P. aeruginosa*, and *Vibrio harveyi* with varying outcomes on plant development, SD, SV, and plant protection (Mandabi et al., 2014).

15.4 WILDFIRES AND MICROBIAL NETWORKS: A LESSER-KNOWN PLANT NEXUS

Over the years, the significance of the plant microbiome as a plant health determinant has become increasingly evident (Babalola et al., 2020). The responsive nature of the plant-associated microbiome toward wildfires or the fire-emerging compounds has also been reported in several studies (Köster et al., 2021; Wani, Akhtar, Sher, et al., 2022). Since wildfires are becoming increasingly prevalent in severity, size, and frequency, understanding their impact on vegetation is imperative (Xu et al., 2020). Plant microbiome confers pest resistance (Gressel, 2018) and nutrition (Rodriguez et al., 2019; Akhtar et al., 2022) and promotes plant growth (Berg et al., 2014); however, fluctuations in plant-associated microbial diversity post-fire is still less well known. *Populus tremuloides* (Aspen), a deciduous native American tree, largely depends on fire to regenerate (Rogers et al., 2020). The regeneration process is driven by vegetative reproduction through root suckering. Characterizing the post-fire microbiome of Aspen will ensure the proper management strategies to promote sustainability. Since the soil microbiome is a major reservoir for endospheric microorganisms dwelling aboveground and belowground (Fernández-González et al., 2019), it is hypothesized that plants that regenerate in burnt sites post-fire will have microbiomes different from the ones growing in unburnt sites. The studies of the impact of wildfire phytomicrobiome are largely confined to the rhizosphere and particularly focus on mycorrhizal symbionts rather than the whole microbiome (Abarenkov et al., 2010). It has been established that mycorrhizal colonization of the rhizosphere is negatively affected by fire in terms of colonization rate and diversity. One meta-analysis report shows that fire reduces mycorrhizal colonization by more than 20%, which reduces the chances of survival for the plant post-fire (Dove and Hart, 2017). On the other hand, fire has also been reported to select growth-promoting bacteria in *Quercus ilex* (Holm oak) rhizosphere (Fernández-González et al., 2019). Metagenomic analyses have shown enrichment of some functional genes involved in nitrogen cycling and nutrient uptake, thus playing its role in plant health (Cobo-Díaz et al., 2015; Wani, Akhtar, Naqash, et al., 2022; Wani, Roy, et al., 2022). Dove *et al.* demonstrated the impact of fire on phytomicrobiome diversity and assembly. There is a dire need to study the impact of fires at the molecular level of plants with a more comprehensive and in-depth impact of the emerging compounds like Kks on plant fitness (Dove et al., 2021).

15.5 PERSPECTIVE: THE WAY FORWARD

Plants that respond to Kk signaling have significant potential to address greater challenges in agriculture by improving nutrient uptake, increasing yields, improving stress tolerance, and fighting parasitic weeds. Hormones and their analogues are potential manipulators in the horticulture, viticulture, and agriculture sector. The PGR market was about 5 billion USD in 2017, and it is expected to double by 2025 (Chesterfield et al., 2020). Auxin (2,4-Dichlorophenoxyacetic acid), a selective herbicide, destroys dicots weeds without having an impact on monocots like crop cereals and lawns (Song, 2014). Cytokines slow down the aging of leaves, flowers, and fruits and improve their shelf life (Miransari and Smith, 2014). GA increases the fruit size and quality and slows down the ripening process, with the global market for GA and its inhibitors reaching 500 million USD annually (Rademacher, 2017). Although IAA and ABA can act synergistically to induce seed dormancy while GA and Kks can improve SG, there is no evidence of whether Kks affects signaling pathways during the SG process. The molecular mechanism underlying the regulation of endogenous plant hormones by Kks during SG is also unclear. Similar in action to SLs, Kks are expected to have applications in agronomy. Even though several studies about their usefulness have been carried out on different plants, the molecular detail of the signaling system remains obscure. There is an utmost need to study the pre- and post-fire changes in the plants that are positively as well as negatively affected to develop a strategy to minimize its negative effects. The similarities to SLs, both in action and structure, also offer Kks an edge because SL synthesis is not realistic for large-scale production.

15.6 CONCLUSION

Kks are simple PGRs that emerge from burning plants, affecting morphological and physiological features of different plant species. They resemble SLs in structure, action, and signaling pathways. Arabidopsis, the most Kk-responsive species, has enabled us to study and understand Kk perception through its signaling cascade. Kks diffuse into the air after plant burning, and then subsequently, Kks settle on the soil surface to induce SG, GD, SV, etc. The Kk perception by Arabidopsis depends largely on KA12, which needs an amino acid triad for its action. Since the response of a model plant like Arabidopsis may vary from those of the horticultural crops, more studies focusing on field crops are required to discover the advanced application of Kks and the potential of plants to carry out the conversion of Kks to different compounds before interacting with KAI2. This will also help in understanding the differences in the response to Kks by different plant species.

REFERENCES

Abarenkov, Kessy, R. Henrik Nilsson, Karl-Henrik Larsson, Ian J. Alexander, Ursula Eberhardt, Susanne Erland, Klaus Høiland, et al. 2010. "The UNITE Database for Molecular Identification of Fungi–Recent Updates and Future Perspectives." *The New Phytologist* 186 (2): 281–85. https://doi.org/10.1111/j.1469-8137.2009.03160.x.

Abisado, Rhea G., Saida Benomar, Jennifer R. Klaus, Ajai A. Dandekar, and Josephine R. Chandler. 2018. "Bacterial Quorum Sensing and Microbial Community Interactions." *MBio* 9 (3): e02331–17. https://doi.org/10.1128/mBio.02331-17.

Abu, Y., J.T. Romo, Y. Bai, and B. Coulman. 2016. "Priming Seeds in Aqueous Smoke Solutions to Improve Seed Germination and Biomass Production of Perennial Forage Species." *Canadian Journal of Plant Science* 96 (4): 551–63. https://doi.org/10.1139/cjps-2015-0229.

Ahmad, Aqeel, Iqra Shahzadi, Samavia Mubeen, Nasim Ahmad Yasin, Waheed Akram, Waheed Ullah Khan, and Tingquan Wu. 2021. "Karrikinolide Alleviates BDE-28, Heat and Cd Stressors in Brassica Alboglabra by Correlating and Modulating Biochemical Attributes, Antioxidative Machinery and Osmoregulators." *Ecotoxicology and Environmental Safety* 213 (April): 112047. https://doi.org/10.1016/j.ecoenv.2021.112047.

Ahmad, Muhammad Zulfiqar, Naveed Ur Rehman, Shuwei Yu, Yuanze Zhou, Basir Ul Haq, Junjie Wang, Penghui Li, Zhixiong Zeng, and Jian Zhao. 2020. "GmMAX2-D14 and -KAI Interaction-Mediated SL and KAR Signaling Play Essential Roles in Soybean Root Nodulation." *The Plant Journal: For Cell and Molecular Biology* 101 (2): 334–51. https://doi.org/10.1111/tpj.14545.

Akeel, Arshiya, M. Masroor Akhtar Khan, Hassan Jaleel, and Moin Uddin. 2019. "Smoke-Saturated Water and Karrikinolide Modulate Germination, Growth, Photosynthesis and Nutritional Values of Carrot (Daucus Carota L.)." *Journal of Plant Growth Regulation* 38 (4): 1387–401. https://doi.org/10.1007/s00344-019-09941-w.

Akhtar, Nahid, Atif Khurshid Wani, Daljeet Singh Dhanjal, and Soumya Mukherjee. 2022. "Insights into the Beneficial Roles of Dark Septate Endophytes in Plants under Challenging Environment: Resilience to Biotic and Abiotic Stresses." *World Journal of Microbiology and Biotechnology* 38 (5): 79. https://doi.org/10.1007/s11274-022-03264-x.

Alabid, Ibrahim, Martin Hardt, Jafargholi Imani, Anton Hartmann, Michael Rothballer, Dan Li, Jenny Uhl, Philippe Schmitt-Kopplin, Stefanie Glaeser, and Karl-Heinz Kogel. 2020. "The N-Acyl Homoserine-Lactone Depleted Rhizobium Radiobacter Mutant RrF4NM13 Shows Reduced Growth-Promoting and Resistance-Inducing Activities in Mono- and Dicotyledonous Plants." *Journal of Plant Diseases and Protection* 127 (6): 769–81. https://doi.org/10.1007/s41348-020-00360-8.

Antala, Michal, Oksana Sytar, Anshu Rastogi, and Marian Brestic. 2019. "Potential of Karrikins as Novel Plant Growth Regulators in Agriculture." *Plants (Basel, Switzerland)* 9 (1): E43. https://doi.org/10.3390/plants9010043.

Aroca, Ricardo, Juan Manuel Ruiz-Lozano, Angel María Zamarreño, José Antonio Paz, José María García-Mina, María José Pozo, and Juan Antonio López-Ráez. 2013. "Arbuscular Mycorrhizal Symbiosis Influences Strigolactone Production under Salinity and Alleviates Salt Stress in Lettuce Plants." *Journal of Plant Physiology* 170 (1): 47–55. https://doi.org/10.1016/j.jplph.2012.08.020.

Asaf, Sajjad, Q.M. Imran, Rahmatullah Jan, Lubna, Amana Khatoon, Jung HeeYoung, and Shafiq-ur-Rehman. 2014. "Plant Derived Smoke Promotes Seed Germination and Alleviates Auxin Stress in Carrot." *Journal of Agricultural and Biological Science* 9 (9): 308–14.

Babalola, Olubukola Oluranti, Ayomide E. Fadiji, Ben J. Enagbonma, Elizabeth T. Alori, Modupe S. Ayilara, and Ayansina S. Ayangbenro. 2020. *The Nexus Between Plant and Plant Microbiome: Revelation of the Networking Strategies. Frontiers in Microbiology*. Vol. 11. Frontiers Media S.A. https://doi.org/10.3389/fmicb.2020.548037.

Bai, Xuegui, Christopher D. Todd, Radhika Desikan, Yongping Yang, and Xiangyang Hu. 2012. "N-3-Oxo-Decanoyl-L-Homoserine-Lactone Activates Auxin-Induced Adventitious Root Formation via Hydrogen Peroxide- and Nitric Oxide-Dependent Cyclic GMP Signaling in Mung Bean." *Plant Physiology* 158 (2): 725–36. https://doi.org/10.1104/pp.111.185769.

Baldrianová, Jana, Martin Černý, Jan Novák, Petr L. Jedelský, Eva Divíšková, and Břetislav Brzobohatý. 2015. "Arabidopsis Proteome Responses to the Smoke-Derived Growth Regulator Karrikin." *Journal of Proteomics* 120 (April): 7–20. https://doi.org/10.1016/j.jprot.2015.02.011.

Baxter, B.J.M., J.E. Granger, and J. van Staden. 1995. "Plant-Derived Smoke and Seed Germination: Is All Smoke Good Smoke? That Is the Burning Question." *South African Journal of Botany* 61 (5): 275–77. https://doi.org/10.1016/S0254-6299(15)30536-6.

Berg, Gabriele, Martin Grube, Michael Schloter, and Kornelia Smalla. 2014. *The Plant Microbiome and Its Importance for Plant and Human Health. Frontiers in Microbiology*. Vol. 5. SEP. Frontiers Research Foundation. https://doi.org/10.3389/fmicb.2014.00491.

Bethke, Paul C., Igor G.L. Libourel, and Russell L. Jones. 2006. "Nitric Oxide Reduces Seed Dormancy in Arabidopsis." *Journal of Experimental Botany* 57 (3): 517–26. https://doi.org/10.1093/jxb/erj060.

Cai, Wenjie, Fuwen Ou, Christian Staehelin, and Weijun Dai. 2020. "Bradyrhizobium Sp. Strain ORS278 Promotes Rice Growth and Its Quorum Sensing System Is Required for Optimal Root Colonization." *Environmental Microbiology Reports* 12 (6): 656–66. https://doi.org/10.1111/1758-2229.12885.

Carbonnel, Samy, Salar Torabi, Maximilian Griesmann, Elias Bleek, Yuhong Tang, Stefan Buchka, Veronica Basso, et al. 2020. "Lotus Japonicus Karrikin Receptors Display Divergent Ligand-Binding Specificities and Organ-Dependent Redundancy." *PLoS Genetics* 16 (12): e1009249. https://doi.org/10.1371/journal.pgen.1009249.

Chesterfield, Rebecca J., Claudia E. Vickers, and Christine A. Beveridge. 2020. "Translation of Strigolactones from Plant Hormone to Agriculture: Achievements, Future Perspectives, and Challenges." *Trends in Plant Science* 25 (11): 1087–106. https://doi.org/10.1016/j.tplants.2020.06.005.

Cobo-Díaz, José F., Antonio J. Fernández-González, Pablo J. Villadas, Ana B. Robles, Nicolás Toro, and Manuel Fernández-López. 2015. "Metagenomic Assessment of the Potential Microbial Nitrogen Pathways in the Rhizosphere of a Mediterranean Forest after a Wildfire." *Microbial Ecology* 69 (4): 895–904. https://doi.org/10.1007/s00248-015-0586-7.

Conn, Caitlin E., and David C. Nelson. 2015. "Evidence That KARRIKIN-INSENSITIVE2 (KAI2) Receptors May Perceive an Unknown Signal That Is Not Karrikin or Strigolactone." *Frontiers in Plant Science* 6: 1219. https://doi.org/10.3389/fpls.2015.01219.

Cook, C.E., L.P. Whichard, B. Turner, M.E. Wall, and G.H. Egley. 1966. "Germination of Witchweed (Striga Lutea Lour.): Isolation and Properties of a Potent Stimulant." *Science (New York, N.Y.)* 154 (3753): 1189–90. https://doi.org/10.1126/science.154.3753.1189.

Danquah, Agyemang, Axel de Zelicourt, Jean Colcombet, and Heribert Hirt. 2014. "The Role of ABA and MAPK Signaling Pathways in Plant Abiotic Stress Responses." *Biotechnology Advances* 32 (1): 40–52. https://doi.org/10.1016/j.biotechadv.2013.09.006.

De Cuyper, Carolien, Sylwia Struk, Lukas Braem, Kris Gevaert, Geert De Jaeger, and Sofie Goormachtig. 2017. "Strigolactones, Karrikins and Beyond." *Plant, Cell & Environment* 40 (9): 1691–703. https://doi.org/10.1111/pce.12996.

Ding, Lina, Jun Cao, Yunfei Duan, Jun Li, Yang Yang, Guoxing Yang, and Yang Zhou. 2016. "Proteomic and Physiological Responses of Arabidopsis Thaliana Exposed to Salinity Stress and N-Acyl-Homoserine Lactone." *Physiologia Plantarum* 158 (4): 414–34. https://doi.org/10.1111/ppl.12476.

Dove, Nicholas C., and Stephen C. Hart. 2017. "Fire Reduces Fungal Species Richness and In Situ Mycorrhizal Colonization: A Meta-Analysis." *Fire Ecology* 13 (2): 37–65. https://doi.org/10.4996/fireecology.130237746.

Dove, Nicholas C., Dawn M. Klingeman, Alyssa A. Carrell, Melissa A. Cregger, and Christopher W. Schadt. 2021. "Fire Alters Plant Microbiome Assembly Patterns: Integrating the Plant and Soil Microbial Response to Disturbance." *New Phytologist* 230 (6): 2433–46. https://doi.org/10.1111/nph.17248.

Downes, Katherine S., Marnie E. Light, Martin Pošta, Ladislav Kohout, and Johannes van Staden. 2013. "Comparison of Germination Responses of Anigozanthos Flavidus (Haemodoraceae), Gyrostemon Racemiger and Gyrostemon Ramulosus (Gyrostemonaceae) to Smoke-Water and the Smoke-Derived Compounds Karrikinolide (KAR_1) and Glyceronitrile." *Annals of Botany* 111 (3): 489–97. https://doi.org/10.1093/aob/mcs300.

Egamberdieva, Dilfuza, Stephan J. Wirth, Abdulaziz A. Alqarawi, Elsayed F. Abd-Allah, and Abeer Hashem. 2017. *Phytohormones and Beneficial Microbes: Essential Components for Plants to Balance Stress and Fitness. Frontiers in Microbiology*. Vol. 8. OCT. Frontiers Media S.A. https://doi.org/10.3389/fmicb.2017.02104.

Elsadek, Mohamed A., and Eltohamy A.A. Yousef. 2019. "Smoke-Water Enhances Germination and Seedling Growth of Four Horticultural Crops." *Plants* 8 (4): 104. https://doi.org/10.3390/plants8040104.

Fernández-González, Antonio J., Pablo J. Villadas, Carmen Gómez-Lama Cabanás, Antonio Valverde-Corredor, Angjelina Belaj, Jesús Mercado-Blanco, and Manuel Fernández-López. 2019. "Defining the Root Endosphere and Rhizosphere Microbiomes from the World Olive Germplasm Collection." *Scientific Reports* 9 (1): 1–13. https://doi.org/10.1038/s41598-019-56977-9.

Flematti, Gavin R., Kingsley W. Dixon, and Steven M. Smith. 2015. "What Are Karrikins and How Were They 'Discovered' by Plants?" *BMC Biology* 13 (December): 108. https://doi.org/10.1186/s12915-015-0219-0.

Flematti, Gavin R., Emilio L. Ghisalberti, Kingsley W. Dixon, and Robert D. Trengove. 2004. "A Compound from Smoke That Promotes Seed Germination." *Science (New York, N.Y.)* 305 (5686): 977. https://doi.org/10.1126/science.1099944.

Flematti, Gavin R., Adrian Scaffidi, Kingsley W. Dixon, Steven M. Smith, and Emilio L. Ghisalberti. 2011. "Production of the Seed Germination Stimulant Karrikinolide from Combustion of Simple Carbohydrates." *Journal of Agricultural and Food Chemistry* 59 (4): 1195–98. https://doi.org/10.1021/jf1041728.

Fuller, Alexandra W., Phoebe Young, B. Daniel Pierce, Jamie Kitson-Finuff, Purvi Jain, Karl Schneider, Stephen Lazar, Olga Taran, Andrew G. Palmer, and David G. Lynn. 2017. "Redox-Mediated Quorum Sensing in Plants." *PloS One* 12 (9): e0182655. https://doi.org/10.1371/journal.pone.0182655.

Ghebrehiwot, H.M., M.G. Kulkarni, K.P. Kirkman, and J. Van Staden. 2008. "Smoke-Water and a Smoke-Isolated Butenolide Improve Germination and Seedling Vigour of Eragrostis Tef (Zucc.) Trotter under High Temperature and Low Osmotic Potential." *Journal of Agronomy and Crop Science* 194 (4). https://doi.org/10.1111/j.1439-037x.2008.00321.x.

Götz-Rösch, Christine, Tina Sieper, Agnes Fekete, Philippe Schmitt-Kopplin, Anton Hartmann, and Peter Schröder. 2015. "Influence of Bacterial N-Acyl-Homoserine Lactones on Growth Parameters, Pigments, Antioxidative Capacities and the Xenobiotic Phase II Detoxification Enzymes in Barley and Yam Bean." *Frontiers in Plant Science* 6: 205. https://doi.org/10.3389/fpls.2015.00205.

Gressel, Jonathan. 2018. "Microbiome Facilitated Pest Resistance: Potential Problems and Uses." *Pest Management Science* 74 (3): 511–15. https://doi.org/10.1002/ps.4777.

Guo, Yongxia, Zuyu Zheng, James J. La Clair, Joanne Chory, and Joseph P. Noel. 2013. "Smoke-Derived Karrikin Perception by the α/β-Hydrolase KAI2 from Arabidopsis." *Proceedings of the National Academy of Sciences of the United States of America* 110 (20): 8284–89. https://doi.org/10.1073/pnas.1306265110.

Gupta, Shubhpriya, Lenka Plačková, Manoj G. Kulkarni, Karel Doležal, and Johannes Van Staden. 2019. "Role of Smoke Stimulatory and Inhibitory Biomolecules in Phytochrome-Regulated Seed Germination of Lactuca Sativa1[OPEN]." *Plant Physiology* 181 (2): 458–70. https://doi.org/10.1104/pp.19.00575.

Hamiaux, Cyril, Revel S.M. Drummond, Bart J. Janssen, Susan E. Ledger, Janine M. Cooney, Richard D. Newcomb, and Kimberley C. Snowden. 2012. "DAD2 Is an α/β Hydrolase Likely to Be Involved in the Perception of the Plant Branching Hormone, Strigolactone." *Current Biology: CB* 22 (21): 2032–36. https://doi.org/10.1016/j.cub.2012.08.007.

Han, Shengcai, Dan Li, Eva Trost, Klaus F. Mayer, A. Corina Vlot, Werner Heller, Michael Schmid, Anton Hartmann, and Michael Rothballer. 2016. "Systemic Responses of Barley to the 3-Hydroxy-Decanoyl-Homoserine Lactone Producing Plant Beneficial Endophyte Acidovorax Radicis N35." *Frontiers in Plant Science* 7: 1868. https://doi.org/10.3389/fpls.2016.01868.

He, Tianhua, Juli G. Pausas, Claire M. Belcher, Dylan W. Schwilk, and Byron B. Lamont. 2012. "Fire-Adapted Traits of Pinus Arose in the Fiery Cretaceous." *The New Phytologist* 194 (3): 751–59. https://doi.org/10.1111/j.1469-8137.2012.04079.x.

Hrdlička, Jakub, Tomáš Gucký, Ondřej Novák, Manoj Kulkarni, Shubhpriya Gupta, Johannes van Staden, and Karel Doležal. 2019. "Quantification of Karrikins in Smoke Water Using Ultra-High Performance Liquid Chromatography-Tandem Mass Spectrometry." *Plant Methods* 15: 81. https://doi.org/10.1186/s13007-019-0467-z.

Jain, Neeru, Manoj G. Kulkarni, and Johannes van Staden. 2006. "A Butenolide, Isolated from Smoke, Can Overcome the Detrimental Effects of Extreme Temperatures during Tomato Seed Germination." *Plant Growth Regulation* 49 (2): 263–67. https://doi.org/10.1007/s10725-006-9136-0.

Jain, N., and Staden J. Van. 2007. "The Potential of the Smoke-Derived Compound 3-Methyl-2H-Furo[2,3-c] Pyran-2-One as a Priming Agent for Tomato Seeds." *Seed Science Research* 17 (3): 175–81. https://doi.org/10.1017/s0960258507785896.

Janssen, Bart J., and Kimberley C. Snowden. 2012. "Strigolactone and Karrikin Signal Perception: Receptors, Enzymes, or Both?" *Frontiers in Plant Science* 3: 296. https://doi.org/10.3389/fpls.2012.00296.

Jin, Guoping, Fang Liu, Hong Ma, Shaoyan Hao, Qian Zhao, Zirui Bian, Zhenhua Jia, and Shuishan Song. 2012. "Two G-Protein-Coupled-Receptor Candidates, Cand2 and Cand7, Are Involved in Arabidopsis Root Growth Mediated by the Bacterial Quorum-Sensing Signals N-Acyl-Homoserine Lactones." *Biochemical and Biophysical Research Communications* 417 (3): 991–95. https://doi.org/10.1016/j.bbrc.2011.12.066.

Kagiyama, Megumi, Yoshinori Hirano, Tomoyuki Mori, Sun-Yong Kim, Junko Kyozuka, Yoshiya Seto, Shinjiro Yamaguchi, and Toshio Hakoshima. 2013. "Structures of D14 and D14L in the Strigolactone and Karrikin Signaling Pathways." *Genes to Cells: Devoted to Molecular & Cellular Mechanisms* 18 (2): 147–60. https://doi.org/10.1111/gtc.12025.

Keeley, Jon E., and C.J. Fotheringham. 1998. "SMOKE-INDUCED SEED GERMINATION IN CALIFORNIA CHAPARRAL." *Ecology* 79 (7): 2320–36. https://doi.org/10.1890/0012-9658(1998)079[2320:SISGIC]2.0.CO;2.

Kochanek, Jitka, Rowena L. Long, Allan T. Lisle, and Gavin R. Flematti. 2016. "Karrikins Identified in Biochars Indicate Post-Fire Chemical Cues Can Influence Community Diversity and Plant Development." *PloS One* 11 (8): e0161234. https://doi.org/10.1371/journal.pone.0161234.

Köster, Kajar, Heidi Aaltonen, Frank Berninger, Jussi Heinonsalo, Egle Köster, Caius Ribeiro-Kumara, Hui Sun, Leho Tedersoo, Xuan Zhou, and Jukka Pumpanen. 2021. "Impacts of Wildfire on Soil Microbiome in Boreal Environments." *Current Opinion in Environmental Science & Health* 22 (August): 100258. https://doi.org/10.1016/j.coesh.2021.100258.

Kulkarni, Manoj G., Glendon D. Ascough, and Johannes Van Staden. 2007. "Effects of Foliar Applications of Smoke-Water and a Smoke-Isolated Butenolide on Seedling Growth of Okra and Tomato." *HortScience* 42 (1): 179–82. https://doi.org/10.21273/HORTSCI.42.1.179.

Lamont, Byron B., Tianhua He, and Zhaogui Yan. 2019. "Evolutionary History of Fire-Stimulated Resprouting, Flowering, Seed Release and Germination." *Biological Reviews of the Cambridge Philosophical Society* 94 (3): 903–28. https://doi.org/10.1111/brv.12483.

Li, Weiqiang, Kien Huu Nguyen, Ha Duc Chu, Chien Van Ha, Yasuko Watanabe, Yuriko Osakabe, Marco Antonio Leyva-González, et al. 2017. "The Karrikin Receptor KAI2 Promotes Drought Resistance in Arabidopsis Thaliana." *PLoS Genetics* 13 (11): e1007076. https://doi.org/10.1371/journal.pgen.1007076.

Lindemann, Andrea, Gabriella Pessi, Amy L. Schaefer, Margrith E. Mattmann, Quin H. Christensen, Aline Kessler, Hauke Hennecke, Helen E. Blackwell, E. Peter Greenberg, and Caroline S. Harwood. 2011. "Isovaleryl-Homoserine Lactone, an Unusual Branched-Chain Quorum-Sensing Signal from the Soybean Symbiont Bradyrhizobium Japonicum." *Proceedings of the National Academy of Sciences of the United States of America* 108 (40): 16765–70. https://doi.org/10.1073/pnas.1114125108.

Liu, Fang, Qian Zhao, Zhenhua Jia, Cong Song, Yali Huang, Hong Ma, and Shuishan Song. 2020. "N-3-Oxo-Octanoyl-Homoserine Lactone-Mediated Priming of Resistance to Pseudomonas Syringae Requires the Salicylic Acid Signaling Pathway in Arabidopsis Thaliana." *BMC Plant Biology* 20 (1): 38. https://doi.org/10.1186/s12870-019-2228-6.

Mandabi, Aviad, Hadas Ganin, Pnina Krief, Josep Rayo, and Michael M. Meijler. 2014. "Karrikins from Plant Smoke Modulate Bacterial Quorum Sensing." *Chemical Communications* 50 (40): 5322–25. https://doi.org/10.1039/C3CC47501H.

Mathesius, Ulrike, Susan Mulders, Mengsheng Gao, Max Teplitski, Gustavo Caetano-Anolles, Barry G. Rolfe, and Wolfgang D. Bauer. 2003. "Extensive and Specific Responses of a Eukaryote to Bacterial Quorum-Sensing Signals." *Proceedings of the National Academy of Sciences of the United States of America* 100 (3): 1444–49. https://doi.org/10.1073/pnas.262672599.

Meng, Yongjie, Feng Chen, Haiwei Shuai, Xiaofeng Luo, Jun Ding, Shengwen Tang, Shuanshuan Xu, et al. 2016. "Karrikins Delay Soybean Seed Germination by Mediating Abscisic Acid and Gibberellin Biogenesis under Shaded Conditions." *Scientific Reports* 6 (1): 22073. https://doi.org/10.1038/srep22073.

Miao, Chunjuan, Fang Liu, Qian Zhao, Zhenhua Jia, and Shuishan Song. 2012. "A Proteomic Analysis of Arabidopsis Thaliana Seedling Responses to 3-Oxo-Octanoyl-Homoserine Lactone, a Bacterial Quorum-Sensing Signal." *Biochemical and Biophysical Research Communications* 427 (2): 293–98. https://doi.org/10.1016/j.bbrc.2012.09.044.

Miransari, Mohammad, and D.L. Smith. 2014. "Plant Hormones and Seed Germination." *Environmental and Experimental Botany* 99 (March): 110–21. https://doi.org/10.1016/j.envexpbot.2013.11.005.

Moshynets, Olena V., Lidia M. Babenko, Sergiy P. Rogalsky, Olga S. Iungin, Jessica Foster, Iryna V. Kosakivska, Geert Potters, and Andrew J. Spiers. 2019. "Priming Winter Wheat Seeds with the Bacterial Quorum Sensing Signal N-Hexanoyl-L-Homoserine Lactone (C6-HSL) Shows Potential to Improve Plant Growth and Seed Yield." *PloS One* 14 (2): e0209460. https://doi.org/10.1371/journal.pone.0209460.

Nelson, David C., Gavin R. Flematti, Emilio L. Ghisalberti, Kingsley W. Dixon, and Steven M. Smith. 2012. "Regulation of Seed Germination and Seedling Growth by Chemical Signals from Burning Vegetation." *Annual Review of Plant Biology* 63: 107–30. https://doi.org/10.1146/annurev-arplant-042811-105545.

Nelson, David C., Gavin R. Flematti, Julie-Anne Riseborough, Emilio L. Ghisalberti, Kingsley W. Dixon, and Steven M. Smith. 2010. "Karrikins Enhance Light Responses during Germination and Seedling Development in Arabidopsis Thaliana." *Proceedings of the National Academy of Sciences of the United States of America* 107 (15): 7095–100. https://doi.org/10.1073/pnas.0911635107.

Nelson, David C., Julie-Anne Riseborough, Gavin R. Flematti, Jason Stevens, Emilio L. Ghisalberti, Kingsley W. Dixon, and Steven M. Smith. 2009. "Karrikins Discovered in Smoke Trigger Arabidopsis Seed Germination by a Mechanism Requiring Gibberellic Acid Synthesis and Light." *Plant Physiology* 149 (2): 863–73. https://doi.org/10.1104/pp.108.131516.

Nelson, David C., Adrian Scaffidi, Elizabeth A. Dun, Mark T. Waters, Gavin R. Flematti, Kingsley W. Dixon, Christine A. Beveridge, Emilio L. Ghisalberti, and Steven M. Smith. 2011. "F-Box Protein MAX2 Has Dual Roles in Karrikin and Strigolactone Signaling in Arabidopsis Thaliana." *Proceedings of the National Academy of Sciences of the United States of America* 108 (21): 8897–902. https://doi.org/10.1073/pnas.1100987108.

Ngoroyemoto, N., S. Gupta, M.G. Kulkarni, J.F. Finnie, and J. Van Staden. 2019. "Effect of Organic Biostimulants on the Growth and Biochemical Composition of Amaranthus Hybridus L." *South African Journal of Botany* 124 (August): 87–93. https://doi.org/10.1016/j.sajb.2019.03.040.

Oracz, Krystyna, Hayat El-Maarouf-Bouteau, Ilse Kranner, Renata Bogatek, Françoise Corbineau, and Christophe Bailly. 2009. "The Mechanisms Involved in Seed Dormancy Alleviation by Hydrogen Cyanide Unravel the Role of Reactive Oxygen Species as Key Factors of Cellular Signaling during Germination." *Plant Physiology* 150 (1): 494–505. https://doi.org/10.1104/pp.109.138107.

Oracz, Krystyna, and Marlena Stawska. 2016. "Cellular Recycling of Proteins in Seed Dormancy Alleviation and Germination." *Frontiers in Plant Science* 7: 1128. https://doi.org/10.3389/fpls.2016.01128.

Ortíz-Castro, Randy, Hexon Angel Contreras-Cornejo, Lourdes Macías-Rodríguez, and José López-Bucio. 2009. "The Role of Microbial Signals in Plant Growth and Development." *Plant Signaling & Behavior* 4 (8): 701–12.

Pazarlar, Sercan, Nedim Cetinkaya, Melike Bor, and Recep Serdar Kara. 2020. "N-Acyl Homoserine Lactone-Mediated Modulation of Plant Growth and Defense against Pseudoperonospora Cubensis in Cucumber." *Journal of Experimental Botany* 71 (20): 6638–54. https://doi.org/10.1093/jxb/eraa384.

Rad, Uta von, Ilona Klein, Petre I. Dobrev, Jana Kottova, Eva Zazimalova, Agnes Fekete, Anton Hartmann, Philippe Schmitt-Kopplin, and Jörg Durner. 2008. "Response of Arabidopsis Thaliana to N-Hexanoyl-DL-Homoserine-Lactone, a Bacterial Quorum Sensing Molecule Produced in the Rhizosphere." *Planta* 229 (1): 73–85. https://doi.org/10.1007/s00425-008-0811-4.

Rademacher, Wilhelm. 2017. "Chemical Regulators of Gibberellin Status and Their Application in Plant Production." In *Annual Plant Reviews Online*, 359–403. American Cancer Society. https://doi.org/10.1002/9781119312994.apr0541.

Rankl, Simone, Benet Gunsé, Tina Sieper, Christoph Schmid, Charlotte Poschenrieder, and Peter Schröder. 2016. "Microbial Homoserine Lactones (AHLs) Are Effectors of Root Morphological Changes in Barley." *Plant Science: An International Journal of Experimental Plant Biology* 253 (December): 130–40. https://doi.org/10.1016/j.plantsci.2016.09.014.

Rodriguez, P.A., M. Rothballer, S.P. Chowdhury, T. Nussbaumer, C. Gutjahr, and P. Falter-Braun. 2019. "Systems Biology of Plant-Microbiome Interactions." *Mole Plant* 12: 804–21. https://doi.org/10.1016/j.molp.2019.05.006.

Rogers, Paul C., Bradley D. Pinno, Jan Šebesta, Benedicte R. Albrectsen, Guoqing Li, Natalya Ivanova, Antonín Kusbach, et al. 2020. "A Global View of Aspen: Conservation Science for Widespread Keystone Systems." *Global Ecology and Conservation* 21 (March): e00828. https://doi.org/10.1016/j.gecco.2019.e00828.

Schenk, Sebastian T., Casandra Hernández-Reyes, Birgit Samans, Elke Stein, Christina Neumann, Marek Schikora, Michael Reichelt, et al. 2014. "N-Acyl-Homoserine Lactone Primes Plants for Cell Wall Reinforcement and Induces Resistance to Bacterial Pathogens via the Salicylic Acid/Oxylipin Pathway." *The Plant Cell* 26 (6): 2708–23. https://doi.org/10.1105/tpc.114.126763.

Schikora, Adam, Sebastian T. Schenk, Elke Stein, Alexandra Molitor, Alga Zuccaro, and Karl-Heinz Kogel. 2011. "N-Acyl-Homoserine Lactone Confers Resistance toward Biotrophic and Hemibiotrophic Pathogens via Altered Activation of AtMPK6." *Plant Physiology* 157 (3): 1407–18. https://doi.org/10.1104/pp.111.180604.

Schuster, Martin, D. Joseph Sexton, Stephen P. Diggle, and E. Peter Greenberg. 2013. "Acyl-Homoserine Lactone Quorum Sensing: From Evolution to Application." *Annual Review of Microbiology* 67: 43–63. https://doi.org/10.1146/annurev-micro-092412-155635.

Shabir, Sumera, Noshin Ilyas, Saira Asif, Muhammad Iqbal, Sidra Kanwal, and Zeshan Ali. 2021. "Deciphering the Role of Plant-Derived Smoke Solution in Ameliorating Saline Stress and Improving Physiological, Biochemical, and Growth Responses of Wheat." *Journal of Plant Growth Regulation* (August). https://doi.org/10.1007/s00344-021-10473-5.

Shah, Faheem Afzal, Jun Ni, Yuanyuan Yao, Hao Hu, Ruyue Wei, and Lifang Wu. 2021. "Overexpression of Karrikins Receptor Gene Sapium Sebiferum KAI2 Promotes the Cold Stress Tolerance via Regulating the Redox Homeostasis in Arabidopsis Thaliana." *Frontiers in Plant Science* 12: 657960. https://doi.org/10.3389/fpls.2021.657960.

Shah, Faheem Afzal, Xiao Wei, Qiaojian Wang, Wenbo Liu, Dongdong Wang, Yuanyuan Yao, Hao Hu, et al. 2020. "Karrikin Improves Osmotic and Salt Stress Tolerance via the Regulation of the Redox Homeostasis in the Oil Plant Sapium Sebiferum." *Frontiers in Plant Science* 11: 216. https://doi.org/10.3389/fpls.2020.00216.

Shiner, Erin K., Kendra P. Rumbaugh, and Simon C. Williams. 2005. "Inter-Kingdom Signaling: Deciphering the Language of Acyl Homoserine Lactones." *FEMS Microbiology Reviews* 29 (5): 935–47. https://doi.org/10.1016/j.femsre.2005.03.001.

Shrestha, Abhishek, Maja Grimm, Ichie Ojiro, Johannes Krumwiede, and Adam Schikora. 2020. "Impact of Quorum Sensing Molecules on Plant Growth and Immune System." *Frontiers in Microbiology* 11 (July): 1545. https://doi.org/10.3389/fmicb.2020.01545.

Shu, Kai, Huawei Zhang, Shengfu Wang, Mingluan Chen, Yaorong Wu, Sanyuan Tang, Chunyan Liu, Yuqi Feng, Xiaofeng Cao, and Qi Xie. 2013. "ABI4 Regulates Primary Seed Dormancy by Regulating the Biogenesis of Abscisic Acid and Gibberellins in Arabidopsis." *PLoS Genetics* 9 (6): e1003577. https://doi.org/10.1371/journal.pgen.1003577.

Sieper, Tina, Sandor Forczek, Miroslav Matucha, Petra Krämer, Anton Hartmann, and Peter Schröder. 2014. "N-Acyl-Homoserine Lactone Uptake and Systemic Transport in Barley Rest upon Active Parts of the Plant." *The New Phytologist* 201 (2): 545–55. https://doi.org/10.1111/nph.12519.

Song, Yaling. 2014. "Insight into the Mode of Action of 2,4-Dichlorophenoxyacetic Acid (2,4-D) as an Herbicide." *Journal of Integrative Plant Biology* 56 (2): 106–13. https://doi.org/10.1111/jipb.12131.

Sparg, S.G., M.G. Kulkarni, M.E. Light, and J. Van Staden. 2005. "Improving Seedling Vigour of Indigenous Medicinal Plants with Smoke." *Bioresource Technology* 96 (12): 1323–30. https://doi.org/10.1016/j.biortech.2004.11.015.

Staden, Johannes van, Shane G. Sparg, Manoj G. Kulkarni, and Marnie E. Light. 2006. "Post-Germination Effects of the Smoke-Derived Compound 3-Methyl-2H-Furo[2,3-c]Pyran-2-One, and Its Potential as a Preconditioning Agent." *Field Crops Research* 98 (2): 98–105. https://doi.org/10.1016/j.fcr.2005.12.007.

Stanga, John P., Steven M. Smith, Winslow R. Briggs, and David C. Nelson. 2013. "SUPPRESSOR OF MORE AXILLARY GROWTH2 1 Controls Seed Germination and Seedling Development in Arabidopsis." *Plant Physiology* 163 (1): 318–30. https://doi.org/10.1104/pp.113.221259.

Sun, Yueming Kelly, Jiaren Yao, Adrian Scaffidi, Kim T. Melville, Sabrina F. Davies, Charles S. Bond, Steven M. Smith, Gavin R. Flematti, and Mark T. Waters. 2020. "Divergent Receptor Proteins Confer Responses to Different Karrikins in Two Ephemeral Weeds." *Nature Communications* 11 (1): 1264. https://doi.org/10.1038/s41467-020-14991-w.

Umehara, Mikihisa, Atsushi Hanada, Satoko Yoshida, Kohki Akiyama, Tomotsugu Arite, Noriko Takeda-Kamiya, Hiroshi Magome, et al. 2008. "Inhibition of Shoot Branching by New Terpenoid Plant Hormones." *Nature* 455 (7210): 195–200. https://doi.org/10.1038/nature07272.

Veliz-Vallejos, Debora F., Giel E. van Noorden, Mengqi Yuan, and Ulrike Mathesius. 2014. "A Sinorhizobium Meliloti-Specific N-Acyl Homoserine Lactone Quorum-Sensing Signal Increases Nodule Numbers in Medicago Truncatula Independent of Autoregulation." *Frontiers in Plant Science* 5: 551. https://doi.org/10.3389/fpls.2014.00551.

Wang, Lei, Bing Wang, Liang Jiang, Xue Liu, Xilong Li, Zefu Lu, Xiangbing Meng, Yonghong Wang, Steven M. Smith, and Jiayang Li. 2015. "Strigolactone Signaling in Arabidopsis Regulates Shoot Development by Targeting D53-Like SMXL Repressor Proteins for Ubiquitination and Degradation." *The Plant Cell* 27 (11): 3128–42. https://doi.org/10.1105/tpc.15.00605.

Wang, Lei, Qian Xu, Hong Yu, Haiyan Ma, Xiaoqiang Li, Jun Yang, Jinfang Chu, et al. 2020. "Strigolactone and Karrikin Signaling Pathways Elicit Ubiquitination and Proteolysis of SMXL2 to Regulate Hypocotyl Elongation in Arabidopsis." *The Plant Cell* 32 (7): 2251–70. https://doi.org/10.1105/tpc.20.00140.

Wang, Lu, Mark T. Waters, and Steven M. Smith. 2018. "Karrikin-KAI2 Signalling Provides Arabidopsis Seeds with Tolerance to Abiotic Stress and Inhibits Germination under Conditions Unfavourable to Seedling Establishment." *The New Phytologist* 219 (2): 605–18. https://doi.org/10.1111/nph.15192.

Wani, Atif Khurshid, Nahid Akhtar, Nafiaah Naqash, Chirag Chopra, Reena Singh, Vineet Kumar, Sunil Kumar, Sikandar I. Mulla, and Juliana Heloisa Pinê Américo-Pinheiro. 2022. "Bioprospecting Culturable and Unculturable Microbial Consortia Through Metagenomics for Bioremediation." *Cleaner Chemical Engineering* (April), 100017. https://doi.org/10.1016/j.clce.2022.100017.

Wani, Atif Khurshid, Nahid Akhtar, Farooq Sher, Acacio Aparecido Navarrete, and Juliana Heloisa Pinê Américo-Pinheiro. 2022. "Microbial Adaptation to Different Environmental Conditions: Molecular Perspective of Evolved Genetic and Cellular Systems." *Archives of Microbiology* 204 (2): 144. https://doi.org/10.1007/s00203-022-02757-5.

Wani, Atif Khurshid, Nahid Akhtar, Reena Singh, Chirag Chopra, Prachi Kakade, Mahesh Borde, Jameel M. Al-Khayri, Penna Suprasanna, and Saurabha B. Zimare. 2022. "Prospects of Advanced Metagenomics and Meta-Omics in the Investigation of Phytomicrobiome to Forecast Beneficial and Pathogenic Response." *Molecular Biology Reports* (September). https://doi.org/10.1007/s11033-022-07936-7.

Wani, Atif Khurshid, Priyanka Roy, Vijay Kumar, and Tahir ul Gani Mir. 2022. "Metagenomics and Artificial Intelligence in the Context of Human Health." *Infection, Genetics and Evolution* (March), 105267. https://doi.org/10.1016/j.meegid.2022.105267.

Waters, Mark T., Gavin R. Flematti, and Steven M. Smith. 2017. "Assaying Germination and Seedling Responses of Arabidopsis to Karrikins." *Methods in Molecular Biology (Clifton, N.J.)* 1497: 29–36. https://doi.org/10.1007/978-1-4939-6469-7_4.

Waters, Mark T., David C. Nelson, Adrian Scaffidi, Gavin R. Flematti, Yueming K. Sun, Kingsley W. Dixon, and Steven M. Smith. 2012. "Specialisation within the DWARF14 Protein Family Confers Distinct Responses to Karrikins and Strigolactones in Arabidopsis." *Development (Cambridge, England)* 139 (7): 1285–95. https://doi.org/10.1242/dev.074567.

Waters, Mark T., Adrian Scaffidi, Yueming K. Sun, Gavin R. Flematti, and Steven M. Smith. 2014. "The Karrikin Response System of Arabidopsis." *The Plant Journal: For Cell and Molecular Biology* 79 (4): 623–31. https://doi.org/10.1111/tpj.12430.

Waters, Mark T., and Steven M. Smith. 2013. "KAI2- and MAX2-Mediated Responses to Karrikins and Strigolactones Are Largely Independent of HY5 in Arabidopsis Seedlings." *Molecular Plant* 6 (1): 63–75. https://doi.org/10.1093/mp/sss127.

Xu, Rongbin, Pei Yu, Michael J. Abramson, Fay H. Johnston, Jonathan M. Samet, Michelle L. Bell, Andy Haines, Kristie L. Ebi, Shanshan Li, and Yuming Guo. 2020. "Wildfires, Global Climate Change, and Human Health." *The New England Journal of Medicine* 383 (22): 2173–81. https://doi.org/10.1056/NEJMsr2028985.

Zhang, Ji-Long, Xiaoting Liu, and Hong-Xing Zhang. 2020. "Interaction Mechanism of the Germination Stimulants Karrikins and Their Receptor ShKAI2iB." *The Journal of Physical Chemistry. B* 124 (44): 9812–19. https://doi.org/10.1021/acs.jpcb.0c06734.

Zheng, Jianshu, Kai Hong, Longjun Zeng, Lei Wang, Shujing Kang, Minghao Qu, Jiarong Dai, et al. 2020. "Karrikin Signaling Acts Parallel to and Additively with Strigolactone Signaling to Regulate Rice Mesocotyl Elongation in Darkness." *The Plant Cell* 32 (9): 2780–805. https://doi.org/10.1105/tpc.20.00123.

Zhou, Jie, Zi-xin Xu, Hui Sun, and Lan-ping Guo. 2019. "Smoke-Isolated Karrikins Stimulated Tanshinones Biosynthesis in Salvia Miltiorrhiza through Endogenous Nitric Oxide and Jasmonic Acid." *Molecules* 24 (7): 1229. https://doi.org/10.3390/molecules24071229.

Zhu, J., P.M. Oger, B. Schrammeijer, P.J. Hooykaas, S.K. Farrand, and S.C. Winans. 2000. "The Bases of Crown Gall Tumorigenesis." *Journal of Bacteriology* 182 (14): 3885–95. https://doi.org/10.1128/JB.182.14.3885-3895.2000.

16 Induction of Dormancy Release in Agricultural Weed Seeds by Plant-Derived Smoke and Smoke-Derived Karrikin 1 (KAR_1)

A Relationship with Plant Hormones

Jan Kępczyński

CONTENTS

16.0 INTRODUCTION: THE STATE OF SEED DORMANCY AND SEED GERMINATION

After harvest, seeds can be dormant (a state called the primary dormancy) or quiescent (i.e., unable to complete germination due to unfavourable conditions). Dormancy induced during maturation means the inability to germinate despite seed viability and conditions (water, temperature, air, light) suitable for the process. Dormancy prevents germination under environmental conditions unfavourable for seedling development, which ensures the species' survival (Finkelstein et al., 2008). Dormancy can be absolute or relative. The absolute dormancy means

DOI: 10.1201/9781003224525-20

the seeds cannot complete germination under any circumstances, whereas the relative dormancy means germination may be completed only within a narrow range of conditions. For the dormant viable seeds to germinate, dormancy must be terminated by appropriate environmental factors or artificial treatments. Under natural conditions, seed dormancy can be released by various factors. Some plant species produce seeds that require variable periods of dry after-ripening, cold or warm stratification, or light. Dormancy release in seeds of many species can also occur as a result of treatment involving a chemical cue such as gibberellins (GAs), ethylene, brassinosteroids, cytokinins, strigolactones, nitric oxide (NO), or hydrogen cyanide (HCN) (Kępczyński and Kępczyńska, 1997; Arc et al., 2013a, 2013b; Corbineau et al., 2014). It is commonly accepted that dormancy status is mainly a result of a balance between the levels of abscisic acid (ABA) and GAs. These hormones are antagonistic with respect to regulating the biosynthetic pathway and signal transduction (Bewley et al., 2013). After their dormancy has been released, viable seeds are able to complete germination, provided conditions for germination are suitable. If the conditions for germination of nondormant seeds are unfavourable for too long, secondary dormancy may ensue. Germination is a process that begins with water being taken up by dry seeds and ends with protrusion of a radicle through the various structures surrounding it. There are three phases differing in the way the water is absorbed. Phase I (imbibition) involves a fast water absorption. During phase II (true germination associated with major metabolic events), the water content is constant or increases only slightly (Bewley et al., 2013). The first two phases can be achieved by both dormant and nondormant seeds. Phase III (growth) can be entered only by nondormant seeds that have completed germination by the radicle emergence.

16.1 THE DISCOVERY OF BIOLOGICAL ACTIVITY OF PLANT-DERIVED SMOKE

The evolution of land plants has made it possible for fires to occur, which, in turn, has enabled some plants to adapt to using fire to their advantage. For example, fire has been well documented as a seed germination stimulant in many species from fire-prone areas such as Southern California, Australia, and South Africa, thus enabling the post-fire regeneration of the vegetation (Light and Van Staden, 2004; Nelson et al., 2012). The beneficial effect of fire depends on various physical and chemical factors. The heat generated by a wildfire can remove seed dormancy by thermal scarification or by a direct stimulation of embryo germination. De Lange and Boucher (1990) were the first to observe that plant-derived smoke or smoke-saturated water (SW) is able to induce seed germination of the fynbos species *Audouinia capitata*. It should be mentioned that fire and smoke had already been used in traditional farming in Australia, India, and South Africa. The interesting information provided by De Lange and Boucher (1990) initiated numerous studies and resulted in publications on the influence of smoke on the germination of seeds of various species from both fire-prone areas and areas where fires do not occur. The role of smoke in seed germination and seedling growth *in vivo* (Brown et al., 1997; Dixon et al., 2009; Kulkarni et al., 2011; Nelson et al., 2012; Kępczyński, 2018; Soós et al., 2019) and *in vitro* (Kępczyński, 2020) was reviewed in several papers The seeds were treated with aerosol smoke or SW obtained by combustion of plant material and bubbling through the water. The discovery of SW being as effective as smoke has doubtless been very important as it led to improved experiments and better practical applications. SW was found to be very active after a substantial dilution with water. Importantly, smoke and SW proved biologically active when produced from leaves of monocotyledons, dicotyledons, and gymnosperms, not only from fire-prone areas. SW can be stored for several years without losing its activity (Brown and Van Staden, 1997). In addition, active smoke can be obtained by the combustion of filter paper, agar, and simple carbohydrates. So far, seeds or seedlings of over 1,300 plant species from about 120 families have been shown to be sensitive to smoke (Soós et al., 2019). Smoke and SW are able to break dormancy in seeds of plants from the major botanical regions of South Africa and the chaparral in the southwestern USA, as well as the kwonogan, Banksia woodlands and Jarrah Forest of southwest Western Australia. Interestingly, not only are the seeds of wild plants from fire-prone areas smoke sensitive but also the seeds of weeds (Kępczyński, 2018) and even vegetables and other crops (Kulkarni et al., 2011; Kępczyński, 2018). Smoke and SW have been used in practical applications to stimulate seed germination. The seeds were treated directly with aerosol smoke for a maximum of one hour and then sown or kept air dry until required (Dixon et al., 2009). Another approach involved soaking the seeds in SW for 6 to 24 hours, followed by drying them before sowing or storing until sowing. There are also examples of smoke or SW application to soil or artificial substrates and various potting media before sowing (Kulkarni et al., 2011). SW applied to seeds germinated in petri dishes or in soil was found to stimulate seedling development even when it did not affect germination. Likewise, foliar application of SW may improve seedling development. SW and smoke were applied to stimulate seed germination in a soil seed bank to restore many native plants, e.g., *Eucalyptus*. Moreover, smoke and SW were used in a newly rehabilitated bauxite mine pit. In addition, SW has been used to manufacture commercial products, such as Kirstenbosch Instant Smoke Plus Seed Primer, containing smoke extracts and other stimulants, and the seed starter Australian Smoke Water, developed by South African and Australian scientists, respectively (Light and Van Staden, 2004).

16.2 THE DISCOVERY OF KARRIKINS IN SMOKE

Since plant-derived smoke contains thousands of compounds, it has been difficult to identify a compound or compounds responsible for smoke's biological activity. As smoke was found to contain ammonia and ethylene (Nelson et al., 2012), these compounds were suspected to be responsible for the stimulatory effect of smoke on seeds. Although the compounds alone did release dormancy in smoke-sensitive seeds, there are also seeds that are sensitive to smoke only. NO, also present in smoke, was considered to be the factor contributing to the beneficial effect of smoke. However, some smoke-sensitive seeds failed to respond to NO, while an NO-specific scavenger did not limit the stimulatory effect of smoke. Moreover, a beneficial effect on seed germination was produced by an NO-free smoke or one containing a low NO concentration only. Thus, it was assumed that a specific signal or signals in smoke other than NO is/are responsible for the induction of germination. Although research on the role of smoke in seed germination and seedling development has been conducted since 1990, it was only in 2004 that a biologically active compound, initially named butanolide, was identified by two research teams, independently, in plant-derived smoke (Van Staden et al., 2004) and in smoke from burnt filter paper (Flematti et al., 2004). At present, the compound is known as karrikinolide or karrikin-1 (KAR_1), the name having been derived from the word *karrik*, meaning *smoke* in the language of the Australian aboriginal Noongar people. The KAR_1 structure comprises three elements only – C, H, and O, and contains a five-member butenolide ring fused to a six-member pyran ring (Figure 16.1). The compound can be produced by burning various materials such as cellulose or pyranose sugar (D-xylose). KAR_1 used in experiments was isolated and purified from SW obtained by burning *Themeda triandra* (Van Staden et al., 2004) or was synthesized. Several methods of KAR_1 synthesis from various compounds including D-xylose have been reported (Figure 16.2). The compound has been demonstrated to stimulate seed germination in numerous smoke-responsive plant species. Interestingly, a positive response to KAR_1 was observed not only in seeds from fire-prone areas but also in agricultural and horticultural plants and weeds from fire-free regions (Dixon et al., 2009; Light et al., 2009; Kępczyński, 2018; Banerjee et al., 2019; Sami et al., 2021). It turned out that a KAR_1 concentration of 6.7×10^{-7} M corresponds to an SW dilution of 1:100 (Dixon et al., 2009). The discovery of KAR_1 and its high activity and possibility of synthesis have created new opportunities for its use in agriculture, horticulture, biotechnology, and research, better than the use of smoke or SW (Dixon et al., 2009; Kępczyński, 2020; Kępczyński and Kępczyńska, Chapter 17 of this volume). KAR_1-containing biochars prepared from green waste or sugarcane by pyrolysis have been

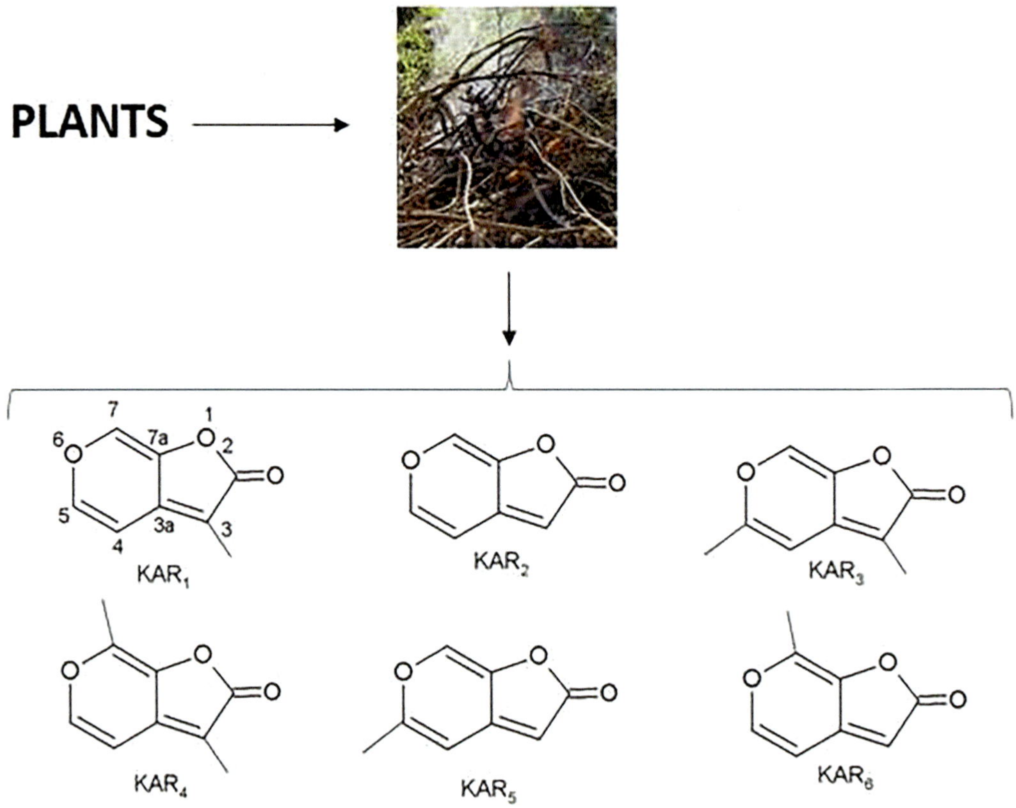

FIGURE 16.1 The structure of karrikins.

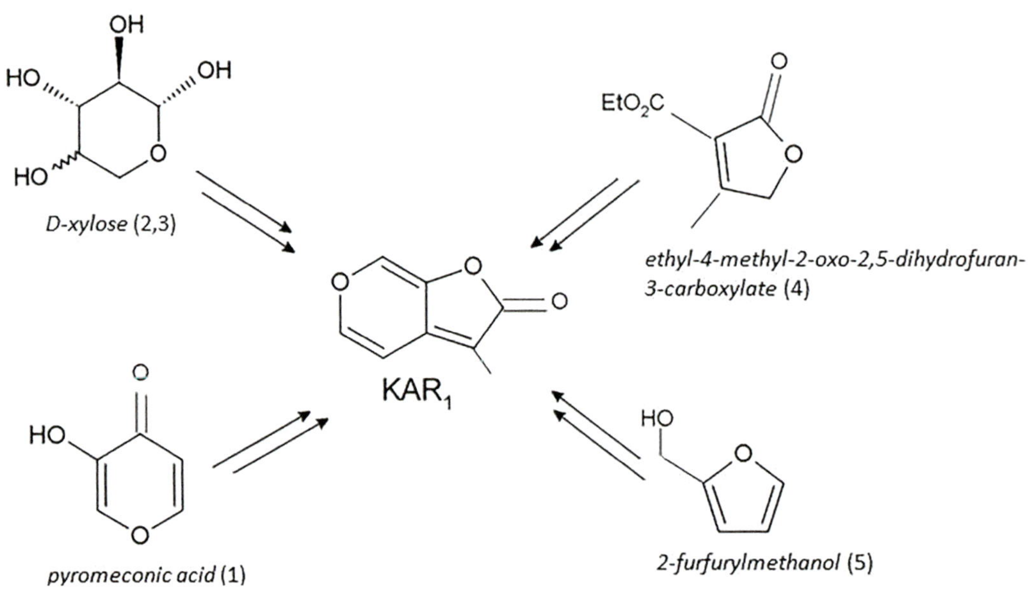

FIGURE 16.2 Synthesis of KAR$_1$ using various compounds.

Source: 1 – Flematti et al., 2005; 2 – Goddard-Borger et al., 2007; 3 – Matsuo and Shindo, 2011; 4 – Sun et al., 2008; 5 – Nagase et al., 2008

shown to have an application potential in agriculture as well (Kochanek et al., 2016). In nature, KAR_1 is produced in an area where wildfire is less intense and can be transferred in smoke for some distance from the source of a fire. After the fire has abated, KAR_1 dissolves in water during the first rain and can then stimulate seed germination. A more intense rain rinses it out to deeper soil layers. There are data indicating that KAR_1 can stay in the soil for over seven years (Flematti et al., 2015). The compound has been reported to be stable at ambient temperature and at 4°C for several years. KAR_1 has been shown experimentally to be very active at extremely low concentrations (on the order of 10^{-10}–10^{-7} M). Just a one-minute imbibition of the seeds of *Emmenanthe penduliflora*, a native Californian species, in a water solution of karrikin was found to be sufficient to stimulate germination (Dixon et al., 2009). The compound proved to be very safe for animals and humans, as it was neither toxic nor genotoxic at concentrations within the 3 x 10^{-10}–10^{-4} M range and is thus safe for use in seed and seedling treatment. Further studies showed the presence of five KAR_1 analogues called karrikins in plant-derived smoke; they are all considered now to be an important family of new plant-growth regulators (Dixon et al., 2009). The compounds are numbered from KAR_2 to KAR_6 (Figure 16.1). Among the various karrikins present in smoke, KAR_1 occurs at the highest concentrations. For this reason, and also because KAR_1 was found to be the most active karrikin, it is regarded as being mainly responsible for the stimulatory effect of smoke. However, there are plant species showing specific responses to individual karrikins. An experiment with *Arabidopsis* seeds demonstrated germination to be stimulated by KAR_1, KAR_2 and KAR_3, the best effect being obtained with KAR_2 (Nelson et al., 2009). On the other hand, KAR_4 stimulated germination of *Lactuca sativa* and *Solanum orbiculatum* seeds but was either inactive with respect to the germination of *Arabidopsis* seeds or inhibited it (Flematti et al., 2007). In addition to karrikins, seed germination was stimulated by glyceronitrile and hydroquinone, and seedling growth was improved by catechol, all the compounds being detected in smoke as well (Flematti et al., 2011, 2013; Kamran et al., 2017). Other experiments revealed the presence in smoke of three compounds: (5RS)-3,4,5-trimethylfuran-2(5H)-one (TMB); 5,5-dimethylfuran-2(5H)-one; and (5RS)-5-ethylfuran-2(5H)-one, which inhibited seed germination and exerted an antagonistic effect toward KAR_1 (Light et al., 2010; Burger et al., 2018). Thus, the effect of smoke should be considered a comprehensive biological response to the interacting compounds present in the smoke.

16.3 THE ROLE OF SMOKE AND KAR_1 IN REGULATING GERMINATION OF WEED SEEDS

The persistence of agricultural weeds in cultivated soil is, at least in part, associated with their dormancy. Dormancy is released in nature during a season that precedes the period suitable for seedling development. Because of seasonal changes in environmental factors, seeds in a soil seed bank, even from one harvest, can be at a different type and degree of dormancy. Thus, dormancy depends on the action of dormancy-releasing and dormancy-inducing factors (Bewley et al, 2013; Kępczyński, 2018). Consequently, seeds can be fully or partially primarily dormant, secondarily dormant, or nondormant. Such seed behavior results in difficulties with a complete elimination of weeds whose seeds can germinate several years after harvest. Weeds are commonly controlled by herbicides. However, the use of herbicides induces the development of herbicide resistance, which necessitates the application of new herbicides. Moreover, due to herbicide resistance, a number of annual weeds persist in soil, enhancing the soil seed bank and the subsequent weed infestation. In addition, the annual use of herbicides leads to environmental pollution. Therefore, it might be useful to treat weed seeds in soil to release their dormancy and stimulate germination to maximize seedling emergence. The emerging seedlings could then be mechanically removed or destroyed by a single application of an herbicide. Such a strategy would allow the soil to be weeded completely and would seriously cut crop cultivation costs. Thus, learning about the mechanism of seed dormancy release under the influence of various factors would allow a new weed-control strategy to be

TABLE 16.1
Effect of Smoke Water and KAR_1 on Seed Germination of Selected Weeds

	Species	Smoke Water	KAR_1	References
Monocotyledones	*Avena fatua*	↑	↑	Stevens et al., 2007; Kępczyński et al., 2010
	Avena sterilis	↑	ND	Adkins and Peters, 2001
	Bromus sterilis	↓	0	Daws et al., 2007
	Hordeum leporinum	↑	↑	Stevens et al., 2007
	Phalaris paradoxa	↑	ND	Adkins and Peters, 2001
	Sorghum halepense	0	↑	Daws et al., 2007
Dicotyledones	*Brassica tournefortii*	↑	↑	Stevens et al., 2007
	Chenopodium album	↑	0	Daws et al., 2007
	Sinapis alba	0	↑	Daws et al., 2007
	Sisymbrium orientale	↓	↑	Stevens et al., 2007
	Stellaria media	↑	↑	Daws et al., 2007
	Malva neglecta	0	↑	Daws et al., 2007

Key: no significant effect – 0, stimulation ↑, inhibition ↓, not determined – ND

developed or the existing one to be improved. Thus, after the stimulatory effect of smoke on seed germination in many plants was discovered in the natural environment, it was logical to embark on research to find out whether seeds of agricultural weeds would also be sensitive to smoke. The effects of SW on the germination of weed seeds were described in only a few papers. Initially, the treatment involved commercial SW. Doherty and Cohn (2000) reported that commercial SW was able to break dormancy in intact and dehulled grain red rice (*Oryza sativa*), a weed seriously affecting rice production. In further studies, Adkins and Peters (2001) applied the commercially available seed starter to seeds of several monocotyledonous and dicotyledonous species (Table 16.1). The strongest effects were found in the monocots *Alopecurus myosuroides*, *Avena sterilis*, *Phalaris paradoxa*, and *Sorghum halepense* (germination increasing by about 40–90%) and the dicot *Malva neglecta* (a 40% germination increase). Subsequent studies involved the effects of SW and KAR_1 (Daws et al., 2007) on seeds of 18 arable weeds and showed SW and KAR_1 to increase the percentage of germinated seeds in as few as two and eight species, respectively. Interestingly, SW stimulated seed germination in two species, *Capsella bursa-pastoris* and *Chenopodium album*, not affected by KAR_1. The effects of SW with KAR_1 on seed germination was compared in nine other weed species (Stevens et al., 2007). Seeds of four species showed a similar sensitivity to SW and KAR_1. Seeds of *Brassica tourneforti* were much more responsive to KAR_1 than to SW, and KAR_1 was able to increase the percentage of germinated seeds up to about 90%, compared to 2% in untreated seeds. It was demonstrated with *Brassica tourneforti* that fully imbibed or fully imbibed and redried seeds were less sensitive to KAR_1 than dry ones, indicating that the compound should be applied in the field prior to the rainfall (Long et al., 2010). Seed germination in *Sisymbrium orientale* and *Arctotheca calendula* was inhibited and not affected by SW, respectively, in contrast to KAR_1, which markedly induced germination in both species (Stevens et al., 2007). Thus, the findings referred to indicate that smoke contains a stimulant and an inhibitor in addition to karrikins, and the response of seeds depends on their sensitivity to specific compounds present in the smoke. The results referred to allow us to conclude that SW could be used to stimulate seed germination in some weeds and seedling development in the soil, followed by weed eradication, before crops are sown. Stevens et al. (2007), who followed seedling emergence in a field experiment, found that doses of 2 to 20g KAR_1/ha were sufficient to enhance seedling emergence from the seed bank in, e.g., *Arctotheca calendula, Brassica tournefortii* and *Raphanus raphanistrum*. It was also demonstrated that soil depth penetration of KAR_1 increased with increasing rainfall and that KAR_1 was effective as a post-rainfall germination inductor of dormant seeds of *B. tournefortii* in sandy soil. The effect of 2–5g KAR_1/ha is similar to the application of 10 tons of SW (Dixon et al., 2009), so doubtless, KAR_1 should be considered as an extremely active factor in the control of a weed seed soil bank. In other words, the extremely high KAR_1 activity as a weed seed germination stimulant, higher than that of compounds known beforehand, offers new opportunities for controlling the seed bank in agricultural soils. It is worth underlining that such activity of KAR_1 presents a great potential for commercial use in agriculture and horticulture.

16.4 *AVENA FATUA* CARYOPSES AS A DORMANCY STUDY MODEL

The annual wild grass *Avena fatua* is the most important agricultural weed worldwide. The grass is one on the world's most widespread and persistent weed, infesting most major cereals such as wheat, barley, oats, rye, and flax in many grooving regions in the world: e.g., North America and Europe, including Great Britain (Simpson, 2007) and Poland (Kępczyński, 2018). After harvest, florets and caryopses of the weed are primarily dormant and, when shed from mother plants, accumulate in soil to form a soil diaspore bank. These dispersal units can remain viable in the soil for at least seven years. Any method that would facilitate a reduction of the soil-stored dispersal unit reservoir is obviously very important. Understanding the mechanism of caryopsis dormancy release is not only important for the broadening of theoretical knowledge but can also be used in developing strategies for controlling this nuisance weed. For this reason, florets and caryopses of the plant have been used as model material in studies of its dormancy (Simpson, 2007; Kępczyński, 2018). Floret dormancy can be associated with the hulls, pericarp, testa, and physiological dormancy of the embryo (Adkins and Peters, 2001; Simpson, 2007). *A. fatua* occurs in various ecotypes differing in the dormancy level or altogether exhibiting a lack of dormancy. Caryopses represent rather the relative dormancy (Kępczyński, 2018). Caryopses of the weed harvested in Poland can germinate at 15°C and 20°C in 40–50% and 0–20%, respectively, depending on the year of harvest (Kępczyński et al., 2010, 2013; Cembrowska-Lech and Kępczyński, 2017). Dormancy in caryopses can be removed by a 12-week floret dry storage at 25°C (Adkins and Peters, 2001; Kępczyński and van Staden, 2012) or by using GA3 (Adkins and Peters, 2001; Kępczyński et al., 2013). The complete caryopsis germination, which takes place after dormancy has been released and suitable conditions provided, can be divided into two stages. The first is related to the emergence of the coleorhiza: i.e., a structure surrounding the radicle, and the second involves the rupture of the coleorhiza by the radicle.

16.5 SMOKE AND KAR_1 IN RELEASING *AVENA FATUA* CARYOPSIS DORMANCY

16.5.1 SW

Early experiments using seed starter and various *Avena fatua* biotypes from western Canada, the northern USA, England, and Australia have shown the strong stimulatory effect of this commercial SW product on germination of partly after-ripened caryopses (Adkins and Peters, 2001). SW, prepared using *Themeda triandra* from South Africa (Kępczyński et al., 2006) or a wild grass growing in Poland (Cembrowska-Lech and Kępczyński, 2017), was found to strongly induce the dormant *A. fatua* caryopses to release dormancy. Thus, as in other experiments, SW was found to be effective at the same dilution level regardless of the biological material used to produce the smoke. SW stimulated germination of both partly and fully dormant caryopses (Adkins and Peters, 2001; Kępczyński et al., 2006; Stevens et al., 2007; Kępczyński, 2018). Dormant caryopses collected in Poland were very sensitive to SW regardless of the harvest year (Kępczyński et al., 2010; Cembrowska-Lech and Kępczyński, 2017). SW at very strong dilutions (e.g., 1:1000) brought about germination of all or almost all caryopses at 20–25°C, resulting in the disappearance of dormancy (Kępczyński et al., 2013; Cembrowska-Lech and Kępczyński, 2017). It was shown that not only dormant caryopses but also dormant florets were very sensitive to SW. Dry storage of florets results in dormancy release as well, the release being more effective in caryopses than in florets (Cembrowska-Lech and Kępczyński, 2017). The sensitivity to SW in both florets and caryopses increased with time of dry storage. Experiments aimed at explaining the role of gibberellins often involve the use of paclobutrazol, a gibberellin biosynthesis inhibitor. The stimulatory effect of SW on the completion of germination of dormant caryopses was not revealed in the presence of paclobutrazol, suggesting that the presence of GAs is essential for SW efficacy. SW was found to enhance the activity of α- and β-amylase and reduce the starch content; thus, it brought about the stored starch mobilization before the radicle protrusion. Moreover, β-tubulin accumulation was found to be enhanced, as was the transition from phase G_1 to S and from S to G_2 in embryos from SW-treated caryopses, also before the radicle protrusion.

16.5.2 KAR_1

As in many seeds, KAR_1 was found to be active in removing caryopsis dormancy and stimulating germination of dormant caryopses (Stevens et al., 2007; Daws et al., 2007; Kępczyński and Van Staden, 2012; Kępczyński et al., 2010, 2013; Long et al., 2010). The effect of KAR_1 was similar regardless of whether the karrikin was isolated from smoke (Kępczyński and van Staden, 2012) or synthesized (Cembrowska-Lech et al., 2015). Concentrations of $3x10^{-9}$ or 10^{-8} M were sufficient for all the caryopses to complete germination, compared to about 10–20% germination in untreated caryopses (Kępczyński et al., 2010; Ruduś and Kępczyński, 2017). Dormant florets were shown to be less sensitive to KAR_1 than dormant caryopses. As in some seeds, KAR_1 does not need to be present all the time to release caryopses from dormancy: a 12- or 24-hour preincubation is sufficient for a highly beneficial effect (Kępczyński et al., 2010, 2013). The caryopsis sensitivity to KAR_1 was demonstrated to decrease with prolonged preincubation in water. The sensitivity to KAR_1 increased with time of dry storage. KAR_1 is not only able to stimulate complete germination of *A. fatua* caryopses under laboratory conditions, but also enhances seedling emergence in soil (Stevens et al., 2007).

16.6 THE RESPONSE OF *AVENA FATUA* CARYOPSES TO KAR_1, WITH A PARTICULAR REFERENCE TO PLANT HORMONES

16.6.1 KAR_1 and Abscisic Acid

Abscisic acid (ABA) is widely recognized as the hormone primarily responsible for the establishment of seed dormancy during maturation and for maintenance of dormancy during seed imbibition (Rodriguez-Gacio et al., 2009; Hilhorst, 2007). The hormone is considered a positive regulator of seed dormancy and a negative one in the case of seed germination. Exogenous ABA deepens seed dormancy and inhibits germination of nondormant seeds. Dormancy release by various factors in seeds of many plants is associated with a reduction in the ABA content (Bewley et al., 2013). The role of ABA in the regulation of dormancy has been extensively studied, mainly with respect to dicotyledonous seeds (Rodriguez-Gacio et al., 2009; Arc et al., 2013b). Much less data is available for monocot seeds, especially weeds. A relationship between KAR_1 and ABA was analyzed in one study only, with seeds of the weed *Arabidopsis* (Nelson, 2009). Both dormant caryopses and those capable of completing germination can absorb water during Phase I. However, only caryopses are capable of completing germination due to KAR_1 treatment that can enter Phase III associated with an increased water uptake (Cembrowska-Lech and Kępczyński, 2016, 2017). Before the KAR_1-effected germination was completed (radicle emergence), a progressively increasing coat rupture and coleorhiza emergence were observed (Cembrowska-Lech and Kępczyński, 2016; Ruduś et al., 2019). The stimulatory effect of KAR_1 was associated with ABA degradation to phaseic acid in embryos (Cembrowska-Lech et al., 2015; Figure 16.3). Further studies showed that KAR_1 reduced

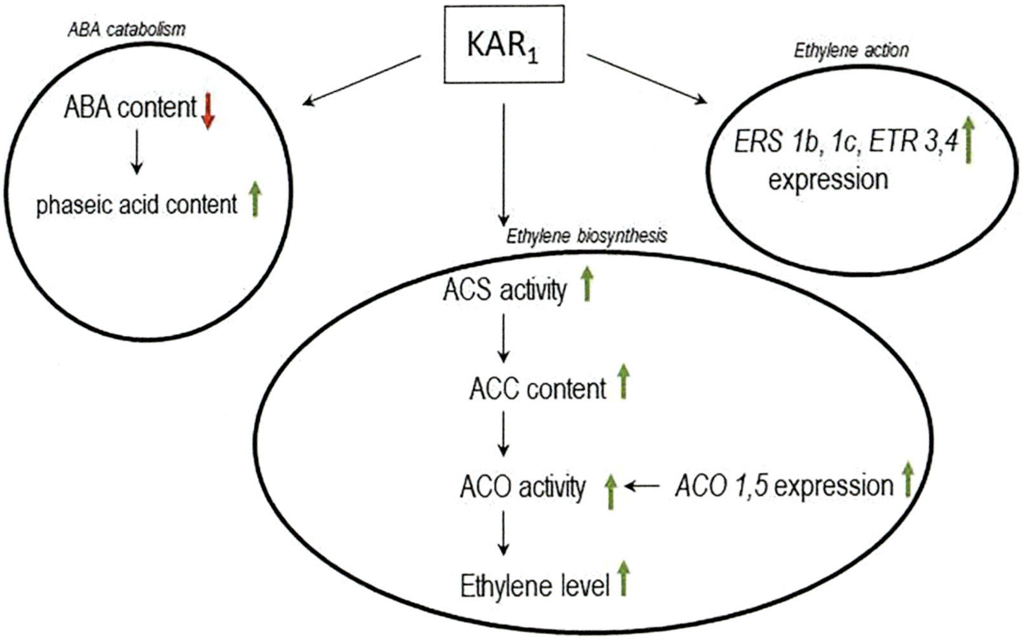

FIGURE 16.3 Relationships between KAR_1, abscisic acid, and ethylene in the release of *A. fatua* caryopsis dormancy.

Key: ACC (1-aminocyclopropane-1-carboxylic acid) = a precursor of ethylene synthesis, ACS = ACC synthase, ACO-ACC oxidase, *ACO* = ACO encoding gene, *ERS*, *ETR* = ethylene receptor genes, ↑ = Stimulation, ↓ = Inhibition

the ABA content to a level similar in the coleorhiza and the radicle (Kępczyński et al., 2021). Furthermore, the coleorhiza turned out to be more sensitive to KAR_1 than the radicle, the opposite being true with respect to ABA sensitivity: the coleorhiza proved less sensitive than the radicle (Kępczyński et al., 2021). The data accumulated indicate that KAR_1-effected induction of dormant caryopses to complete germination is mediated by a reduction of the ABA contents in the embryo, coleorhiza, and radicle.

16.6.2 KAR_1 and Gibberellins

Gibberellins (GAs) act antagonistically to ABA, promote dormancy release, and are required for germination (Kucera et al., 2005; Bewley et al., 2013). Unlike ABA, GAs are considered hormones negative toward dormancy induction and maintenance and positive toward dormancy termination and germination (Feurtado and Kermode, 2007). Exogenous gibberellins are known to release seed dormancy in many plants and to stimulate germination (Kucera et al., 2005). Thus, it was interesting to compare the effects of KAR_1 and GA_3 on the germination of dormant *A. fatua* caryopses. The comparison (Kępczyński et al., 2013) showed both compounds to be very active inductors of caryopsis germination. As in other experiments (Adkins and Peters, 2001; Daws et al., 2007; Stevens et al., 2007), GA_3 induced germination of dormant *A. fatua* caryopses harvested in Poland (Kępczyński et al., 2013). However, KAR_1 was much more effective in this regard; for example, to get the same result at 20°C, it was sufficient to apply KAR_1 at a concentration about 1,000 times lower than that at which GA_3 was applied. Three inhibitors of GAs synthesis – paclobutrazol, ancymidol, and flurprimidol (Yamaguchi et al., 2007) – were used to study the role of endogenous gibberellins in releasing dormancy by KAR_1. When used simultaneously with KAR_1, all the inhibitors prevented dormancy release (Kępczyński et al., 2013). The inhibition was reversed by applying exogenous GA_3, which probably replaced the endogenous gibberellins. Thus, it was suggested that the biosynthesis/presence of gibberellins is essential for KAR_1 to be effective. Since a higher concentration of GA_3 than KAR_1 is required to obtain a similar effect, it may be thought that karrikin does not act via gibberellins, but rather requires their presence to act. Moreover, although the most common effect of the GAs inhibitors used is likely to be related to gibberellin content reduction, it should be kept in mind that the inhibitors may also increase the ABA concentration (Yamaguchi et al., 2007).

16.6.3 KAR_1 and Ethylene

Ethylene is commonly regarded as a factor important for regulation of seed dormancy in many plant species (Kępczyński and Kępczyńska, 1997; Matilla and Matilla-Vazquez, 2008; Corbineau et al., 2014). It stimulates germination of primarily and secondarily dormant seeds as well as that of nondormant seeds incubated in nonoptimal conditions. The hormone is produced from S-adenosyl-methionine (SAM) which is converted to 1-aminocyclopropane-1-carboxylic acid (ACC) by ACC synthase, whereupon ACC is oxidized by ACC oxidase to ethylene (Yang and Hoffman, 1984). To explore the role of ethylene, ethephon, which liberates ethylene in the cell, was used in addition to exogenous ethylene. Moreover, ACC, a direct ethylene biosynthesis precursor, was used, as well as ethylene biosynthesis inhibitors and inhibitors of ethylene binding to receptors. Neither ethylene nor ethephon and ACC was able to remove primary dormancy in *A. fatua* caryopses (Kępczyński et al., 2006; Kępczyński and van Staden, 2012). It was thus concluded that ethylene alone was not able to release caryopsis dormancy. The stimulatory effect of KAR_1 on germination of dormant caryopses was blocked by 2,5-norbornadiene (NBD), a competitive, reversible inhibitor of ethylene binding to the receptors, which points to the necessity of ethylene being bound to its receptors for the KAR_1 effect to materialize in the germination of dormant caryopses. It was also demonstrated that the inhibitory effect of NBD can be counteracted by exogenous ethylene. Consequently, the dormancy release by KAR_1 requires that ethylene be active. The importance of ethylene action was confirmed by using 1-MCP, a non-reversible ethylene binding receptor, which also counteracted the stimulatory effect of KAR_1. Aminoethoxyvinylglycine (AVG), an ethylene biosynthesis inhibitor that inhibits the ACC synthase activity, reinforced the NBD inhibition of the stimulatory effect of KAR_1 on dormancy release (Kępczyński and van Staden, 2012), thus confirming that for KAR_1 to release dormancy, endogenous ethylene is required. KAR_1 was found to stimulate ethylene production prior to the radicle protrusion, which was associated with an increase in ACC synthase activity, ACC content, and the ACC oxidase activity (Ruduś et al., 2019; Figure 16.3). It was reported that *AfACS6* was unaffected, but *AfACO1 and AfACO5* were upregulated by KAR_1, suggesting non-transcriptional and transcriptional activation of ACC synthase and ACC oxidase, respectively. Thus, the response to KAR_1 was associated with regulation of sensitivity to ethylene through stimulation of ethylene receptor synthesis.

16.6.4 KAR_1 and Oxidative Homeostasis

Reactive oxygen species (ROS) such as $O_2^{\bullet -}$, H_2O_2, and hydroxyl radicals (·OH) are known to play an essential role in seed dormancy (El-Maarouf-Boutau and Bailley, 2008; Diaz-Vivancos et al., 2013). When their concentrations are too high, ROS act adversely: they cause damage, loss of vigor, aging, and even death. If the concentrations are too low, seeds are dormant; it is only after a concentration suitable for germination (the "oxidative window") has been reached that the seeds can complete germination (Bailey et al., 2008). The ROS level is controlled by the opposite processes of production and scavenging by enzymatic and non-enzymatic antioxidants. Enzymatic scavenging is accomplished by superoxide dismutase (SOD) and catalase (CAT) as well as by enzymes engaged in the ascorbate-glutathione (AsA-GSH) cycle, involved in indirect or direct ROS scavenging:

ascorbate peroxidase (APX), monodehydroascorbate reductase (MR), dehydroascorbate reductase (MDHAR), and glutathione reductase (GR) (Noctor and Foyer, 1998; Foyer and Noctor, 2011). Non-enzymatic antioxidants such as α-tocopherol, reduced glutathione (GSH), and ascorbate (AsA) are active as well. Studies on the seeds of several plant species have shown that dormancy can be released by exogenous ROS or ROS-generating compounds (El-Maarouf-Bouteau and Bailley, 2008; Diaz-Vivancos et al., 2013). Several studies have demonstrated the AsA-GSH cycle as being mainly responsible for the scavenging of H_2O_2. Using AsA as the electron donor, APX caused H_2O_2 detoxification; AsA is regenerated by MR and DHAR with the participation of GSH and NADPH. GSH is oxidized to oxidized glutathione (GSSG) and may be reduced to GSH by GR (Li et al., 2010). As in many other seeds, dormancy of *A. fatua* caryopses was released by H_2O_2 and by aminotriazole (AT), a compound known to increase H_2O_2 content due to inhibition of the catalase activity (Cembrowska-Lech et al., 2015). Stimulation of dormant *A. fatua* caryopsis germination by the $O_2^{\bullet-}$ generating methylviologene also indicates the participation of that ROS in dormancy release. In addition, $O_2^{\bullet-}$ involvement in the caryopsis response to KAR_1 was confirmed by the use of diphenyleneiodonium (DPI), an inhibitor of NADPH oxidase that catalyzes the production of $O_2^{\bullet-}$, which dampens the stimulatory effect of KAR_1. The stimulatory effect of KAR_1 was associated with reduced $O_2^{\bullet-}$ and H_2O_2 contents and with increased activities of SOD and CAT in embryos, so it was concluded that the stimulatory effect of KAR_1 was associated with its regulating the ROS concentration such that it is adjusted to nontoxic and signaling levels (Figure 16.4). Further studies revealed that the maintenance of ROS homeostasis also involves the ascorbate-glutathione (AsA-GSH) cycle (Cembrowska-Lech and Kępczyński, 2016). An increase in the content of AsA, a non-enzymatic antioxidant associated with the appearance of an additional isoform, APX, under the influence of stimulators used, demonstrates the importance of AsA in generating the response to KAR_1. A similarly essential role of AsA is shown by the reduction of the stimulatory effect of

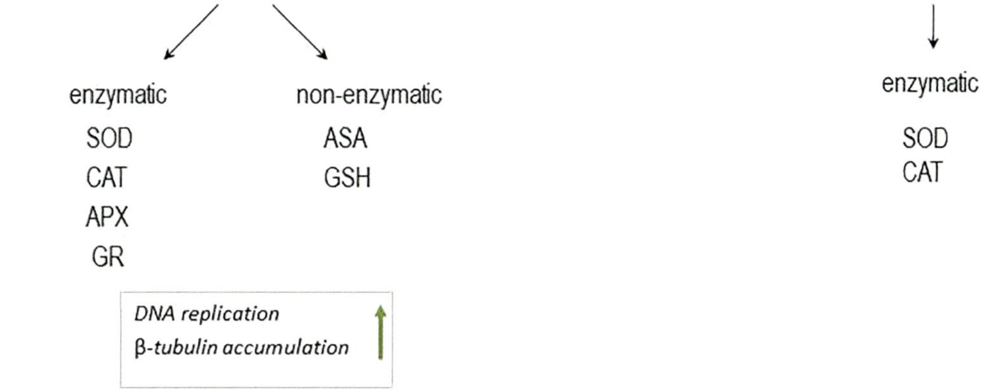

FIGURE 16.4 A summary of the effect of KAR_1 on reactive oxygen species and antioxidants in the embryo and aleurone layers of *A. fatua* caryopses.

Key: H_2O_2 = hydrogen peroxide, $O_2^{\bullet-}$ = superoxide anion, SOD = superoxide dismutase, CAT = catalase, APX = ascorbate peroxidase, GR = glutathione reductase, ASA = ascorbate, GSH = reduced glutathione, ↑ = Stimulation, ↓ = Inhibition

KAR_1 by the AsA biosynthesis inhibitor. KAR_1 was found to increase the contents of both AsA and DHA, but the increase was higher in the case of AsA than in DHA. The content of GSH was increased, but that of GSSG was reduced due to the treatment with the stimulant mentioned earlier. KAR_1 increased the AsA/DHA and GSH/GSSG ratios, indicating that a KAR_1-induced increase in the two ratios allows the excessive ROS to be scavenged and dormancy to be released. Moreover, a stronger effect on the GSH/GSSG ratio compared to that on the AsA/DHA one demonstrates the higher importance of the former in the stimulation of complete germination by KAR_1.

Germination and post-germination processes in grasses such as barley, wheat, and *A. fatua* are associated with the gibberellin-induced synthesis of hydrolytic enzymes, mainly α-amylases, in the aleurone layer and their secretion into the starchy endosperm where they are responsible for mobilization of the stored reserves (starch hydrolysis) for the growing embryo (Bewley et al., 2013). After their function has been terminated, aleurone cells undergo the programmed cell death (PCD), stimulated by gibberellins and inhibited by ABA (Bethke et al., 1999). Experiments showed that ROS, especially H_2O_2, were responsible for PCD induction in GA_3-treated cells (El-Maarouf-Bouteau and Bailley, 2008). GA_3 was found to inhibit the activity of scavenging enzymes such as CAT, APX, and SOD, which resulted in an increase in the H_2O_2 content and the PCD induction. The activity of α-amylase and the ROS-antioxidant status in the *A. fatua* aleurone layers were found to be regulated by both KAR_1 and GA_3 (Cembrowska-Lech et al., 2015). It has been demonstrated that not only KAR_1 and GA_3 but also H_2O_2, MV, and AT, all releasing dormancy, are able to stimulate the α-amylase activity in the aleurone layers of *A. fatua*. Both KAR_1 and GA_3 increased the contents of H_2O_2 and $O_2^{\bullet -}$, which correlated to decreased activities of SOD and CAT (Figure 16.4). Induction of PCD in the aleurone layer is probably associated with an increase in the ROS concentration induced by the compounds mentioned.

16.6.5 KAR_1 and Cell Cycle Activity

Currently, the role of cell cycle in seed dormancy release is still not entirely clear. Cells in radicle tips of dry dormant seeds of *Lycopersicon esculentum* (de Castro et al., 2001) and *Hordeum vulgare* (Gendreau et al., 2012), as well as in *A. fatua* caryopses (Cembrowska-Lech and Kępczyński, 2016; Kępczyński et al., 2021), have been found to be arrested at phase G_1. Like SW, KAR_1 decreased the percentage of nuclei at G_1 but increased their percentages at S and G_2 (Cembrowska-Lech and Kępczyński, 2016, 2017). KAR_1 caused accumulation of β-tubulin prior to, or coincident with, DNA replication in the radicle tip. Since KAR_1 caused transition from G_1 to S and from S to G_2, which occurred a few hours before the coleorhiza emergence and radicle protrusion, it was concluded that the induction of dormancy release and germination of *A. fatua* caryopses by KAR_1 involves activation of the cell cycle (Figure 16.3).

16.7 SUMMING UP AND PERSPECTIVES

This chapter discussed the recent advances in research on the role of smoke and KAR_1 detected in smoke in seed dormancy release, with a particular emphasis on agricultural weed seeds. In addition to KARs, smoke has been found to contain both seed germination stimulants and inhibitors. Looking to the future, it can be assumed that more biologically active compounds will be isolated from smoke. It should be emphasized that both smoke/SW and KAR_1 have found practical applications: e.g., in the priming of seed crops, weed control, and ecological restoration. Considering the *A. fatua* caryopses as a model for a better understanding of the regulation of dormancy release and germination by KAR_1, it will be necessary to expand the research on KAR_1 effects on the contents of individual gibberellins, on the expression of genes encoding enzymes in their biosynthetic and catabolic pathways, on genes encoding enzymes active in the ABA biosynthetic and catabolic pathways, and on genes coding the enzymes responsible for weakening the coleorhiza, which enables the radicle to break out. Further studies involving *A. fatua* caryopses should focus on the relationship between KAR_1 and strigolactones since research on *Arabidopsis thaliana* showed that karrikin and strigolactones display a similar signaling mechanism.

REFERENCES

Adkins, Steve W., and Peters, Nicholas C.B. "Smoke derived from burnt vegetation stimulates germination of arable weeds." Seed Science Research 11 no. 3 (2001): 213–222. https://doi.org/10.1079/SSR200177

Arc, Erwann, Galland, Marc, Godin, Beatrice, Cueff, Gwendal, and Rajjou, Loic. "Nitric oxide implication in the control of seed dormancy and germination." Frontiers in plant science 4 (2013a): 346. https://doi.org/10.3389/fpls.2013.00346

Arc, Erwann, Sechet, Julien, Corbineau, Françoise, Rajjou, Loïc, and Marion-Poll, Annie. "ABA crosstalk with ethylene and nitric oxide in seed dormancy and germination." Frontiers in Plant Science 4 (2013b): 63. https://doi.org/10.3389/fpls.2013.00063

Bailey, Christophe, El-Maarouf-Bouteau, Hyat, and Corbineau, Francoise. "From intracellular signaling networks to cell death: the dual role of reactive oxygen species in seed physiology." Comptes Rendus Biologus 331 (2008): 806–814. https://doi.org/10.1016/j.crvi.2008.07.022

Banerjee, Aditya, Tripathi, Durgesh K., and Roychoudhury, Aryadeep. "The karrikin 'calisthenics': Can compounds derived from smoke help in stress tolerance?" Physiologia Plantarum 165 (2019): 290–302. http://dx.doi.org/10.1111/ppl.12836

Bethke, Paul C., Lonsdale, Jennifer E., Fath, Angelika, and Jones, Russell L. "Hormonally regulated programmed cell death in barley aleurone cells." Plant Cell 11 (1999): 1033–1045. https://doi.org/10.2307/3870796

Bewley, J. Derek, Bradford, Kent, J., Hilhorst, Henk W.M., and Nonogaki, Hiro. *Seeds: Physiology of development, germination and dormancy*. New York: Springer, 2013. https://doi.org/10.1017/S0960258513000287

Brown, Neville A.C., and van Staden, Johannes. "Smoke as a germination cue: A review." Plant Growth Regulation 22 (1997): 115–124. https://doi.org/10.1023/A:1005852018644

Burger, Barend V., Pošta, Martin, Light, Marnie E., Kulkarni, Manoj G., Viviers, Marlize Z., and van Staden, Johannes. "More butenolides from plant-derived smoke with germination inhibitory activity against karrikinolide." South African Journal of Botany 115 (2018): 256–263. https://doi.org/10.1016/j.sajb.2018.01.023

Cembrowska-Lech, Danuta, and Kępczyński, Jan. "Gibberellin-like effects of KAR_1 on dormancy release of *Avena fatua* caryopses include participation of non-enzymatic antioxidants and cell cycle activation in embryos." Planta 243 (2016): 531–548. https://doi.org/10.1007/s00425-015-2422-1

Cembrowska-Lech, Danuta, and Kępczyński, Jan. "Plant derived smoke induced activity of amylases, DNA replication and β-tubulin accumulation before radicle protrusion of dormant *Avena fatua* L. Caryopses." Acta Physiologiae Plantarum 39 (2017): 39. http://dx.doi.org/10.1007/s11738-016-2329-x

Cembrowska-Lech, Danuta, Koprowski, Marek, and Kępczyński, Jan. "Germination, induction and dormant *Avena fatua* caryopses by KAR_1 and GA_3 involving the control of reactive oxygen species (H_2O_2 and O_2-) and enzymatic antioxidants (superoxide dismutase and catalase) both in the embryo and the aleurone layers." Journal of Plant Physiology 176 (2015): 169–179. https://doi.org/10.1016/j.jplph.2014.11.010

Corbineau, Francoise, Qiong, Xia, Christophe, Bailly, and Hayat El-Maarouf-Bouteau. "Ethylene, a key factor in the regulation of seed dormancy" Frontiers in Plant Science 5 (2014): 539. https://doi.org/10.3389/fpls.2014.00539

Daws, Matthew I., Davies, Jennifer, Pritchard, Hugh W., Brown, Neville A. C., and van Staden, Johannes. "Butenolide from plant-derived smoke enhances germination and seedling growth of arable weed species." Plant Growth Regulation 51 (2007): 73–82. https://doi.org/10.1007/s10725-006-9149-8

de Castro, Renato D., Bino, Raoul J., Jing, Hai-Chun, Kieft, Henk, and Hilhorst, Henk. "Depth of dormancy in tomato (Lycopersicon esculentum Mill) seeds is related to the progression of the cell cycle prior to the induction of dormancy." Seed Science Research 11 (2001): 45–54.

De Lange, Johannes H., and Boucher, C. "Autoecological studies on Audouinia capitata (Bruniaceae). I. Plant-derived smoke as a germination cue." South African Journal of Botany 56 no. 6 (1990): 700–703. https://doi.org/10.1016/S0254-6299(16)31009-2

Diaz-Vivancos, Pedro, Barbara-Espin, Gregorio, Hernandez, and Josef A. "Elucidating hormonal/ROS network during seed germination: Insights and perspectives." Plant Cell Reports 32 (2013): 1491–1502. http:/doi.org/10.1007/s00299-13-1473-7

Dixon, Kingsley W., Merritt, David J., Flematti Gavin R., and Ghisalberti Emilio L. "Karrikinolide-a phytoreactive compound derived from smoke with applications in horticulture, ecological restoration and agriculture." Acta Horticulturae 813 (2009): 155–170. https://doi.org/10.17660/ActaHortic.2009.813.20

Doherty, Lucia C., Cohn, and Marc A. "Seed dormancy in red rice (Oryza sativa) XI Commercial liquid smoke elicits germination." Seed Science Research 10 (2000): 515–421. https://doi.org/10.17/s096025800000465

El-Maarouf-Bouteau, Hyat, Bailly, and Christophe. "Oxidative signaling in seed germination and dormancy." Plant Signal Behaviour 3 (2008): 175–182. https://doi.org/10.4161/psb.3.3.5539

Feurtado, Allan J., Kermode, and Allison R. "A merging of paths: Abscisic acid and hormonal cross-talk in the control of seed dormancy maintenance and alleviation." In *Seed development, dormancy and germination*. Hoboken, NJ: Blackwell Publishing Ltd, 2007, 176–223.

Finkelstein, Ruth, Reeves, Weendy, Ariizumi, Tohru, and Steber, Cawille. "Molecular aspects of seed dormancy." Annual Review Plant Biology 59 (2008): 387–415. https://doi.org/10.1146/annurev.arplant.59.032607.092740

Flematti, Gavin R., Dixon, Kingsley W., and Smith, Steven M. "What are karrikins and how were they 'discovered' by plants?" BMC Biology 13 no. 1 (2015): 108. https://bmcbiol.biomedcentral.com/articles/10.1186/s12915-015-0219-0

Flematti, Gavin, R., Ghisalberti, Emilio L., Dixon, Kingsley W., and Trengove, Robert D. "A compound from smoke that promote seed germination." Science 305 no. 5686 (2004): 977. https://doi.org/10.1126/science.1099944

Flematti, Gavin R., Ghisalberti, Emilio L., Dixon, Kingsley W., and Trengove, Robert D. "Synthesis of the seed germination stimulant 3-methyl-2H-furo[2,3-c]pyran-2-one." Tetrahedron Letters 46 no. 34 (2005): 5719–5721. https://doi.org/10.1016/j.tetlet.2005.06.077

Flematti, Gavin, R., Goddard-Borger, Merritt, David J., Ethan D., Ghisalberti, Emilio L., Dixon, Kingsley W., and Trengove, Robert D. "Preparation of 2H-furo[2,3-c]pyran-2-one derivatives and evaluation of their germination-promoting activity." Journal of Agricultural Food Chemistry 55 (2007): 2189–2194. https://doi.org/10.1021/jf0633241

Flematti, Gavin R., Merritt, David J., Piggott, Matthew J., et al. "Burning vegetation produces cyanohydrins that liberate cyanide and promote seed germination." Nature Communications 2 no. 1 (2011): 360. http://dx.doi.org/10.1038/ncomms1356

Flematti, Gavin R., Waters, Mark T., Scaffidi, Adrian et al. "Karrikin and cyanohydrin smoke signals provide clues to new endogenous plant signaling compounds." Molecular Plant 6 no. 1 (2013): 29–37. https://doi.org/10.1093/mp/sss132

Foyer, Christine H., and Noctor, Graham. "Ascorbate and glutathione: the hearth of the redox hub." Plant Physiology 155 (2011): 2–18. https://doi.org/10.1104/pp.110.167569

Gendreau, Emmanuel, Cayla, Thibaud, and Corbineau, Francoise. "S phase of the cell cycle: A key phase for the regulation of thermodormancy in barley grain." Journal of Experimental Botany 63 (2012): 5535–5543. https://doi.org/10.1093/jxb/ers204

Goddard-Borger, Ethan D., Ghisalberti, Emilio L., and Stick, Robert V. "Synthesis of the germination stimulant 2H-furo[2,3-c]pyran-2-one and analogous compounds from carbohydrates." European Journal of Organic Chemistry 23 (2007): 3925–3934. http://dx.doi.org/10.1002/ejoc.200700334

Hilhorst, Henk W.M. "Definitions and hypotheses of seed dormancy." In *Seed development, dormancy and germination*. Hoboken: Blackwell Publishing Ltd, 2007, 50–71.

Kamran, Muhhamad, Khan, Abdul L., Ali, Liaqat et al. "Hydroquinone: A novel bioactive compound from plant-derived smoke can cue seed germination of lettuce." Frontiers in Chemistry 5 (2017): 30. https://dx.doi.org/10.3389%2Ffchem.2017.00030

Kępczyński, Jan. "Induction of agricultural weed seed germination by smoke and smoke-derived karrikin (KAR_1), with a particular reference do *Avena fatua* L." Acta Physiologiae Plantarum 40 no. 5 (2018): 87. http://dx.doi.org/10.1007/s11738-018-2663-2

Kępczyński, Jan. "Progress in utilizing plant-derived smoke water and smoke-derived KAR_1 in plant tissue culture." Plant Cell, Tissue and Organ Culture (PCTOC) 140 (2020): 271–278. https://doi.org/10.1007/s11240-019-01739-8

Kępczyński, Jan, Białecka, Bożena, Light, Marnie E., and van Staden, Johannes. "Regulation of *Avena fatua* seed germination by smoke solution, gibberellin A_3 and ethylene." Plant Growth Regulation 49 (2006): 9–16. https://doi.org/10.1007/s10725-006-0008-4

Kępczyński, Jan, Cembrowska, Danuta, and van Staden, Johannes. "Necessity of gibberellin for stimulatory effect of KAR_1 on germination of dormant *Avena fatua* L. Caryopses." Acta Physiologiae Plantarum 35 (2013): 379–387. http://dx.doi.org/10.1007/s11738-012-1080-1

Kępczyński, Jan, Cembrowska, Danuta, and van Staden, Johannes. "Releasing primary dormancy in *Avena fatua* L. Caryopses by smoke-derived butenolide." Plant Growth Regulation 62 (2010): 85–91. https://doi.org/10.1007/s10725-010-9493-6

Kępczyński, Jan, and Kępczyńska, Ewa. "Ethylene in seed dormancy and germination." Physiologiae Plantarum 101 no. 4 (1997): 720–726. https://doi.org/10.1111/j.1399-3054.1997.tb01056.x

Kępczyński, Jan, and van Staden, Johannes. "Interaction of karrikinolide and ethylene in controlling germination of dormant *Avena fatua* L. Caryopses." Plant Growth Regulation 67 (2012): 185–190. https://doi.org/10.1007/s10725-012-9675-5

Kępczyński, Jan, Wójcik, Agata, and Dziurka, Michał. "*Avena fatua* caryopsis dormancy release is associated with changes in KAR_1 and ABA sensitivity as well as with ABA reduction in coleorhiza and radicle." Planta (2021): 253–52. https://doi.org/10.1007/s00425-020-03562-4

Kochanek, Jitka, Long, Rowena L., Lisle, Allan, T., Flematti, and Gavin, R. "Karrikins Identified in biochars indicate post-fire chemical cues can influence community diversity and plant development." Plos One 11 (2016): e0161234. https://doi.org/10.1371/journal.pone.0161234

Kucera, Bright, Cohn, Marc A., and Leubner-Metzger, Gerhard. "Plant hormone interactions during seed dormancy release and germination." Seed Science Research 15 (2005): 281–307. https://doi.org/10.1079/SSR2005218

Kulkarni, Manoj G., Light, Marnie E., and Van Staden, Johannes. "Plant-derived smoke: Old technology with possibilities for economic applications in agriculture and horticulture." South African Journal of Botany 77 no. 4 (2011): 972–979. https://doi.org/10.1016/j.sajb.2011.08.006

Li, Yingli, Liu, Yufang, and Zhang, Jianguang. "Advances in the research on the AsA-GSH cycle in horticultural crops." Frontiers Agriculture of China 4 (2010): 84–90. https://doi.org/10.1007/s11703-009-0089-8

Light, Marnie E., Burger, Ben V., Staerk, Dan, Kohout, Ladislav, and Van Staden, Johannes. "Butenolides from plant-derived smoke: Natural plant growth regulators with antagonistic actions on seed germination." Journal of Natural Products 73 no. 2 (2010): 267–269. https://doi.org/10.1016/j.sajb.2008.10.004

Light, Marnie E., Daws, Matthew I., and Van Staden, Johannes. "Smoke-derived butenolide: Towards understanding its biological effects." South African Journal of Botany 75 (2009): 1–7. https://doi.org/10.1021/np900630w

Light, Marnie E., and van Staden, Johannes. "The potential of smoke in seed technology." South African Journal of Botany 70 no. 1 (2004): 97–101. http://dx.doi.org/10.1016/S0254-6299(15)30311-2

Long, Rowena L., Williams, Kimberlyn, Griffiths, Erin M., et al. "Prior hydration of Brassica tournefortii seeds reduces the stimulatory effect of karrikinolide on germination and increases seed sensitivity to abscisic acid." Annals of Botany 105 (2010): 1063–1070. https://doi.org/10.1093/aob/mcq061

Matilla, Angel J.M., and Matilla-Vazquez M. A. "Involvement of ethylene in seed physiology." Plant Science 175 (2008): 87–97. https://doi.org/10.1016/j.plantsci.2008.01.014

Matsuo, Kazumasa, and Shindo, Mitsuru. "Efficient synthesis of karrikinolide via Cu(II)-catalyzed lactonization." Tetrahedron 67 no. 5 (2011): 971–975. https://doi.org/10.1016/j.tet.2010.11.108

Nagase, Ryohei, Katayama, Mayumi, Mura, Hiroaki, Matsuo, Noritada, and Tanabe, Yoo. "Synthesis of the seed germination stimulant 3-methyl-2H-furo[2,3-c]pyran-2-ones utilizing direct and regioselective Ti-crossed aldol addition." Tetrahedron Letters 49 no. 29–30 (2008): 4509–4512. http://dx.doi.org/10.1016%2Fj.tetlet.2008.05.052

Nelson, David C., Flematti, Gavin R, Ghisalberti, Emilio L, Dixon, Kingsley W., and Smith, Steven M. "Regulation of seed germination and seedling growth by chemical signals from burning vegetation." Annual Review of Plant Biology 63 (2012): 107–130. https://doi.org/10.1146/annurev-arplant-042811-105545

Nelson, David C., Riseborough Julie-Anne, Flematti, Gavin R., et al. "Karrikins discovered in smoke trigger Arabidopsis seed germination by a mechanism requiring gibberellic acid synthesis and light." Plant Physiology 149 no. 2 (2009): 863–873. https://doi.org/10.1104/pp.108.131516

Noctor, Graham, and Foyer, Christine H. "Ascorbate and Glutathione: Keeping Active Oxygen under Control." Annual Review of Plant Physiology and Plant Molecular Biology 49 (1998): 249–279. https://doi.org/10.1146/annurev.arplant.49.1.249

Rodriguez-Gacio, María C., Matilla-Vázquez, Miguel A., and Matilla, Angel J. "Seed dormancy and ABA signaling: the breakthrough goes on." Plant Signaling & Behavior 4 no. 11 (2009): 1035–1048. https://doi.org/10.4161/psb.4.11.9902

Ruduś, Izabela, Cembrowska-Lech, Danuta, Jaworska, Anna, and Kępczyński, Jan. "Involvement of ethylene biosynthesis and perception during germination of dormant *Avena fatua* Lcaryopses induced by KAR_1 or GA_3." Planta 249 no. 3(2019):719–738. https://doi.org/10.1007/s00425-018-3032-5

Ruduś, Izabela, and Kępczyński, Jan. "Exogenous putrescine increase the responsiveness of thermodormant *Avena fatua* L. Caryopses to karrikinolide and gibberellic acid." Acta Physiologiae Plantarum 39 no. 2 (2017): 49. http://dx.doi.org/10.1007/s11738-016-2344-y

Sami, Abdul, Zhu, Z.H., Zhu T.X., et al. "Influence of KAR_1 on the plant growth and development of dormant seeds by balancing different factors." International Journal of Environmental Science and Technology (2021). https://doi.org/10.1007/s13762-021-03282-6

Simpson, Graham M. *Seed dormancy in grasses*. Cambridge: Cambridge University Press, 2007. https://doi.org/10.1017/CBO9780511721816

Soós, Vilmos, Badics, Eszter, Incze, Norbert, and Balázs, Ervin. "Fire-borne life: A brief of Smoke-induced germination." Natural Product Communications 14 no. 9 (2019): 1934578X1987292. https://doi.org/10.1177%2F1934578X19872925

Stevens, Jason C., Merritt, David J., Flematti Gavin R., Ghisalberti Emilio L., and Dixon, Kingsley W. "Seed germination of agricultural weeds is promoted by the butenolide 3-methyl-2H-furo{2,3-c}pyran-2-one under laboratory and field conditions." Plant and Soil 298 (2007): 113–124. https://doi.org/10.1007/s11104-007-9344-z

Sun, Kingmo, Chen, Yuzong, Wagerle, Ty, Linnstaedt, David et al. "Synthesis of butenolides as seed germination stimulants." Tetrahedron Letters 49 no. 18 (2008): 2922–2925. https://doi.org/10.1016/j.tetlet.2008.03.024

Van Staden, Johannes, Jäger, Anna K., Light, Marnie E., and Burger, Barend V. "Isolation of the major germination cue from plant-derived smoke." South African Journal of Botany 70 no. 4 (2004): 654–659. https://doi.org/10.1016/S0254-6299(15)30206-4

Yamaguchi, Shinjiro, Kamiya, Yuji, and Nambura, Eiji. "Regulation of ABA and GA levels during seed development and germination in Arabidopsis." In *Seed development, dormancy and germination*. Hoboken: Blackwell Publishing Ltd, 2007, 223–247.

Yang, Shang F., and Hoffman, Neil E. "Ethylene biosynthesis and its regulation in higher plants." Annual Review of Plant Physiology 35 (1984): 155–189. https://doi.org/10.1146/annurev.pp. 35.060184.001103

17 Application of Smoke-Water and KAR_1 in Plant Biotechnology in vito Cultures

Jan Kępczyński and Ewa Kępczyńska

CONTENTS

17.0 INTRODUCTION

As a result of a considerable advancement of knowledge, the 1970s witnessed the emergence of a new interdisciplinary field of studies, biotechnology. It deals with the use of living cells, plants, or their parts in practical applications, such as propagation of organisms, according them desired characteristics, and production of specific compounds. Biotechnology is defined as a discipline that integrates agricultural and engineering sciences to obtain products and services by using organisms, cells, or their parts and molecular analogues. The discovery of biological activity produced by plant-derived smoke and the subsequent isolation of KAR_1 – a compound present in the smoke and responsible mainly, or at least to a large extent, for this activity – directed subsequent research not only to studies on the mechanism of KAR_1 activity but also to developing technologies facilitating practical applications of both smoke and KAR_1, with particular attention being paid to the latter (Light et al., 2009; Nelson et al., 2012; Morffy et al., 2016; Kępczyński, 2020). Both smoke and smoke-saturated water or smoke water (SW) have been used for seed treatment in soil, for priming seeds before sowing, and for treating seedlings by spraying or watering them using appropriate dilutions of SW. Likewise, KAR_1, which is active at extremely low concentrations, has been and will probably continue to be used instead of smoke in practical applications: e.g., *in vivo* treatment of seeds and seedlings. This chapter will focus on the use of SW and KAR_1 in plant and cell tissue cultures as a basis of plant biotechnology (Figure 17.1).

17.1 PLANT CELL AND TISSUE CULTURE

Development of plant biotechnology has been made possible by advances in *in vitro* plant culture methods; *in vitro* cultures are now an important component of the discipline. In turn, *in vitro* plant cultures have been made possible by the detection and subsequent application of plant hormones or

DOI: 10.1201/9781003224525-21

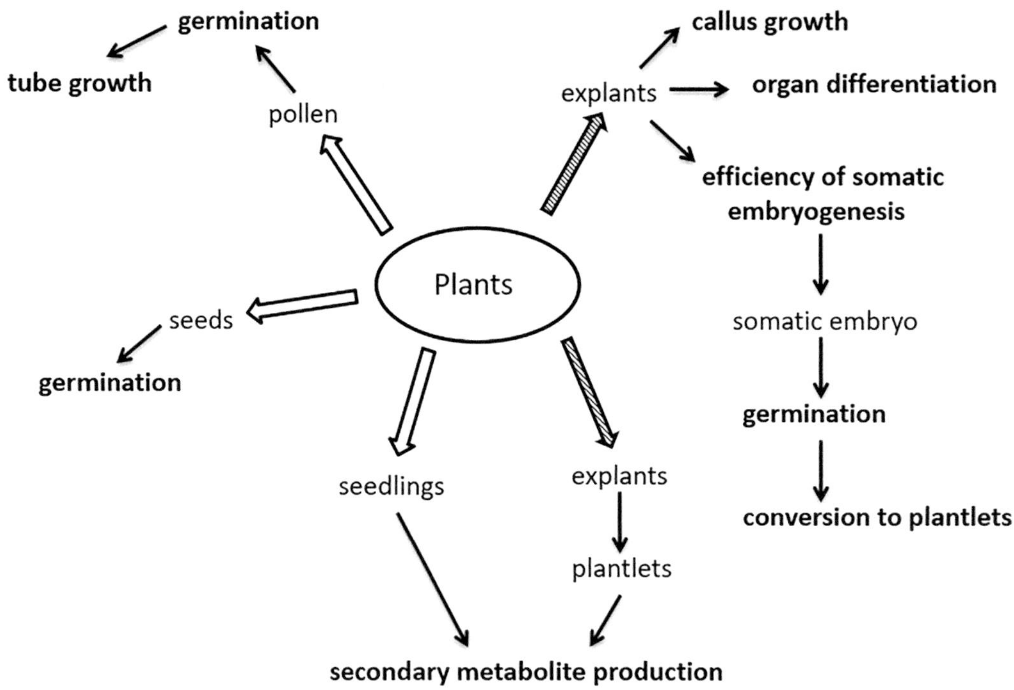

FIGURE 17.1 Diagram summarizing the stimulating effect of smoke water and KAR_1 on different processes (in bold) using pollen, seeds, seedlings, and explants.

their synthetic analogues. They are key to regulation of *in vitro* growth and development of organs and whole plants. The most frequently used plant hormones include cytokinins and auxins, which, when applied in appropriate proportions, enable induction of the callus tissue on explants, callus proliferation, induction of organogenesis in shoots and roots, and somatic embryogenesis (Moshkov et al., 2008). *In vitro* cultures involve the use of other hormones as well. For example, abscisic acid is used to stimulate the synthesis of storage reserves in somatic embryos, to induce desiccation tolerance, and to ease the entry into the quiescent state (Ruduś et al., 2006). The hormone is also used as an anti-transpirant in acclimatization of tissue-raised plants. The use of gibberellins is less known. There are conflicting data showing either stimulation or inhibition by exogenous gibberellins of somatic embryo production. Ethylene plays a particularly important role in *in vitro* growth and organogenesis since it is produced and accumulated to physiologically active concentrations in airtight culture vessels (Kępczyński et al., 2006). Although plant hormones and other plant growth regulators are widely used for mass propagation of plant material in *in vitro* cultures, the search for new agents is still ongoing. Therefore, research aimed at evaluating the effectiveness of SW and KAR_1 in regulation of processes in *in vitro* cultures is necessary as the compounds mentioned show a high *in vivo* activity.

17.1.1 Pollen Germination and Tube Growth

Wildfires and smoke play an important role in ecosystems. Smoke from burning plant material not only induces seed dormancy release and germination (see Chapter 16) but also stimulates bulb and rhizome sprouting (Soós et al., 2019). Data on about 400 plant species from South Africa and Australia show that fire can stimulate flowering and, thus, seed production (Lamont and Downes, 2011). Orchids are the most numerous group that respond to fire by flowering. It turns out that fire can act directly to stimulate flowering, but smoke can elicit such response as well. Kelley (1993) found smoke to be responsible for induction of flowering in the fire-sensitive fynbos geophyte *Cyrtanthus ventricosus*. Likewise, *Watsonia borbonica*, responsive to fire, can be induced to flower by SW (Kulkarni et al., 2007). Since smoke has been proven to induce flowering, it was interesting to find out if it was also effective for pollen germination of plants from fire-prone areas. It is important to bear in mind that pollen quality is important as high pollen quality ensures fast germination, enabling tube growth and the resultant successful fertilization. The *in vivo* pollen germination as well as tube growth depend on both genome information and external factors. Research on the physiology of pollen involved, *inter alia*, assessing the influence of hormones on the germination and growth of the pollen tube. Cytokinins, auxins, and gibberellic acid have been found to stimulate *in vitro* pollen germination and tube growth in numerous plant species. Papenfus et al. (2014) showed SW to stimulate pollen germination and tube growth in three species of the Amaryllidaceae (*Clivia gardenii*, *Cyrtanthus mackenii*, and *Scadoxus multiflorus*) from fire-prone areas. Moreover, germination of *C. gardenii* and *S. multiflorus* was also enhanced by SW applied in combination with nutrients. KAR_1 was also able to stimulate pollen germination of all the species tested and to enhance tube growth in *C. gardenii* and *S. multiflorus*. The application of KAR_1 in the medium increased tube elongation only in *S. multiflorus*. Research on the influence of SW and KAR_1 also involved pollen of species from other plant families (Kumari et al., 2015) and demonstrated that SW stimulates pollen germination and tube growth in *Kniphofia uvaria* and *Paeonia lactiflora*, whereas KAR_1 stimulated both germination and tube elongation in *Aechmea kertesziae*, *Aloe maculata*, *Kniphofia uvaria*, *Nematanthus crassifolius*, and *Tulbaghia simmleri*. Adding SW or KAR_1 to the medium changed the pollen response to SW and/or KAR_1. Kumari et al. (2015) concluded that since SW or KAR_1 increases pollen germinability and tube growth potential, they can be used to increase the number of pollinated flowers, thus increasing fertilization and, consequently, the seed yield.

17.1.2 Efficiency of Somatic Embryogenesis

Zygotic embryogenesis should be considered the most important process in the plant life cycle; its starts with egg fertilization and ends with the formation of a mature embryo. In nature, however, embryogenesis without fertilization, so-called somatic embryogenesis (SE), can occur as well. This process can be defined as a developmental pathway leading to differentiation of somatic cells into zygotic embryo-like structures called somatic embryos or artificial seeds. Thus, on account of the totipotency of plant cells, SE, an asexual process not involving gamete fusion, leads to the formation of embryos. This multi-step process, depending chiefly on auxins and cytokinins as well as on ethylene, abscisic acid, jasmonic acid, and gibberellins includes two main phases: induction and differentiation (Mikuła et al., 2022). SE is an excellent model system used in research on embryogenesis; it also allows the use of diploid embryos for commercial mass plant production and haploid embryos to obtain haploid plants for breeding purposes. Therefore, research aimed at optimizing conditions for SE to increase the efficiency of the process is highly desirable. This research involves, in addition to the previously applied hormones and growth regulators, SW. Senaratna et al. (1999) were the first to experimentally study the influence of SW on SE using a *Pelargonium hortorum* culture. Explants from seedling hypocotyls were incubated in a thidiazuron-enriched induction medium for three days and were then transferred to a medium without thidiazuron to obtain callus and somatic embryos. When SW was applied to explants before transfer to the induction medium, the total number of somatic embryos increased, with about twice the number of embryos per explant being observed. When used during induction, SW increased the number of somatic embryos to a similar amount as well. Effects of SW on the number of cotyledonary embryos, the most advanced stage of embryo development, was also assessed. Regardless of whether SW was applied to the hypocotyl before or during the induction phase, it did increase the number of cotyledonary embryos by a factor of about five. The data provided showed SW to have markedly increased the SE efficiency and embryo quality. SE, a unique process in the plant life cycle, has been widely applied in biotechnology as a basic tool for mass production of forest trees. SE was induced in some *Pinus* species using vegetative shoot apices. Malabadi and

Nataraja (2007) examined effects of SW on SE using vegetative shoot apices of three genotypes of *Pinus wallichiana*, an important native tree of the Himalayan region. SW applied to the induction medium containing 2.4-dichlorophenoxyacetic acid, 1-naphthaleneacetic acid and benzyladenine or to the maturation medium increased the number of somatic embryos to a genotype-dependent level. In the most sensitive genotype, SW applied to the induction medium increased the number of cotyledonary embryos about 3.5 times. Data reported by Ma et al. (2006) on an experiment involving KAR_1 used in the induction medium showed that KAR_1 accelerated the development of torpedo embryos on explants from coleoptiles, shoots, and leaves of *Baloskion tetraphyllum*. Effects of SW on induction of the secondary embryogenesis using hypocotyl-derived embryos of *Brassica napus* were investigated as well (Abdollahi et al., 2012). Secondary embryogenesis, also termed a repetitive embryogenesis, is the formation of embryos on primary embryos, usually on cotyledons or hypocotyls. SW was applied prior to or during germination of microspore-derived embryos. In both cases, the secondary embryogenesis was markedly enhanced (Abdollahi et al., 2012).

17.1.3 Somatic Embryo Germination and Conversion to Plantlets

SE makes it possible to produce a huge number of embryos from which to obtain homogenous plants. However, somatic embryos can be used for the production of plants if they are at an appropriate developmental stage and when they are able to germinate and convert into plantlets. Experiments with *P. wallichiana* showed that SW induced germination of cotyledonary embryos, the inducement being expressed as a five- to ten-fold increase in the number of germinated embryos, depending on the genotype (Malabadi and Nataraja, 2007). The beneficial effect of SW was also evident when the number of surviving seedlings was determined. Without SW, as few as 1 to 3 seedlings were obtained, whereas SW made it possible to obtain about 30 seedlings. KAR_1 also had a beneficial effect on the germination of the somatic embryos and development of plantlets of *Baloskion tetraphyllum* (Ma et al., 2006). Namely, KAR_1 increased the frequency of root formation by somatic embryos and the frequency of plantlet development. SW was also used for treating *Brassica napus* microspore-derived embryos (Ghazanfari et al., 2012), which increased the percentage of plantlets' regeneration from embryos. An increase in root and shoot length was also recorded. In addition, the beneficial effects of SW were related to a decline in callogenesis.

17.1.4 Seed Germination

The most important information on the effects of SW and KAR_1 on seed germination in plants from fire-prone areas as well as in weeds and cultivated plants has already been published (Light et al., 2009; Nelson et al., 2012; Kępczyński, 2018), so it is not surprising that experiments have been carried out on the influence of SW and KAR_1 on *in vitro* seed germination. In particular, the two agents were used to assess their suitability for improving seed germination in orchids. Orchids are very important because of their medicinal use, and also, because of their beautiful flowers of various shapes and colors, they are widely used in ornamental horticulture, as both cut and potted flowers. In the commercial mass production of orchids, they are propagated using different *in vitro* cultures of, e.g., shoot tips, nodal segments, callus, or seeds. To improve seed germination, SW was used as the seed-sowing basal medium for a few orchid species, including *Vanda parviflora*, a native epiphytic orchid that requires a symbiotic relationship with mycorrhizal fungus to germinate in nature. SW turned out to stimulate germination of asymbiotic seeds, differentiation of protocorms, and plant regeneration (Malabadi et al., 2008). Similarly, SW applied to seeds of the epiphytic orchid *Xenikophyton smeeanum* also produced a high percentage of germination and plantlet recovery (Malabadi et al., 2011). Another experiment compared the effects of SW and KAR_1 on germination in the epiphytic orchid *Ansellia africana* (Papenfus et al., 2016). In contrast to KAR_1, SW increased seed germination and the formation of protocorms. The effects of SW on seed germination in *Tulbaghia ludwigiana* and *T. violacea*, popular garden plants, were also tested (Aremu et al., 2014). SW was shown to accelerate seed germination and root production in *T. ludwigiana* only. SW was also used to stimulate *Aloe arborescens* seeds, not particularly widely used for traditional and *in vitro* propagation because of their usually poor germination. In this case, KAR_1 was able to increase the seed germination percentage too (Espinosa-Leal and Garcia-Lara, 2020). In another experiment, various karrikins were used to test their suitability for improving *in vitro* germination of seeds of *Balsamorhiza deltoidea* and *B. sagittata*, plants native to America that grow in fire-prone areas and are used as food and in medicine (Monthony et al., 2020). Seed germination in both species was enhanced by KAR_2, but not by KAR_1. An experiment with *Baloskion tetraphyllum* showed that KAR_1 accelerated germination and increased the total percentage of seed germination (Ma et al., 2006).

17.1.5 Callus Proliferation and Organ Differentiation

Callus, which consists of an undifferentiated or poorly differentiated tissue, is formed under natural conditions and in *in vitro* cultures as a result of wound-caused induction. In *in vitro* cultures, wounding is associated with excising an explant from the plant organism. Calluses are used in bioassays to determine cytokinin activity and to prepare cell suspensions, as well as to induce indirect organogenesis. Soybean callus was used in a bioassay to test the SW activity. The results showed that KAR_1 stimulated callus growth by enhancing cell division without affecting the cell size (Jain et al., 2008). Thus, it was correctly concluded that KAR_1 was characterized by cytokinin-like activity. When used to treat mung bean cuttings, KAR_1 was observed to stimulate rooting, thus demonstrating auxin-like activity. In another experiment, when SW was introduced to a suspension of tomato primary roots, the root growth was stimulated, and the frequency of lateral roots was increased (Taylor and Van Staden, 1998). In an experiment with bananas explants, SW – in contrast to KAR_1 – increased the number of leaves and pseudostems (Aremu et al., 2012).

17.1.6 Production of Secondary Metabolites

In addition to synthesizing compounds participating in basic metabolism, regulating it by controlling the acquisition of nutrients necessary for basic metabolic pathways, plants also synthesize tens of thousands of various compounds referred to as secondary metabolites. These compounds include plant hormones, with a regulatory function as well as substances enabling plant survival in nature; there are also compounds the role of which is still unclear. Many compounds classified as secondary metabolites or their derivatives are used in pharmacology, human diet, and animal nutrition. Deriving secondary metabolites from plants grown *in vivo* is difficult, as it often requires large areas for cultivation because the target crop occurs at a low density in association with a regular crop; in addition, crops are dependent on environmental conditions. Moreover, specialized compound isolation and purification technologies are required. Although technologies for the production of many of them have been developed, methods for increasing production efficiency are still being pursued. For these reasons, research is being carried out on the use of *in vitro* cultures, cell suspensions, and organ cultures for the production of secondary metabolites. The use of SW and or KAR_1 to increase the production of certain secondary metabolites – e.g., phenolic compounds and terpenoids – has been dealt with in a few studies only.

17.1.6.1 Phenolic Compounds

Phenolic compounds serve various very important functions; notably, they protect plants against infections by microorganisms, against herbivores, and against the noncarcinogenic effects of ultraviolet radiation. Moreover, they attract pollinators and are involved in allelopathy. There are many instances of medicinal uses of phenolic acids. Rosmarinic acid, a polyphenol occurring in several plants, acts as an antioxidant as well as an antimicrobial, immunomodulatory, antidiabetic, antiallergic, anti-inflammatory, and hepato- and renal-protection agent. Salvianolic acid is a strong antioxidant, used in cardiovascular and cerebrovascular diseases, aging, and liver fibrosis. Experiments involving explants from banana plantlets to determine the effects of SW and KAR_1 on secondary metabolites showed SW to be capable of increasing the total content of phenolics in roots and KAR_1 to increase the total phenolics in both plantlet leaves and roots (Aremu et al., 2012). The total flavonoid level in leaves and roots was found to be increased by both SW and KAR_1. In another experiment, hairy roots induced by *Agrobacterium rhizogenes* in *Salvia miltiorrhiza*, a plant important in traditional Chinese medicine, were treated with SW and KAR_1. Here, both SW and KAR_1 increased the content of salvianolic and rosmarinic acids (Zhou et al., 2018b). Aremu et al. (2014) examined the *in vitro* effects of SW on the contents of phenolics, flavonoids, and condensed tannins in seedlings of *Tulbaghia ludwigiana* and *T. violacea*, plants used in medicine and horticulture. The concentrations of all the compounds were demonstrated to be increased by the SW treatment in *T. ludwigiana* seedlings only. Experiments in which SW and KAR_1 were used to stimulate the production of secondary metabolites in *Eucomis autumnalis*, a plant popular in African folk medicine and applied against various diseases, showed SW and KAR_1 to increase the content of protocatechuic and p-hydrobenzoic acids in plantlets induced on leaf explants (Aremu et al., 2019). Likewise, levels of caffeic, coumaric, and cinnamic acids were also enhanced by SW and KAR_1. In addition, both SW and KAR_1 increased the antioxidant activity in plantlets.

17.1.6.2 Terpenoids-Tanshinones

In addition to salvianolic and rosmarinic acids, *Salvia miltiorrhiza* synthesizes other pharmacological compounds, tanshinones. Roots of the plant are used to relieve pain and remove blood stasis. Tanshinones were reported as potentially having a wide range of antitumour effects with respect to several human tumour cell lines. Zhou et al. (2018a) applied SW or KAR_1 to hairy root cultures obtained by using *A. rhizogenes* for leaf infection. Both agents increased the tanshinone content. Other experiments demonstrated that KAR_1 enhanced the level of jasmonic acid and stimulated production of NO, which was able to increase the concentration of tanshinone I. Jasmonic acid synthesis inhibitors antagonized the effects of KAR_1 and NO (Zhou et al., 2019a,b), and an NO scavenger decreased the stimulatory effect of KAR_1 on the tanshinone I content. Thus, both jasmonic acid and NO are involved in the response to KAR_1.

17.2 SUMMING UP AND PERSPECTIVES

Detection of the biological activity of plant-derived smoke and, later, KAR_1 resulted in great interest in the practical application of these agents under *in vivo* conditions and in explaining the mechanism of KAR_1 activity. So far, most studies have investigated the influence of these agents on releasing seed dormancy and seedling emergence and development, as well as their application for seed priming. A similar biological material was used in studies aimed at explaining the mechanism of activity. On the other hand, there is relatively little information on the use of SW and KAR_1 in tissue culture–based plant biotechnology. There are a few examples of applying SW/KAR_1 in *in vitro* cultures, so it is necessary to extend the scope of the research by using more plant species. It would also be worthwhile to compare the effectiveness of KAR_1 with that of other karrikins. Moreover, it would be interesting to broaden the extent of the studies by including research on interactions between KAR_1 and plant hormones. Examples of the successful use of SW or KAR_1 as elicitors also indicate that it is worth checking the effects of these substances on production of secondary metabolites other than those investigated so far. *In vitro* cultures provide ample opportunity for further research into the mechanism of karrikin activity.

REFERENCES

Abdollahi, Mohammad R., Ghazanafari, Pouriya, Corral-Martínez, Patricia, Moieni, Ahmad, and Seguí-Simarro, Jose M. "Enhancing secondary embryogenesis in Brassica napus by selecting hypocotyl-derived embryos and using plant-derived smoke extract in culture medium." Plant Cell Tissue and Organ Culture 110 no. 2 (2012): 307–315. http://dx.doi.org/10.1007/s11240-012-0152-7

Aremu, Adeyemi O., Bairu, Michael W., Finnie, Jeffrey F., and van Staden, Johannes. "Stimulatory role of smoke water and karrikinolide on the photosynthetic pigment and phenolic contents of micropropagated 'Williams" bananas." Plant Growth Regulation 67 no. 3 (2012): 271–279. http://dx.doi.org/10.1007/s10725-012-9685-3

Aremu, Adeyemi O., Masondo Nqobile A., Gruz, Jiri, Dolezal, Karel, et al. "Potential of smoke-water and one of its active compounds (karrikinolide, KAR_1) on the phytochemical and antioxidant activity of euconis autumnalis." Antioxidants 8 no. 611 (2019). http://dx.doi.org/10.3390/antiox8120611

Aremu, Adeyemi O., Masondo, Nqobile A., and van Staden, Johannes. "Smoke-water stimulates secondary metabolites during in vitro seedling development in Tulbaghia species." South African Journal of Botany 91 (2014): 49–52. http://dx.doi.org/10.1016/j.sajb.2013.12.001

Espinosa-Leal, Claudia A., and Garcia-Lara, Silverio. "In vitro germination and initial seedling development of krantz aloe by smoke-saturated water and seen imbibition." HortTechnology 30 no. 5 (2020). https://doi.org/10.21273/HORTTECH04643-20

Ghazanfari, Pouriya, Abdollahi, Mohammad R., Moieni, Ahmad, and Moosavi, Sayyed S. "Effect of plan-derived smoke extract on in vitro plantlet regeneration from rapeseed (Brassica napus L. cv. Topas) microspore-derived embryos." International Journal of Plant Production 6 no. 3 (2012): 309–324.

Jain, Neeru, Stirk, Wendy A., and van Staden, Johannes. "Cytokinin-and auxin-like activity of a butenolide isolated from plant-derived smoke." South African Journal of Botany 74 (2008): 327–331. http://dx.doi.org/10.1016/j.sajb.2007.10.008

Kelley, Jon E. "Smoke-induced flowering in the fire-lily Cyrtanthus ventricosus." South African Journal of Botany 59 (1993): 638. https://doi.org/10.1016/S0254-6299(16)30681-0

Kępczyński, Jan. "Induction of agricultural weed seed germination by smoke and smoke-derived karrikin (KAR_1), with a particular reference do *Avena fatua* L". Acta Physiologiae Plantarum 40 no. 5 (2018): 87. http://dx.doi.org/10.1007/s11738-018-2663-2

Kępczyński, Jan. "Progress in utilizing plant-derived smoke water and smoke-derived KAR_1 in plant tissue culture." Plant Cell, Tissue and Organ Culture (PCTOC) 140 (2020): 271–278. https://doi.org/10.1007/s11240-019-01739-8

Kępczyński, Jan, Nemoykina, Anna, and Kępczyńska, Ewa. "Ethylene and in vitro rooting of rose shoots." Plant Growth Regulation 50 (2006): 23–28. https://doi.org/10.1007/s10725-006-9122-6

Kulkarni, Manoj G., Ascough, Glendon, and van Staden, Johannes. "Improved flowering of a South African Watsonia with smoke treatments." South African Journal of Botany 73 no. 2 (2007): 298–298. https://doi.org/10.1016/j.sajb.2007.02.074

Kumari, Aloka, Papenfus, Heino B., Kulkarni, Manoj G., Pošta, Martin, and van Staden, Johannes. "Effect of smoke derivatives on in vitro pollen germination and pollen tube elongation of species from different plant families." Plant Biology 17 no. 4 (2015): 825–830. http://dx.doi.org/10.1111/plb.12300

Lamont, Byron B., and Downes, Katherine. "Fire-stimulated flowering among resprouters and geophytes in Australia and South Africa." Plant Ecology 212 no. 12 (2011): 2111–2125. https://doi.org/10.1007/s11258-011-9987-y

Light, Marnie E., Daws, Matthew I., and van Staden, Johannes. "Smoke-derived butenolide: towards understanding its biological effect." South African Journal of Botany 75 (2009): 1–7. https://doi.org/10.1016/j.sajb.2008.10.004

Ma, Guoha, Bunn, Eric, Dixon, Kingsley, and Flematti, Gavin. "Comparative enhancement of germination and vigor in seed and somatic embryos by the smoke chemical 3-methyl-*2H*-furo [2,3-*C*] pyran-2-one in Baloskion tetraphyllum (Restionaceae)." In Vitro Cell Development Biology 42 no. 3 (2006): 305–308. http://dx.doi.org/10.1079/IVP2006758

Mikuła, Anna, Gaj, Małgorzata, Grzyb, Małgorzata, Hazubska-Przybył, Teresa, Kępczyńska, Ewa, et al. "Polish contribution to global research on somatic embryogenesis." Acta Societatis Botanicorum Poloniae 91(9115) (2022): 1–25.

Malabadi, Ravindra B., da Silva, Jaime A.T., Mulgund, Gangadhar S. "Smoke-saturated water influences in vitro seed germination of Vanda parviflora Lindl." Seed Science and Biotechnology 2 (2008): 65–69.

Malabadi, Ravindra B., and Nataraja, K. "Smoke-saturated water influences somatic embryogenesis using vegetative shoot apices of mature trees of Pinus wallichiana A.B. Jacks." Journal of Plant Sciences 2 no. 1 (2007): 45–53. http://dx.doi.org/10.3923/jps.2007.45.53

Malabadi, Ravindra B., Vijaykumar, S, da Silva, Jaime A.T., Mulgund, Gangadhad S., and Nataraja, K. "In vitro seed germination of an epiphytic orchid Xenikophyton smeeanum (Reichb. f.) by using smoke-saturated-water as natural growth promoter." International Journal of Biological Technology 2 no. 2 (2011): 35–41.

Monthony, Adrian S., Baethke, Kathy, Erland, Lauren A. E., and Murch, Susan J. "Tools for conservation of Balsamorhiza deltoidea and Balsamorhiza sagittata: Karrikin and thidiazuron-induced growth." In Vitro Cellular & Developmental Biology – Plant (2020). https://doi.org/10.1007/s11627-019-10052-0

Morffy, Nicholas, Faure, Lionel, and Nelson, David C. "Smoke and hormone mirrors: Action and evolution of karrikin and strigolactone signaling." Trends in Genetics 32 (2016): 176–188. https://doi.org/10.1016/j.tig.2016.01.002

Moshkov, Igor E., Novikova, Galina V., Hall, Michael A., and George, Edwin F. "Plant growth regulators III: Gibberellins, ethylene, abscisic acid, their analogues and inhibitors; miscellaneous compounds." In George, Edwin F., Hall, Michael A., and de Klerk, Geert-Jan (eds). *Plant propagation by tissue culture*. Springer, Dordrecht, 2008, 227–282. https://doi.org/10.1007/978–1 - 4020–5005–3_7

Nelson, David C., Flematti, Gavin R., Ghisalberti, Emilio L., Dixon, Kingsley W., et al. "Regulation of seed germination and seedling growth by chemical signals from burning vegetation." Annual Review of Plant Biology 63 (2012): 107–130. https://doi.org/10.1146/annurev-arplant-042811-105545

Papenfus, Heino B., Kumari, Aloka, Kulkarni, Manjo G., Finnie, Jeffrey F., and van Staden, Johannes. "Smoke-water enhances in vitro pollen germination and tube elongation of three species of Amaryllidaceae." South African Journal of Botany 90 (2014): 87–92. https://doi.org/10.1016/J.SAJB.2013.10.007

Papenfus, Heino B., Naidoo, Devashan, Pošta, Martin, Finnie, Jeffrey F., and van Staden, Johannes. "The effects of smoke derivatives on in vitro seed germination and development of the leopard orchid Ansellia Africana." Plant Biology 18 (2016): 289–294. https://doi.org/10.1111/plb.12374

Ruduś, Izabela, Kępczyńska, Ewa, and Kępczyński, Jan. "Comparative efficacy of abscisic acid and methyl jasmonate for indirect somatic embryogenesis in Medicago sativa L." Plant Growth Regulation 48 (2006): 1–11. https://doi.org/10.1007/s10725-005-5136-8

Senaratna, Tissa, Dixon, Kingsley, Bunn, Eric, and Touchell, Darren. "Smoke-saturated water promotes somatic embryogenesis in geranium." Plant Growth Regulation 28 (1999): 95–99. https://doi.org/10.1023/A:1006213400737

Soós, Vilmos, Badics, Eszter, Incze, Norbert, and Balazs, Ervin. "Fire-borne life: A brief review of smoke-induced germination." Natural Product Communications (2019). https://doi.org/10.1177/1934578X19872925

Taylor, Joslyn L.S., and van Staden, Johannes. "Plant-derived smoke solutions stimulate the growth of Lycopersicon esculentum roots in vitro." Plant Growth Regulation 26 (1998): 77–83. https://doi.org/10.1023/A:1006088109106

Zhou, Jie, Ran, Zhifang, Liu, Qian, Xu, Zixin, et al. "Jasmonic acid serves as a signal role in smoke-isolated butanolide-induced tanshinones biosynthesis in Salvia miltiorrhiza hairy root." South African Journal of Botany 121 no. 11 (2019a): 355–359. http://doi.org/10.1016/j.sajb.2018.10.029

Zhou, Jie, Ran, Zhifang, Ran, Xu Zixin, Liu, Qian, et al. "Effect of smoke-water and smoke-isolated karrikinolide on tanshinones production in Salvia miltiorrhiza hairy roots." South African Journal of Botany 119 (2018a): 265–270. https://doi.org/10.1016/j.sajb.2018.09.005

Zhou, Jie, Xu, Zixin, Ran, Zhifang, Fang, Lei, and Guo, Lanping. "Effects of smoke-water and smoke-derived butanolide on accumulation of phenolic acids in cultures hairy roots of Salvia miltiorrhiza bung". Bangladesh Journal of Botany 47 no. 3 (2018b): 479–485. http://dx.doi.org/10.3329/bjb.v47i3.38715m

Zhou, Jie, Xu, Zixin, Sun, Hui, and Guo, Lan-Ping. "Smoke-isolated karrikins stimulated tanshinones biosynthesis in Salvia miltiorrhiza through endogenous nitric oxide and jasmonic acid." Molecules 24 no. 7 (2019b): 1229. http://dx.doi.org/10.3390/molecules24071229

18 The Role of Karrikins in Addressing Pressing Global Challenges

Plant Abiotic Stress Tolerance, Biodiversity Restoration, and Karrikin Commercial Realities

Jitka Kochanek, Georgina McGregor, Kenneth A. Tryggestad, Gavin R. Flematti, Yunwei Wang, and Santi Krisantini

CONTENTS

18.0 INTRODUCTION

Agriculture, ecological restoration, and other plant-related sectors are facing critical global challenges. Anthropogenic climate change, pollution, and soil degradation subject crops and natural environments to abiotic stresses with increasing intensity and frequency and are often simultaneous and/or sequential (Rivero et al., 2022). While the world's population rapidly approaches 10 billion (forecast for circa 2057; United Nations, Department of Economic and Social Affairs, Population Division 2019), climatic extremes such as heat waves and droughts, pollutants such as heavy metals, and declining soil health are exacerbating pest and disease stressors and increasingly threatening crop yields. This perfect storm imperils food and fibre supply and, therefore, future global food security and resource availability (Rivero et al., 2022; Li et al., 2021). Alarmingly, 5–50% yield reductions are predicted by 2100 for major cereal crops if climate change is not slowed (Arora, 2019). In parallel, large-scale ecosystem conservation/restoration is urgently required to protect biodiversity and create environmental resilience to better withstand climate change and related public health threats, including future pandemics (Dinerstein et al., 2020). Hence, enhancing plant stress tolerance and effective ecosystem restoration are urgent targets and the focus of this review.

As sessile organisms, plants cannot escape their surrounding environment but have instead evolved a sophisticated arsenal of sensing and adaptive mechanisms that modify their development and growth to cope when their surroundings are harmful and thrive when conditions are favourable (Kochanek et al., 2011). To this end, almost every plant function is coordinated by phytohormones, which are endogenous signaling molecules that regulate plant development, growth, defense, and

DOI: 10.1201/9781003224525-22

more (Li et al., 2021; Vanstraelen and Benkova, 2012). Plant growth regulators (PGRs) are naturally occurring or synthetic compounds widely utilized in modern agriculture/horticulture/viticulture, most commonly to control plant developmental and metabolic processes by modulating the balance of endogenous phytohormones (Rademacher, 2015). Increasingly, PGRs are also being recognized for their potential role in ameliorating damage from abiotic stress and/or improving stress tolerance (Li et al., 2021). This chapter reviews the emerging PGR karrikinolide (KAR_1) and its analogues for their capacity to alleviate plant stress and enhance biodiversity restoration success. Unlike previous literature, this review additionally explores the barriers to commercial karrikin applications and suggests future research priorities to overcome such barriers.

Karrikins are a family of butenolide (furanone) containing compounds fused to a pyran ring, which are derived in nature from charred plant materials and smoke released by burning vegetation. Karrikins provide environmental signals in the form of chemical cues that stimulate germination from seed banks for many wild, fire-following plant taxa (Flematti et al., 2009). Extraordinarily, the karrikin perception system is also fundamental in broad plant functioning across angiosperms (Nelson et al., 2009), likely owing to the evolution of vascular plants at the peak of global fire events (Waters et al., 2013). Since the discovery of karrikinolide or KAR_1 (Flematti et al., 2004) and its availability through chemical synthesis (Flematti et al., 2005), six karrikins have been identified in smoke derived from combustion of plant vegetation (KAR_1 to KAR_6; Flematti et al., 2009). KAR_1 is the most bioactive and abundant germination stimulant, although KAR_2 is more effective for *Arabidopsis* (Nelson et al., 2010) and possibly certain crops (Meng et al., 2016), and KAR_3 is equally effective for certain fire-following taxa (Flematti et al., 2007). The model plant *Arabidopsis thaliana* has been used to demonstrate karrikin function in plants, whereby karrikin is believed to interact with a putative α/ß-hydrolase receptor protein, KAI2, which induces a conformational change (Xu et al., 2016; Guo et al., 2013; Waters et al., 2012), transducing a signal to the F-box protein MAX2 that degrades various growth-repressing proteins, resulting in plant developmental responses (Yang et al., 2019; Stanga et al., 2013; Nelson et al., 2011).

Such responses include stimulated seed germination for wild plants from fire-prone environments but also for non-fire-following species (Flematti et al., 2004) such as *Arabidopsis* (Nelson et al., 2009), certain crops (Jain and Van Staden, 2007) and weeds (Long et al., 2011; Stevens et al., 2007), and enhanced seed germination and/or seedling development under suboptimal light and/or in darkness (Merritt et al., 2006), including enhancing the light response that accelerates the transition to photoautotrophy (i.e., photomorphogenesis via hypocotyl de-etiolation and accelerated cotyledon expansion) (Bursch et al., 2021, Nelson et al., 2010), enhanced root growth and development (Swarbreck, 2021; Villaecija-Aguilar et al., 2019; Carbonnel et al., 2020), enhanced phloem formation and meristem growth (Wallner et al., 2017), and enhanced tolerance to abiotic stress (Kucukakyuz and Catav, 2021; Kochanek et al., 2016). Crosstalk with endogenous phytohormones, such as gibberellic acid (GA), abscisic acid (ABA) (Shah et al., 2020; Meng et al., 2016), cytokinin (CK) (Aremu et al., 2016), and ethylene (ET) (Swarbreck, 2021), assists in plant developmental regulation under both abiotic stress and non-stress conditions (Yang et al., 2019).

The focus of this review is on the potential role of karrikins in overcoming two of the most pressing global challenges of our time: the restoration of ecosystem biodiversity to decelerate species loss (Section 18.1) and the enhancement of plant abiotic stress tolerance to provide food and fibre security for a rapidly expanding global population under increasingly extreme conditions (Section 18.2). Uniquely, this review also explores the barriers to commercial karrikin usability and applicability (Section 18.3), concluding with a summary of research gaps and future prospects to maximize the potential impact of karrikins for food/fibre security and ecosystem functioning (Section 18.4). This review focuses exclusively on pure karrikins, while smoke water/pyrolytic products/extracts are not reviewed since such products contain complex organic compounds/toxins (Flematti et al., 2009), which can be permissive or inhibitory (Soos et al., 2012); hence, benefits to plants are often species specific and can be hit and miss (Kochanek et al., 2022). Karrikins for crop benefit alone were recently reviewed (Antala et al., 2020) and hence are not covered here.

18.1 KARRIKINS FOR ECOSYSTEM BIODIVERSITY RESTORATION

The restoration/conservation of nature is crucial to combat the dual crises of biodiversity loss and climate change (Dinerstein et al., 2020), and wild plant seeds underpin large-scale plant reintroduction success since direct seeding is substantially more cost effective than seedling transplants (Raupp et al., 2020; Perez et al., 2019; Merritt and Dixon, 2011). However, seed-based restoration success is low; ≤ 11% of sown seeds result in viable seedlings. This is due, in part, to the lack of "on-demand" dormancy release pre-treatments, such as effective KAR_1 delivery (for broader factors, see Ceccon et al., 2016; Merritt and Dixon, 2011). Seed dormancy occurs in 50–90% of wild plant species, and chemical stimulants, such as KAR_1, can release the most common seed dormancy classes, including physiological, morpho-physiological, and combinational dormancy (i.e., seeds with embryos displaying low growth potential but which can germinate with stimulants) (Kildisheva et al., 2020).

It has long been known that many wild plant seeds germinate after fire (Keeley et al., 1985). Smoke was specifically recognized as a seed-germination cue by Brown (1993) while KAR_1 was discovered by Flematti et al. (2004) and shown to germinate seeds of Australian, South African, and North American wild species and lettuce (*Lactuca sativa* L. cv. Grand Rapids). KAR_1 has since been demonstrated to enhance seed germination of wild plants from broad regions (e.g., China, the USA and Australia; see Table 18.1) (Zhou et al., 2014; Flematti et al., 2007; Kildisheva et al., 2019; Commander et al., 2008) and broad families (e.g., Asteraceae, Solanaceae and Poaceae) (Merritt et al., 2006; Commander et al., 2008; Berto et al., 2021). In most cases, KAR_1 results in higher seed germination than smoke water (Flematti et al., 2004; Commander et al., 2008), other karrikins (Flematti et al., 2007), and, for many Australian species, gibberellic acid (GA_3) (Erickson et al., 2016), although in other studies, GA_3 outperformed KAR_1: e.g., certain Australian *Solanum* species (Commander et al., 2008) and USA Great Basin species (Kildisheva et al., 2019). One clear advantage of karrikins is that they are known to concomitantly promote seed germination and healthy seedling growth (or photomorphogenesis) (Bursch et al., 2021) while gibberellins can cause seedling abnormalities, e.g., skotomorphogenesis or etiolated growth (Kusnetsov et al., 2020); hence, they require cautious use.

To deliver compounds to seeds, commercial companies utilize seed priming and coating. Seed priming, used since the 1800s (Hegarty, 1978), is the hydration of seeds until pre-germination metabolism begins, but then seeds are redried and stored before germination (Farooq et al., 2019). Seed coating, used since the 1930s (Kaufman, 1991), covers seeds with materials to assist with their handling, germination, and/or establishment (Pedrini et al., 2017). Similar protocols were tested by Flematti et al. (2004), who soaked seeds in KAR_1 solution, dried them, and sowed them into propagation media, observing positive effects on wild plant seed germination in all but one species tested. Similarly, Berto et al. (2021) observed that Australian native spinifex seeds had their seed germination increased and time to germinate accelerated when primed in a KAR_1 solution within specially made aerated cylinders. Nonetheless, priming on agar containing KAR_1 outperformed the KAR_1 solution treatment, particularly after flash-flaming co-removed outer seed coverings (Table 18.1; control seed germination was 0–6.4%; flaming + KAR_1-solution-priming 38–62%; flaming + KAR_1-agar-priming 74–79%). Seed coating reduced germination and was not recommended. The same experiment should be repeated in soil since Berto et al. (2021) germinated seeds only on agar. By contrast, Kildisheva (2019), who used the same KAR_1-solution-priming protocol, observed <20% spinifex seed germination when sown into soil (Tables 18.1 and 18.2). Most other studies have germinated seeds/grown seedlings on filter paper/agar in a petri dish rather than in soil/in-field (Table 18.1); hence, positive seed germination outcomes using KAR_1 should be viewed with caution. Indeed, substantially more research is required to develop protocols to deliver KAR_1 to wild plants within soil. The wild plant restoration sector generally acknowledges that current seed treatments/handling are inadequate, and improved germination fractions, synchronicity of seedling emergence, survival, and growth rates in-field are urgently needed to manage seed supply and conservation

TABLE 18.1
Karrikin Effects on Wild Plant Germination and/or Growth for Ecosystem Biodiversity Restoration Purposes (Studies are grouped by application and growing methods to illustrate the field and/or commercial relevance of each study.) Created with BioRender.com.

Application Method	Summary	Results – KAR_1 Effects on Plants (Relative to Untreated Controls without KAR_1)
Seeds primed in a Petri dish with filter paper wetted with $H_2O \pm KAR_1$ and rewetted with H_20 as required. Seedlings grown in the same Petri dish. 	KAR_1 improved seed germination of Australian native *Conostylis aculeata* R. Br. and *Stylidium affine* Sonder and lettuce (*Lactuca sativa* L. cv. Grand Rapids).	**0.67 nM–6.7 µM KAR_1** (Flematti et al., 2004) • ↑ seed germination of lettuce and both native species. • Mostly outperformed smoke water.
	KAR_1 improved seed germination of the Chinese native species, *Aristolochia debilis* Sieb. Et Zucc. But GA_3 outperformed KAR_1.	**0.1 µM KAR_1** (1 and 10 nM also tested; Zhou et al., 2014) • ↑ seed germination with KAR_1 (46% in the dark at all doses, >90% in the light with KAR_1 at 100 nM, approximately 80% with 10 nM, *c.* 59% with 1 nM). GA_3 outperformed KAR_1 – higher germination at optimal concentration in darkness.
	KAR_1 improved seed germination of Australian native *Solanum orbiculatum*, USA native *Emmenanthe penduliflora* Benth, and lettuce cv. Grand Rapids.	**6.7 nM–6.7 µM KAR_1** (Flematti et al., 2007) • ↑ seed germination of lettuce and native species at all concentrations. *S. orbiculatum* displayed the greatest improvement in germination (control <5%; all KAR_1 dosages ≥90%). • KAR_1 outperformed most 3-methyl, 3-desmethyl and nitrogen analogues across species.
	↑ seed germination for 5 out of 7 Australian native ephemeral Asteraceae species.	**KAR_1** (Merritt et al., 2006) Detailed in Table 18.2 since suboptimal light and temperatures tested.
Seeds soaked for 24 hours, dried in a laminar flow cabinet, and sown into a propagation mix. 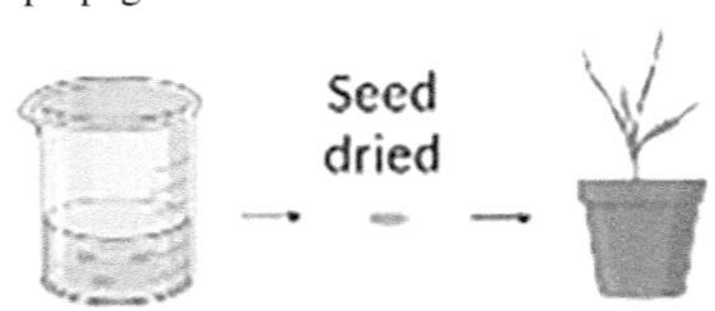	KAR_1 improved germination for 6 Australian, 2 South African, and 4 North American native species.	**67 nM KAR_1** (Flematti et al., 2004) • ↑ seed germination for 6 Australian, 2 South African, and 3 North American species. • No effect on seed germination for 1 South African species. 0.67 µM KAR_1 ↑ seed germination for 1 North American species.

Cylinders with 1.5 L liquid hydro-primed (H_2O) or KAR_1-primed (H_2O + KAR_1) seeds. Primed seeds were sown onto water agar (0.7% w/v) in a Petri dish. 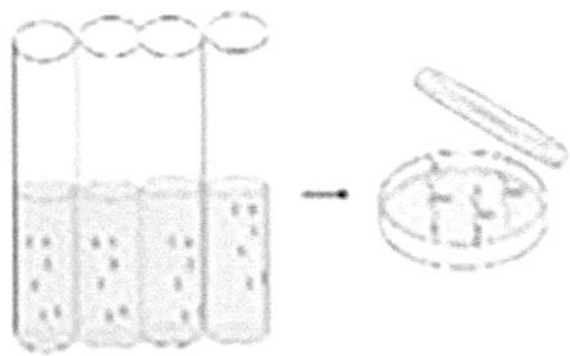	2 Australian native species tested. Priming on agar (KAR_1-agar) improved seed germination more effectively than KAR_1 priming for *Triodia wiseana*. KAR_1 had no effect on or reduced *Rytidosperma caespitosum* seed germination.	**0.67 µM KAR_1** (Berto et al., 2021) *Germination % reported was maximum achieved and not average seed germination. *Triodia wiseana* • ↑ seed germination with hydro-priming (up to +41%), KAR_1-agar (+40%) & KAR_1-priming (+24%) • ↑↑ seed germination with flaming when combined with hydro-priming (+50%), KAR_1-agar (+73%), KAR_1-priming (+55%) • Seed coating had no effect or reduced seed germination *Rytidosperma caespitosum* • ↑ seed germination with hydro-priming (19%). • ↓ time to germinate with hydro-priming (4 d), KAR_1-agar (2 d) and KAR_1-priming (2.5 d). • ↓ seed germination with flaming (-22%) and flaming + KAR_1-agar (-14%). • No effect of KAR_1-agar (-flaming) or KAR_1-priming (±flaming) on seed germination. Additional treatments: ± flash flaming to remove unwanted appendages, ± seed coating[a,b] Optimal seed germination temperatures were used for each species. *Germination variation for *T wiseana*: control 0–6.4%; hydro-primed 25–47%; KAR_1-primed 27–30%; KAR_1-agar 17–46%; flaming + hydro-primed 21–57%; flaming + KAR_1-primed 38–62%; flaming + KAR_1-agar 74–79%.
A "KAR_1-agar" treatment delivered KAR_1 to seeds and seedlings within water agar (0.7% w/v) in a Petri dish. 	KAR_1 improved seed germination in the native Australian spinifex grass, *Triodia pungens*.	**KAR_1** (Kildisheva, 2019) Detailed in Table 18.2 since osmotic stress and mining waste soil tested.
KAR_1 delivered via water agar (0.7% w/v) or MS media 	8 Australian native *Triodia* collections tested[b] (common name Spinifex, Poaceae family) KAR_1 improved seed germination of 7 collections with florets removed and 6 with florets present. KAR_1 outperformed GA_3	**0.67 µM KAR_1** (control was untreated seeds/florets + seeds; Erickson et al., 2016) • ↑ seed germination with KAR_1 in *T. augusta* (bare seeds > seeds with florets attached), *T. basedowii* (seeds), *T. brizoides* (seeds >> florets), *T. pungens* (seeds >> florets), *T. sp. Shovelanna Hill* (seeds), *T. wiseana* (both collections, seeds >> florets), *T. epactia* (florets). • ↑↑ seed germination when dry after-ripening (DAR) was additionally used with KAR_1 for *T. augusta* (seeds and florets), *T. basedowii* (seeds), *T. brizoides* (seeds and florets, although seeds only after 2 months DAR), *T. pungens* (florets), *T. wiseana* (florets, both collections). • ↓ seed germination with DAR for *T. epactia* (seeds) & *T. wiseana* (seeds, both collections). • KAR_1 outperformed GA_3 for all collections except *T. sp. Shovelanna Hill* (seeds, similar germination improvements).

(Continued)

TABLE 18.1 (Continued)

Application Method	Summary	Results – KAR_1 Effects on Plants (Relative to Untreated Controls without KAR_1)
	KAR_1 improved seed germination of 9 desert perennial species (across 5 families) from Australia's Great Sandy Desert; 25 species (58%) had physiological dormancy. KAR_1 did not promote the germination of tree seeds.	**1 μM KAR_1** (Commander et al., 2017) • ↑ seed germination with KAR_1 in the Goodeniaceae (1 species), Poaceae (3 *Triodia* species), Proteaceae (3 *Grevillea* species), Asteraceae (1 species) and Myrtaceae (1 species) out of 25 physiologically dormant species. • KAR_1 outperformed GA_3 in the seed germination of 3 Proteaceae and 3 Poaceae. • KAR_1 widened the range of temperatures over which 1 Asteraceae species germinated. • GA_3 outperformed KAR_1 in the seed germination of 3 Goodeniaceae, 1 Poaceae, and 1 Solanaceae species. • KAR_1 did not promote the germination of tree seeds
	26 USA Great Basin native species tested. KAR_1 enhanced seed germination in 5 species out of 16 with physiological dormancy (PD). GA_3 outperformed KAR_1.	**1 μM KAR_1** (control was untreated seed; Kildisheva et al., 2019) • ↑ seed germination with KAR_1 in 5 species (*Erigeron linearis*, *Erigeron pumilus*, *Ipomopsis fricana*, *Lupinus arbustus*, *Potentilla glandulosa*). Improvement ranged from 9–26%. • ↓ seed germination for 1 species (*Dieteria canescens*). • Germination slowed for 2 species (*Agoseris heterophylla* and *Sphaeralcea munroana*). • GA_3 outperformed KAR_1 for 10 species (GA_3 improved germination of 12 species). • While the amount of germination with both stimulants was generally temperature mediated, both expanded the thermal germination envelope into warmer temperatures.
	KAR_1 had no effect on seed germination of the endangered African orchid *Ansellia fricana* Lindl.	**1–100 nM KAR_1** (Papenfus et al., 2016) • No effect on seed germination.

Seeds were soaked for 24 h in H_2O, 0.67 µM KAR_1, smoke water (SW), or 2.89 mM GA_3. Primed seeds were sown onto water agar (0.7% w/v) in a Petri dish. 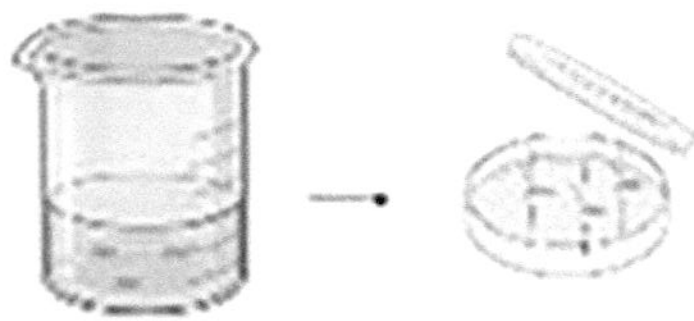	Seed germination was improved by KAR_1 for 5 out of 8 Australian native *Solanum* species tested.	**0.67 µM KAR_1** (Commander et al., 2008) Low (26/13°C) and high (33/18°C) temperature tests for all species. • ↑ seed germination at low temp for *S. cunninghamii* and *S. phlomoides* and both temps for *S. centrale*, *S. orbiculatum*, and *S. dioicum*. • No effect on seed germination for *S. chippendalei*, *S. diversiflorum*, and *S. sturtianum* (only GA_3 improved germination). • Seed germination behaved similarly for KAR_1 and GA_3 for 4 species, GA_3 outperformed KAR_1 for 3 species, KAR_1 outperformed GA_3 for 1 species. • KAR_1 outperformed SW for 4 species. Temperature gradient (10, 15, 20, 25, 30°C) for *S. orbiculatum* (SO) and *S. centrale* (SC). • ↑ seed germination at all temps for SO but only at 10–20°C for SC. • KAR_1 outperformed GA_3 for SO at 10–15°C; GA_3 outperformed KAR_1 for SC (all temps). • KAR_1 outperformed SW for both species and at most temps.

Notes: [a]Seed coating: 22 g (8% w/v) of polyvinyl alcohol (Selvol 205, Sekisui Specialty Chemicals America, Dallas, TX, USA) and 90 g of calcium carbonate ($CaCO_3$; Omyacarb, Omya Australia Pty. Limited, Lindfield, NSW, Australia).

[b]Gibberellic acid (GA_3, 289 µM) and dry after-ripening (storage at 30°C/50% RH over 12–24 months) were additional treatments (Berto et al., 2021; Kildisheva, 2019).

Abbreviations: karrikinolide, KAR_1; gibberellic acid, GA_3.

outcomes (Hancock et al., 2020). Another impediment to using karrikins for restoration is cost, estimated at \$20,000 g^{-1} KAR_1 upon upscale (personal experience, GRF) and, at present, commercially costing approximately \$1,000 for 5 mg (Toronto Research Chemicals, Ontario, Canada). Initially, it was anticipated that KAR_1 would be applied directly to the soil seed bank using 1–20 g KAR_1 ha^{-1} (versus 10 t ha^{-1} of smoke water to obtain the same results) (Rokich and Dixon, 2007), but this option is clearly cost prohibitive. Novel seed enhancement technologies and multidisciplinary problem resolution are needed (Svejcar et al., 2022) and are discussed in Section 18.4.

18.2 KARRIKINS FOR PLANT ABIOTIC STRESS TOLERANCE

"Climate time is upon us", and among the suggested mitigation strategies against magnified abiotic/biotic stressors that threaten crops and wild plants are chemical/hormone applications (Rivero et al., 2022). Karrikins were first explored as stress mitigators in 2006 (Table 18.2), when KAR_1 promoted normal tomato seedling development under extreme chilling and heat (higher seedling weight, size, and vigour at 10 and 40°C) (Jain et al., 2006) and promoted wild Australian plant seed germination under suboptimal light and temperatures (Merritt et al., 2006). Subsequently, karrikins have protected plants against stressors such as salinity (Table 18.2, Shah et al., 2021; Shah et al., 2020), heavy metals (Sardar et al., 2021; Kochanek et al., 2016; Kucukakyuz and Catav, 2021), osmotic stress (Kildisheva, 2019; Masondo et al., 2018; Ghebrehiwot et al., 2008), shading (KAR_2 in soybeans, Meng et al., 2016), suboptimal light (KAR_2 more effective in Arabidopsis) (Nelson et al., 2010), and temperature extremes (KAR_2 in Arabidopsis) (Wang et al., 2018a). Further, KAR_1 has improved the quality of aged and immature seeds, resulting in higher seedling vigour indices (Demir et al., 2018; Jain and Van Staden, 2007) and improved helpful plant attributes, such as increased stem thickness, which reduces crop lodging and improves yields (Ghebrehiwot et al., 2013; Kulkarni et al., 2008).

Suppression of oxidative stress through reduced lipid peroxidation, increased antioxidant activity (Shah et al., 2021; Sardar et al., 2021; Sunmonu et al., 2016) and enhanced stomatal conductance, gas exchange capability, photosynthetic capacity, and root growth (Swarbreck, 2021; Sardar et al., 2021; Akeel et al., 2019) are other KAR_1-enhanced mechanisms that can assist plant stress tolerance. Another curious phenomenon is seed germination delay (Meng et al., 2016) and exacerbated hypocotyl de-etiolation (Figure 18.1) (Kochanek, 2016) following KAR_1 application under stressful conditions (e.g., shading, supra-optimal temperatures). Meng et al. (2016) determined that gibberellin biosynthesis suppression and enhanced abscisic acid biogenesis are responsible. Indeed gibberellins (GA) are elevated by abiotic stress in bolting-prone crops, which manifests as stem elongation, early flowering, poor-quality head formation, and reduced yields (e.g., lettuce and brassicas) (Fukuda et al., 2009, 2012; Guo et al., 2004). Thus, we hypothesise that GA suppression following KAR_1 applications could delay post-stress bolting in sensitive crops, noting this requires validation via long-term growth trials. As for revegetation studies, past studies have demonstrated abiotic stress tolerance largely in laboratory conditions (Table 18.2); thus, future research must utilize long-term soil/field studies to generate harvest indices and accurately quantify abiotic stress tolerance improvements and, therefore, commercial viability. The following section dissects additional barriers to the commercial use of karrikins.

18.3 COMMERCIAL REALITIES OF KARRIKIN USE IN PLANT-RELATED SECTORS

Two clear disadvantages of using karrikins are their high cost (≥ \$20,000 g^{-1} KAR_1, GRF) and the lack of reliable field application protocols. Cost can be overcome if minute KAR_1 dosages are applied via seed since low nanomolar quantities can be effective for many plant species (e.g., Shah et al., 2021; Flematti et al., 2004). Indeed, we have shown that KAR_1 applied via seeds has stronger effects on plant quality attributes than root drenching or shoot sprays (Kochanek, 2021). Commercial seed priming and coating techniques apply compounds via a solution followed by drying and storing

TABLE 18.2

Karrikin Effects on Abiotic and Biotic Plant Stress Resilience. Studies are grouped by application and growing methods to illustrate field and/or commercial relevance for each study. Created with BioRender.com.

Application Method	Summary of KAR_1 Effect	Detailed Results – KAR_1 Effects on Plants Relative to Untreated Controls (without KAR_1)
Seeds primed in a Petri dish with filter paper wetted with water ± KAR_1 ± treatment solutions and rewetted with H_20 or treatment solutions. Seedlings grown in the same Petri dish.	Resilience to temperature extremes (10–40°C) in tomato, where 20–30°C is optimal.	**100 nM KAR_1** (Jain et al., 2006) • ↑ normal seedling development at temperature extremes (10 and 40°C) with KAR_1 versus only radicle protrusion in control (without KAR_1). • ↑ seedling growth at all temperatures – i.e., ↑ weight and vigour index, longer seedlings.
	Enhanced germination under varying light wavelengths, in darkness and/or under suboptimal temperatures for 5 Australian native ephemeral Asteraceae species (*Angianthus tomentosus*, *Genphosis tenuissima*, *Myriocephalus guerinae*, *Podolepis canescens*, and *Rhodanthe citrina*). 2 species had germination suppressed or not affected by KAR_1. KAR_1 usually outperformed smoke water; GA_3 often outperformed KAR_1.	**0.67 µM KAR_1** (Merritt et al., 2006) • ↑ seed germination in white light for 2 species (400–700 nM), 5 species under 570, 640 and/ or 720 nm (most in red light at 640 nm) and 3 species in darkness. • ↓ seed germination in red light for 1 species (640 nM). • KAR_1 generally outperformed smoke water. 6.7, 67 & 670 nM KAR_1 tested for 3 species under winter (18/7°C), spring/autumn (26/13°C), or summer (33/18°C) conditions ± light. • ↑ seed germination in darkness under all conditions for all species. • ↑ seed germination under light in winter conditions for all species and under spring/autumn and winter conditions for *M. guerinae*. • ≥ 67 nM KAR_1 generally enhanced germination more than 6.7 nM KAR_1. • GA_3 generally outperformed KAR_1 for 2 species (strongly for *M. guerinae*).

(Continued)

TABLE 18.2
(Continued)

Application Method	Summary of KAR_1 Effect	Detailed Results – KAR_1 Effects on Plants Relative to Untreated Controls (without KAR_1)
	Resilience to heavy metal stress (cadmium) in coriander (*Coriandrum sativum* L.).	**1 μM KAR_1** most enhanced stress tolerance (10^{-4} and 10^{-8} M KAR_1 also tested; Sardar et al., 2021). • ↑ seed germination, plant chlorophyll concentration, and gas exchange in stressed and non-stressed plants. • ↑ relative water content (+62%), soluble sugar content (+53%), proline content, and antioxidant activity (SOD, POD, and CAT) in stressed plants. • ↑ metal tolerance value in stressed plants. ↓ lipid peroxidation (↓ MDA & H_2O_2) and electrolyte leakage in stressed plants (63–69%).
	Suppression of oxidative stress in beans and maize	**10 nM KAR_1** (Sunmonu et al., 2016) • ↓ lipid peroxidation in bean roots and shoots and maize coleoptiles and roots. • ↑ antioxidant activity in maize endosperm and bean cotyledons (especially SOD and CAT). • ↑ root and shoot elongation, possibly due to ROS depletion in seedlings. • ↑ degradation of starch from bean cotyledons and maize endosperms, ↑ mobilization to other seedling parts (e.g., roots). • ↑ amylase activity in seedling parts, which may improve osmotic adjustment.
	Resilience to osmotic (PEG-6000) and temperature stress (20–40°C, 30/15 °C) in tef (*Eragrostis tef* (Zucc.) Trotter).	**10 nM KAR_1** (Ghebrehiwot et al., 2008) • ↑ 16-day-old seedling length and vigour at ≥25°C (except at 25°C under constant light or darkness). • ↑ water imbibition under osmotic stress (-0.15 to -0.49 Mpa at 25°C). • ↑ seed germination and 7-day-old seedling length at extreme osmotic stress (-0.49 Mpa).
	Reversed the effects of aquaporin inhibitors ($HgCl_2$, $ZnCl_2$ at 10–30 μM) in tomato.	**0.1 μM KAR_1** (Jain et al., 2008) • ↑ water imbibition, root length, and seedling moisture content (as effectively as Hg scavenging agents, ME at 500 μM and DTT at 250 μM, for $HgCl_2$).

Resilience to osmotic (PEG-6000), salinity and low temperature stress (10, 15, 10/15°C) in the African leafy vegetable *Ceratotheca triloba* (Bernh.) Hook. F.	**1 μM KAR_1-priming** (versus water-priming; Masondo et al., 2018) • ↑ seed germination at low temperatures (10/15, 15°C). • ↑ shoot and root length, vigour index at 15°C. • ↑ germination under osmotic stress and for unstressed plants (0, -0.05, -0.15 Mpa PEG-6000). • ↑ root length and vigour index under mild osmotic stress (-0.05 Mpa). • ↓ shoot length and vigour index in unstressed plants (0 Mpa) and under mild salinity stress (5 mM). • No effect on seed germination at 10°C (0%) and under salinity stress (≥ 5 mM NaCl at 25°C). • No effect on shoot or root growth, vigour index under severe salinity stress (≥15 mM NaCl).
Resilience to salinity stress (NaCl) in wheat.	**1 nM KAR_1** (Shah et al., 2021) • ↑ seed germination. • ↑ 7-day-old seedling root and shoot growth • ↑ antioxidant encoding genes (e.g., SOD, CAT, POD, APX). Post-germination: ABA up- & GA downregulation.
KAR_2 had a negative effect on seed germination and post-germination growth under shaded conditions by increasing ABA and decreasing GA biogenesis.	**1 μM KAR_2** (Meng et al., 2016) • ↓ seed germination under shaded conditions. KAR_2 positively affected abscisic acid (ABA) biogenesis but negatively regulated gibberellin (GA) biosynthesis. • Paclobutrazol, a GA biogenesis inhibitor, mimicked the KAR_2 treatment (i.e., inhibitory effect on soybean seed germination). • No effect on soybean seed germination under dark and white light conditions.

(Continued)

TABLE 18.2
(Continued)

Application Method	Summary of KAR_1 Effect	Detailed Results – KAR_1 Effects on Plants Relative to Untreated Controls (without KAR_1)
Seeds primed in 2.5 L cylinders with H_20 or 1 μM KAR_1 for 10 h & dried for 48 h. Experiment 1 seeds sown into petri dish with filter paper rewetted with H_20 or PEG-8000; Experiment 2 seeds sown into pots containing topsoil + mining overburden waste (1:1, v:v). Harvest at seedling establishment (harvest 1), during active growth (2) or at the end of the growing season (3). 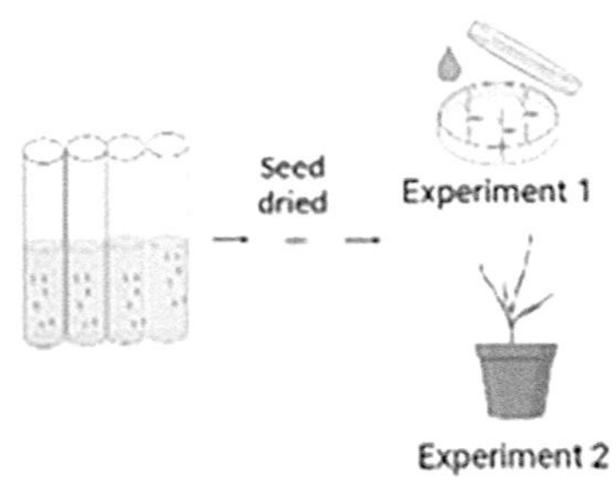 	Resilience to osmotic stress (PEG-8000 at -0.25, -0.5, and -0.75 Mpa), drought (irrigated with 12, 24, 48 mm to simulate low, medium and high Pilbara rainfall, respectively) and mining waste soil[b] in the native Australian spinifex grass, *Triodia pungens*. Regardless of treatment[c], seedling losses between the germination and emergence phase posed a major limitation to plant recruitment, and maximum emergence was <20% even with the best KAR_1 priming treatment.	**1 μM KAR_1** (no water priming control utilized, Kildisheva, 2019) Osmotic stress (Experiment 1) • ↑ seed germination at 0 Mpa (+61%), -0.25 Mpa (+30%) and -0.5 Mpa (+15%). • Seed coating[a] + KAR_1 priming (relative to KAR_1 priming alone): ↓ seed germination at 0 Mpa, had similar outcomes at -0.5 and -0.25 Mpa and improved germination at -0.75 Mpa. Mining waste and drought stress (Experiment 2) • ↑ seedling emergence at low (+4%), medium (+13%), and high (+6%) rainfall (control 0–2%). • However, max emergence was <20% even with the best KAR_1 priming treatment. • ↑ seedling size with med (harvest 1) and high rainfall (harvest 2) • ↑ root biomass with med rainfall (harvest 2, same as seed coating[a] + KAR_1 priming); otherwise, no effect on roots. • Seed coating[a] + KAR_1 priming: ↓ seedling emergence at med and high rainfall (-2–4%) relative to KAR_1 priming alone; ↑ seedling size at low, med, and high (harvest 1); low and high (harvest 2), and high (harvest 3) rainfall relative to no treatment; often outperformed KAR_1 priming for plant size. • Seed coating[a] alone reduced or had no effect on plant emergence, root biomass, or root size relative to the untreated control.
Seeds primed on agar or Phytagel (± KAR ± MS) and seeds/seedlings grown in the same media or transferred for growth into plant-growing media. 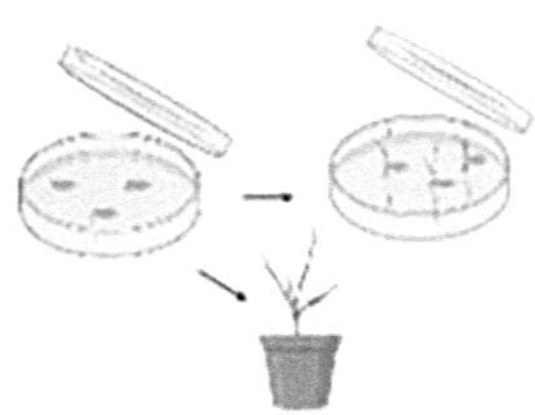	Resilience to salinity (NaCl) and drought stress (mannitol) in the Chinese tallow tree (*Sapium sebiferum* L.).	**1 nM KAR_1** (Shah et al., 2020) • ↑ seed germination in unstressed and stressed plants. • ↑ survival rate, plant height, biomass accumulation, maximum photochemical efficiency, and root growth (↑ weight and length of taproot; ↑ later root number) in stressed plants. • ↑ antioxidative enzyme activity (SOD, POD, CAT, APX) in stressed plants. • ↓ lipid peroxidation (↓ MDA & H_2O_2) and electrolyte leakage in stressed plants. • ↑ organic and amino acids that control redox homeostasis under stress. • ↓ endogenous ABA in all plants, ABA signaling genes altered under stress • Upregulation of stress-related genes (e.g., SOS1, DREB2A, WRKY33, ERF6).

	Enhanced light responses during germination and seedling development in *Arabidopsis*; positive influence on photomorphogenesis in *Arabidopsis*, lettuce (*Lactuca sativa* L. cv. Grand Rapids) and the weed *Brassica tournefortii*.	**1 µM KAR$_1$** (Nelson et al., 2010) • ↑ light-dependent seed germination at a range of low light fluences in *Arabidopsis*. Hypothesis: KAR$_1$ may promote emergence from the soil seed bank by enhancing light perception and sensitivity for some species (i.e., seeds can germinate at depth and/or with intermittent light exposure; may not be a broad mechanism across species). • ↓ hypocotyl elongation in *Arabidopsis* (at 1–10 µM KAR$_1$), lettuce, and *B. tournefortii* (0.1–10 µM KAR$_1$) and ↑ cotyledon expansion in *Arabidopsis* across a range of light fluences. • KAR$_2$ more effective than KAR$_1$ in *Arabidopsis*.
	Resilience to salinity (NaCl) and temperature stress in Arabidopsis.	**100 nM KAR$_2$** (Wang et al., 2018a) • ↓ seed germination during stress; ↑ under favourable conditions. • ↓ hypocotyl elongation regardless of stress. • Hypothesis: KAI2 may be an environmental sensor that modulates dormancy/germination so plants grow under favourable conditions, enacted by ABA and GA signaling pathways.
Seeds primed in water ± KAR$_1$ in a beaker/tube or within a petri dish + filter paper. Post-priming seed dried and resown into paper rolls, pots, trays, or petri dish with filter paper for seedling growth.	Better quality capsicum plants from immature and mature seeds.	**0.1 µM KAR$_1$** (Demir et al., 2018) • ↑ seed germination, emergence, seedling weight (all seeds). • ↑ rapid emergence, larger seedlings from immature seeds, KAR$_1$ priming "upgraded" the immature seed lot and outperformed hydro-priming. • ↓ ROS hence reduced damage to cell membranes (all seeds). • ↑ enzyme activity, ↑ CAT activity resulted in fewer abnormal seedlings. *Seeds were dried for 24 hours between priming and resowing.
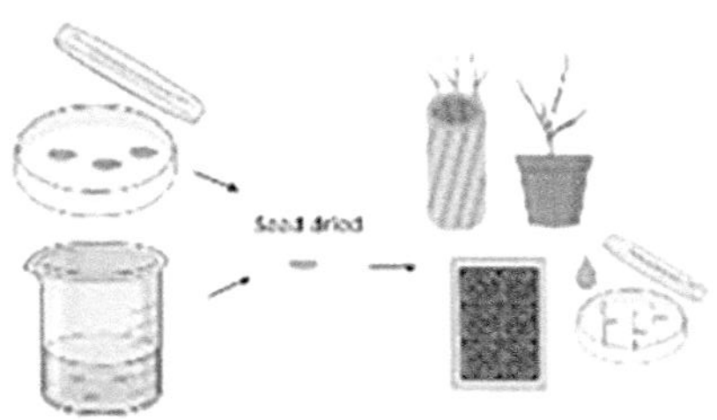	Higher vigour index for tomato seedlings grown from aged seeds for KAR$_1$-primed seeds compared to water-primed and unprimed seeds. Resilience to salinity (NaCl), osmotic (PEG-6000) and temperature stress (10–35°C) in tomato.	**Priming with 0.1 µM KAR$_1$** (Jain and Van Staden, 2007) Vigour index: • ↑ seedling vigour for seeds primed with KAR$_1$ and dried, then stored at room temperature; however, 13% decline in vigour for KAR$_1$-primed seeds, 27% decline for water-primed seeds after 4 months storage. • Accelerated radicle emergence (0% in unprimed, 12% in water-primed, 22% in KAR$_1$-primed seeds at 36 hours post water imbibition). • ↑ root and shoot length, seedling weight. • ↑ seedling vigour for 24 h KAR$_1$-primed and stored seeds than for seeds exposed continuously to KAR$_1$ on filter paper for 7 d. Salinity, osmotic and temperature stress: • ↑ seed germination, seedling vigour, and weight for stressed and unstressed plants.

(Continued)

TABLE 18.2
(Continued)

Application Method	Summary of KAR_1 Effect	Detailed Results – KAR_1 Effects on Plants Relative to Untreated Controls (without KAR_1)
	Reduced lodging and improved growth under water-stress conditions for tef.	**10 nM KAR_1** (drenched at germination initiation; Ghebrehiwot et al., 2013) • ↑ stem thickness results in ↑ lodging resistance (which ↑s grain yields). • ↑ plant height. • No effect on grain yield or dry biomass. • ↑ chlorophyll-to-carotenoid ratio.
Plants drenched with H_20 ± KAR_1 within pots.	Increased stem thickness and plant size, faster fruiting, and increased harvest index in tomato.	**1 nM KAR_1** (irrigated twice weekly, grown until harvest; Kulkarni et al., 2008) • ↑ plant height, number leaves, stem thickness, fruiting rate (i.e., fruited earlier), harvest index (total fruit yield per aboveground biomass). • No effect on fruit yield per plant, fruit weight or size, or nutritional value of fruit.
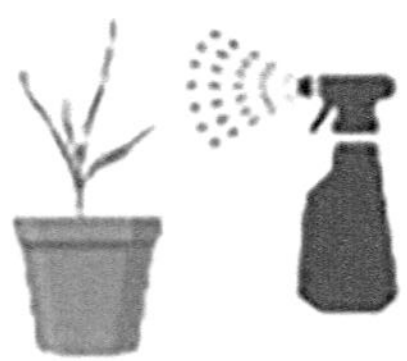 Foliar spray with H_20 ± KAR_1 ± surfactant and grown within pots.	Seedling vigour and size increase in tomato but no increase in stem thickness. No improvements in okra growth.	**0.1 µM KAR_1** (sprayed at 8, 12, 16, and 20 days after sowing; Kulkarni et al., 2007) • No effect on okra seedling growth or size, except ↓ root dry weight. • ↑ tomato shoot and root weight, number of leaves and leaf area, seedling vigour, growth rate.

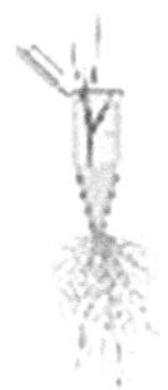 Seedling roots grew into nutrient solution ± KAR_1	Resilience to heavy metal stress (boron) in wheat.	**100 nM KAR_1** (Kucukakyuz and Catav, 2021) • ↑ root growth. • ↓ boron accumulation in seedlings. • Antioxidant enzymes (GR, POD, SOD) low in boron-stressed seedlings.

Notes: [a]Seed coating: 22 g (8% (w/v)) of polyvinyl alcohol (Selvol 205, Sekisui Specialty Chemicals America, Dallas, TX, USA) and 90 g of calcium carbonate ($CaCO_3$) (Omyacarb, Omya Australia Pty. Limited, Lindfield, NSW, Australia). [b]Mining overburden waste substrate + topsoil are used in post-mining restoration in the Pilbara (Kildisheva, 2019). [c]Gibberellic acid (GA_3, 289 µM) and dry after ripening (storage at 30°C/50% RH over 12–24 months) were additional treatments.

Abbreviations: MDA, malondialdehyde; H_2O_2, hydrogen peroxide; SOD, superoxide dismutase; CAT, catalase; POD, peroxidase; APX, ascorbate peroxidase; GR, glutathione reductase; ROS, reactive oxygen species; KAI2, KARRIKIN INSENSITIVE2; ABA, abscisic acid; PEG, polyethylene glycol; ME, β-mercaptoethanol; DTT, dithiothreitol.

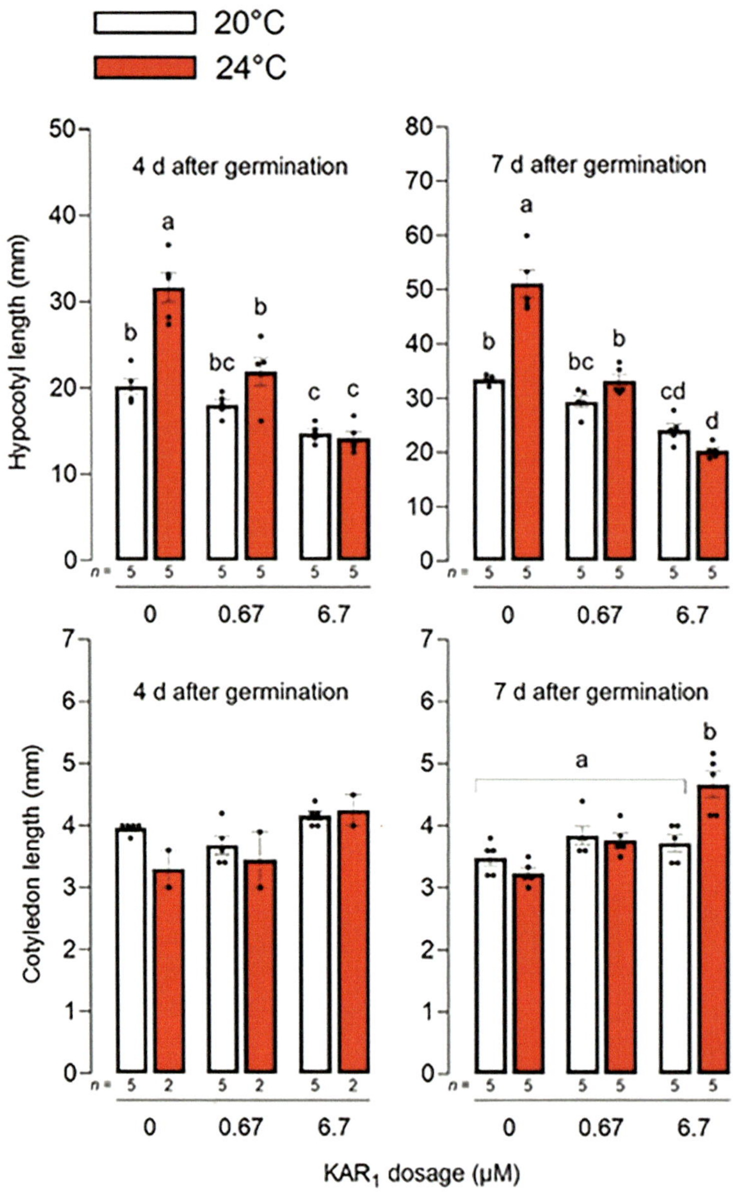

FIGURE 18.1 The effects of KAR_1 priming on lettuce (*Lactuca sativa* L. var. Nicky) seedling traits when grown at optimal and supra-optimal temperatures. Seeds were sown and seedlings grown in darkness at 20°C (optimal, white bars) or 24°C (supra-optimal, red bars) in 90 mm Petri dishes on 1% w/v agar with 0 (control), 0.67 or 6.7 μM KAR_1. Quantification of seedling cotyledon expansion and hypocotyl etiolation was at 4 and 7 days after germination. Nested analysis of variance and means tested the effect of KAR_1 (0, 0.67 or 6.7 μM) within temperature (Temp, 20 or 24°C) on lettuce seedling growth: 4 d hypocotyl, Temp: $F_{1,20} = 28.9$, $P < 0.001$, KAR (Temp): $F_{4,20} = 33.8$, $P < 0.001$; 4 d cotyledon, Temp: ns, KAR (Temp): $F_{4,11} = 3.7$, $P = 0.037$; 7 d hypocotyl: Temp: $F_{1,20} = 35$, $P < 0.001$, KAR (Temp): $F_{4,20} = 88$, $P < 0.001$; 7 d cotyledon: Temp: ns, KAR (Temp): $F_{4,20} = 14.5$, $P < 0.001$. Letters represent means that are different to one another with a *post-hoc* Tukey test ($\alpha = 0.05$). The exact sample size (n) for each block (true biological replication) is given as a discrete number in each panel. At 4 d after germination cotyledons had not emerged in most 24°C blocks. Transformation did not improve homogeneity of variance so data are presented untransformed. Statistical analysis/graphing were carried out using GRAPHPAD PRISM, Version 9.3.1 (GraphPad Software, San Diego, CA, USA) and MINITAB, Version 17 (Minitab Inc., State College, PA, USA).

the seeds prior to sowing at a later time. However, both technologies have limitations. Firstly, the drying of seeds post-treatment compromises priming effectiveness (Mehboob et al., 2018; Farooq et al., 2010) and plant benefits generally disappear approximately two weeks after sowing if seeds have been conventionally primed (Farooq et al., 2019). Secondly, drying seeds post-treatment can severely compromise seed quality if seeds are stored, especially for high-quality seed lots (Farooq et al., 2019; Wang et al., 2018b; Powell et al., 2000).

In contrast to these commercial techniques, priming on agar dosed with karrikins is currently the most promising protocol for delivery to plants, being more effective than conventional seed priming, coating (Berto et al., 2021), root drenching, or foliar sprays (Kochanek, 2021). For example, KAR_1 seed applications enhanced okra growth (van Staden et al., 2006) while foliar sprays did not (Kulkarni et al., 2007). Seed coating appears to be detrimental for seed germination and is not recommended for wild plant restoration (Berto et al., 2021). KAR_1 delivery to seeds via agar may outperform conventional priming because agar, compared to a solution, may better control seed hydration and minimize soaking injury (i.e., ensure excessively rapid water uptake does not damage cell membranes, etc. Roos and Pollock, 1971). Alternatively, the better performance may be because seeds were kept continuously moist post-priming in these studies (rather than being dried and stored prior to sowing). Indeed, it has long been known that post-priming dehydration may alter seed coat properties, produce secondary dormancy, or cause fatal injury to the embryo if at a critical development stage (Hegarty, 1978), and it is believed that sowing seeds moist post-priming is the best way to maximize longer-term growth benefits (Mehboob et al., 2018; Farooq et al., 2010).

In a bid to understand the effects of KAR_1 seed priming when seeds are redried after priming versus sown wet immediately after priming, we performed an experiment that compared priming on agar with and without post-priming drying. Seed longevity, seedling quality, and KAR_1 effectiveness after three storage durations (zero, one, or two months; Figure 18.2) were quantified for lettuce, selected for its extremely short-lived seeds (Roos and Davidson, 1992). Results showed that "wet" treatments (seed sown while moist, pink bars) maintained seed longevity while "wet-dry-wet" treatments did not (white bars). Specifically, "wet" seed germination remained at 100% across treatments (± KAR_1; Figure 18.2a, b) while "wet-dry-wet" seed germination was reduced by 30–50% for the control (without KAR_1) after one to two months of storage and by 10–20% for KAR_1-dosed seeds after two months' storage. These results imply that lettuce seed longevity was harmed by redrying seeds post-priming.

An unexpected result was that subsequent seedling quality rapidly deteriorated for "wet-dry-wet" seeds regardless of KAR_1 treatments while seedling quality was largely maintained within "wet" treatments and further enhanced by KAR_1 (Figure 18.2c, d, e). For example, radicle length for redried seeds was reduced by approximately 70% after one to two months storage (approximately 13 mm after zero months storage, <5 mm after one to two months storage; Figure 18.2c), and cotyledons were about half the size of "wet" seedlings at one to two months' storage (Figure 18.2d). Also, KAR_1 did not improve any seedling quality parameter for "wet-dry-wet" seeds relative to the control. By contrast, 0.67 μM KAR_1 resulted in 40% longer radicles than the control for "wet" treated seeds, and these radicles were comparable in length after two months to after zero months storage (Figure 18.2c) while radicle length of "wet" control seeds was reduced by approximately 25–40% after one to two months compared to zero months' storage. Similarly, after two months storage, 0.67 μM KAR_1 also increased cotyledon size by 20%, and both KAR_1 doses reduced hypocotyl length by ≥25% relative to the control for "wet" seeds. This implies accelerated photomorphogenesis with KAR_1 dosing for wet treated seeds since KAR_1 resulted in seedlings with larger cotyledons and hypocotyl de-etiolation after two months' storage when seeds were sown wet. By contrast, KAR_1 had no effect on seedling quality if seeds were redried after priming.

These results agree with Jain and Van Staden (2007), who showed that primed and redried seeds stored for 4 months had 13% reduced seedling vigour for KAR_1 primed seeds and 27% reduced vigour for water-primed seeds relative to fresh seeds (i.e., not stored). By contrast, Turner et al.

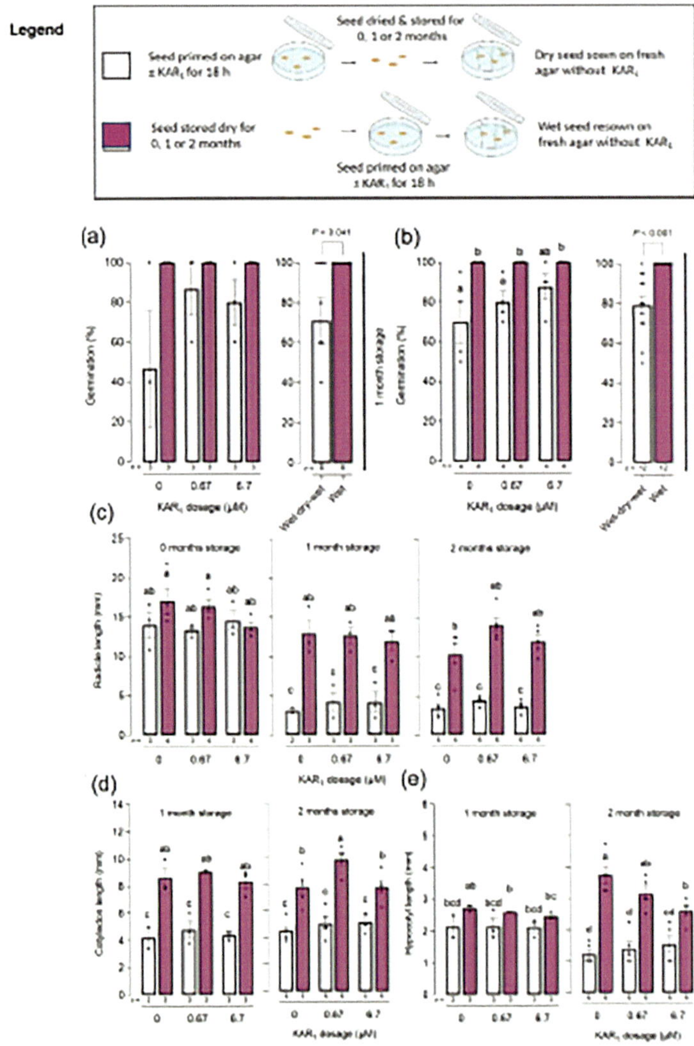

FIGURE 18.2 The effects of KAR_1 priming and post-priming seed drying on lettuce (*Lactuca sativa* L. var. Nicky) seed longevity and seedling quality after 0, 1 or 2 month(s) of storage. Seeds were either primed for 18 h, dried and stored for 0, 1 or 2 month(s) at room temperature and subsequently sown dry onto fresh media (white bars, 'wet-dry-wet' treatment) or seeds were stored for 0, 1 or 2 month(s) at room temperature and then primed for 18 h and sown wet onto fresh media (pink bars, 'wet' treatment). Seed priming used 1% w/v agar with KAR_1 at 0 (control), 0.67 or 6.7 μM for 18 h, while growing-on media used agar + half strength MS media + charcoal. Quantification of seed germination after a, 1 month or b, 2 months of storage. c, Quantification of seedling c, radicle length after 0, 1 or 2 months storage and d, cotyledon expansion and e, hypocotyl etiolation after 1 or 2 month(s) storage. Statistical analysis/graphing were carried out using GRAPHPAD PRISM, Version 9.3.1 (GraphPad Software, San Diego, CA, USA), MINITAB, Version 17 (Minitab Inc., State College, PA, USA) and BioRender.com. The letters represent means that are different to one another with a *post-hoc* Tukey test ($\alpha = 0.05$). For a, b, General linear model analysis of variance was used to compare the effects of priming method (wet-dry-wet vs. wet) and KAR_1 dose (0, 0.67 or 6.7 μM) and their interaction, with block as a random factor, while for c-e, storage duration (0, 1 and 2 month(s)) was additionally compared. The exact sample size (*n*) for each block (true biological replication) is given as a discrete number in each panel. Significant differences for a, seed germination at 1 month: priming method: $F_{1,10} = 5.5$, $P = 0.041$, b, at 2 months: priming method: $F_{1,15} = 43$, $P < 0.001$; c, radicle length: storage: $F_{2,46} = 77.5$, $P < 0.001$, priming method: $F_{1,46} = 153$, $P < 0.001$, storage × priming: $F_{2,46} = 21$, $P < 0.001$; d, cotyledon length: priming method: $F_{1,31} = 216$, $P < 0.001$, KAR_1: $F_{2,31} = 5.0$, $P = 0.013$; e, hypocotyl length: priming method: $F_{1,31} = 93$, $P < 0.001$, storage × priming: $F_{1,31} = 21$, $P < 0.001$, KAR_1 × priming: $F_{2,31} = 5.2$, $P = 0.012$. Proportions were √arcsine transformed prior to analysis otherwise transformation did not improve homogeneity of variance so data are presented untransformed. Priming and incubation was at room temperature (Jain and Van Staden, 2007) and exact storage timing was 3 days (shown as 0 months), 5 weeks (1 month) and 8 weeks (2 months).

(2013) found that seeds that were stored for 36 months and then primed (analogous to the "wet-primed" treatment mentioned earlier) versus seeds stored for 36 months after priming (analogous to "redried") showed few germination differences between treatments for seeds of 18 Australian wild plants. They did not, however, measure seedling quality and may not have used species with seeds as inherently short lived as lettuce. Hence, we recommend more studies compare wet-dried-wet to continuously wet seed priming protocols since most studies that observe strong karrikin effects use continuously wet protocols alone (e.g., Bursch et al., 2021, Berto et al., 2021). Given our results, we also recommend that consideration be given to sowing seeds immediately or soon after priming to maximize seed longevity and to maintain positive karrikin seedling effects, at least for species with short-lived seeds such as lettuce. A caveat is that moist seeds pose handling and pathogen complications so cannot be stored for > one week (Burrell et al., 1980). And while agar priming is a time-honoured method of controlling seed hydration (Hegarty, 1978) and has so far yielded the best outcomes for karrikin delivery (Berto et al., 2021; Commander et al., 2017), it has little practicality for large-scale field applications. Clearly, more work is needed to resolve such issues before reliable KAR_1 application protocols can be recommended to industry.

18.4 CONCLUSIONS AND FUTURE PERSPECTIVES

Climate change, pollution, and soil degradation are increasing the frequency and intensity of stressors that threaten crop yields and wild plant biodiversity. Chemical/hormone applications to plants are among the mitigation strategies for lessening the impact of such stressors (Rivero et al., 2022), and an emerging group of plant growth–regulating compounds called karrikins may have a role in addressing such pressing global challenges. This review has confirmed that karrikins have potential to promote plant resilience against abiotic stressors, such as sub- and supra-optimal temperatures, drought, and heavy metal and salinity stress. Additionally, as seed pretreatments, karrikins are stimulants that have the potential to release seed dormancy "on demand" for many wild plants across broad ecological niches and taxa, thus promising improved seed-based biodiversity restoration outcomes. However, two critical caveats uncovered by this review are the high cost of karrikins and the lack of reliable karrikin application protocols for field use. Priming on agar dosed with KAR_1 and sowing seeds while moist (or at least soon after priming) are currently the most promising protocol to deliver karrikins to plants, being more effective than conventional seed priming, coating, and root drenching/foliar sprays. However, seed priming on agar has little practicality for large-scale field applications, and sowing seeds moist poses handling and pathogen complications, and such seeds cannot be stored. Research should be targeted to resolve such issues: for example, by developing novel seed-enhancement technologies that deliver KAR_1 as effectively as agar but in a field setting. Further, karrikins' effects on plants must be validated in long-term soil/in-field studies as such research is currently lacking in the literature.

ACKNOWLEDGMENTS

We gratefully acknowledge the support of the Australian Institute for Bioengineering and Nanotechnology at the University of Queensland for financially supporting the writing of this review.

REFERENCES

Akeel A, Khan MMA, Jaleel H, Uddin M. 2019. Smoke-saturated water and karrikinolide modulate germination, growth, photosynthesis and nutritional values of carrot (Daucus carota L). *Journal of Plant Growth Regulation* 38(4): 1387–1401.

Antala M, Sytar O, Rastogi A, Brestic M. 2020. Potential of karrikins as novel plant growth regulators in agriculture. *Plants-Basel* 9(1).

Aremu AO, Plackova L, Novak O, Stirk WA, Dolezal K, Van Staden J. 2016. Cytokinin profiles in ex vitro acclimatized Eucomis autumnalis plants pre-treated with smoke-derived karrikinolide. *Plant Cell Reports* 35(1): 227–238.

Arora NV. 2019. Impact of climate change on agriculture production and its sustainable solutions. *Environmental Sustainability* 2: 95–96.

Berto B, Ritchie AL, Erickson TE. 2021. Seed-enhancement combinations improve germination and handling in two dominant native grass species. *Restoration Ecology* 29(1).

Brown NAC. 1993. Promotion of germination of fynbos seeds by plant-derived smoke. *New Phytologist* 123(3): 575–583.

Burrell NJ, Knight GP, Armitage DM, Hill ST. 1980. Determination of the time available for drying rapeseed before the appearance of surface molds. *Journal of Stored Products Research* 16(3–4): 115–118.

Bursch K, Niemann ET, Nelson DC, Johansson H. 2021. Karrikins control seedling photomorphogenesis and anthocyanin biosynthesis through a HY5-BBX transcriptional module. *Plant Journal* 107(5): 1346–1362.

Carbonnel S, Das D, Varshney K, Kolodziej MC, Villaecija-Aguilar JA, Gutjahr C. 2020. The karrikin signaling regulator SMAX1 controls *Lotus japonicus* root and root hair development by suppressing ethylene biosynthesis. *Proceedings of the National Academy of Sciences of the United States of America* 117(35): 21757–21765.

Ceccon E, Gonzalez EJ, Martorell C. 2016. Is direct seeding a biologically viable strategy for restoring forest ecosystems? Evidences from a meta-analysis. *Land Degradation & Development* 27(3): 511–520.

Commander LE, Golos PJ, Miller B, Merritt DJ. 2017. Seed germination traits of desert perennials. *Plant Ecology* 218(9): 1077–1091.

Commander LE, Merritt DJ, Rokich DP, Flematti GR, Dixon KW. 2008. Seed germination of Solanum spp. (Solanaceae) for use in rehabilitation and commercial industries. *Australian Journal of Botany* 56(4): 333–341.

Demir I, Ozden E, Yildirim KC, Sahin O, Van Staden J. 2018. Priming with smoke-derived karrikinolide enhances germination and transplant quality of immature and mature pepper seed lots. *South African Journal of Botany* 115: 264–268.

Dinerstein E, Joshi AR, Vynne C, Lee ATL, Pharand-Deschenes F, Franca M, Fernando S, Birch T, Burkart K, Asner GP, et al., 2020. A "Global Safety Net" to reverse biodiversity loss and stabilize Earth's climate. *Science Advances* 6(36).

Erickson TE, Shackelford N, Dixon KW, Turner SR, Merritt DJ. 2016. Overcoming physiological dormancy in seeds of *Triodia* (Poaceae) to improve restoration in the arid zone. *Restoration Ecology* 24: S64–S76.

Farooq M, Usman M, Nadeem F, Rehman HU, Wahid A, Basra SMA, Siddique KHM. 2019. Seed priming in field crops: potential benefits, adoption and challenges. *Crop & Pasture Science* 70(9): 731–771.

Farooq M, Wahid A, Ahmad N, Asad SA. 2010. Comparative efficacy of surface drying and re-drying seed priming in rice: Changes in emergence, seedling growth and associated metabolic events. *Paddy and Water Environment* 8(1): 15–22.

Flematti GR, Ghisalberti EL, Dixon KW, Trengove RD. 2004. A compound from smoke that promotes seed germination. *Science* 305(5686): 977–977.

Flematti GR, Ghisalberti EL, Dixon KW, Trengove RD. 2005. Synthesis of the seed germination stimulant 3-methyl-2H-furo[2,3-c]pyran-2-one. *Tetrahedron Letters* 46(34): 5719–5721.

Flematti GR, Ghisalberti EL, Dixon KW, Trengove RD. 2009. Identification of alkyl substituted 2H-Furo[2,3-*c*] pyran-2-ones as germination stimulants present in smoke. *Journal of Agricultural and Food Chemistry* 57(20): 9475–9480.

Flematti GR, Goddard-Borger ED, Merritt DJ, Ghisalberti EL, Dixon KW, Trengove RD. 2007. Preparation of 2H-furo[2,3-c]pyran-2-one derivatives and evaluation of their germination-promoting activity. *Journal of Agricultural and Food Chemistry* 55(6): 2189–2194.

Fukuda M, Matsuo S, Kikuchi K, Mitsuhashi W, Toyomasu T, Honda I. 2009. The endogenous level of GA(1) is upregulated by high temperature during stem elongation in lettuce through LsGA3ox1 expression. *Journal of Plant Physiology* 166(18): 2077–2084.

Fukuda M, Matsuo S, Kikuchi K, Mitsuhashi W, Toyomasu T, Honda I. 2012. Gibberellin metabolism during stem elongation stimulated by high temperature in lettuce. *Acta Horticulturae* 932: 359–364.

Ghebrehiwot H, Kulkarni M, Bairu M, Van Staden J. 2013. Plant-derived aerosol-smoke and smoke solutions influence agronomic performance of a traditional cereal crop, tef. *Experimental Agriculture* 49(2): 244–255.

Ghebrehiwot HM, Kulkarni MG, Kirkman KP, Van Staden J. 2008. Smoke-water and a smoke-isolated butenolide improve germination and seedling vigour of *Eragrostis tef* (Zucc.) trotter under high temperature and low osmotic potential. *Journal of Agronomy and Crop Science* 194(4): 270–277.

Guo DP, Shah GA, Zeng GW, Zheng SJ. 2004. The interaction of plant growth regulators and vernalization on the growth and flowering of cauliflower (*Brassica oleracea* var. botrytis). *Plant Growth Regulation* 43(2): 163–171.

Guo YX, Zheng ZY, La Clair JJ, Chory J, Noel JP. 2013. Smoke-derived karrikin perception by the alpha/beta-hydrolase KAI2 from Arabidopsis. *Proceedings of the National Academy of Sciences of the United States of America* 110(20): 8284–8289.

Hancock N, Gibson-Roy P, Driver M, Broadhurst L. 2020. *The Australian native seed sector survey report.* Australian Network for Plant Conservation, Canberra.

Hegarty TW. 1978. The physiology of seed hydration and dehydration, and the relation between water stress and the control of germination: A review. *Plant, Cell and Environment* 1: 101–119.

Jain N, Ascough GD, Van Staden J. 2008. A smoke-derived butenolide alleviates $HgCl_2$ and $ZnCl_2$ inhibition of water uptake during germination and subsequent growth of tomato – Possible involvement of aquaporins. *Journal of Plant Physiology* 165(13): 1422–1427.

Jain N, Kulkarni MG, van Staden J. 2006. A butenolide, isolated from smoke, can overcome the detrimental effects of extreme temperatures during tomato seed germination. *Plant Growth Regulation* 49(2–3): 263–267.

Jain N, Van Staden J. 2007. The potential of the smoke-derived compound 3-methyl-2H-furo[2,3-c]pyran-2-one as a priming agent for tomato seeds. *Seed Science Research* 17(3): 175–181.

Kaufman G. 1991. Seed coating: A tool for stand establishment: A stimulus to seed quality. *Horttechnology* 1: 98–102.

Keeley JE, Morton BA, Pedrosa A, Trotter P. 1985. Role of allelopathy, heat and charred wood in the germination of chaparral herbs and suffrutescents. *Journal of Ecology* 73(2): 445–458.

Kildisheva OA. 2019. *Improving the outcomes of seed-based restoration in cold and hot deserts: An investigation into seed dormancy, germination, and seed enhancement.* Doctor of Philosophy, The University of Western Australia.

Kildisheva OA, Erickson TE, Madsen MD, Dixon KW, Merritt DJ. 2019. Seed germination and dormancy traits of forbs and shrubs important for restoration of North American dryland ecosystems. *Plant Biology* 21(3): 458–469.

Kildisheva OA, Dixon KW, Silveira FAO, Chapman T, Di Sacco A, Mondoni A, Turner SR, Cross AT. 2020. Dormancy and germination: Making every seed count in restoration. *Restoration Ecology* 28: S256–S265.

Kochanek J. 2016. *Internal report for Horticulture Innovation Australia project VG15021.* North Sydney: Horticulture Innovation Australia.

Kochanek J. 2021. *Sowing success through transformational technologies. Final report for Hort Innovation project VG15021.* North Sydney: Horticulture Innovation Australia.

Kochanek J, Long RL, Lisle AT, Flematti G. 2016. Karrikins identified in biochars indicate post-fire chemical cues can influence community diversity and plant development. *Plos One* 11(8): e0161234.

Kochanek J, Soo RM, Martinez C, Dakuidreketi A, Mudge AM. 2022. Biochar for intensification of plant-related industries to meet productivity, sustainability and economic goals: A review. *Resources, Conservation and Recycling* 179: 106109.

Kochanek J, Steadman KJ, Probert RJ, Adkins SW. 2011. Parental effects modulate seed longevity: Exploring parental and offspring phenotypes to elucidate pre-zygotic environmental influences. *New Phytologist* 191(1): 223–233.

Kucukakyuz K, Catav SS. 2021. Physiological effects of smoke-water and karrikinolide on wheat seedlings grown under boron stress. *Russian Journal of Plant Physiology* 68(3): 552–558.

Kulkarni MG, Ascough GD, Van Staden J. 2007. Effects of foliar applications of smoke-water and a smoke-isolated butenolide on seedling growth of okra and tomato. *Hortscience* 42(1): 179–182.

Kulkarni MG, Ascough GD, Van Staden J. 2008. Smoke-water and a smoke-isolated butenolide improve growth and yield of tomatoes under greenhouse conditions. *Horttechnology* 18(3): 449–454.

Kusnetsov VV, Doroshenko AS, Kudryakova NV, Danilova MN. 2020. Role of phytohormones and light in de-etiolation. *Russian Journal of Plant Physiology* 67(6): 971–984.

Li N, Euring DJ, Cha JY, Lin Z, Lu MZ, Huang LJ, Kim WY. 2021. Plant hormone-mediated regulation of heat tolerance in response to global climate change. *Frontiers in Plant Science* 11.

Long RL, Stevens JC, Griffiths EM, Adamek M, Powles SB, Merritt DJ. 2011. Detecting karrikinolide responses in seeds of the Poaceae. *Australian Journal of Botany* 59(7): 609–619.

Masondo NA, Kulkarni MG, Finnie JF, Van Staden J. 2018. Influence of biostimulants-seed-priming on *Ceratotheca triloba* germination and seedling growth under low temperatures, low osmotic potential and salinity stress. *Ecotoxicology and Environmental Safety* 147: 43–48.

Mehboob N, Minhas WA, Nawaz A, Shahzad M, Ahmad F, Hussain M. 2018. Surface drying after seed priming improves the stand establishment and productivity of maize than seed re-drying. *International Journal of Agriculture and Biology* 20(6): 1283–1288.

Meng YJ, Chen F, Shuai HW, Luo XF, Ding J, Tang SW, Xu SS, Liu JW, Liu WG, Du JB, et al., 2016. Karrikins delay soybean seed germination by mediating abscisic acid and gibberellin biogenesis under shaded conditions. *Scientific Reports* 6.

Merritt DJ, Dixon KW. 2011. Restoration seed banks-a matter of scale. *Science* 332(6028): 424–425.

Merritt DJ, Kristiansen M, Flematti GR, Turner SR, Ghisalberti EL, Trengove RD, Dixon KW. 2006. Effects of a butenolide present in smoke on light-mediated germination of Australian Asteraceae. *Seed Science Research* 16(1): 29–35.

Nelson DC, Flematti GR, Riseborough JA, Ghisalberti EL, Dixon KW, Smith SM. 2010. Karrikins enhance light responses during germination and seedling development in Arabidopsis thaliana. *Proceedings of the National Academy of Sciences of the United States of America* 107(15): 7095–7100.

Nelson DC, Riseborough JA, Flematti GR, Stevens J, Ghisalberti EL, Dixon KW, Smith SM. 2009. Karrikins discovered in smoke trigger Arabidopsis seed germination by a mechanism requiring gibberellic acid synthesis and light. *Plant Physiology* 149(2): 863–873.

Nelson DC, Scaffidi A, Dun EA, Waters MT, Flematti GR, Dixon KW, Beveridge CA, Ghisalberti EL, Smith SM. 2011. F-box protein MAX2 has dual roles in karrikin and strigolactone signaling in Arabidopsis thaliana. *Proceedings of the National Academy of Sciences of the United States of America* 108(21): 8897–8902.

Papenfus HB, Naidoo D, Posta M, Finnie JF, Van Staden J. 2016. The effects of smoke derivatives on in vitro seed germination and development of the leopard orchid Ansellia africana. *Plant Biology* 18(2): 289–294.

Pedrini S, Merritt DJ, Stevens J, Dixon K. 2017. Seed Coating: science or marketing spin? *Trends in Plant Science* 22(2): 106–116.

Perez DR, Gonzalez F, Ceballos C, Oneto ME, Aronson J. 2019. Direct seeding and outplantings in drylands of Argentinean Patagonia: estimated costs, and prospects for large-scale restoration and rehabilitation. *Restoration Ecology* 27(5): 1105–1116.

Powell AA, Yule LJ, Jing HC, Groot SPC, Bino RJ, Pritchard HW. 2000. The influence of aerated hydration seed treatment on seed longevity as assessed by the viability equations. *Journal of Experimental Botany* 51(353): 2031–2043.

Rademacher W. 2015. Plant growth regulators: Backgrounds and uses in plant production. *Journal of Plant Growth Regulation* 34(4): 845–872.

Raupp PP, Ferreira MC, Alves M, Campos EM, Sartorelli PAR, Consolaro HN, Vieira DLM. 2020. Direct seeding reduces the costs of tree planting for forest and savanna restoration. *Ecological Engineering* 148.

Rivero RM, Mittler R, Blumwald E, Zandalinas SI. 2022. Developing climate-resilient crops: Improving plant tolerance to stress combination. *Plant Journal* 109: 373–389.

Rokich DP, Dixon KW. 2007. Recent advances in restoration ecology, with a focus on the Banksia woodland and the smoke germination tool. *Australian Journal of Botany* 55(3): 375–389.

Roos EE, Davidson DA. 1992. Record longevities of vegetable seeds in storage. *Hortscience* 27(5): 393–396.

Roos EE, Pollock BM. 1971. Soaking injury in lima beans. *Crop Science* 11(1): 78–81.

Sardar R, Ahmed S, Yasin NA. 2021. Seed priming with karrikinolide improves growth and physiochemical features of Coriandrum sativum under cadmium stress. *Environmental Advances* 5: 100082.

Shah FA, Ni J, Tang CG, Chen X, Kan WJ, Wu LF. 2021. Karrikinolide alleviates salt stress in wheat by regulating the redox and K+/Na+ homeostasis. *Plant Physiology and Biochemistry* 167: 921–933.

Shah FA, Wei X, Wang QJ, Liu WB, Wang DD, Yao YY, Hu H, Chen X, Huang SW, Hou JY, Lu R, Liu C, Ni J, Wu L. 2020. Karrikin improves osmotic and salt stress tolerance via the regulation of the redox homeostasis in the oil plant *Sapium sebiferum*. *Frontiers in Plant Science* 11: 216.

Soos V, Sebestyen E, Posta M, Kohout L, Light ME, Van Staden J, Balazs E. 2012. Molecular aspects of the antagonistic interaction of smoke-derived butenolides on the germination process of Grand Rapids lettuce (*Lactuca sativa*) achenes. *New Phytologist* 196(4): 1060–1073.

Stanga JP, Smith SM, Briggs WR, Nelson DC. 2013. Suppressor of more axillary growth2 1 controls seed germination and seedling development in Arabidopsis. *Plant Physiology* 163(1): 318–330.

Stevens JC, Merritt DJ, Flematti GR, Ghisalberti EL, Dixon KW. 2007. Seed germination of agricultural weeds is promoted by the butenolide 3-methyl-2H-furo[2,3-c]pyran-2-one under laboratory and field conditions. *Plant and Soil* 298(1–2): 113–124.

Sunmonu TO, Kulkarni MG, Van Staden J. 2016. Smoke-water, karrikinolide and gibberellic acid stimulate growth in bean and maize seedlings by efficient starch mobilization and suppression of oxidative stress. *South African Journal of Botany* 102: 4–11.

Svejcar LN, Brown VS, Ritchie AL, Davies KW, Svejcar TJ. 2022. A new perspective and approach to ecosystem restoration: A seed enhancement technology guide and case study. *Restoration Ecology* 30: e13615.

Swarbreck SM. 2021. Phytohormones interplay: Karrikin signalling promotes ethylene synthesis to modulate roots. *Trends in Plant Science* 26(4): 308–311.

Turner SR, Steadman KJ, Vlahos S, Koch JM, Dixon KW. 2013. Seed treatment optimizes benefits of seed bank storage for restoration-ready seeds: The feasibility of prestorage dormancy alleviation for mine-site revegetation. *Restoration Ecology* 21(2): 186–192.

United Nations, Department of Economic and Social Affairs, Population Division. 2019. World Population Prospects 2019: Highlights (ST/ESA/SER.A/423).

van Staden J, Sparg SG, Kulkarni MG, Light ME. 2006. Post-germination effects of the smoke-derived compound 3-methyl-2H-furo[2,3-c]pyran-2-one, and its potential as a preconditioning agent. *Field Crops Research* 98(2–3): 98–105.

Vanstraelen M, Benkova E. 2012. Hormonal interactions in the regulation of plant development. *Annual Review of Cell and Developmental Biology* 28(28): 463–487.

Villaecija-Aguilar JA, Hamon-Josse M, Carbonnel S, Kretschmar A, Schmid C, Dawid C, Bennett T, Gutjahr C. 2019. SMAX1/SMXL2 regulate root and root hair development downstream of KAI2-mediated signalling in Arabidopsis. *Plos Genetics* 15(8).

Wallner ES, Lopez-Salmeron V, Belevich I, Poschet G, Jung I, Grunwald K, Sevilem I, Jokitalo E, Hell R, Helariutta Y, et al., 2017. Strigolactone – and karrikin-independent SMXL proteins are central regulators of phloem formation. *Current Biology* 27(8): 1241–1247.

Wang L, Waters MT, Smith SM. 2018a. Karrikin-KAI2 signalling provides Arabidopsis seeds with tolerance to abiotic stress and inhibits germination under conditions unfavourable to seedling establishment. *New Phytologist* 219(2): 605–618.

Wang WQ, He AB, Peng SB, Huang JL, Cui KH, Nie LX. 2018b. The effect of storage condition and duration on the deterioration of primed rice seeds. *Frontiers in Plant Science* 9.

Waters MT, Nelson DC, Scaffidi A, Flematti GR, Sun YKM, Dixon KW, Smith SM. 2012. Specialisation within the DWARF14 protein family confers distinct responses to karrikins and strigolactones in Arabidopsis. *Development* 139(7): 1285–1295.

Waters MT, Scaffidi A, Flematti GR, Smith SM. 2013. The origins and mechanisms of karrikin signalling. *Current Opinion in Plant Biology* 16(5): 667–673.

Xu YQ, Miyakawa T, Nakamura H, Nakamura A, Imamura Y, Asami T, Tanokura M. 2016. Structural basis of unique ligand specificity of KAI2-like protein from parasitic weed Striga hermonthica. *Scientific Reports* 6.

Yang T, Lian YK, Wang CY. 2019. Comparing and contrasting the multiple roles of butenolide plant growth regulators: Strigolactones and karrikins in plant development and adaptation to abiotic stresses. *International Journal of Molecular Sciences* 20(24).

Zhou JF, da Silva JAT, Ma GH. 2014. Effects of smoke water and karrikin on seed germination of 13 species growing in China. *Central European Journal of Biology* 9(11): 1108–1116.

Index

Note: Page numbers in *italic* indicate a figure and page numbers in **bold** indicate a table on the corresponding page.

T

W